Microwave Engineering and Systems Applications

Microwave Engineering and Systems Applications

Edward A. Wolff
Roger Kaul

WILEY

A WILEY-INTERSCIENCE PUBLICATION
JOHN WILEY & SONS
New York • Chichester • Brisbane • Toronto • Singapore

Library of Congress Cataloging-in-Publication Data

Wolff, Edward A.
  Microwave engineering and systems applications.

  "A Wiley-Interscience publication."
  Includes bibliographies and index.
  1. Microwave devices. 2. Microwaves. I. Kaul,
Roger. II. Title.
TK7876.W635 1988   621.381′3   87-25223
ISBN 0-471-63269-4

Printed in the United States of America

10 9 8 7 6 5 4 3 2

Contributors

J. Douglas Adam (Chapter 10, co-author), Westinghouse Research Laboratories, Pittsburgh, Pennsylvania

David Blough (Chapter 22, co-author), Westinghouse Electric Co., Baltimore, Maryland

Michael C. Driver (Chapter 16), Westinghouse Research Laboratories, Pittsburgh, Pennsylvania

Albert W. Friend (Chapter 5), Space and Naval Warfare Systems Command, Washington, D.C.

Robert V. Garver (Chapters 6, 9, and 12; Chapter 10, co-author), Harry Diamond Laboratories, Adelphi, Maryland

William E. Hosey (Chapter 17; Chapter 22, co-author), Westinghouse Electric Co., Baltimore, Maryland

Roger Kaul (Chapters 2, 3, 4, 8, 11, 13, 14, 18, 19, 20, 21; Chapters 7, 15 co-author), Litton Amecom, College Park, Maryland (Presently at Harry Diamond Laboratories)

David A. Leiss (Chapters 7 and 15, co-author), EEsof Inc., Manassas, Virginia

Preface

This book had its beginnings when Richard A. Wainwright, Cir-Q-Tel President, asked Washington area microwave engineers to create a course to interest students in microwave engineering and prepare them for positions industry was unable to fill. Five of these microwave engineers, H. Warren Cooper, Albert W. Friend, Robert V. Garver, Roger Kaul, and Edward A. Wolff, responded to the request. These engineers formed the Washington Microwave Education Committee, which designed and developed the microwave course. Financial support to defray course expenses was provided by Bruno Weinschel, President of Weinschel Engineering.

The course was given for several years to seniors at the Capitol Institute of Technology in Laurel, Maryland. It was also given to seniors at the University of Maryland in College Park, Maryland and was given as a continuing education course to practicing engineers on several successive weekends at the University of Maryland Adult Education Center. The students had studied electromagnetic theory, so the emphasis of the course was placed on engineering and the types of problems encountered by practicing engineers. Lengthy proofs were only referenced so that the application of microwave techniques could be emphasized.

Once the course was designed and available textbooks were examined, it appeared to the Committee that it would need its own textbook to give a balanced, systems oriented presentation of modern microwave engineering. The course made extensive use of expert guest lecturers, and the notes used by the course lecturers provide the basis for this book. The course lasted two semesters or three quarters. The students were given extensive opportunity to use the SUPER-COMPACT computer program for the design of their own microwave circuits. This text should provide the basics for a beginning microwave engineer, a reference book to the fundamentals of microwave engineering for practicing engineers, and a text for short-course classes.

This textbook is divided into four parts. Part I (Chapters 2–4) provides the motivation for studying microwaves and describes the three major systems that use microwaves: communications, radar, and electronic warfare. These chapters show the importance of system parameters, such as noise temperature, bandwidth, and circuit losses which the microwave engineer must consider in circuit design.

Part II (Chapters 5–17) provides information on the design of various microwave components used for microwave generation, transmission, control, and detection. The components discussed include transmission lines, transmission line components, filters, ferrites, antennas, diodes, amplifiers, oscillators, vacuum tubes, and monolithic microwave integrated circuits.

Part III (Chapters 18–20) presents the measurement techniques needed to verify the performance of the components fabricated for a microwave subsystem. If a laboratory accompanies this course, Part III can be studied after Chapter 6 and can be taught simultaneously with Part II.

Part IV (Chapters 21 and 22) describes the design procedures for interconnecting microwave components into receiver and transmitter subsystems. These subsystems are the major microwave parts of communications, radar, or electronic warfare systems. The subsystem performance dictates the system performance presented in Part I and depends on the capabilities of the components described in Part II.

The creation of a comprehensive text— including applications and requirements, components, circuits, measurements, and subsystems—within the bounds of a single book has made it necessary to present engineering relations and equations without always including their often lengthy derivations. When derivations are omitted, references are given to sources in the literature.

Numerous periodicals and books provide a source of additional material on microwaves. After World War II and the development of radar, the MIT Radiation Laboratory Series of books became the "Bible" and main source of educational material. Soon to follow was the IEEE (formerly IRE) Transactions on Microwave Theory and Techniques, which began in 1953. (A cumulative index was published in June 1981.) In 1958, Theodore S. Saad began publication of *The Microwave Journal*. This was followed in 1962 by Hayden's regular publication of *Microwaves* (in a square format) and in 1971 by Weber Publications' *Microwave System News*. Add to this the digests from the International Solid State Circuits Conferences, the International Microwave Symposium, the *IEEE Journal of Solid State Circuits*, selected issues of the *IEEE Transactions on Electron Devices*, and *Proceedings of the IEEE*, and you can fill more than 6 meters of bookshelf. In more recent years, there have also been European and Japanese publications that should be included in any comprehensive microwave library. In 1963, A. F. Harvey attempted to condense all of the useful microwave information into one encyclopedic book.* It served as a good starting point for any new research endeavors because the researcher was not forced to scan all the old journals. The *Advances in Microwaves* series, edited by Dr. Leo Young (published from 1966 to 1971), presents in-depth articles on numerous microwave subjects, and the two-volume *Microwave Engineers' Handbook*, edited by Theodore Saad (published in 1971), provides excellent reference material for practicing engineers. *Introduction to Microwave Theory and Measurement*, written by Algie L. Lance in 1964, has been the most popular course text. Because all these books are at least ten years old, they do not include modern developments (such as monolithic integrated circuits) and generally do not emphasize the systems aspects found in this textbook.

We acknowledge the contributions Melvin Zisserson made to many of the chapters. Patricia Campbell meticulously typed the manuscript.

EDWARD A. WOLFF
ROGER KAUL

January, 1988

---

* A. F. Harvey, *Microwave Engineering*, Academic Press, London, 1963.

Contents

Major Symbols, Abbreviations, and Acronyms

A	ampere
A	amplitude, antenna area, area
A_e	antenna effective area
A_{eu}	unity gain antenna effective area
A_v	video pulse amplitude
AC	alternating current
AFC	automatic frequency control
AM	amplitude modulation
ANA	automatic network analyzer
B	magnetic flux density
B_{IF}	intermediate frequency bandwidth
B_n	noise bandwidth
B_S	source susceptance
BIT	built-in test
BPF	bandpass filter
BPSK	binary phase shift keyed
BPT	bipolar transistor
BW	bandwidth
b	bit
C	camouflage factor, capacitance, coupling factor
C_c	case capacitance
C_{dg}	drain-gate capacitance

C_{ds}	drain-source capacitance
C_f	fringe capacitance
C_j	junction capacitance
C_p	package capacitance
C_u	capacitance per unit length
C_V	voltage coupling ratio
CAD	computer-aided design
CC-TWT	cavity-coupled traveling wave tube
CDMA	code division multiple access
CFA	crossed field amplifier
C/N	carrier-to-noise ratio
CSL	central spectrum line
CW	continuous wave
c	velocity of light
D	diameter, directivity
DC	direct current
DR	dynamic range
DSO	dielectric stabilized oscillator
DUT	device under test
d	diameter
dB	decibel
dBc	decibels above carrier
dBm	decibels above 1 mW
dBW	decibels above 1 W
E	electric field vector
E_b	energy per bit
E_g	energy gap
E_i	incident voltage
E_r	reflected voltage
E_t	transmitted voltage
ECCM	electronic counter-countermeasures
ECM	electronic countermeasures
EIRP	effective isotropic radiated power
EMC	electromagnetic compatibility
EMI	electromagnetic interference
ESM	electronic support measures
EW	electronic warfare
e	electron charge
F	Farad

F	focal length, noise figure
F_{min}	minimum noise factor
F_n	noise factor
FDMA	frequency division multiple access
FET	field-effect transistor
FIT	fault isolation test
FM	frequency modulation
FPA	final power amplifier
FSK	frequency shift keyed
f	frequency
f_c	carrier frequency
f_{diff}	difference frequency
f_{if}	intermediate frequency
f_{in}	input frequency
f_{lo}	local oscillator frequency
f_{op}	operating frequency
f_{out}	output frequency
f_{res}	resonant frequency
G	Gauss, giga-
G	conductance, gain
G_A	available power gain
G_a	antenna gain
G_r	receiver antenna gain
G_S	source conductance
G_T	transducer power gain
G_t	transmitter antenna gain
GaAs	gallium arsenide
g	gram
g_m	transconductance
H	magnetic field vector
H	Henry
H_c	coercive force
HEMT	high electron mobility transistor
HVPS	high voltage power supply
Hz	Hertz
h	height, Planck's constant
I	current, isolation
IF	intermediate frequency
IFM	instantaneous frequency measurement

IGFET	insulated gate field-effect transistor
IL	insertion loss
Im	imaginary part of complex number
IMD	intermodulation distortion
IMN	input matching network
IMPATT	impact ionization avalanche transit-time
IPA	intermediate power amplifier
i	current
$\mathbf{J}$	current density vector
JFET	junction field-effect transistor
J/S	jammer-to-signal ratio
K	Kelvin
K	perveance
k	Boltzmann's constant, kilo-, wave number
$\mathbf{k}$	unit vector in propagation direction
L	inductance, length, loss
L_p	package inductance
L_system	system loss
L_w	wire inductance
LNA	low noise amplifier
LO	local oscillator
LVPS	low voltage power supply
M	mega-
M_s	saturation magnetization
MESFET	metal-semiconductor field-effect transistor
MIC	microwave integrated circuit
MMIC	monolithic microwave integrated circuit
MOCVD	metalorganic chemical vapor deposition
MODFET	modulation-doped field-effect transistor
MOPA	master oscillator power amplifier
MOS	metal-on-semiconductor
MSG	maximum signal gain
MTI	moving target indicator
m	meter, milli-
m	modulation index
m_e	electron mass
N	noise, number
N_B	number of pulses
N_i	input noise power

N_o	output noise power
NF	noise figure
n	nano-
n	unit normal vector
Oe	Oersted
OMCVD	organometallic chemical vapor deposition
OMT	orthomode transducer
P_{ant}	power at antenna
P_{avg}	average power
P_D	power dissipated
P_{DC}	direct-current power, input power
P_{in}	incident power, input power
P_L	load power
P_{LIM}	limited power
P_n	noise power
P_{out}	output power
P_r	received power
P_T	transmitter output power
P_t	transmitter output power
PCU	protection and control unit
PD_{inc}	incident power density
PD_{rec}	received power density
PD_{rerad}	reradiated power density
PD_t	transmitted power density
PIN	p-i-n semiconductor
PM	phase modulation
PMS	port modeling and synthesis
PPM	periodic permanent magnet
PPM-TWT	periodic permanent magnet traveling wave tube
PRF	pulse repetition frequency
PSK	phase shift keyed
p	pico-
Q	resonance factor
Q_{rad}	radiation resonance factor
QPSK	quadrature phase shift keyed
R	range, resistance
R_b	bit rate
R_d	drain resistance
R_{ff}	far field range

R_{in}	input resistance
R_j	junction resistance
R_L	load resistance
R_{max}	maximum range
R_p	parallel resistance
R_{rad}	radiation resistance
R_s	series resistance, source resistance
Re	real part of complex number
RF	radio frequency
RWR	radar warning receiver
rms	root mean square
S	signal, surface
S_i	input signal power
S_{min}	minimum detectable signal
S_o	output signal power
S_{11}	S-parameter
S_{12}	S-parameter
S_{21}	S-parameter
S_{22}	S-parameter
SAW	surface acoustic wave
SCPC	single channel per carrier
SI	international system of units
SMA	subminiature connector
S/N	signal-to-noise ratio
SOJ	stand-off jammer
SSJ	self-screening jammer
STC	sensitivity time control
s	second
s	distance
T	temperature, transmission coefficent
T_A	antenna temperature
T_c	Curie temperature
T_{eq}	equivalent noise temperature
T_i	input temperature
T_j	junction temperature
T_m	modulation period
T_o	reference temperature
T_p	pulse period
T_R	round trip time delay

T_{rcvr}	receiver noise temperature
TDMA	time division multiple access
TDR	time domain reflectometer
TE	transverse electric wave
TEGFET	two-dimensional electron gas field-effect transistor
TEM	transverse electromagnetic wave
TM	transverse magnetic wave
TRF	tuned radio frequency
TV	television
TWT	traveling wave tube
t	time
t_o	pulse width
t_r	rise time
t_{rise}	rise time
UHF	ultra high frequency
V	Volt
V	velocity, voltage, volume
V_B	breakdown voltage
V_{gs}	gate-source voltage
V_I	incident voltage
V_{max}	maximum voltage
V_{min}	minimum voltage
V_R	reflected voltage
V_{sat}	saturation voltage
V_T	transmitted voltage
VCO	voltage controlled oscillator
VHF	very high frequency
VHSIC	very high speed integrated circuit
VPE	vapor phase epitaxy
VSWR	voltage standing wave ratio
VTO	voltage tuned oscillator
v	velocity, voltage
W	Watt
W_{eff}	effective width
Wb	Weber
w	width
X	reactance
X_{in}	input reactance
X_j	junction reactance

X_L	load reactance
XMTR	transmitter
x	distance
$\mathbf{x}$	unit vector in x direction
Y	admittance
Y_{in}	input admittance
Y_o	characteristic admittance
YIG	yttrium iron garnet
YTF	YIG tuned filter
$\mathbf{y}$	unit vector in y direction
Z	impedance
Z_{in}	input impedance
Z_L	load impedance
Z_o	characteristic impedance, wave impedance
Z_{oe}	even mode characteristic impedance
Z_{oo}	odd mode characteristic impedance
Z_{out}	output impedance
Z_S	source impedance
Z_T	terminating impedance
Z_w	branch line characteristic impedance
$\mathbf{z}$	unit vector in z direction
α	attenuation
α_c	conductive attenuation
β	phase constant
Γ	reflection coefficient
Γ_g	generator reflection coefficient
Γ_{in}	input reflection coefficient
Γ_L	load reflection coefficient
Γ_l	load reflection coefficient
Γ_{out}	output reflection coefficient
Γ_S	source reflection coefficient
Γ_T	termination reflection coefficient
γ	complex propagation constant, gyromagnetic ratio
Δ	matrix determinant
δ	insertion loss, skin depth
ϵ	permittivity
ϵ_{eff}	effective permittivity
ϵ_o	free space permittivity

ϵ_r	relative permittivity
η	efficiency, electron charge/mass ratio, isolation
θ	angle, helix pitch angle
θ_B	Bragg angle
θ_T	thermal resistance
λ	wavelength
λ_c	cutoff wavelength
λ_g	waveguide wavelength
λ_o	free space wavelength
μ	micro-, permeability, cathode to electrode voltage ratio
μ_c	cathode to electrode cutoff voltage ratio
μ_o	free space permeability
μ_r	relative permeability
ρ	efficiency, radiation efficiency, VSWR
σ	conductivity, radar cross-section, standard deviation
σ^o	radar cross-section per unit area
τ	pulse width
ϕ	angle, depression angle, work function
χ_m	magnetic susceptibility
ω	angular frequency
∇	del operator

COMMONLY-USED CONSTANTS

$k = 1.38 \times 10^{-23}$ Joule/°K

$kT = 4 \times 10^{-21}$ Watts/Hertz   when $T = 290$ °K

$c = 3 \times 10^8$ m/s

$\mu_0 = 4\pi \times 10^{-7}$ Henry/m

$\epsilon_0 = 8.85 \times 10^{-12}$ Farad/m

m_e = electron mass = 9.1×10^{-31} kg

e = electron charge = 1.6×10^{-19} Coulombs

h = Planck's constant = 6.6×10^{-34} Joule seconds

Z_0 = intrinsic impedance of free space = 377 ohms

1 Introduction

This book is an introduction to microwave engineering. Microwave engineering is the branch of electrical engineering that deals with the transmission, control, detection, and generation of radio waves whose wavelength is short compared to the physical dimensions of the system. Wavelengths less than 30 cm (corresponding to frequencies in excess of 10^9 Hz), but greater than 0.03 mm (10^{13} Hz) are considered microwaves. Millimeter waves are a subset of microwaves in the 10–0.03 mm range. Microwaves are bounded on the long wavelength side by radio waves and on the short wavelength side by infrared waves. The relationship between wavelength and frequency is

$$f\lambda = c = 3 \times 10^8 \text{ m/s} \tag{1.1}$$

where frequency f is in units of Hertz, wavelength λ is in meters, and c is the speed of light. This relation applies in a vacuum (free space); in a dielectric medium, the wavelength λ is shortened. Unless specified otherwise, the relation between frequency and wavelength assumes a free space condition.

1.1 EARLY HISTORY OF MICROWAVES (1)

The history of microwaves is embodied in the evolution of electromagnetic waves. One of the earliest talks on this subject was Michael Faraday's (1791–1867) impromptu presentation to the Friday Night Lectures of the Royal Society of London in 1846 entitled "Thoughts on Ray Vibrations." [Faraday's talk was given because Charles Wheatstone (1802–1875) was afraid to give his scheduled presentation entitled "Electromagnetic Chronoscope" before this prestigious audience.] Faraday's idea included the propagation of magnetic disturbances by means of transverse vibrations. Later James Clerk Maxwell (1831–1879) credited this idea for helping him determine the electromagnetic theory of light.

Maxwell was primarily a theoretician and began to translate Faraday's experimentally generated ideas into mathematical relations (1864). The resulting four relations express the fundamental laws for the electromagnetic spectrum. These relations along with Newton's laws, the three laws of thermodynamics, the quantum theory, and the theories of relativity form the basis for our current understanding of the physical universe. The Maxwell relations appear to be quite fundamental in that they are not modified by relativity as Newton's relations are when particle velocities approach the speed of light.

Heinrich Rudolf Hertz (1857–1894) confirmed Maxwell's predictions via experiments conducted between 1886 and 1888. Hertz used an oscillating electric spark near 10 cm wavelength to induce similar oscillations in a distant wire loop. Although this was the first microwave-like experiment, the first developments in the microwave region of the spectrum were carried out by Guglielmo Marconi (1874–1937) in the 20th century. Marconi built parabolic antennas to demonstrate wireless telegraphic communications via Hertzian waves up to 550 MHz (5.5×10^8 Hz). Marconi is considered the first "microwave engineer" because of these trendsetting experiments demonstrating the use of ever higher frequencies.

More detailed development of the microwave technology involving waveguides (hollow conductors for propagation of microwave frequencies above cutoff frequencies determined by dimensions and dielectric properties) was done by George C. Southworth in 1930. In preliminary experiments he also observed standing waves (1920) that proved to be functions of the dimensions of the experimental apparatus. In 1933 Southworth and his colleagues at AT&T sent and received telegraph messages over 6 m of enclosed transmission line (waveguide).

The next major advance in microwave technology was the development of a continuous wave source called the klystron in 1937. This microwave vacuum tube was invented at Stanford by Russell and Sigurd Varian and William Hansen. This invention was helped by the financial support from the Sperry Gyroscope Company and its desire to develop an instrument landing system for aircraft during poor weather conditions. Concurrently, companies (AT&T, ITT, Marconi) were supporting microwave developments primarily for use in communications.

Based on these developments before World War II, radio detection and ranging (radar) motivated an immense increase in microwave developments. Most of the microwave devices used today had their early development during WWII in the British and U.S. war laboratories. These developments are well documented in a 28 volume set written by members of the MIT Radiation Laboratory.

1.2   ADVANTAGES OF MICROWAVES

The advantages of microwaves over other regions of the electromagnetic spectrum arise from the short wavelength monochromatic radiation and from the width of the spectrum available for use. The short wavelength monochromatic radiation results in high directivity and resolving power of microwave antennas. At longer wavelengths, much larger physical structures would be required for the same directivity and resolving power. Here resolving power is defined as the ability of the antenna to resolve two objects with different reflectivities. By contrast, the infrared spectrum would be expected to be even more directive and possess higher resolving power, but some infrared sources are not monochromatic (emit only one coherent frequency) and suffer from significant attenuation when propagating through the atmosphere except in certain narrow "windows" or regions of low attenuation. Thus, for some applications microwaves give the best overall system performance, having properties of high directivity and resolving power and low atmospheric attenuation.

The second major advantage of microwaves is the wide frequency spectrum available for communications. For example, the frequency range between a 10 and 1 cm radiator is 27,000 MHz (27 GHz), or more than 50 times the combined frequency bandwidth of the AM radio, FM radio, and television allocations. In addition modern microwave sources and systems are readily modulated (modified to carry information) at large bandwidths so that all modulation formats and high data rates are possible.

1.3   APPLICATIONS OF MICROWAVES

The most important applications of microwaves are in communications and radar. In the communications application, radio relay systems for telephone and television are the largest markets, with Earth–space (satellite) communications growing rapidly. Numerous special communications applications such as point-to-point communications have replaced the use of telephone lines in metropolitan regions where the lines are very expensive to install. Within a few years, direct broadcast satellites will open an immensely competitive microwave market unknown heretofor.

Radar provided the major incentive for the development of microwave technology because only this region of the spectrum could provide the required resolution with antennas of reasonable size. More details of the communication and radar applications will be found in Chapters 2 and 3.

A relatively recent application of microwave technology has been electronic warfare. As the name implies, this is a military application of microwaves involving specialized broadband receivers and high-power jamming transmitters. The receivers are used to monitor the enemy's transmissions passively, primarily for intelligence purposes. The jammer transmitters are used to confuse and deceive the enemy's received signals thus making their systems less effective. More information on this application of microwaves is given in Chapter 4.

The most familiar consumer application of microwaves is the microwave oven. This application uses a minimum of sophisticated microwave circuitry but has revolutionized the technique for heating foods and other products without convectively heating their entire surroundings.

Microwaves have also flourished in basic research and science applications. Atomic clocks use microwave resonance interactions with either ammonia or cesium molecules to provide extremely stable oscillating frequencies. In a similar application, the fine structure of the hydrogen atom known as the Lamb shift was discovered with microwaves at 1420 MHz. This line has been observed in stellar radiation with masers (low-noise microwave amplifiers) by radio astronomers. Finally, microwave spectroscopy is used to study the structure of numerous molecules and crystals.

1.4 OVERVIEW OF MODERN MICROWAVE ENGINEERING

To compose a book of finite length, the material must be tailored to what the reader needs to know and already knows. Too many microwave books start out with Maxwell's equations, immediately immersing the reader in a highly complex and mathematical marathon. Microwaves is more exciting than that. The physical concepts are really the most important aspect of the art and certainly more interesting. The mathematics should follow after as required. We have chosen to begin with the more exciting system applications of microwaves.

Microwave theory and techniques of today permit circuits to be modeled on the computer in great detail. The circuit can be built up "on paper" (really in the computer memory) and exercised to ascertain its performance over wide frequency bands, temperature ranges, and variations in dielectric constants, mechanical dimensions, and components. The computer can account for nuisance problems like irregularities due to discontinuities, propagation velocities that may vary with frequency, and small attenuations that accumulate to have a large effect on overall design. Once the model on the computer is made to perform satisfactorily, a working model can be built up and tested. The computer comes to the rescue again. Now the measurements can be carried out automatically with great precision using a computer-controlled network analyzer. If any difference exists between the experimental model and the computer model, it can be reconciled by experimenting on the computer model. When the troublesome design parameter is isolated, a working model is modified and the modifications are noted for future designs. Before powerful computers and computer programs were available, the design cycle was much slower and more empirical. The mathematical concepts used today did not even exist in the 1950s. It was not possible to combine discrete components and transmission lines in the same formulation. Both types of elements are now easily handled by *ABCD* matrices, and all measurements are made in *S*-parameters (see Chapter 6). Mathematical manipulation of these two types of matrices provides all of the computational power required to solve most of the microwave analysis and design problems today using computers.

The computer has the added convenience that it can store for immediate recall a large collection of tables. Thus, when a large computer program such as SUPER-COMPACT® (see Chapter 7) is used, all of the lookup information is already on file and automatically called up by the program as needed.

What do microwave engineers need to know today in order to carry out their work? There can be no substitute for going back and learning all that the literature has to offer. It is most interesting to go back by seeking out the root of the design being pursued. Frequently the original authors gave a better overview of the design than do those who have built upon their work. This is accomplished by looking up references and references in the references, and so on. But as a beginning point, a book for new microwave engineers must dwell on the types of technology being used in the design of devices now. The microwave state-of-the-art today makes use of semiconductor devices for generating, controlling, amplifying, and detecting microwave signals. The more advanced systems are made using microwave integrated circuits (MIC) (see Chapter 8). The information processing at radio frequency (RF) is done on transmission line structures using a high dielectric substrate (to keep it small) and transverse electromagnetic (TEM) transmission lines (see Chapter 5), typically microstrip. The trend is toward higher frequencies (10 GHz and higher) in order to make the circuits their smallest (shorter wavelengths). The present art of MIC uses an alumina substrate, and the components are mounted on it as chips. Future art will make use of monolithic structures (see Chapter 16) in which the transmission lines and components are all made on gallium arsenide substrate much as other integrated circuits are made. In order to accomplish designs of this type, the computer models must be highly refined so that few remakes are required. Monolithic MIC must grow out of conventional MIC. Thus, learning the art of conventional MIC is a good first step in preparation for carrying out future microwave work.

1.5 MICROWAVE UNITS

Microwave engineering progressed rapidly during World War II. Because the early documentation used a mixture of English, CGS, and MKS units, this convention has continued into today's literature and practice. For example, the inside dimensions of a common-sized waveguide are 0.4 by 0.9 in., whereas two common 50-ohm coaxial air lines measure 3.5 and 7 mm for the inside dimension of the outer conductor. This book will use the International System (SI) of units and parenthetically reference the common usage units where appropriate.

REFERENCES

1. T. S. Saad, National Lecturer's Talk, IEEE Microwave Theory and Techniques Society, 1972.

Part One
Systems

2 | Microwave Communications Systems

2.1 INTRODUCTION

Communication has become more electronic since Samuel Morse sent the first public telegraph message in 1844. As the amount of transmitted information increases, the bandwidth or frequency spectrum over which the information is transferred must increase.

Communication via radio or microwaves began in 1895, when Guglielmo Marconi invented the wireless telegraph. Radio communications uses large portions of the radio spectrum. As bandwidth requirements and the cost of laying coax cables in cities rose in the 1950s, many communicators turned to microwaves. Today nearly every long-distance telephone call, television program, and data link involves a microwave link as part of the system. In this context microwaves are defined as the frequency range from 1 GHz to the frequencies where optical communications dominate (several thousand gigahertz).

Telephone, radio, TV, and data use both terrestrial and satellite links (see Table 2.1). These links must be along line-of-sight paths free of intervening obstructions between the transmitter and receiver. The microwave antennas on relay towers are usually located on hilltops with a clear view to the previous and following tower of the link. Earth–space paths to geostationary satellites positioned above the equator also must be free of obstructions but are frequently located in valleys to provide shielding from terrestrial transmitters that might interfere with the reception of the weak satellite signals. The desired signal power must exceed the combined noise and interference power by an amount specified by the signal-to-noise (S/N) ratio. The lower

Graph of Probability of Error Rates for Selected Binary Coding Systems

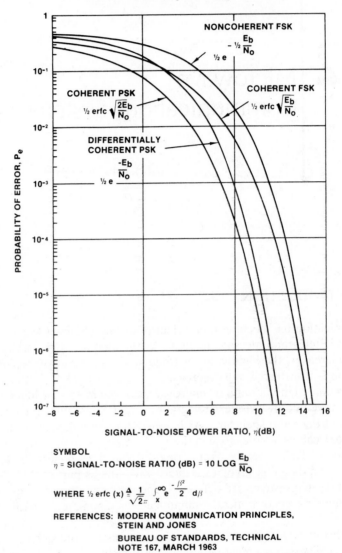

SYMBOL

η = SIGNAL-TO-NOISE RATIO (dB) = $10 \log \frac{E_b}{N_0}$

WHERE $\frac{1}{2}$ erfc (x) $\triangleq \frac{1}{\sqrt{2\pi}} \int_{x}^{\infty} e^{-\frac{\beta^2}{2}} d\beta$

REFERENCES: **MODERN COMMUNICATION PRINCIPLES, STEIN AND JONES**

BUREAU OF STANDARDS, TECHNICAL NOTE 167, MARCH 1963

Figure 2.1. *Probability of error versus signal-to-noise ratio for selected binary coded systems. Reprinted with permission of the publisher. (From* The Handbook of Digital Communications *1979. All rights reserved.)*

the S/N, the more difficult it is to reconstruct the desired data information. The presence of noise in digital data communication introduces errors (increases the bit error rate). This error rate is plotted in Fig. 2.1 for several types of modulation. The quality of analog signals is subjective and measured by human reactions to television signals as shown in Fig. 2.2.

Table 2.1   Typical Microwave Transmission System Characteristics

	Typical Digital Microwave Relay Systems Frequencies (GHz)	Type Transmission
Japan	3.6–4.2	8 QPSK—24 Mb/s
	10.7–17.7	4 QPSK—200 Mb/s per channel
	14.4–15.25	4 QPSK—200 Mb/s per channel
	17.7–20.2	4 QPSK—200 Mb/s per channel
United States	3.7–4.2	4 QPSK—20 Mb/s (TD-2)
	5.9–6.4	8 QPSK—43.3 Mb/s (DATRON)
	10.7–11.7	4 QPSK—40.15 Mb/s
	10.7–11.7	4 QPSK—70 Mb/s (MAV-12D)
	17.7–20.2	4 QPSK—274 Mb/s
England	30.67	ASK—100 Mb/s
	29.92	
Italy	13	4 QPSK—35 Mb/s

System	Date Introduced	Development of Bell Network Maximum Length (Miles)	Two-way Channels	Circuits/Channel Voice Video	Frequency Band in GHz
TD-2	1948	4000	5	480	3.7–4.2
TH-1	1959	4000	8	1860 or 1	5.9–6.4
TD-3	1966	4000	12	1200 or 1	3.7–4.2
TD-2B	1968	4000	12	1200 or 1	3.7–4.2
TH-3	1969	4000	8	1800 or 1	5.9–6.4
TD-2C	1972	4000	12	1500 or 1	3.7–4.2
TD-2	1973	4000	12	1500 or 1	3.7–4.2
TN-1	1974	1000	12	1800 or 1	10.7–11.7

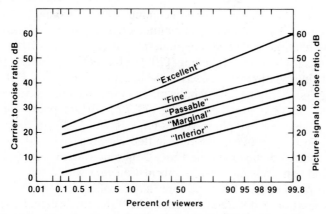

Figure 2.2. *Video signal-to-noise ratio versus viewer evaluation. Reprinted with permission of the publisher. (From* The Handbook of Digital Communications *1979. All rights reserved.)*

2.2   THE DECIBEL

The decibel parameter describes the ratio of two quantities. Typically, these are the output and input power of an amplifier, called the power gain of the amplifier. Mathematically,

$$\text{Power gain} = \frac{P_{out}}{P_{in}} \tag{2.1}$$

which in decibels is

$$\text{Gain} = 10 \log_{10}\left(\frac{P_{out}}{P_{in}}\right) \text{ dB} \tag{2.2}$$

The power gain in decibels is equal to 10 times the logarithm base 10 of the power gain (power ratio).

The decibel is one-tenth of a bel, a unit named after Alexander Graham Bell. It is the unit of change in audio level that is discernable by the ear (approximately).

Some values of decibels should be easy to remember. For example, if $P_{out}/P_{in} = 10$, then $\log_{10}(10) = 1$ and the gain is 10 dB. If $P_{out}/P_{in} = 2$, the gain is 3 dB. With the amplifier off, $P_{out}/P_{in} = 0.001$, the gain (loss) is 10 $\log_{10}(0.001) = -30$ dB.

2.3   MODULATION TECHNIQUES

The information is transmitted over a communication channel by modulating the transmitter. Typical analog communication modulation forms are frequency modulation (FM) and amplitude modulation (AM). Digital communication links use phase-shift keying (PSK) or frequency-shift keying (FSK).

Several abbreviations were used in Table 2.1 and Fig. 2.1. QPSK is quadrature phase-shift keying in which the phase of the transmitted signal is given one of four values (e.g., 0°, 90°, 180°, or 270°). Binary phase shift keying (BPSK) signals have two phase values (e.g., 0° and 180°). In coherent FSK there is coherence between the transmitted carrier signal and the receiver oscillator used to demodulate the FSK. Coherent systems are usually more efficient and can transmit a given information rate with a lower carrier-to-noise ratio for the same bit error rate than incoherent or differentially coherent systems. For differentially coherent PSK, the receiver oscillator is coherent for a block of PSK data, but this coherence may be lost between blocks of data.

The signal-to-noise ratio in Fig. 2.1 is expressed in terms of the energy in a bit (the smallest element of information) of the signal and the energy in the noise accompanying the signal into the receiver's detector stage. The energy bit (E_b) can be expressed as the power received in 1 s divided by the bit rate (R_b). If the symbol (letter or number) rate is specified, the number of bits per symbol is needed to compute the bit rate. Mathematically,

$$P_{received} = E_b R_b \tag{2.3}$$

2.4 SATELLITE TRANSMISSION TECHNIQUES

For satellite and other communication systems where more than one signal is carried by the link, multiple access techniques are used. These include

FDMA: frequency division multiple access
SCPC: single carrier per channel
TDMA: time division multiple access
CDMA: code division multiple access or spread spectrum

A summary of the multiple access methods is shown in Fig. 2.3.

The FDMA and SCPC techniques use assigned frequency channels to separate the signals. This is the method used for most commercial satellite communications (see Fig. 2.4). A new system being developed will use TDMA to separate the information using the access code shown in Fig. 2.3b. The information is contained in the frame immediately following the access code. This is shown in Fig. 2.5 for a satellite communication system. The advantage of TDMA over SCPC is that more ground stations can share a single satellite transponder. Note in Fig. 2.5 how the downlink signals are time phased to get to the three ground stations. These stations could be the #1 through #3 of the frame in Fig. 2.3.

CDMA or spread spectrum is useful for military communications systems because of its immunity to enemy jamming. Jamming is the transmission of interfering signals to increase the bit error rate or decrease the signal-to-noise ratio. CDMA requires a high bandwidth because many bits of information are sent for each symbol. The signal is reconstructed from the noise using coherent detection techniques with a minimal impact from an interfering signal not coherent with the desired transmission. The coherent–noncoherent discrimination is sufficient so that several desired transmissions with different codes can use the same frequency band with minimal interference to each other.

14

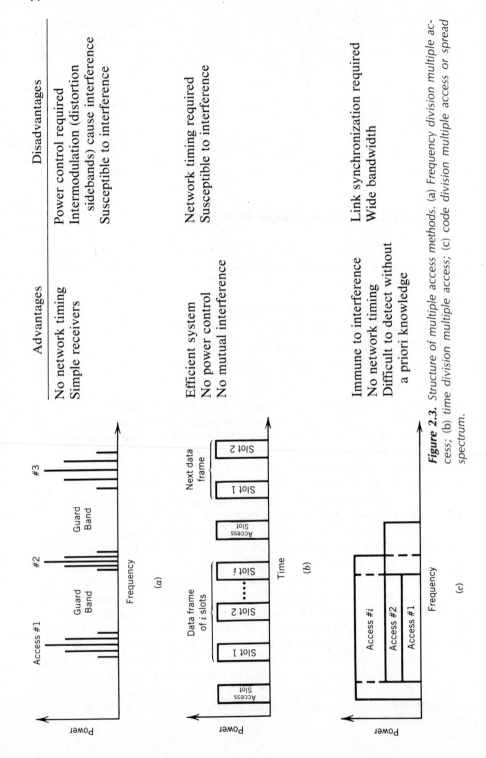

Advantages	Disadvantages
No network timing Simple receivers	Power control required Intermodulation (distortion sidebands) cause interference Susceptible to interference
Efficient system No power control No mutual interference	Network timing required Susceptible to interference
Immune to interference No network timing Difficult to detect without a priori knowledge	Link synchronization required Wide bandwidth

Figure 2.3. Structure of multiple access methods. (a) Frequency division multiple access; (b) time division multiple access; (c) code division multiple access or spread spectrum.

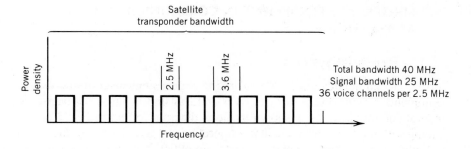

Satellite
transponder bandwidth

Power density

2.5 MHz

3.6 MHz

Frequency

Total bandwidth 40 MHz
Signal bandwidth 25 MHz
36 voice channels per 2.5 MHz

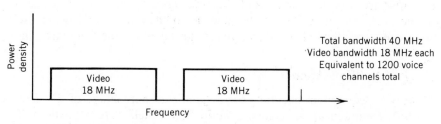

Power density

Video
18 MHz

Video
18 MHz

Frequency

Total bandwidth 40 MHz
Video bandwidth 18 MHz each
Equivalent to 1200 voice
channels total

Figure 2.4. *Frequency division multiple access (FDMA) technique.*

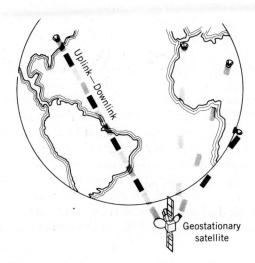

Uplink—Downlink

Geostationary
satellite

Figure 2.5. *Time division multiple access (TDMA) technique. Reprinted with permission of the publisher. (From* The Handbook of Digital Communications 1979. *All rights reserved.)*

2.5 MICROWAVE EQUIPMENT IN COMMUNICATIONS APPLICATIONS

2.5.1. Terrestrial Systems

Terrestrial communications systems are the largest single user of microwave equipment. A typical repeater (receiver and transmitter) used for transcontinental communications is shown in block diagram format in Fig. 2.6. Beginning with the transmitter oscillator operating at 2 GHz (typical), the signal is amplified in a class C bipolar transistor amplifier, multiplied in a varactor (variable reactance) multiplier, and filtered before being emitted by the antenna. In the receiver a filter, isolator (unidirectional transmission line) and a downconverter consisting of a mixer and an intermediate frequency (IF) amplifier are used to amplify and shift the signal frequencies down to frequencies better suited for filtering and demodulation. A microwave (solid state Gunn) oscillator is used to provide the signal for the frequency shift.

2.5.2 Satellite Systems

Microwave equipment accounts for a significant percentage of the cost of satellite communication systems. Here both satellite transponders (frequency-shifting repeaters) and ground stations require microwave devices.

Aboard the satellite several types of transponders are used (see Fig. 2.7). A transponder is a receiver, frequency converter, and transmitter designed to pass the uplink signal through to the downlink. If the frequency were not converted, the receiver would hear the satellite transmitter and a condition of "lock up" would occur (similar to positive feedback).

Satellite antennas are designed to radiate to fixed portions of the Earth on the downlink. The Intelsat V antenna patterns are shown in Fig. 2.8. Spot beams are positioned to the regions with the highest data rate requirements, while zone beams cover the continents with high telecommunication requirements and hemisphere beams cover the remainder of the region. These complex satellite antennas are used to provide the undulating contours shown in Fig. 2.8. Future antennas will use multibeam phase-shifted designs (phased arrays). The antennas will be similar to the phased array antennas used in radar systems, except they will be lighter and use less power. These components and subsystems are discussed in more detail in later chapters.

A sketch of a large Earth station is shown in Fig. 2.9; the corresponding system block diagram is in Fig. 2.10. Earth stations are complex because it is desirable to do the data processing on the ground. As technology develops, it is expected more complexity will be shifted into the satellite.

Figure 2.6. *Transmitter and receiver block diagrams for a transcontinental repeater.* (a) *Transmitter;* (b) *receiver.*

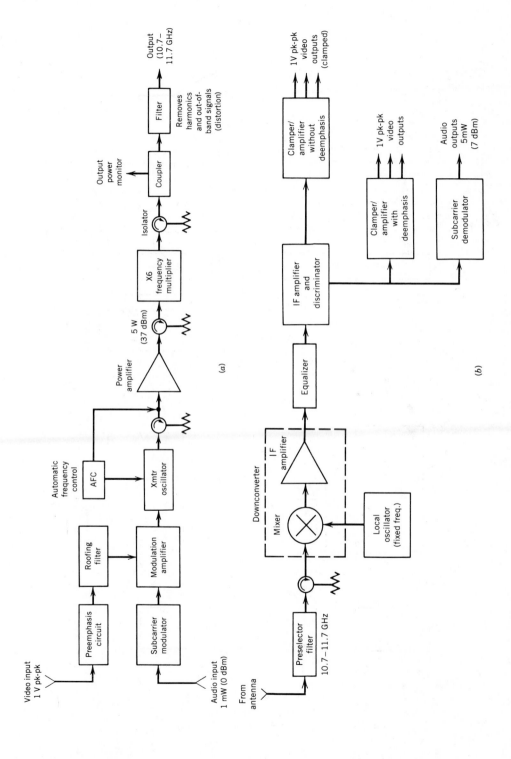

Video input 1 V pk-pk

Preemphasis circuit

Roofing filter

Modulation amplifier

Subcarrier modulator

Audio input 1 mW (0 dBm)

Automatic frequency control

AFC

Xmtr oscillator

Power amplifier

5 W (37 dBm)

X6 frequency multiplier

Isolator

Output power monitor

Coupler

Filter

Removes harmonics and out-of-band signals (distortion)

Output (10.7– 11.7 GHz)

(a)

From antenna

Preselector filter

10.7–11.7 GHz

Downconverter

Mixer

IF amplifier

Local oscillator (fixed freq.)

Equalizer

IF amplifier and discriminator

Clamper/ amplifier without deemphasis

1V pk-pk video outputs (clamped)

Clamper/ amplifier with deemphasis

1V pk-pk video outputs

Subcarrier demodulator

Audio outputs 5mW (7 dBm)

(b)

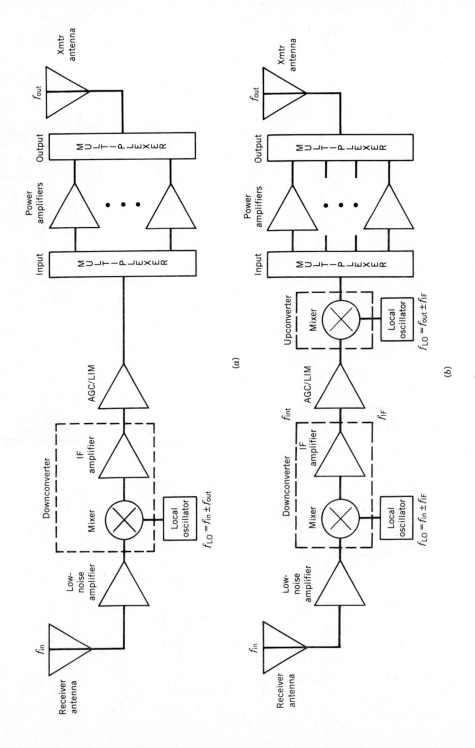

(a)

(b)

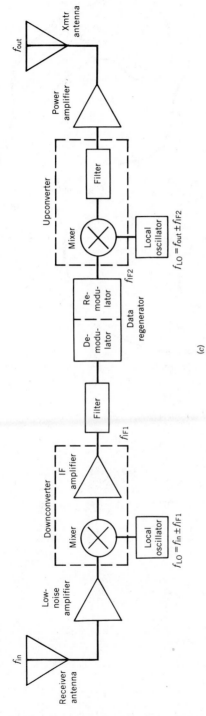

Figure 2.7. Types of satellite transponders. (a) Single conversion transponders; (b) double conversion transponder; (c) data regenerative transponder. IF, intermediate frequency; AGC, automatic gain control; LIM, limiter; LO, local oscillator.

Figure 2.8. Antenna pattern for the Atlantic Intelsat V Satellite. Reprinted with permission of the publisher. (From The Handbook of Digital Communications 1979. All rights reserved.)

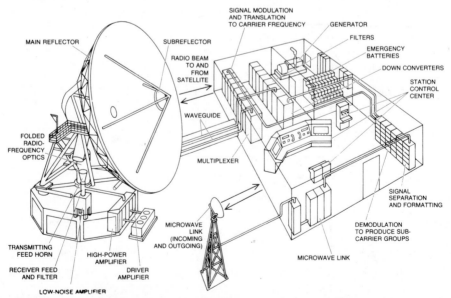

Figure 2.9. A typical large Earth station layout. Reprinted with permission of the publisher. (From The Handbook of Digital Communications 1979. All rights reserved.)

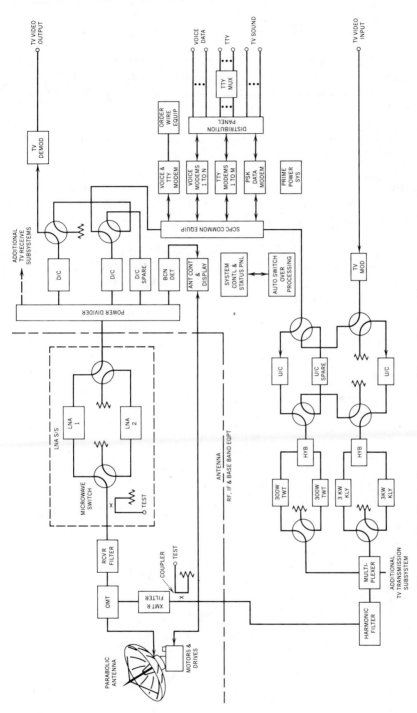

Figure 2.10. Block diagram of a typical large Earth station. OMT, orthomode transducer; LNA, low-noise amplifier; S/S, subsystem; D/C, downconverter; U/C, upconverter; BCN, beam control network; TWT, Traveling wave tube; KLY, klystron; TTY, teletype. Reprinted with permission of the publisher. (From The Handbook of Digital Communications 1979. All rights reserved.)

2.6 THE LINK EQUATION

Analysis of the performance of a terrestrial or satellite communication system requires the determination of the signal and noise levels at the receiver detector to estimate the quality of the transmission. The technique for estimating the signal level is the link equation.

The power of a typical television signal (see Fig. 2.11) from the time it enters one Earth station, makes a round trip through the satellite transponder, and finally leaves another Earth station covers a dynamic range from less than 10^{-17} W to the effective equivalent of nearly 10^9 W (a range from -170 dBW—decibels above 1 W—to more than 80 dBW). The largest losses occur on the Earth–satellite path. The largest gains are produced by the antennas at the earth stations. On the uplink a 30-m-diameter antenna boosts the signal density by a factor of about 1 million, i.e., from an output power of a few hundred watts to an effective isotropically radiated power (EIRP) of several megawatts. This EIRP represents the power that would be required at the Earth station if the signal were being radiated equally in all directions instead of being focussed.

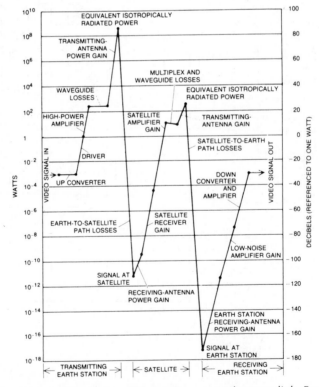

Figure 2.11. *Gains and losses for a TV signal on an Earth–space link. Reprinted with permission of the publisher. (From* The Handbook of Digital Communications *1979. All rights reserved.)*

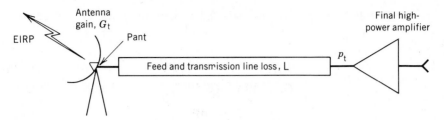

Figure 2.12. *Earth station EIRP parameters. L, total loss from amplifer to antenna feed; EIRP = G_tP_t/L; EIRP(dBW) = P_t(dBW) − L(dB) + G_t(dB).*

2.6.1   Effective Isotropic Radiated Power

The uplink EIRP is computed using the simplified block diagram in Fig. 2.12. The three important parameters are the power output of the high-power amplifier P_t, the waveguide and feed losses L, and the transmitter antenna gain G_t. For this formulation, the loss L is the ratio of the power at the output of the amplifier P_t to the power at the output of the feed horn P_{ant},

$$L = \frac{P_t}{P_{ant}} \tag{2.4}$$

greater or equal to 1. The EIRP is then

$$EIRP = \frac{P_t G_t}{L} \tag{2.5}$$

or expressed in dBW (decibels referenced to 1 W),

$$EIRP(dBW) = P_t(dBW) - L(dB) + G_t(dB) \tag{2.6}$$

Note that the loss and gain are expressed in decibels because they are ratios and are not referenced to any specific power level.

Consider an example of the computation of the EIRP based on the gain–loss curve in Fig. 2.11. Assume that 20% of the power from the final amplifier is lost in the waveguide and feedhorn. Using the relation $L = P_t/P_{ant}$ and the fact the $P_{ant} = 0.8\,P_t$, results in

$$L = \frac{P_t}{P_{ant}} = \frac{P_t}{0.8\,P_t} = 1.25 \tag{2.7}$$

Also given the fact that the gain $G_t = 1,000,000 = 10^6$, the EIRP for a 500 W transmitter is

$$EIRP = \frac{P_t G_t}{L} = \frac{(500)(10^6)}{1.25} = 4 \times 10^8 \text{ W} \tag{2.8}$$

Working the same numbers in decibels and decibels referenced to 1 W results in

$$P_t = 500 \text{ W} = 27 \text{ dBW} \qquad (2.9a)$$

$$G_t = 10^6 = 60 \text{ dB} \qquad (2.9b)$$

$$L = 1.25 = 1 \text{ dB} \qquad (2.9c)$$

and using the relation given earlier

$$\text{EIRP(dBW)} = P_t(\text{dBW}) - L(\text{dB}) + G_t(\text{dB}) \qquad (2.10a)$$

$$= 27 \text{ dBW} - 1 \text{ dB} + 60 \text{ dB} \qquad (2.10b)$$

$$= 86 \text{ dBW} \qquad (2.10c)$$

Using the inverse of the decibel relation

$$X(\text{dB}) = 10 \log_{10}(Y) \qquad (2.11)$$

Namely,

$$Y = 10^{[X(\text{dB})/10]} \qquad (2.12)$$

yields

$$P_t = 10(86 \text{ dB}/10) = 10^{8.6} = 4 \times 10^8 \quad \text{W} \qquad (2.13)$$

which agrees with the number calculated earlier.

2.6.2 Antenna Gain

In the previous example is was stated the 30-m-diameter parabolic reflector antenna has a gain of 10^6. The antenna gain is estimated by the relation

$$G = \frac{4\pi A_e}{\lambda^2} = \frac{4\pi A \rho}{\lambda^2} \qquad (2.14)$$

where

A_e = the effective area in meters squared

A = the physical area of the antenna's reflector = $\pi d^2/4$

d = the diameter of the antenna in meters

ρ = the efficiency of the antenna

λ = the wavelength of the radiation in meters = c/f

c = the speed of light in meters/second = 3×10^8 m/s

f = the frequency in hertz

Typical efficiencies for antennas range from 0.6 to 0.65.

2.6.3 Antenna Beamwidth

The 3 dB ($\frac{1}{2}$ power) beamwidth of a parabolic antenna is useful when estimating the pointing accuracy required of an antenna system. The relation is

$$\text{beamwidth} = \frac{65\lambda}{d} \text{ deg} \qquad (2.15)$$

where λ and d are in the same units as defined for Eq. 2.14.

2.6.4 Waveguide and Feed Losses

Waveguide and feed losses rob power from the output of the final amplifier and introduce additional noise into the receiver. Generally these losses are less than 2 dB.

2.6.5 Space Loss

The remainder of the uplink loss is termed space loss due to the spreading of the RF energy as it propagates through space. Space loss can be derived using the power density concept. The power density from a transmitter measured 1 m from an isotropic antenna (radiates equally in all directions) is four times as large as the power density measured 2 m from the same antenna. At the larger distance, the transmitted power is distributed over a sphere with four times the surface area. Writing this mathematically for an isotropic antenna

$$\text{Transmitted power density} = PD_t \qquad (2.16)$$

$$= \frac{P_t}{4\pi R^2} \text{ W/m}^2$$

where P_t is in Watts and R is in meters. The power received at a range R by a receiver with a unity gain antenna can be found from Eq. 2.14:

$$G = \frac{4\pi A_e}{\lambda^2} = 1 \qquad (2.17)$$

Solving Eq. 2.17 for the effective area of a unity gain antenna gives

$$A_{eu} = \frac{\lambda^2}{4\pi} \tag{2.18}$$

This is the area of the surface of the sphere with radius R that delivers power into a receiver antenna with unity gain. Therefore, the power received by a unity gain system is

$$P_r = PD_t A_{eu} \tag{2.19}$$

where

P_r = the received power in Watts

PD_t = the transmitted power density in Watts per square meter

A_{eu} = the effective area of the unity gain antenna in square meters

Substituting Eqs. 2.16 and 2.18 into 2.19 gives

$$P_r = \frac{P_t}{4\pi R^2} \frac{\lambda^2}{4\pi} = \left(\frac{\lambda}{4\pi R}\right)^2 P_t \tag{2.20}$$

which depends on the signal frequency because of the λ^2 dependence. The ratio of the transmitted power to the received power is the space loss ratio:

$$\text{Space loss ratio} = \frac{P_t}{P_r} = \left(\frac{4\pi R}{\lambda}\right)^2 \tag{2.21}$$

The space loss is usually expressed in decibels as

$$\text{Space loss} = 20 \log_{10}\left(\frac{4\pi R}{\lambda}\right) \text{ dB} \tag{2.22}$$

The numbers associated with space loss ratio and space loss are large. For example, for the range from the synchronous orbit to the satellite's suborbital point (R = 35,860 km) and an operating frequency of 4 GHz the space loss ratio is

$$\text{Space loss ratio} = \left(\frac{4\pi R}{\lambda}\right)^2$$

$$= \left(\frac{4(3.14)(3.586 \times 10^7)}{0.075}\right)^2 \tag{2.23}$$

$$= 3.61 \times 10^{19}$$

which means that the received power at the satellite, assuming it had a unity gain antenna, would only be $1/(3.61 \times 10^{19})$ of the EIRP transmitted. The space loss in decibels is

$$\text{Space loss} = 10 \log_{10}(3.61 \times 10^{19})$$
$$= 196 \text{ dB}$$

(2.24)

2.6.6   Earth Station *G/T*

The weakest signal portion of a satellite path usually occurs on the downlink. For the example shown in Fig. 2.11, the received power at the ground is about 10^{-17} W. At this point, the effects of noise are most important.

The G/T parameter is a measure of the Earth station's ability to receive weak signals in noise. G is the antenna gain of Eq. 2.14, and T is an effective temperature used to describe the amount of noise power received.

The available thermal noise power P_n generated by a receiver of bandwidth B_n (measured in Hertz) at a temperature T (measured in degrees Kelvin) is

$$P_n = kTB_n \quad \text{W}$$

(2.25)

where k is Boltzmann's constant $(1.38 \times 10^{23}$ J/deg). The noise bandwidth for most receivers with sharp IF filters is nearly equal to the 3-dB bandwidth of the receiver. If $T = 290°$K, which corresponds to a room temperature near 62°F, the factor kT is 4×10^{-21} W/Hz bandwidth.

2.6.7   Receiver Noise Factor/Figure

Any receiver adds noise to the signal above that which would be added by an ideal receiver operating at $T = T_o = 290°$K = standard temperature. The ratio of the noise out of a nonideal receiver to that predicted from an ideal receiver is termed the noise factor:

$$F_n = \text{noise factor} = \frac{\text{Noise from practical receiver}}{\text{Noise from ideal receiver}}$$
$$= \frac{N_o}{kT_iB_nG_r}$$

(2.26)

where

$$N_o = \text{noise output from practical receiver}$$

$$G_r = \text{receiver gain}$$

$$T_i = \text{equivalent input temperature}$$

This equation can be rearranged into another familiar form using the relations G_r = output signal power/input signal power = S_o/S_i and N_i the input noise power = kT_iB_n to give

$$F_n = \frac{S_i/N_i}{S_o/N_o} \tag{2.27}$$

The noise factor may be interpreted as a measure of the signal-to-noise ratio degradation as the signal passes through the receiver.

The noise factor F_n can be related to an equivalent noise temperature T_{eq} for the receiver referenced to the standard temperature T_o = 290°K

$$T_oF_n = T_o + T_{eq} \tag{2.28}$$

or rearranging

$$T_{eq} = T_{rcvr} = T_o(F_n - 1) \tag{2.29}$$

In this relation the F_n must be expressed as a ratio and not in decibels. In this text, when F_n is expressed in decibels it is called noise figure.

2.6.8   Other Contributions to T

Figure 2.13 shows several other contributions to the T-parameter including the noise added by the losses in the feedhorn and transmission lines connecting the antenna to the receiver and the effective losses due to noise entering the sidelobes of the antenna. These two terms are

$$\text{Noise temperature due to waveguide and feedhorn} = \frac{L-1}{L}T_o \tag{2.30}$$

$$\text{Noise due to antenna sidelobes} = \frac{T_A}{L} \tag{2.31}$$

Sometimes losses incurred during propagation through the troposphere or ionosphere must be considered (2).

2.6.9   Effective Noise Temperature

The total effective noise temperature is the sum of the effective temperature contributions, namely

$$T = T_{\text{antenna sidelobes}} + T_{\text{feed losses}} + T_{\text{rcvr}} \tag{2.32}$$

$$= \frac{T_A}{L} + \frac{L-1}{L}T_o + T_{\text{rcvr}} \text{ in degrees Kelvin}$$

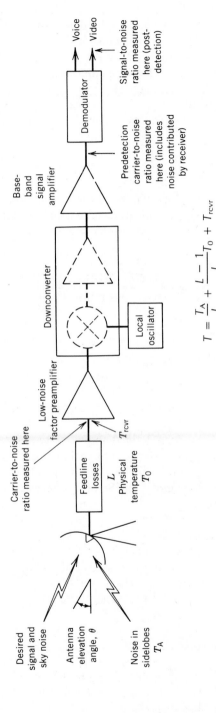

Figure 2.13. *Parameters of the Earth Station G/T.*

$$T = \frac{T_A}{L} + \frac{L-1}{L}T_0 + T_{rcvr}$$

The values of the loss in waveguide and the feedhorn are the same order of magnitude as those described for the transmitter, typically 1–2 dB.

The noise temperature introduced by the sidelobes of the antenna and the atmospheric background to which the antenna is pointed is more difficult to estimate. This last term is called the sky noise temperature. Typical values for this contribution are 10° or less for antenna elevation angles exceeding 10° or 15° above the horizon. The sky noise temperature is plotted in Fig. 2.14.

2.6.10 Signal-to-Noise Ratio

The goal of these computations has been the calculation of the signal-to-noise ratio to estimate the communications signal quality. The signal-to-noise ratio measured at the output of the baseband amplifier shown in Fig. 2.13 is

$$\frac{S}{N} = \frac{(\text{Satellite EIRP})(G_a)}{(\text{Space loss ratio})kTB_n} \qquad (2.33)$$

where

$$S = \text{the signal power}$$

$$N = \text{the noise power}$$

$$G_a = \text{the receiving station antenna gain}$$

The other parameters have been defined above. Note that G_a is *not* the gain of the satellite's antenna, since that gain is included in the EIRP.

The data from Fig. 2.11, and estimates for the quality of the receiver and the antenna can be used to compute the S/N for the television example given earlier.

Example 2.1. Assume the satellite EIRP is 300 W, $G_a = 10^6$, the space loss ratio is 3.61×10^{19} (computed earlier), $T = 600°\text{K}$, and the noise bandwidth is 4.5×10^6 Hz to assure that the full bandwidth of the signal is received. The signal-to-noise ratio using Eq. 2.33 is S/N = 223 = 23.5 dB. From Fig. 2.2 most viewers would consider the signal to be inferior.

2.7 S/N versus C/N versus E_b/N_o

The signal-to-noise ratio (S/N) usually refers to the ratio of the power in the signal to the power in the noise measured after the demodulator of the receiver (the point labeled video in Fig. 2.13). The carrier-to-noise ratio (C/N) is measured at the antenna terminals before the receiver noise is added

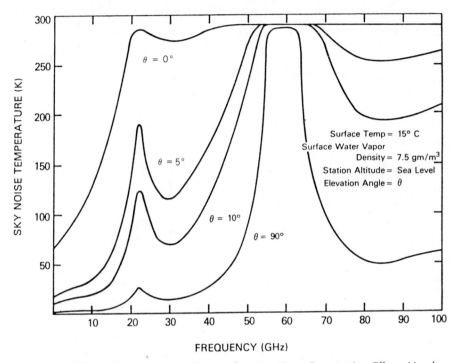

Figure 2.14. *Sky noise temperature due to clear air. (From* Propagation Effects Handbook, *Ref. 2.2.)*

to currupt the information. Both of these terms are dependent on the bandwidth of the signal and the receiver because this same bandwidth is required to pass the signal to the demodulator.

When the link equation is written in terms of E_b/N_o, the bandwidth dependence is lost because N_o is the noise density per Hertz (the noise power measured per Hertz of bandwidth). This is a convenient terminology for computing the bit error rate in digital communication links.

2.8 CALCULATION OF EIRP

The S/N can be used to estimate the EIRP required for a given bit error rate on a digital channel. For example, for the NASA Tracking and Data Relay Satellite System (TDRSS) the S-band (about 2 GHz) single access link is made up of two parts–the user-to-TDRS link and the TDRS-to-ground (White Sands, NM) link. In this system, the user is another satellite orbiting in a nongeostationary orbit. Because of this double hop, noise introduced into the user–TDRS link is retransmitted and combined with the noise in the TDRS-to-ground link. The distribution of the S/N's between the links is

Table 2.2   Typical Link Budgets for the Communications Technology Satellite (CTS)

Typical U.S. User Video Transmit Ground Terminal

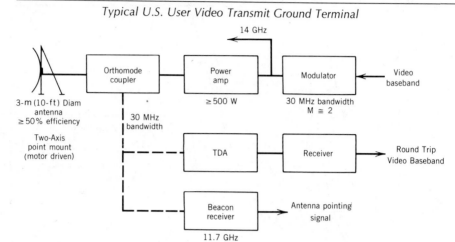

Sample Uplink Calculation for Lewis Ground Station
(Uplink frequency, 14.2 GHz)

Characteristic	Spacecraft Receiver Noise Temperature (K)	
	1315	2315
Terminal:		
Transmitter power (1250.0 W), dBW	30.97	30.97
Feed loss, dB	−2.00	−2.00
Antenna gain 4.88 m (16.0 ft), 0.31° half-power beam width (HPBW), dB	54.53	54.53
Effective Isotropic Radiated Power (EIRP), dBW	83.50	83.50
Antenna pointing error (0.05°), dB	±0.26	−0.26
Margin, dB	−3.00	−3.00
Propagation loss (23 074 statute miles; latitude, 41.4°; relative longitude, 35.1°), dB	−207.22	−207.22
Atmospheric loss (0.100% outage; CCIR Rainfall Region 2), dB	−2.23	−2.23
Polarization loss, dB	−0.25	−0.25
Spacecraft:		
Feed loss, dB	−0.00	−0.00
Antenna gain [0.70 m by 0.70 m (2.3 ft by 2.3 ft); 2.15° by 2.15° HPBW], dB	37.68	37.68
Antenna pointing error (0.38°), dB	−0.31	−0.31
Received carrier power, dBW	−92.03	−92.03
Noise power density, dBW/Hz	−197.41	−194.96
Bandwidth, dB (Hz) (27.0 MHz)	−123.10	−120.04
Carrier-power/receiver-noise ratio, dB	31.02	28.56

Typical U.S. User Video Receive Ground Terminal

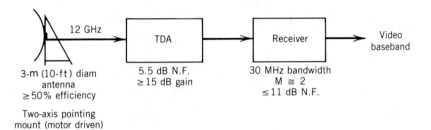

Sample Downlink Calculation for Lewis Ground Station
(Downlink frequency, 12.1 GHz)

Characteristic	Spacecraft Receiver Noise Temperature (K)	
	1315	2315
Spacecraft:		
Output tube power (200 W), dBW	23.01	23.01
Feed loss, dB	−0.00	−0.00
Antenna gain [0.70 m by 0.70 m (2.3 ft by 2.3 ft); 2.52° by 2.521° + HPBW], dB	36.28	36.28
Effective Isotropic Radiated Power (EIRP), dBW	59.29	59.29
Antenna pointing error (0.38°), dB	−0.22	−0.22
Margin, dB	−3.00	−3.00
Propagation loss (23 074 statute miles; latitude, 41.4°; relative longitude, 35.1°), dB	−205.81	−205.81
Atmospheric loss (0.100% outage; CCIR Rainfall Region 2), dB	−1.52	−1.52
Polarization loss, dB	−0.25	−0.25
Terminal:		
Feed loss, dB	−1.00	−1.00
Antenna gain [4.88 m (16.0 ft); 0.30° HPBW], dB	53.12	53.12
Antenna pointing error (0.05°), dB	−0.18	−0.18
Received carrier power, dBW	−99.58	−99.58
Noise power density (T = 800 K), dBW/Hz	−199.57	−199.57
Bandwidth, dB (Hz) (27.0 MHz)	74.31	74.31
Terminal receiver noise power, dBW	−125.26	−125.26
Uplink noise contribution (C/N, 31.02; 28.6 dB), dB	0.95	1.80
Terminal net noise power, dBW	124.14	123.45
Terminal carrier-power/receiver-noise ratio, dB	24.56	23.87
FM improvement (M = 2.00), dB	21.58	21.58
Noise weighting factor (CCIR), dB	10.20	10.20
Preemphasis improvement, dB	2.40	2.40
Signal/noise ratio, dB	58.91	58.05

beyond the scope of this text. The results are (for an acceptable bit error rate)

User-to-TDRS	S/N = 13.3 dB
TDRS-to-ground	S/N = 21.7 dB
Overall	S/N = 12.7 dB

(2.34)

The *S*-band link operates between the TDRS and one of its users (e.g., the Space Telescope). The 2.2555-GHz BPSK signal covers a maximum range of 4.3×10^4 km (26,800 statute miles). The TDRS bandwidth is 18.4 MHz and the *S*-band receiving antenna on TDRS is 4.9 m (15 ft) in diameter. The space loss ratio is

$$\text{SLR} = \left(\frac{4\pi R}{\lambda}\right)^2 = \left(\frac{4\pi Rf}{c}\right)^2 = 1.64 \times 10^{19} \qquad (2.35)$$

corresponding to 192.2 dB. The G/T for the satellite receiver is 8.57 dB (corresponding to a ratio of 7.19), i.e., $G/T = 7.19$. Assuming the gain of the 4.9-m-diameter satellite antenna is

$$G_a = \frac{4\pi A}{\lambda^2}\rho = \left(\frac{\pi df}{c}\right)^2 \rho = 7400 \qquad (2.36)$$

corresponding to 38.7 dB, the noise temperature of the satellite receiver is

$$T = G/7.19 = 1030 \text{ °K} \qquad (2.37)$$

This noise temperature arises because of (1) the warm earth (a blackbody radiator) is in the antenna's field of view, (2) transmission line losses between the antenna and the receiver's input stage, and (3) the noise in the receiver. G_a/T could be used directly in the link equation, but it is instructive to see its two components separately. An S/N of 13.3 dB (ratio of 21.4) is required for this link.

Substituting into the link equation (Eq. 2.33)

$$\text{EIRP} = (\text{S/N})(\text{SLR})kB_n(T/G_a)$$
$$= (21.4)(1.64 \times 10^{19})(1.38 \times 10^{-23})(1.84 \times 10^7)(7.19)^{-1} \quad (2.38)$$
$$= 12400 \text{ W}$$

corresponding to 40.9 dBW.

On the Space Telescope the *S*-band antenna is 42 in. (1.07 m) in diameter. The resulting antenna gain is

$$G_t = \rho\left(\frac{\pi df}{c}\right)^2 = 400 \quad (26 \text{ dB}) \qquad (2.39)$$

The required transmitter power at the antenna is

$$\text{EIRP}/G_t = 31 \text{ W} \tag{2.40}$$

in order to maintain an acceptable bit error rate for the maximum data rate allowed on this link. Since the Space Telescope will require less bandwidth because it uses less than the maximum data rate, a lower EIRP is required to maintain the overall S/N at the satellite and ground station.

2.9 ACTUAL LINK CALCULATIONS

The link budgets considered above are simplified from those required for actual system design. Actual link budget calculations for the 12/14 GHz television links on the Communications Technology Satellite are shown in Table 2.2.

2.10 COMMUNICATION SYSTEM COMPONENTS

From the link equation it is clear that there are several tradeoffs between the power level of the transmitter, antenna gains, and receiver amplifier noise figure. There is no unique solution to the design of a communication system. The design depends on the costs and constraints of the various components. Many of these components are discussed in this book. Included are the waveguides, filters, low-noise amplifiers, antennas, and high-power amplifiers.

PROBLEMS

1. What is the gain in decibels of an amplifier if the power gain is 4?

2. What is the power gain ratio of a 5-dB amplifier?

3. Calculate the gain of a $d = 30$-m diameter antenna operating at a frequency of 4×10^9 Hz (4 GHz), $d = 15$ m at 4 GHz, and $d = 30$ m at 2 GHz. Express the answers as gain ratios and decibels.

4. Calculate the 3-dB beamwidth for the three antenna–frequency combinations in Problem 3.

5. What percentage of the output power of the final amplifier is lost in a 2-dB loss feed system?

6. Compute the space loss ratio and space loss for a terrestrial microwave link with tower spacing of 60 km at 4 GHz.

7. Compute the space loss ratio and space loss for an equatorial ground

station directly below a geostationary satellite with a 12-GHz downlink. What is the frequency dependence of the space loss ratio?

8. What is the effective area of a unity gain antenna operating at 4 and 12 GHz? What is the frequency dependence of A_{eu}? What is the gain of an antenna whose effective area is $\lambda^2/4\pi$?

9. What is the S/N for the audio channel if the EIRP for audio is 30 W and the bandwidth is 10^5 Hz for the link in Example 1?

10. If 30 W of EIRP are dedicated to the audio, only 270 W are available for video. Compute the video S/N for this case using the parameters in Example 1.

11. Derive the link equation (Eq. 2.33) for E_b/N_o rather than S/N.

12. The spacing between two repeater towers in a 4-GHz microwave telephone link is 50 km. If the voice is digitized at 6000 bits per second and a bit error rate of 10^{-5} ($E_b/N_o = 10$ dB) is desired, what is the minimum EIRP for this channel? Other parameters of the link are

    Diameter of circular receiving antenna = 2 m
    Total receiver temperature = $T = 700\ °K$
    Diameter of circular transmitting antenna = 3 m
    Efficiency of antennas = 0.6 $k = 1.38 \times 10^{-23}$ J/deg

    Why is the answer so small?

13. During the Saturn flyby the *Voyaguer* spacecraft sent back digital data at 7200 bits per second. The link operated at 10 GHz with a 3.66 m-diameter satellite antenna. The signals were received by the Deep Space Network ground stations with 67-m-diameter parabolic antennas. The effective noise temperature of these receivers is 25 °K (maser front end mounted at the focal point). Assume the other noise contributions are 25 °K. The distance to Saturn is 1.45×10^9 km! Assume the antenna efficiencies are 63%. What was the minimum EIRP and the power into the *Voyaguer* antenna to yield a data bit error rate of 10^{-5}? The actual transmitter operated at 12 W (low-power mode) and 21 W (high-power mode). Why is your answer lower than either of these numbers?

REFERENCES

1. C. L. Cuccia, ed., *The Handbook of Digital Communications*, EW Communications, Palo Alto, CA, 1979.
2. L. J. Ippolito, R. D. Kaul, and R. G. Wallace, *Propagation Effects Handbook for Satellite Systems Design*, 3rd ed. NASA Reference Publication 1082(03), June 1983.

3 | Microwave Radar Systems

3.1 INTRODUCTION

Radar (radio detection and ranging) uses radio energy to extend the operator's ability to detect and monitor objects. A radar can observe through media (fog, rain, clouds) opaque to the human eye when it operates at microwave and millimeter wave frequencies not greatly attenuated by these media. Radars transmit a known waveform and detect the scattered energy from objects (called targets) and the natural surrounding background (called clutter). Selective waveforms and signal processing are used to differentiate between desired objects and clutter. The waveforms allow measurement of target parameters such as radial range, velocity, and size. The nonradial components of position and velocity are usually sensed by the radar antenna. This chapter explains how the radar transmitter, receiver, and antenna design establish the radar performance.

Radar has become a familiar safety device for all air travelers (the primary commercial application) and for surveillance, tracking, and weapons delivery systems (military applications). The commercial systems are frequently adaptations of military systems.

The U.S. military uses a formal equipment designation system for its electronic equipment beginning with the letters AN (formerly meant Army/Navy, but now respresent the Joint Electronics Type Designation System). Following the AN/ are three letters denoting platform, type, and function, respectively, followed by sequentially assigned numbers and possibly (. . .) with modification sequence delineated by a letter starting with A. A typical radar designation is AN/MPQ-46, a mobile continuous wave (CW) tracker

37

and illuminator for the Hawk missile system. The definition of the letters is given in Table 3.1. The designations AN/MPQ-46( ) or AN/MPQ-( ) refers to systems not yet assigned formal modification or sequence numbers. The ( ) are referred to as "bowlegs." Commercial radars are given designators such as the ARSR-3 (air route surveillance radar).

The equipment layout sketch in Fig. 3.1 gives an indication of the number of cabinets of equipment needed for the ARSR-3 radar. Dual systems operate simultaneously on opposite polarizations and slightly different frequencies. This allows operation with one system shut down for repairs. The mean time to failure for one side of this radar exceeds 1500 hr (3 months). The avail-

Table 3.1 Joint Electronics-Type Designation System Definitions

Platform Installation

A	Piloted aircraft	S	Water
B	Underwater mobile, submarine	U	General utility
D	Pilotless carrier	V	Vehicular (ground)
F	Fixed ground	W	Water surface and underwater combination
G	General ground use		
K	Amphibious	Z	Piloted–pilotless airborne vehicle combination
M	Mobile (ground)		
P	Portable		

Equipment Type

A	Invisible light, heat, radiation	N	Sound in air
C	Carrier	P	Radar
D	Radiac	Q	Sonar and underwater sound
G	Telegraph or teletype	R	Radio
I	Interphone and public address	S	Special or combinations of types
J	Electromechanical or inertial wire covered	T	Telephone (wire)
K	Telemetering	V	Visual and visible light
L	Countermeasures	W	Armament
M	Meteorological	X	Facsimile or television
		Y	Data processing

Equipment Function or Purpose

B	Bombing	N	Navigational aids
C	Communications	Q	Special or combination of purposes
D	Direction finder, reconnaissance, and/or surveillance		
E	Ejection and/or release	R	Receiving, passive detecting
G	Fire control or searchlight directing	S	Detecting and/or range and bearing, search
H	Recording and/or reproducing	T	Transmitting
K	Computing	W	Automatic flight or remote control
M	Maintenance and/or test assemblies	X	Identification and recognition

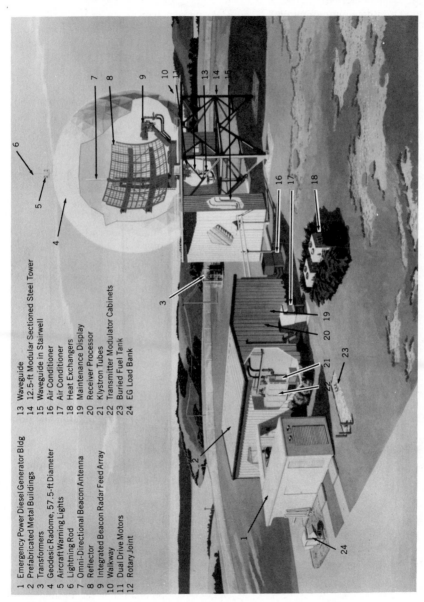

1 Emergency Power Diesel Generator Bldg
2 Prefabricated Metal Buildings
3 Transformers
4 Geodesic Radome, 57.5-ft Diameter
5 Aircraft Warning Lights
6 Lightning Rod
7 Omni-Directional Beacon Antenna
8 Reflector
9 Integrated Beacon Radar Feed Array
10 Walkway
11 Dual Drive Motors
12 Rotary Joint
13 Waveguide
14 12.5-ft Modular Sectioned Steel Tower
15 Waveguide in Stairwell
16 Air Conditioner
17 Air Conditioner
18 Heat Exchangers
19 Maintenance Display
20 Receiver Processor
21 Klystron Tubes
22 Transmitter Modulator Cabinets
23 Buried Fuel Tank
24 EG Load Bank

Figure 3.1. ARSR-3 Diagram. (Courtesy Westinghouse Electric Corp.)

Table 3.2   Impact of User Requirements on Microwave Design

User Requirement	Impact on Microwave Subsystems
Type of target (scattering cross section) and maximum/minimum range	Antenna size Transmitter power Transmitter types: solid state or tube Waveform
Ability to observe close-in targets	Sensitivity time control (STC) provided by PIN-diode attenuator in RF or IF
Coverage	Antenna rotation (azimuth and/or elevation) Mechanically versus electronically scanned antenna
Number of targets observed simultaneously	Mechanically versus electronically scanned antenna
Minimum spacing of targets	Antenna size Operating frequency (higher frequencies yield more spatial resolution) Pulse width (modulation frequency)
Operating weather conditions	Operating frequency Increase transmitter power to overcome attenuation due to weather
Site restrictions	Transmitter and antenna size Transmitter efficiency (affects size of power conditioning circuitry) Blank sector (turn off transmitter at specified azimuth angles) Reduce ground clutter (use phase stable transmitter, higher operating frequency)
Reliability/availability	Redundant transmitters, power supplies, receivers Switchover box
Maintainability	Replacable subsystems (consider interface specification tolerances to eliminate retuning circuits) Connectors positioned for ease of operation
Electromagnetic compatability	Receiver has high dynamic range to minimize spurious signals Transmitter pulse shaping to restrict bandwidth Transmitter filtering to avoid spurious signals
Weight	Efficient transmitter reduces supply weight Lightweight antenna affects performance
Cost/delivery schedule	Use known, proven designs Introduce new technologies where needed Begin development/testing of "untried" components early

ability exceeds 99.6% because of the built-in redundancy and a preventive maintenance time of less than 1/2 hr per week.

The microwave portion of the radar is located in cabinets in the building and is connected via a waveguide to the adjoining antenna. The microwave portion of modern radar does not represent the largest volume or cost for these systems. The sophisticated signal processing and the software are more expensive.

Ultimately the user requirements define the system and the microwave design. Typical user requirements and their impact on the microwave portion of the radar are given in Table 3.2. Clearly the impacts are many, varied, and highly interrelated.

Table 3.3   Parameters of the ARSR-3*

Minimum target size	2 m^2
Maximum range	370 km (200 nautical miles)
Operating frequency range	1.25–1.35 GHz
Probability of detection	80% (one scan)
Probability of false alarm	10^{-6}
Pulse width	2 s
Resolution	500 m
Pulse repetition frequency	310–365 Hz
Lowest blind speed	600 m/s (1200 knots)
Peak power	5 MW
Average power	3600 W
Receiver noise figure	4 dB
Moving Target Indicator (MTI) improvement	39 dB (3-pulse canceler, 50 dB with 4-pulse canceler)
Antenna size	12.8 m (42 ft) × 6.9 m (22 ft)
Beamwidth (azimuth/ elevation)	1.25/40°
Polarization	Horitontal, vertical, and circular
Antenna gain	34 dB (34.5 dB lower-beam, 33.5 dB upper beam)
Antenna scan time	12 s
EIRP (on-axis effective isotropic radiate power)	101 dBW (12,500 MW)
Average power/effective aperture/scan time product (antenna efficiency = 60%)	$2300 \text{ kWm}^2\text{s}$

* Courtesy Westinghouse Electric Corp.

Table 3.4   Radar Frequency Bands

Band Designation	Nominal Frequency Range	Specific Radiolocation (Radar) Bands Based on ITU Assignments for Region 2 (North and South America)
VHF	30–300 MHz	138–144 MHz
		216–255
UHF	300–3000 MHz	420–450 MHz
		890–942
L	1000–2000 MHz	1215–1400 MHz
S	2000–4000 MHz	2300–2500 MHz
		2700–3700
C	4000–8000 MHz	5250–5925 MHz
X	8000–12000 MHz	8500–10680 MHz
K_u	12.0–18 GHz	13.4–14.0 GHz
		15.7–17.7
K	26.5 GHz	24.05–24.25 GHz
K_a	26.6–40 GHz	33.4–36.0 GHz
EHF	30–300 GHz	

Selected parameters and user requirements for the ARSR-3 radar are given in Table 3.3. The frequency range was selected for all-weather operation in the L-band (see Table 3.4 for radar band designations) radar band to operate at 370 km range with an 80% probability of detection and 10^{-6} false alarm rate (one false report per 10^6 pulses) against a target with a backscatter area of at least 2 m² (allows detection of small general aviation aircraft). The transmitter pulse is generated by a phase-coherent klystron and associated pulse modulator. The moving target indicator (MTI) feature reduces detection of stationary objects by detecting the presence of Doppler shift in the received signal compared to the transmitted frequency.

The microwave system engineer must consider the options and select the "optimum" technique for the design. This technique is bounded by factors defined by the application and interrelated by the physics of radar systems.

3.2   APPLICATIONS AND BASIC TYPES

Radar applications include altimetry, air surveillance, and weapon guidance. These three provide the basis for the description of different types of microwave radar hardware. Others are the over-the-horizon radar for monitoring the movement of ships and planes in the high-frequency (3–30 MHz) skip zone, synthetic aperture radar for high-resolution radar mapping from aircraft or spacecraft, navigation radars for ships (AN/SPS-53) and terrain avoidance, and following radars for aircraft and missiles.

3.2.1 Radar Altimeter

Altimeters are low weight, simple, low cost radars for measurement of altitude above the Earth's surface. These systems use two general types of radar: FM-CW and pulse.

3.2.1.1 FM–CW Altimeter. In the FM–CW altimeter system, the frequency difference between the FM transmitted signal and the received signal delayed by the round-trip transit time to the target is measured. For a stationary target the waveforms shown in Fig. 3.2a are obtained for a linear triangular FM modulation between $+\Delta f$ and $-\Delta f$. The time between the two identical frequencies is related to the round-trip delay $T_R = 2R/c$, where R is the range and c is the speed of light. If the slope of the ramp is df/dt, then at a given time the difference frequency is

$$f_{\text{diff}} = \frac{df}{dt} T_R = \frac{2R}{c}(df/dt) \qquad (3.1)$$

This difference is independent of the operating frequency (unlike Doppler frequency).

The system block diagram for an FM–CW altimeter (AN/APN-22) is shown in Fig. 3.2b. Here a coupler feeds about 10 mW of local oscillator (LO) power to a mixer. The nonlinear process in the mixer (Chapter 13) generates the audio difference signal (plus others) sent to the audio amplifier. The input filtering of the audio amplifier attenuates all other signals passing through the mixer. The limiter (which provides a constant amplitude output signal as in an FM radio) drives the frequency counter and the indicator in the cockpit. The frequency counter measures $N = f_{\text{diff}}T_m$ complete cycles during one modulation period T_m. The resulting range is found from Eq. 3.1 to be

$$R = \frac{cf_{\text{diff}}}{2(df/dt)} = \frac{cN}{2T_m(df/dt)} \qquad (3.2)$$

Relating T_m and Δf to df/dt yields $(df/dt) = 4\Delta f/T_m$ (see Fig. 3.2a); so

$$\Delta R = \frac{cN}{8\Delta f} \qquad (3.3)$$

Since the counter can have an error ΔN of at least one count, the error in range is

$$\Delta R = \frac{c\Delta N}{8\Delta f} \qquad (3.4)$$

For the APN-22 with $\Delta f = 35$ MHz; the range error is at least 1 m.

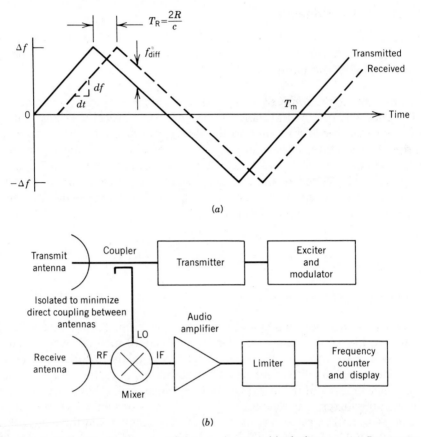

Figure 3.2. *FM–CW altimeter waveforms and system block diagram. (a) Frequency–time diagram for an FM–CW radar; (b) System block diagram.*

The APN-22 uses a 1.5-W output CW magnetron operating near 4.3 GHz. The modulation frequency ($1/T_m$) is 120 Hz. The antenna beamwidth is 60° to assure a return from the ground while the aircraft banks or climbs. This large beamwidth and low power provides adequate signal because of the large scattering cross section of the target.

3.2.1.2   *Pulse Altimeter.* Pulse radar altimeters use the leading edge of a short pulse (pulse length 20–200 ns) to measure the time difference between the transmitted and received pulses. The pulse altimeter waveforms are shown in Fig. 3.3*a*. An estimate of the error in measuring the position of the leading edge is ΔT_R for a video pulse of amplitude A_v with risetime t_{rise} after video filtering. The value of ΔT_R can be estimated by equating the slope of the noisy signal to that of the ideal pulse; i.e.,

$$\frac{n(t)}{\Delta T_R} = \frac{A_v}{t_{rise}} \quad \text{or} \quad \Delta T_R = n(t)\left(\frac{t_{rise}}{A_v}\right) \tag{3.5}$$

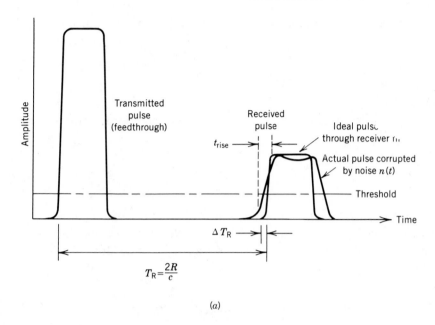

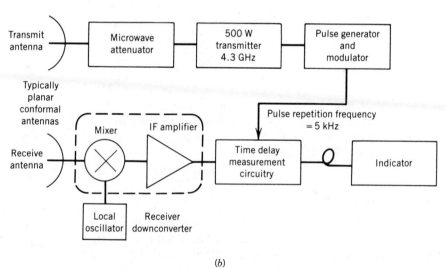

Figure 3.3. *Pulsed altimeter waveforms and system block diagram. (a) Altimeter pulses;* (b) *System block diagram.*

Since $n(t)$ is the assumed Gaussian noise voltage, the ΔT_R will be distributed similarly. Taking the expected value of ΔT_R yields

$$\sqrt{\overline{(\Delta T_R)^2}} = \sigma_{T_R} = \frac{t_{rise}}{A_v/\sqrt{\overline{n^2}}} \tag{3.6}$$

Since $\bar{n}^2$ is the noise power N and A_v^2 is the signal power S, Eq. 3.6 becomes

$$\sigma_{T_R} = \frac{t_{\text{rise}}}{(S/N)^{1/2}} \tag{3.7}$$

in the IF.

In the video circuitry following the detector an additional factor of 2 arises (1) yielding

$$\sigma_{T_R} = \frac{t_{\text{rise}}}{(2S/N)^{1/2}} \tag{3.8}$$

If the IF bandwidth B_{IF} allows a risetime t_{rise}, they are related by $t_{\text{rise}} = (B_{\text{IF}})^{-1}$. Substituting the pulse energy $E = S\tau$, where τ is the pulse width, and $N = N_0 B_{\text{IF}}$ due to thermal noise per unit bandwidth N_0, yields the relation

$$\sigma_{T_R} = \frac{\tau^{1/2}}{2^{1/2} B_{\text{IF}} E^{1/2}/N_0^{1/2} B_{\text{IF}}^{1/2}} = \left(\frac{\tau}{2B_{\text{IF}}E/N_0}\right)^{1/2} \tag{3.9}$$

Relating σ_{T_R} to σ_R via $R = cT_R/2$ yields

$$\sigma_R = \left(\frac{c}{2}\right)\sigma_{T_R} \tag{3.10}$$

The system block diagram for a unit similar to the APN-22 is shown in Fig. 3.3*b*. The microwave attenuator is used to reduce power at low altitudes for receiver front-end overload protection. Systems like these yield accuracies of 2 m or 2% (whichever is greater) up to 20-km altitudes.

3.2.2   Surveillance Radars

Surveillance radars are the largest high-production volume radars manufactured. These radars survey large areas, and detect out to their horizon of 100–400 km depending on the altitude of the target. Some are able to track while scanning to monitor the movement of some targets while searching for others.

The parameters of a typical surveillance radar are given in Table 3.3. These radars use high-power transmitters and large complex antennas. A single antenna is used for both transmit and receive. A duplex circuit directs the transmitter power to the antenna while directing the return signal to the receiver. A limiter protects the front end of the receiver from burnout due to the leakage of the transmitter through the duplexer. Good design is required to prevent high voltage breakdown (arcing) in the transmission lines and rotary joints in the antenna pedestals while maintaining low loss and good performance.

3.2.3 Weapons Guidance Radar

Weapons guidance radar is used to extend the eyes of a pilot or ground operator. A typical radar is the AN/APG-63 used on the F15 aircraft to locate and track hostile aircraft at long ranges and to look down and direct weapons via a clutterfree display. This system uses high-reliability hybrid and integrated circuits and four standard module sizes. The radar is a pulse–Doppler unit operating near 10 GHz. Various situation dependent waveforms and pulse repetition frequencies (PRF) are available in part because of the use of a traveling-wave tube amplifier in the radar transmitter. For long range, a high PRF is used to maximize the energy on the target. The high PRF causes an ambiguity in the range measurement as described below. A low PRF allows unambiguous range measurements and better clutter suppression. Clutter suppression is possible in this lightweight (224 kg) radar because digital signal processing rather than Doppler filters is used.

If the period between transmitted pulses is short, an ambiguity develops because the return signal cannot be identified uniquely with a single transmitted pulse. Consider the time diagram in Fig. 3.4a. If the target is less than $c/[2(PRF)]$ from the radar, the return pulse appears as shown at time A. If the range exceeds $c/2(PRF)$, the return pulse appears at time B. The radar is unable to determine if the round-trip time is between pulse 1 and B or pulse 2 and B. Thus the range is ambiguous. The maximum unambiguous range is $c/2(PRF)$. If more range is required, the PRF is reduced until the echo returns before the next pulse or a "staggered" PRF is used (explained below).

If the radar is turned on at $t = 0$ and the first echo is observed, the ambiguity is not present. This mode of operation is usually not used because the received signal grows slowly as the beam scans across ("paints") the target and then decreases to zero. The transmitter is operating all the time. What is done in many systems is to change the PRF during the period that the target is in the beam (staggered PRF) and note if the relative position of the transmitted and received pulses changes. If they change, the range is ambiguous, and some additional algebra is needed to compute the range.

Obviously the radar cannot receive while the transmitter pulse is on. This limits the range of the radar because targets at ranges near $c/2(PRF)$ are masked out. Also targets whose range is less than $c\tau/2$, where τ is the pulse width, are masked out because the receiver needs time to recover from (dissipate) the transmitter energy in the receiver.

Some radars use a phased array antenna to direct the beam either in azimuth, elevation, or both. One example is the AN/TPS-59 long-range three-dimensional (measures range, azimuth, and elevation) air surveillance radar. The front view of the radar is shown in Fig. 3.5a, and Fig. 3.5b (2) shows

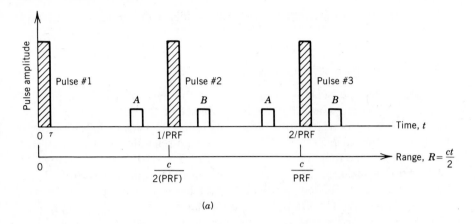

(a)

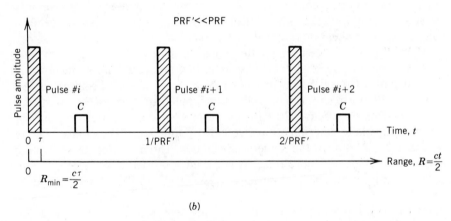

(b)

Figure 3.4. *Pulse diagram demonstrating the unambiguous range and minimum range.*
(a) Short-range target; (b) long-range target.

a schematic of the hardware that feeds energy to each of 54 rows. The radar
operates in the 1.215–1.4 GHz band with a 90-W solid-state transmitter module driving each row. The phased array allows the energy from each row to
be combined in lossless space rather than a complex beam forming network.
The radar array rotates mechanically in azimuth and electronically scans
the pencil beam between 0° and 19° in elevation.

   The phased array rows are fed from three column feeds. The RF exciter
is phase shifted, amplified using the module shown in Fig. 3.6, filtered,
distributed by the feed in each row, and radiated. Both the azimuth-difference and sum signals pass through filters, transmit/receive switches, pream-
plifiers, and phase shifters before being combined in their respective column
feeds. The elevation-difference signals are generated in the column feeds.

(a)

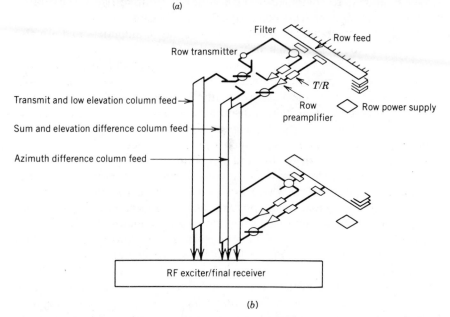

(b)

Figure 3.5. *AN/TPS-59 phased array radar. Notes: Entire array rotates at 6 revolutions per minute. Array leans to place boresight near middle of ±9.5° electronic elevation scan. (a) Photo of AN/TPS-59 in operating position. (Courtesy General Electric Co.); (b) antenna array block diagram. (From Ref. 2, © 1975 IEEE)*

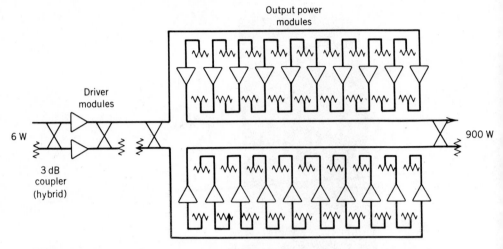

Figure 3.6. *Schematic of a power module within a row transmitter. (From Ref. 2, © 1975 IEEE)*

The sum and azimuth-difference outputs are derived in two 3-dB couplers etched in a stripline circuit. Stripline (explained in Chapter 8) is a planar form of coaxial-like transmission line.

The power module in Fig. 3.6 is located on each of the 54 rows. The two driver modules are combined in 3-dB couplers to drive two serial 10-way power dividers that provide equal signal levels to each amplifier input. The 10-way power combiners sum the outputs of the 10 amplifiers and are also combined in the coupler that drives each row distribution network. The divider and combiner provide isolation between stages so that failure of one amplifier only degrades the output by its power loss. This graceful degradation is a highly desirable feature of multiple-module solid-state radar transmitters compared to single tube transmitters. As a result, the solid-state radar has over 99.9% availability and a mean time between failures of 1400 hr. Normal preventive maintenance allows time to replace defective modules. The mean time to repair for this radar is 40 min.

Since solid-state devices do not provide significantly more power when pulsed than operated CW, these radars use long pulses. Long modulated pulses provide sufficient energy on long-range targets and can be pulse compressed upon receive to provide adequate range resolution. A popular waveform for pulse compression is linear chirp (FM). The key component in the pulse compression radar (Fig. 3.7a) is the pulse compression filter (Fig. 3.7b). A simple surface acoustic wave device is shown, but others (3) are also used. Surface acoustic waves are a set of compression and rarefaction surface waves excited by the thin metal input interdigital transducer. The surface waves travel outward from the input transducer. When the waves are aligned under the output transducer, the charge induced on the surface of the piezoelectric material is matched to the finger spacing and a pulse

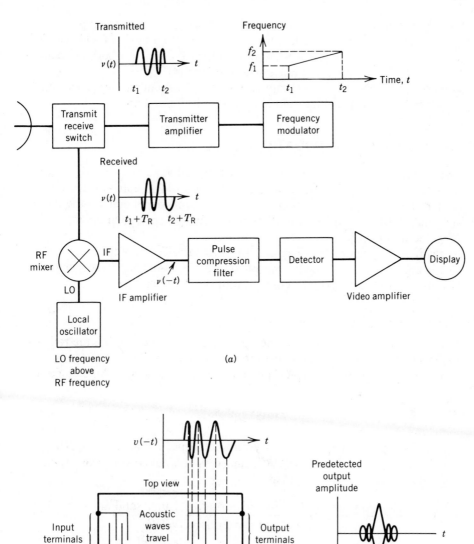

Figure 3.7. Block diagram and operation of a pulse-compression radar. Note: The pulse-compression filter may also be excited by a pulse to generate the FM waveform. (a) Block diagram of a linear FM pulse-compression radar; (b) surface acoustic wave (SAW) interdigital pulse-compression filter.

and sidelobes appear at the output terminals. The energy in the input $v(-t)$ during the period $t_2 - t_1$ has been compressed into a pulse of width about $(f_2 - f_1)^{-1}$. For example, a SAW device with $f_2 - f_1 = 500$-MHz bandwidth yielded a compressed pulse width of 3 ns (4).

Another method for pulse compression uses phase-coded subpulses within a long pulse. Additional information is given in Refs. 3 and 5.

3.3 RADAR RANGE EQUATION

The performance of the radar system is evaluated via a power budget equation similar to the communications link equation. The equation for radars (called the radar range equation) has several forms depending on the target size and clutter. In this section the surveillance (termed searchlight) radar range equation is derived.

Many of the relations in Chapter 2 are used here. The new concepts unique to the radar problem include the radar cross section of a target and the importance of the target size compared to the beam size.

3.3.1 Radar Cross Section

The cross section of a target is the equivalent area intercepting the radar's radiated power, which when scattered equally in all directions (4π steradians), produces a received signal equal to that of the target. Algebraically,

$$PD_{\text{rerad}} = \sigma \frac{PD_{\text{inc}}}{4\pi} \qquad (3.11)$$

where PD_{rerad} is the power density reradiated per steradian (W), PD_{inc} is the incident power density (W/m^2), and σ is the equivalent cross section (m^2). Rearranging Eq. 3.11 yields

$$\sigma = \frac{\text{Power reflected toward surface per unit solid angle}}{\text{Incident power density}/4\pi} \qquad (3.12)$$

For simple shapes, σ can be computed analytically, but most practical objects are measured at specified frequencies, polarizations, and target aspect angles. Table 3.5 from Ref. 6 gives numbers suitable for preliminary calculations.

3.3.2 Basic Radar Relations

The incident power density at range R from a radar with transmitter output power P_t and antenna gain G_t is

$$PD_{\text{inc}} = \frac{P_t G_t}{4\pi R^2} \qquad (3.13)$$

and the total power reradiated from the target is

$$P_{\text{rerad}} = 4\pi PD_{\text{rerad}} = \frac{P_t G_t \sigma}{4\pi R^2} \tag{3.14}$$

The power density at the receiving antenna is derived assuming the target to be the transmitter of P_{rerad} watts with a unity gain antenna. The power density is

$$PD_{\text{rec}} = \frac{P_{\text{rerad}}}{4\pi R^2} = \frac{P_t G_t \sigma}{(4\pi R^2)^2} \tag{3.15}$$

The signal power at the feed of the receiving antenna is

$$P_r = PD_{\text{rec}} A_{e,r} = \frac{P_t G_t \sigma A_{e,r}}{(4\pi)^2 R^4} \tag{3.16}$$

where $A_{e,r}$ is the effective area of the receiving antenna. Thus the received power for a surveillance radar varies as the inverse fourth power of the range. Doubling the radar range requires 16 times the transmitted power.

By substituting

$$G_t = \frac{4\pi A_{e,t}}{\lambda^2} \quad \text{and} \quad G_r = \frac{4\pi A_{e,r}}{\lambda^2} \tag{3.17}$$

other forms of the simple radar range relation can be derived. For most surveillance radars, the same antenna is used for transmit and receive, so

$$G = G_t = G_r \tag{3.18}$$

If the minimum detectable signal is $S_{\min}$, the maximum range of the radar is

$$R_{\max} = \left[\frac{P_t G_t A_{e,r} \sigma}{(4\pi)^2 S_{\min}} \right]^{1/4} \tag{3.19}$$

This simple relation does not predict the range performance accurately. The other factors needed are the signal-to-noise ratio for the output of the receiver and the effect of integrating n_B pulses. The n_B is determined by the number of pulses on the target as the antenna scans the beamwidth from one -3 dB azimuth angle through boresight to the other -3 dB angle. The result is

$$n_B = \frac{(3 \text{ dB beamwidth})(\text{PRF})}{\text{Scan rate in degrees/s}} = \frac{[\theta_B(\text{deg})](\text{PRF})}{\theta_{\text{scan}}(\text{deg}/s)} \tag{3.20}$$

Table 3.5 Approximate Microwave Radar Cross Sections

	σ (m^2)
Aircraft	
Small, single engine	1–2
Small fighter and four-passenger jet	2–4
Large fighter	6–10
Medium bomber (airliner)	20
Large bomber (airliner)	40
Jumbo jet	100
Missile (unmanned, winged)	0.5
Ship	
Ship (zero grazing angle)	$52f^{1/2}$(MHz)$D^{3/2}$(D = displacement in kilotons)
Ship (higher grazing angles)	D
Cabin cruiser	10
Small pleasure boat	2
Rowboat	0.02
Vehicles	
Pick-up truck	200
Automobile	100
Bicycle	2
Living Objects	
Person	1
Bird	10^{-2}
Insect	10^{-5}

Adapted from Ref. 6.

The signal-to-noise ratio is introduced via the definition of noise factor (Eq. 2.27), namely,

$$F_n = \frac{S_i/N_i}{S_0/N_0} = \frac{\text{Input signal-to-noise ratio}}{\text{Output signal-to-noise ratio}} \tag{3.21}$$

Substituting $N_i = kT_0B_n$, the minimum detectable signal is

$$S_{min} = kT_0B_nF_n\left(\frac{S_0}{N_0}\right)_{min} \tag{3.22}$$

Typical $(S_0/N_0)_{min}$ ratios that yield a reasonable probability of detection and low probability of false alarm are 10:1 to 40:1 (10–16 dB).

The final factor needed to make the system performance determination more accurate is the loss, L_{system}, which attenuates both the transmitted signal and the received signal. This is a number greater than 1 that is the sum of one-way losses from the point where the transmitter power is measured to the antenna and back to the reference plane where the receiver noise factor is measured. Typical hardware contributors are the transmission line, connectors, rotary joint, and duplexer (a component designed to route the transmitter power to the antenna while protecting the receiver front end, transmit/receive switch). L_{system} values range from 2 to 10 (3–10 dB). Propagation (attenuation in the atmosphere especially during heavy rain) loss can have a significant effect on two-way loss at frequencies above 8 GHz (7).

Introducing the effect of n_B pulses in making the detection decision in a perfect processor, the noise factor, and the system losses between the antenna and the point where F_n is measured yields a relation that more accurately predicts performance:

$$R_{max} = \left[\frac{P_t G_t A_{e,R} \sigma n_B}{(4\pi)^2 k T_0 B_n F_n (S_0/N_0)_{min} L_{system}} \right]^{1/4} \tag{3.23}$$

The average power P_{avg} of a radar determines its detection range. The relation between P_t and P_{avg} involves the duty cycle (percentage of time the transmitter is on):

$$P_{avg} = P_t(\text{duty cycle}) = P_t \frac{\tau}{T_P} = P_t \tau(\text{PRF}) \tag{3.24}$$

Substituting into Eq. 3.23 yields a common form of the radar range equation for use in surveillance radar performance calculations where

$$G = G_t = \frac{4\pi A_{e,r}}{\lambda^2}$$

$$\tag{3.25}$$

$$R_{max} = \left[\frac{P_{avg} G^2 \lambda^2 \sigma n_B}{(4\pi)^3 k T_0 (B_n \tau) F_n (S_0/N_0)_{min} L_{system}(\text{PRF})} \right]^{1/4}$$

For most receiver designs the noise bandwidth B_n is equal to the inverse pulse width, so $B_n \tau = 1$.

For the surveillance radar, the target receives energy from only a portion of the beam, thus the R^{-4} relation between transmitted and received power levels. If the entire beam energy is reflected, for example with an altimeter or clutter, a different relation is obtained and the optimum receiver detection process may be affected (8).

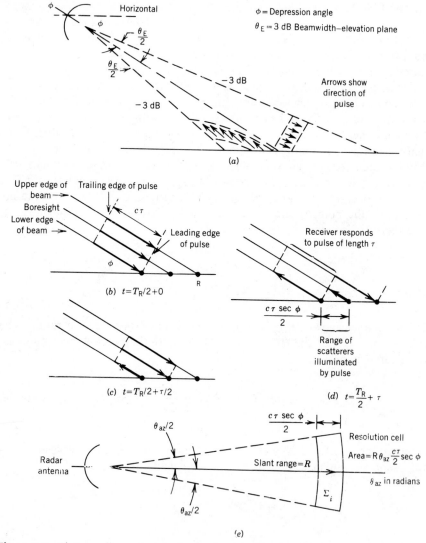

Figure 3.8. *Clutter cell area for a pulsed radar. (a) Side view in elevation plane; (b, c, d) detail of scatter region at different times in elevation plane; (e) top view in azimuthal plane.*

3.3.3   Simple Radar Relations for Clutter

Consider sea clutter where $\overline{\sigma}_i$ is the time averaged cross section of the ith scatterer. From the radar relation (Eq. 3.16 with $G = G_t = 4\pi A_{e,r}/\lambda^2$)

$$P_r = \frac{P_t G^2 \lambda^2}{(4\pi)^3 R^4} \sum_i \overline{\sigma}_i \qquad (3.26)$$

where the summation applies to all scatterers in the resolution cell of the radar. The resolution cell is determined by the 3-dB azimuthal and elevation angles and/or the range resolution of the radar. A typical cell for a small depression angle ϕ is shown in Fig. 3.8. Here the range cell determines the total area

$$R\theta_{az}\left(\frac{c\tau}{2}\right)\sec\phi$$

where τ is the pulse length and ϕ is the depression angle as shown in Fig. 3.8*e*. The factor of 2 in the area relation is explained in the time sequence series in Figs. 3.8*b–d*. Note that if the receiver bandwidth is not matched to the pulse length τ, the area must be adjusted accordingly. Defining σ^0 as the average cross section per unit area,

$$\sum_i \overline{\sigma}_i = \sigma^0 R\theta_{az}\left(\frac{c\tau}{2}\right)\sec\phi \tag{3.27}$$

The power received from the clutter is proportional to R^{-3}; namely,

$$P_{r\text{ clutter}} = \frac{P_t G^2 \lambda^2 \sigma^0 \theta_{az} c\tau \sec\phi}{2(4\pi)^3 R^3} \tag{3.28}$$

If the depression angle is large, the $P_{r\text{ clutter}}$ is proportional to R^{-2} (see Problem 11). For an ocean, σ^0 depends on sea state, wind direction, polarization, frequency, and depression angle (9).

3.4   RADAR HARDWARE

For most radars the microwave receiver is similar to those discussed in Chapter 4. However, the transmitters and antennas are significantly different for different applications. The transmitter and its power supply (modulator) are usually the reliability-limiting microwave subsystem.

The performance requirements usually establish the required antenna design. If a reflector antenna with a single feed is used, a single transmitter is the likely candidate. For high power and high efficiency the microwave magnetron, traveling-wave tube or crossed-field amplifier are the logical candidates. For moderate power levels, parallel banks of solid-state devices, power combined to provide one output port, are an alternative. Tubes are more efficient in converting prime power to microwaves, but solid-state devices, when properly designed, are more reliable and provide better system availability because of the "graceful degradation" feature when a few fail.

3.4.1 Vacuum Tubes for Radar Applications

The magnetron is a DC-pulsed oscillator whose oscillations are not coherent from pulse to pulse. Thus, unless combined with a coherent oscillator that retains the phase information between pulses, this transmitter is not usable in phase coherent systems. Output power up to 25 MW peak at 1.3 GHz using 260 A and 52,000 V are available commercially. The design of the power supply to provide that much energy in a controlled pulse is a major task.

Crossed-field amplifiers look similar to magnetron oscillators but operate as amplifiers. The name crossed field describes the RF interaction region containing crossed electric and magnetic fields (also true in a magnetron). These tubes have lower voltage (30–50 kV), high efficiency (40–80%), low gain (6–12 dB), and moderate bandwidth (typically 10%). A 1.3-GHz unit provides 1.8 MW with 9 dB gain. Since the tube operates in a saturated mode, output amplitude is constant, but the phase can be controlled for phase coherent systems. Crossed-field amplifiers, when unbiased, pass the driver power through with less than 2 dB loss. Thus when power level variation is needed for observing close-in targets, the tube acts as a lossy transmission line. The AN/TPN-25 (precision approach radar) operates using the high-power mode to compensate for attenuation in heavy rain at 9 GHz. Traveling wave tubes are used in the driver stage.

The traveling wave tube (TWT) can be used in coherent systems. TWTs have wide bandwidth and can be backed off for linear (amplitude modulation) systems with a decrease in efficiency. These tubes have high gains (30–60 dB) thus requiring only a low-power driver stage (solid state) since TWT amplifier output powers are 100–750 kW. The TWT generally has lower efficiency (30–40%) than the other tubes. Like magnetrons they required 10–50-kV power supplies.

3.4.2 Solid-State Devices for Radar Applications

If the antenna is a phased array, the logical technique for combining power is "space combining." This involves numerous low-power sources phase controlled to provide coherent wavefronts from the antenna. One low-power source is the solid-state module (a transceiver, or *trans*mitter/*receiver* combination) rather than a single large source divided in a network and recombined in space (phased array lens antennas use the latter technique).

Solid-state modules using transistors similar to the AN/TPS-59 have provided excellent reliability performance. A test performed over 374,000 module-hours yielded a mean time between failures of 55,000 hr for the transmitter (10). The efficiency of solid-state devices is lower, so the prime power requirements are higher. Modules producing 3 W at 10 GHz (11) use monolithic microwave integrated circuits (these MMICS are gallium arsenide circuits containing both passive and active elements produced using techniques similar to digital electronics) have been built.

3.4.3 Conclusions

Radar transmitter components will continue to improve, and both solid state and tubes will be used. Nongridded TWTs and klystrons (a relative of the TWT) have achieved 100,000 hr MTBFs (12). Their efficiency remains high (50%) thus requiring less prime power than their solid-state competitors. Above 18 GHz the tube is the dominant source for high-power applications because power transistors are not available and tube research at MMW is quite active. Solid-state transmitters have the advantage of graceful degradation and lower-voltage (but higher-current) operation. The radar system designer will select from these two basic approaches. Reference 13 is an excellent book for further study.

PROBLEMS

1. Use the radar equation (Eqs. 3.16 and 3.17 combined) to determine the peak transmitter power and antenna physical area that makes the cost of the ARSR-3 radar a minimum. Use data from Table 3.3 with an antenna efficiency of 60% and S_{min} about 4×10^{-13} W. Assume antennas cost $1,000/m^2 and transmitters cost 70¢/W. The cost of the receiver and peripherals is $250,000. Compare your answer to the actual system. (A similar problem was suggested by M. I. Skolink, 1981.)

2. What is the total cost of the "optimum" ARSR-3 system in Problem 1.

3. Use the parameters in Table 3.3 to compute the average power needed to observe a 1 m^2 target at 370 km.

4. Redo Problem 3 for a radar operating at 3 GHz with an antenna area of 38.3 m^2.

5. How does the performance of a surveillance radar vary with the antenna aperture–frequency product and the average power–antenna aperture squared product.

6. Make a plot of FM–CW altimeter difference frequency versus reasonable aircraft range for a FM ramp of 16.8 GHz/s.

7. Compute the standard deviation in range for a pulse altimeter with $\tau = 100$ ns, $E/N_0 = 14$ dB, $B_{IF} = 100$ MHz $= t_{rise}^{-1} = (10\tau)^{-1}$

8. Compute the shortest and longest unambiguous range for a radar with a 1 μs pulse width and a PRF $= 1$ kHz.

9. Surface acoustic wave (SAW) velocities are near 3.6×10^5 cm/s. Compute the center-to-center spacing between fingers on the 1.3-GHz output transducer. Most SAW devices use lower frequencies.

10. Use Eq. 3.25 to predict the maximum range for the ARSR-3 in Table 3.3 for the minimum target size of 2 m². Assume the system losses are 6 dB.

11. Derive the power received from clutter if the resolution cell is defined by the elevation and azimuth beamwidths (companion relation to Eq. 3.28).

12. Derive the signal-to-clutter ratio relations for the two cases where the pulse width determines the clutter resolution cell (Eq. 3.28) and the case in Problem 11. In most cases the clutter "noise" swamps the thermal noise in the receiver so clutter limits the performance. Hence clutter rejection techniques such as MTI were developed.

13. Sea clutter in X-band is about -30 dB for grazing angles from 10 to 40°. Prepare a graph of the power returned from the clutter as a function depression angle ϕ ($10° \le \phi \le 40°$) for an aircraft altitude of 1 km. If the aircraft is to see (signal 10 dB above clutter) a cabin cruiser (Table 3.4), what clutter rejection ratio is required at $\phi = 10°$, antenna diameter $= 1$ m and pulse length $= 1$ μs.

14. A 10-GHz phased array is being built using the 3-W MMIC amplifiers. Assuming a 1 m² antenna and elements are needed every half-wavelength, estimate the number of modules, the prime power consumed by the array (assume 20% efficiency), and the amount of heat to be dissipated.

15. A 94-GHz air–air missile guidance radar antenna is 15 cm in diameter. Estimate the maximum range of operation against a small fighter if antenna efficiency is 60%, noise figure $= 10$ dB, system losses $= 10$ dB, pulse length $= 100$ ns, PRF $= 1$ kHz, $n_B = 20$, $(S_0/N_0)_{min} = 10$ dB, and peak power $= 1$ W. How accurately must the missile be pointed to have the target in the main beam when launched?

REFERENCES

1. A. J. Mallinckrodt and T. E. Sollenberger, "Optimum Pulse-Time Determination," *IRE Trans. Inf. Th.,* Vol. PGIT-3, March 1954, p. 151.
2. C. M. Lain and E. J. Gersten, "AN/TPS-59 Overview," IEEE 1975 Intl. Radar Conf., Arlington, VA, April 21–23, *IEEE Publ.* 75 CHO 938-1 AES, p. 527.
3. M. I. Skolnik, *Introduction to Radar Systems,* 2nd ed., McGraw-Hill, New York, 1980, sec. 11.5.
4. M. Lipka, "Pulse Compression Filter and Wideband Receiver Evaluation," 1975 IEEE Intl. Radar Conf., Arlington VA, April 21–23, *IEEE Publ.* 75 CHO 938-1 AES, p. 283; R. J. Purdy, "Signal Processing Linear Frequency Modulated Signals," Ch. 10 in *Radar Technology,* E. Brookner, ed., Artech House, Dedham, MA, 1979.

5. E. C. Farnett, T. B. Howard, and G. H. Stevens, "Pulse Compression," in *Radar Handbook,* M. I. Skolnik, ed., McGraw-Hill, New York, 1970, Chap. 20.

6. M. I. Skolnik, *Introduction to Radar Systems,* 2nd ed., McGraw-Hill, New York, 1980, Sec. 2.7.

7. M. I. Skolnik, *Introduction to Radar Systems,* 2nd ed., McGraw-Hill, New York, 1980, Secs. 2.12, 2.13 and Chap. 12.

8. M. I. Skolnik, *Introduction to Radar Systems,* 2nd ed., McGraw-Hill, New York, 1980, Chap. 13.

9. M. I. Skolnik, "Sea Echo," *Radar Handbook,* M. I. Skolnik ed., McGraw-Hill, New York, 1970, Chap. 26.

10. B. C. Dodson, Jr., "L Band Solid State Array Radar Overview," *Radar Technology,* E. Brookner, ed., Aertech House, Dedham, MA, 1979, Chap. 19.

11. R. G. Freitag, J. E. Degenford, D. C. Boire, M. C. Driver, R. A. Wickstrom, and C. D. Change, "Wideband 3W Amplifier Employing Cluster Matching," B. E. Spielman, ed., 1983 IEEE Micro. Millimeter Wave Monolithic Circuits Symp. Dig., Boston, MA, May 31–June 1, p. 62.

12. A. D. LaRue, "High Efficiency Klystron CW Amplifier for Space Application," IEEE Intl. Elect. Dev. Meeting, Washington, DC, Dec. 1976, p. 385.

13. G. W. Ewell, *Radar Transmitters,* McGraw-Hill, New York, 1981.

4 | Electronic Warfare Systems

4.1 INTRODUCTION

Modern military weaponry is frequently directed toward the target using electronic detection. This detection process provides faster response times than a human operator can provide, thus increasing the weapon's effectiveness. If the detection process is interrupted or confused with intentionally generated noise, the weapon's effectiveness is reduced. Electronics has neutralized the mechanical power of the weapon. The process of disrupting the electronic performance of a weapon is called electronic warfare, or EW.

Electronic warfare is generally conducted in the transmission media between the transmitter and the receiver. It is the battle for control of the electromagnetic spectrum in a region. The control of the spectrum can either mask the intended communication with noise (jamming) or substitute false information (deceptive jamming) for the desired signal. A repeater jammer that sends false signals to a radar is an example of a deceptive jammer.

The broad activity of EW is subdivided into three modes: electronic support measures (ESM), electronic countermeasures (ECM), and electronic counter–countermeasures (ECCM). The activity of intercepting enemy communications and using the information to plan counter activities is termed electronic support measures. The information can be used immediately for threat recognition or stored for future reference to anticipate enemy action during a conflict. An example of this latter mode of ESM is an enemy ship "shadowing" or monitoring a fleet when it is on maneuvers. The types of activities carried out in the ESM mode of EW are shown in Fig. 4.1. Sophisticated receivers and recording/signal processing systems are the principal hardware implementations of this type of EW.

Electronic countermeasures involves the generation of intentional noise to decrease the signal-to-noise ratio or the generation of false targets to deceive the enemy. This denies the enemy effective use of the spectrum. Active ECM involves the generation of broadband noise (barrage jammers) or deceptive (smart) jammers that confuse the enemy's detection or communication system. Passive ECM involves the use of chaff (a highly microwave reflective material dispensed over large volumes to present a huge target) or decoys such as expendable repeater jammers that appear to be targets.

ECCM actions are taken to ensure the use of the electromagentic spectrum by one side when the other side is conducting ECM. Typical actions taken for ECCM are listed in Fig. 4.1. The implementation of these actions in hardware frequently requires sophisticated microwave equipment. For example, overpowering the jammer may require high-power tubes, avoiding the jammer's frequency may require broadband frequency-hopping subsystems, preventing receiver overload may require programmable microwave attenuators, and signal discrimination may require high-speed sophisticated modulators. Each of these ECCM techniques involves unique microwave hardware designs. Frequently off-the-shelf components require redesign before incorporation into these systems.

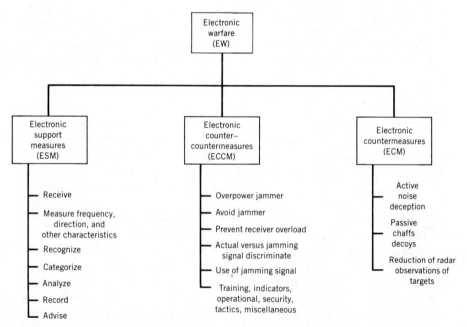

Figure 4.1. *Modes of electronic warfare. (From LeRoy B. VanBrunt, WMEC Course, 1981.)*

4.2   ECM SCENARIOS*

Two illustrative scenarios will be described to serve as a medium for development of the basic power budget equations. The scenarios are the self-screening jammer (SSJ) attacking a radar-protected missile site and a CW barrage stand-off jammer (SOJ) screening the attack aircraft. These scenarios may take place individually or simultaneously as shown in Fig. 4.2.

In Chapter 3 the radar equation was derived. The received power from the target is

$$P_r = \left(\begin{array}{c}\text{Trans.}\\\text{EIRP}\end{array}\right)\left(\begin{array}{c}\text{Path}\\\text{loss}\end{array}\right)\left(\begin{array}{c}\text{Target}\\\text{cross}\\\text{section}\end{array}\right)\left(\begin{array}{c}\text{Path}\\\text{loss}\end{array}\right)\left(\begin{array}{c}\text{Rcvr}\\\text{ant}\\\text{gain}\end{array}\right) \tag{4.1}$$

$$= (P_t G_t) \left(\frac{\lambda^2}{(4\pi R)^2}\right)\left(\frac{4\pi\sigma}{\lambda^2}\right)\left(\frac{\lambda^2}{(4\pi R)^2}\right) G_r \tag{4.2}$$

where

$$P_t = \text{transmitter peak power}$$

$$G_t = G = G_r = \text{radar antenna gain}$$

$$R = \text{range}$$

$$\lambda = \text{wavelength}$$

$$\sigma = \text{scattering cross section}$$

The objective of both the SSJ and the SOJ is to put enough disruptive energy into the radar antenna so that it will overwhelm the power backscattered from the target. Assuming the jammer knows the radar's operating frequency, the SSJ need only overcome the space loss on a one-way path directly into the main beam of the radar's antenna. For the SOJ scenario, however, the attack aircraft may be in the mainlobe, whereas the SOJ must introduce noise into the radar antenna's sidelobes. For this SOJ case, the jammer's EIRP must compensate for the radar's sidelobe level. The critical parameter is the P_{noise}/P_r or jammer-to-signal (J/S) ratio. For a given scenario, there is usually a range where J/S is favorable for the attacking aircraft and a range where the radar will "see" through (also called burnthrough) the noise. In this chapter the effects of integrating numerous pulses in the radar receiver and the ability to discriminate between a target and noise based on the correlation/decorrelation of the signal will not be described (see 1, 2). This chapter deals only with the single-pulse J/S ratio.

---

* Extensive use of the link and radar equations and concepts developed in Chapters 2 and 3 will be used. If this material has not been studied, reference to it may be necessary to appreciate the equations presented here.

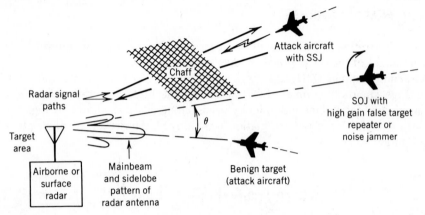

Figure 4.2. *Typical radar jamming scenario. SSJ, self-screen jammer; SOJ, stand-off jammer. (From LeRoy B. VanBrunt, WMEC Course, 1981.)*

4.2.1   Self-Screening Jammer Power Relation

The SSJ EIRP over a bandwidth BW is simply the transmitted power output times the antenna gain $P_{\text{tSSJ}}\,G_{\text{tSSJ}}$. The one-way space loss is $\lambda^2/(4\pi R)^2$ and the radar antenna's gain G yields a receiver noise power of

$$P_{\text{noise, SSJ}} = P_{\text{tSSJ}}G_{\text{tSSJ}}\frac{\lambda^2}{(4\pi R)^2}\,G\left(\frac{\text{BW}_{\text{radar}}}{\text{BW}_{\text{SSJ}}}\right) \tag{4.3}$$

where R is the range from radar to SSJ. For a typical attack airplane there are restrictions on the amount of transmitter power and the size of the antenna to allow room for other armaments. Typical transmitter powers are 1000 W with antenna gains of 10 dB at S-band frequencies. Because the SSJ does not know the radar's operating frequency exactly, it typically jams over a 30-MHz bandwidth. As a result of the radar using only a 1-MHz bandwidth, a scaling factor $\text{BW}_{\text{SSJ}}/\text{BW}_{\text{radar}}$ in favor of the radar is obtained. The resulting noise-to-signal ratio is

$$\begin{aligned}
\frac{P_{\text{noise}}}{P_{\text{signal}}} &= \frac{P_{\text{tSSJ}}G_{\text{tSSJ}}\left(\dfrac{\lambda^2}{(4\pi R)^2}\right)G\text{BW}_{\text{radar}}}{P_t G^2\left(\dfrac{\lambda^2}{(4\pi R)^2}\right)^2\left(\dfrac{4\pi\sigma}{\lambda^2}\right)\text{BW}_{\text{SSJ}}} \\[2mm]
&= \frac{P_{\text{tSSJ}}G_{\text{tSSJ}}\,4\pi R^2}{P_t G\sigma}\left(\frac{\text{BW}_{\text{radar}}}{\text{BW}_{\text{SSJ}}}\right) \\[2mm]
&= \frac{J}{S}
\end{aligned} \tag{4.4}$$

Normally $P_{tSSJ} < P_t$, $G_{tSSJ} < G$, $BW_{radar} < BW_{SSJ}$, so for short ranges $S > J$ and the jammer is ineffective. But for large ranges $J > S$ and the jammer is effective in masking the return signal from the skin of the aircraft. The minimum J/S for which the target is obscured is called the camouflage factor C. For pulsed and CW radars $C = 0$–6 dB, whereas for pulse-coded radars $C = 10$–30 dB is required.

4.2.2 Stand-Off Jammer Power Relation

For the SOJ scenario (refer to Fig. 4.2) the angle θ is important because the main beam of the radar will be on the attack aircraft while the jammer is forced to inject power into the radar antenna sidelobes. Thus the gain of the radar antenna toward the SOJ is typically 20 dB below G. The noise introduced into the radar receiver is

$$P_{noise, SOJ} = P_{tSOJ}G_{tSOJ}G_{SL}\left(\frac{\lambda^2}{(4\pi R_{SOJ})^2}\right)\left(\frac{BW_{radar}}{BW_{SOJ}}\right) \tag{4.5}$$

where $P_{tSOJ}G_{tSOJ}$ is the EIRP of the SOJ. Typical values are $P_{tSOJ} = 5000$ W, $G_{tSOJ} = 20$ dB, and $G_{SL} = G/100$. The same BW scaling factor as the SSJ case applies here so

$$\frac{P_{noise}}{P_{signal}} = \frac{P_{tSOJ}G_{tSOJ}G_{SL}\left(\frac{\lambda^2}{(4\pi R_{SOJ})^2}\right)BW_{radar}}{P_tG^2\left(\frac{\lambda^2}{(4\pi R)^2}\right)^2\left(\frac{4\pi\sigma}{\lambda^2}\right)BW_{SOJ}}$$

$$= \frac{4\pi P_{tSOJ}G_{tSOJ}G_{SL}R^4 BW_{radar}}{P_t G^2 R_{SOJ}^2 \sigma BW_{SOJ}} \tag{4.6}$$

$$= \frac{J}{S}$$

Typically, R_{SOJ} is greater than the range of the missile sited for target defense near the radar. The radar operator knows the direction of the SOJ but not its range. The missile could home on the jammer signal, but this is ineffective because the jammer platform remains out of range of the missile.

4.2.3 Other Considerations

Several other scenarios, such as repeater jammers, could be analyzed, but each involves the same basic considerations. In the above two scenarios some implicit simplifying assumptions have been made. These assumptions include:

1. The jammer's polarization is identical to the radar.
2. The jammer signal is significantly higher than the radar receiver noise.
3. The jammer and radar system losses are negligible.
4. Reflections of the microwave signal from the Earth or sea surface are negligible (no multipath effects).

4.3. HARDWARE IMPLEMENTATION

The block diagram for a typical CW noise jammer system is shown in Fig. 4.3. The key microwave components are the power amplifier string, the oscillator power combiner, and the tuned oscillators. The entire system is triggered by a broadband radar warning receiver (RWR) designed to recognize the presence and frequency of a radar. The RWR then communicates this information to a computer, which decides if action is to be taken and controls the jamming process.

The power amplifiers are similar to those used in radar and communication applications described earlier. The tuned oscillator designs usually use voltage controlled oscillators (VCOs) similar to those described in Chapter 13. The transmitter part of the jammer will not be described further in this text since it is similar to airborne radar transmitters.

The design of the receiver could take several forms. Typically, receiver types for this application are crystal video, compressive (also called microscan), instantaneous frequency measurement, and channelized configurations. Some new systems are using the acousto-optic receiver because it

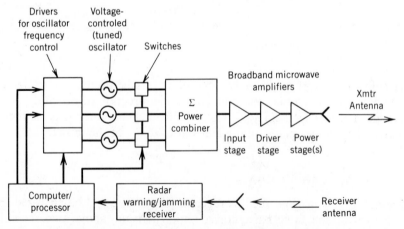

Figure 4.3. *Block diagram of a CW noise jamming system.*

provides small size and weight while being able to measure signal parameters and direction of arrival on a single pulse. A summary of the key parameters for these five receiver configurations is given in Table 4.1 (3). An introductory description of each of these receiver configurations follows to better delineate the microwave componentry used in their respective implementations. This presentation provides additional motivation for the following chapters.

4.3.1 Crystal Video Receiver

The crystal video receiver uses a broadband RF filter followed by an RF preamplifier to increase the sensitivity of a crystal detector. The detector output drives a logarithmic video amplifier that accommodates a wide input dynamic range. The log video amplifier output has significantly less dynamic range for driving the logic circuitry following the receiver. Tunable RF crystal video receivers similar to the design in Fig. 4.4 are used today because of their added flexibility and high probability of intercepting the desired signal. Low cost, size, and complexity are key features of this configuration.

4.3.2 Compressive Receiver

The compressive filter (a filter whose time delay is proportional to the input frequency) is the key component of the compressive receiver configuration. When combined with a swept frequency local oscillator (LO) signal whose sweep rate df/dt is the negative of the compressive filter characteristic, the output of the filter is a spike for a constant input frequency, as shown in Fig. 4.5. The microwave scanning (microscan) LO inputs a chirp to the constant frequency input pulse, which results in the energy throughout the pulse arriving at the output of the filter simultaneously. The video detector creates the envelope of the filter output (similar to an AM radio detector) for the following logic circuitry. By knowing the position of the pulse during the scan time, the RF frequency is determined. This scanning process is analogous to a chirped radar receiver.

4.3.3 Instantaneous Frequency Measurement Receiver

The instantaneous frequency measurement (IFM) receiver uses a delay line discriminator to measure the frequency of the incoming signal. The discriminator concept (similar to the detection process in an FM receiver) is shown in Fig. 4.6. Following RF amplification and limiting, the RF input is split, one signal passing directly to a mixer (correlator) LO input and the other signal passing through a delay line of physical length L to the RF input of the mixer. This length of line introduces a phase shift ϕ (radians) $= 2\pi L/\lambda$ $= 2\pi L f/v$. Thus as the frequency increases the phase difference $\Delta \phi$ between the two channels increases linearly. Knowing the phase difference, the fre-

Table 4.1 Comparison of Wideband Receiver Configurations

Parameter / RCVR Type[a]	Sensitivity	Dynamic Range	Frequency Accuracy	Simultaneous Signal Capability	TOA Availability	Instantaneous Bandwidth	Processing Complexity	Hardware Complexity	Cost	Stage of Development
CVR	Low	Low	Poor	Poor	Excellent	Excellent	Low	Low	Low	Mature
A/O	High	Low	Good	Excellent	Poor	Fair	Med	Med	Med	Early
Channelized	High	Med	Good	Good	Excellent	Fair	High	High	High	Emerging
IFM	Med	High	Excellent	Poor	Excellent	Excellent	Med	Med	Med	Mature
Compressive	High	Med	Good	Good	Poor	Good	High	High	High	Emerging

[a] CVR, crystal video receiver; A/O, acousto-optic; IFM, instantaneous frequency measurement; TOA, time of arrival.
(Reprinted with permission of Microwave Journal, from the February 1981 issue, © 1981 Horizon House-Microwave, Inc.)

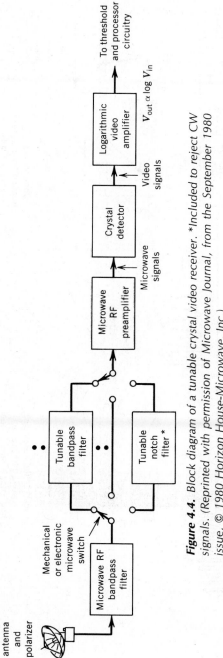

Figure 4.4. Block diagram of a tunable crystal video receiver. *Included to reject CW signals. (Reprinted with permission of Microwave Journal, from the September 1980 issue. © 1980 Horizon House-Microwave, Inc.)

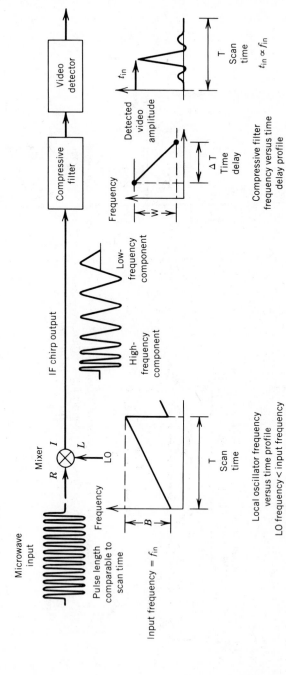

Figure 4.5. Key concepts in a compressive receiver. (Reprinted with permission of Microwave Journal, from the September 1980 issue, © 1980 Horizon House-Micro-wave, Inc.)

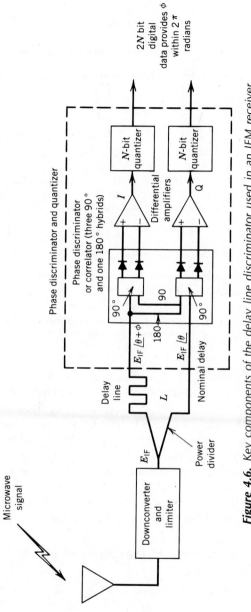

Figure 4.6. Key components of the delay line discriminator used in an IFM receiver. L, additional delay line length; $E_{IF} = A \cos \omega t$; $I = k E_{IF}^2 \cos \phi$; $Q = k E_{IF}^2 \sin \phi$.

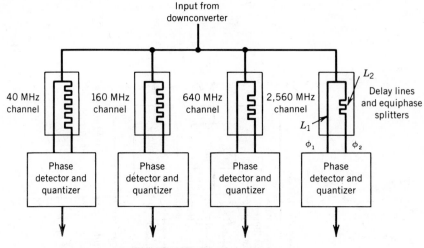

Figure 4.7. Multiple channel delay line discriminator in an IFM receiver for 2–4 GHz. $\phi_1(f_L) = 2\pi n L_1 f_L / v$; $\phi_2(f_L) = 2\pi(n + 1)L_2 f_L / v$; $\phi_1(f_H) = 2\pi n L_1 f_H / v$; $\phi_2(f_H) = 2\pi(n + 1)L_2 f_H / v$; f_L = low frequency; f_H = high frequency; $\Delta L = L_2 - L_1 = v/(f_H - f_L)$ = $v/\Delta f$; v = velocity of propagation in delay lines. (Reprinted with permission of Microwave Journal, from the September 1980 issue, © 1980 Horizon House-Microwave, Inc.)

quency can be inferred. For a typical system, if a 10-MHz accuracy over a 2-GHz bandwidth is desired, the phase accuracy must exceed 1.8°. This is a stringent requirement, so multiple delay lines are used as shown in Fig. 4.7. The 2560-MHz wide channel provides guard bands around the 2-GHz channel to assure that ambiguities are avoided. To expand this unit for wider frequency coverage (2–18 GHz), channelized downconverters operated by appropriate local oscillators are used, as shown in Fig. 4.8. The overall performance is shown in Table 4.2. Because of its large bandwidth the receiver is also used to cue other superheterodyne analysis receivers (such as the narrowband channelized receiver) to the appropriate frequency.

Table 4.2 Typical IFM Receiver Performance Characteristics

Instantaneous IFM bandwidth	4 GHz
Total frequency coverage	2–18 GHz
Frequency accuracy	2 MHz
Dynamic range	60 dB
Sensitivity (S/N = 1)	−70 dBm
Minimum pulse width	100 ns
Pulse on CW ratio[a]	6 dB

[a] A pulse 6 dB stronger than a CW signal will be measured with some accuracy degradation.

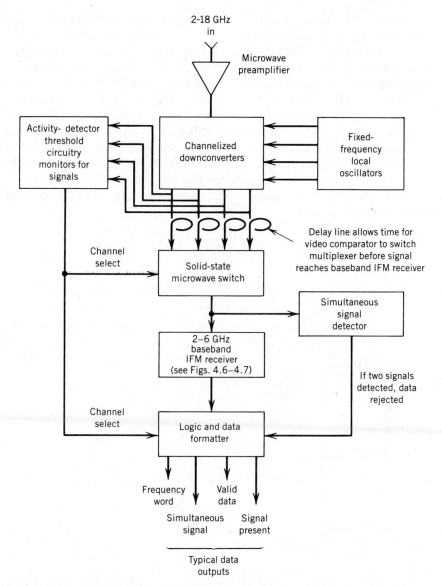

Figure 4.8. *Two-18-GHz IFM receiver. (Reprinted with permission of Microwave Journal, from the September 1980 issue, © 1980 Horizon House-Microwave, Inc.)*

4.3.4 Channelized Receiver

The channelized receiver has many of the desirable properties of the crystal video and IFM receivers with the sensitivity and frequency resolution of a superheterodyne configuration. Unfortunately this improved performance is gained only at higher cost and complexity.

The block diagram of a channelized receiver covering the 2–18 GHz range is shown in Fig. 4.9. By using a group of decreasing bandwidth filter banks (2 GHz, 200 MHz, and 20 MHz) and the appropriate downconverters, the signal appears in one or two neighboring detectors. Neighboring filters are excited because the power spectrum density of a pulsed signal is distributed in frequency. Also if the signal frequency is at the crossover frequency between two 20-MHz filters, both detectors are excited equally. By rapidly switching between LO frequencies, a wide bandwidth can be covered but not as quickly as the IFM or crystal video receiver. Clearly, the number of PIN switches, LO sources, etc., in a channelized receiver exceeds those in any of the earlier designs. The acousto-optic receiver is being developed to reduce the number of components while retaining many of the features of the channelized receiver.

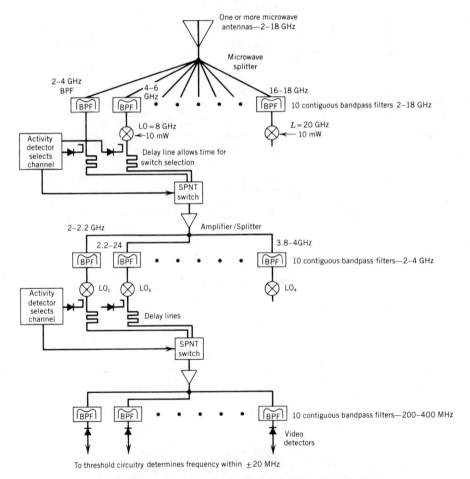

Figure 4.9. *Block diagram of a channelized receiver.*

4.3.5 Acousto-Optic Receiver

This receiver uses the properties associated with the interaction of mono-chromatic light with a microwave frequency acoustic beam as shown in Fig. 4.10. The light diffracted at the Bragg angle θ_B depends on the wavelength of the acoustic beam through the relation $\theta_B = \sin^{-1}(\lambda/2\Lambda)$. Downconverted signals (typically centered at 1 GHz with 500-MHz bandwidth) excite the transducer, which compresses and expands the crystal changing the dielectric constant in a traveling acoustic wave. This varying dielectric constant is analogous to a moving diffraction grating. By placing a segmented optical detector in the diffracted beam, the input frequency can be inferred. This process is analogous to the narrowest (20 MHz) filters in Fig. 4.9, except that only one optical filter is required.

The overall arrangement of an acousto-optic receiver is shown in Fig. 4.11. The array of photosensors is processed and displayed as required. If multiple transducers are positioned on the crystal, a simultaneous measure of the frequency and direction of arrival is possible (4). The overall performance

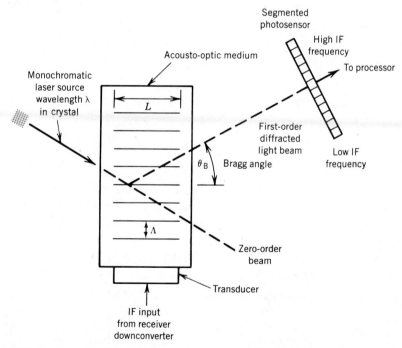

Figure 4.10. Conceptual layout of a Bragg cell in an acousto-optic receiver. Bragg's law, $\sin \theta_B = \lambda/2\Lambda$; diffraction relation, $L \gg 2\Lambda^2/\lambda$. Λ, Wavelength of acoustic energy (if induced); λ, wavelength of light in crystal. (Reprinted with permission of Microwave Journal, from the September 1980 issue, © 1980 Horizon House-Microwave, Inc.)

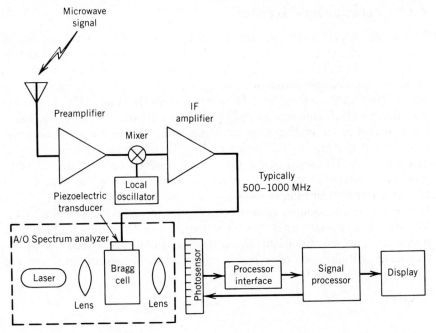

Figure 4.11. *Block diagram of an acousto-optic receiver. (Reprinted with permission of Microwave Journal, from the September 1980 issue, © 1980 Horizon House-Microwave, Inc.)*

characteristics for an advanced acousto-optic receiver are given in Table 4.3. Compared to the channelized receiver, the major differences are reduced dynamic range, size, and weight.

4.3.6   Data Processing

The data processing for complex receivers can handle millions of pulses per second (see Fig. 4.12). Future systems are expected to use the results of the very high-speed integrated circuit (VHSIC) program (5) or GaAs logic cir-

Table 4.3   Typical A/O Receiver Performance Characteristics

Instantaneous RF bandwidth	500 MHz
Dynamic range	40 dB
Minimum discernable signal	−90 dbM
Frequency resolution/accuracy	20 MHz
Time-of-arrival/pulse width resolution	1 μs

(Reprinted with permission of Microwave Journal, from the September 1980 issue, © 1980 Horizon House-Microwave, Inc.)

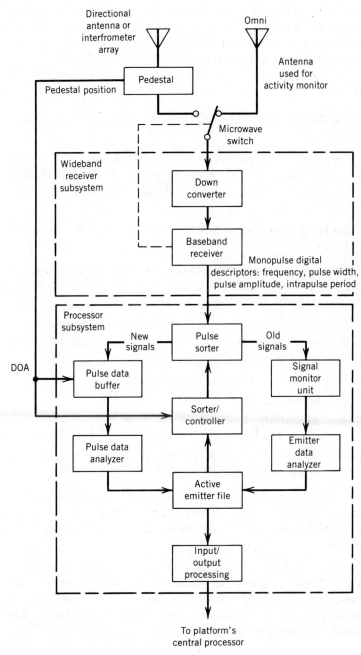

Figure 4.12. *Block diagram of a wideband ESM receiving system and processor system. (Reprinted with permission of Microwave Journal, from the February 1981 issue, © 1981 Horizon House-Microwave, Inc.)*

cuitry (6). The design of many of these circuits requires consideration of high-frequency signal propagation of pulses on a line. Here the line between analog microwaves and extremely high-speed digital logic becomes fuzzy. The topics to follow in this text apply to both technologies.

PROBLEMS

1. Using Eq. 4.4, derive the SSJ burnthrough range $R_{B,SSJ}$ as a function of $J/S = C$.

2. For the following parameters, compute $R_{B,SSJ}$. $P_{tSSJ} = 1000$ W, $G_{tSSJ} = 10$ dB, $C = 0$ dB, $P_t = 10^6$ W, $= 10$ m^2, $G = 36$ dB, BW$_{radar}$ $= 1$ MHz, BW$_{SSJ} = 30$ MHz. Compute the $R_{B,SSJ}$ if P_{tSSJ} is doubled.

3. Using Eq. 4.6, derive the SOJ burnthrough range $R_{B,SOJ}$ as a function of $J/S = C$.

4. Compute $R_{B,SOJ}$ using the following typical parameters: $P_{tSOJ} = 5$ kW, $G_{tSOJ} = 20$ dB, $G_{SL}(dB) = G(dB) - 20$ dB, BW$_{SOJ} = 30$ MHz, and the other values in Problem 4.2. Compute $R_{B,SOJ}$ if P_{tSOJ} is doubled.

5. What is the minimum range for the target defense missiles to assure interception with the SOJ for $C = C'$ (a constant). Neglect the distance associated with altitude differences.

6. For the values quoted above, $R_{B,SSJ} < R_{B,SOJ}$.

   a. Compute $P_{noise,SOJ}/P_{noise,SSJ}$.

   b. If $P_{noise,SOJ} = P_{noise,SSJ}$, $R_{SOJ} = 10\ R_{SSJ}$, $G_{tSSJ}/G_{tSOJ} = 0.1$, $G/G_{SL} = 100$, BW$_{SOJ}$ = BW$_{SSJ}$, compute P_{tSOJ}/P_{tSSJ}.

7. Derive the delay line length relation $\Delta L = v/\Delta f$ using the notation in Fig. 4.7 for an IFM.

8. To avoid ambiguities, the frequency span of the channels is usually increased by 20%. For the four-channel IFM in Fig. 4.7, estimate the mass of the SiO$_2$ delay lines assuming $v = 0.8c$ and the cable weighs 20 g/m.

9. Compute the mass of the delay lines in the channelized receiver of Fig. 4.9 if 20 ns is needed for switching the *SPNT* switches. Use cable parameters above. Estimate the prime power for the local oscillators alone assuming 10% efficiency and 10 mW to each mixer. The power supply efficiency is 50%.

REFERENCES

1. L. V. Blake, "Prediction of Radar Range," in *Radar Handbook*, M. I. Skolnik, ed., McGraw-Hill, New York, 1970.

2. L. B. VanBrunt, *Applied ECM,* EW Engineering, Dunn Loring, VA, 1978.

3. C. B. Hofmann and A. R. Baron, "Wideband ESM Receiving Systems," *Microwave Journal,* Pt. 1, February 1981 and Pt. 2 March, 1981.

4. R. A. Coppock, R. F. Croce, and W. L. Regier, "Acousto-Optic Processor Feasibility Model for Simultaneous Spatial and Spectral Analysis," GTE Sylvania Rpt. No. M1685, Mountain View, CA, June 30, 1978.

5. L. W. Sumney, "VHSIC: A Status Report," *IEEE Spectrum,* Vol. 19, No. 12, December 1982, p. 34.

6. R. C. Eden, ed., "Very Fast Solid-State Technology," Special Issue, *Proc. IEEE,* Vol. 70, No. 1, January 1982.

Part Two
Microwave Transmission, Control, Detection, and Generation

5 | Transmission Line Concepts

The purpose of this chapter is to help you understand transmission lines intuitively. Almost everything at microwave frequencies and above has some transmission line properties. Therefore, it is extremely important to understand not only the formulas, but what they mean.

By the time you finish this chapter, you will have reviewed basic transmission lines and the Smith chart. You will also have been introduced to an advanced version of the Smith chart; microwave engineers use this chart constantly.

It is assumed that you already have some background in electromagnetic theory and an elementary knowledge of transmission lines. You may want to review some of your earlier work in these areas. The problems at the end of the chapter will help you in deciding what material to review.

5.1 ELECTROMAGNETIC WAVES

It is helpful to have some kind of real model when attempting to visualize wave phenomena. Ocean waves, ripple tanks, rubber bands, etc., have been used for this purpose. One of the best models is a child's toy, called a "Slinky®." It can be purchased for a few dollars and be used to illustrate almost every kind of wave motion. A Slinky is like a big floppy spring. One end can be attached to a chair or desk or held firmly by a second experimenter, which (or who) acts as a fixed *termination*, or short. The first experimenter can demonstrate a traveling wave by jerking the Slinky sideways

® Registered trademark of James Industries, Inc. Holidaysburg, PA.

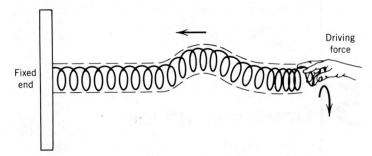

Figure 5.1. *Traveling wave demonstration using a "Slinky."*

with a quick motion. A sideways kink will travel down the Slinky and, if the far end is firmly held, reflect and come back, as illustrated in Fig. 5.1. The velocity v of the envelope of the wave relates the wavelength λ to the frequency f by the relation

$$f\lambda = v$$

Another kind of wave, a standing wave, can be demonstrated by shaking the free end of the Slinky up and down rythmically, as shown in Fig. 5.2. The experimenter then will notice that at certain frequencies of shaking the Slinky assumes a steady-state condition with one or more regions that oscillate up and down separated by points (nodes) that don't move. It is possible to have any number of nodes, depending on the rate of shaking. Obviously, there is a certain practical limit of shaking. Usually it is possible to demonstrate zero nodes, one node, two nodes, and even three nodes. This is a standing wave. It shakes up and down but doesn't go any place.

The waves demonstrated above are transverse waves. That is, the wave motion is sideways. Notice that it is possible to have two kinds of transverse motion, up and down or left and right. Any other kind of transverse motion can be represented as combinations of up and down or left and right waves. Up and down waves are said to be vertically polarized and left and right waves to be horizontally polarized. This property of polarization is associated with transverse waves.

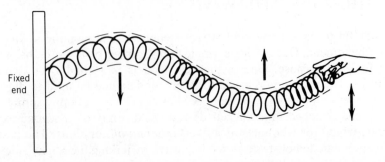

Figure 5.2. *Standing wave demonstration.*

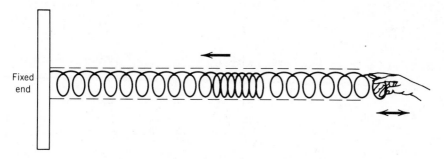

Figure 5.3. Longitudinal waves.

Not all waves are polarized. Instead of shaking (or jerking) the Slinky sideways, give it a jerk (or shake it) along its axis. The experimenter will see regions where the coils of the Slinky are closer together or farther apart moving along it. In order to come closer together or be farther apart, the coils must move longitudinally along the axis of the Slinky. These kind of waves are called longitudinal waves and are illustrated in Fig. 5.3.

This does not exhaust all the possibilities for different kinds of waves. A third kind of wave can be demonstrated by quickly twisting the Slinky. A twisted region will move down the Slinky and reflect off the far end. This kind of wave is a torsional wave. It is illustrated in Fig. 5.4.

Electromagnetic waves are normally transverse waves in free space and air. (If you are not certain why this is so review an introductory text on electromagnetic theory, e.g., Ref. 2, 5–12, 14). The electric and magnetic fields are perpendicular to the direction of propagation. They are also polarizable, as are all transverse waves. In certain kinds of transmission lines, they are also transverse. These are called TEM lines for *t*ransverse *e*lectric and *m*agnetic fields. A TEM wave propagating in free space is illustrated in Fig. 5.5.

Electromagnetic waves may have *longitudinal* components in come kinds of transmission lines. In fact, they must have them in waveguides. We will discuss the conditions when this is true in more detail below. In media other than air and space electromagnetic waves may have not only transverse and

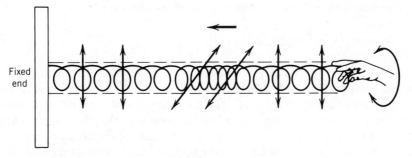

Figure 5.4. Torsional waves.

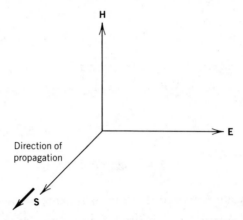

Figure 5.5. Field configuration for a TEM wave and coordinate system.

longitudinal components but also torsional components. The most general description of a wave in complex media uses tensors. We will not discuss such complex waves in detail here, but it is important to understand that they exist and are used in useful devices. In a later chapter one kind of material that makes use of more complex waves is discussed, namely ferrites.

All waves in any kind of medium obey the same equation, called the wave equation. For a one-dimensional wave in a simple medium the wave equation simplifies to

$$\frac{\partial^2 \psi}{\partial x^2} = \frac{1}{v^2} \frac{\partial^2 \psi}{\partial t^2} \tag{5.1}$$

where

ψ = the physical property that is waving (voltage, current, pressure, displacement, etc.)

t = time

x = distance transverse to direction of propagation for transverse waves, etc.

v = velocity of propagation

Equation 5.1 is a partial differential equation. (If you are not familiar with partial differentials now would be a good time to look them up. Even though what follows can be understood without them, you will understand waves better if you know what they are.) The equation says that the rate of change of the physical property with respect to distance equals the rate of change of the physical property with respect to time divided by the velocity squared. Equations of this form can be derived for the electromagnetic field from

Table 5.1   Types of Waves and Media

Type of Wave	Physical Medium
Transverse	Electromagnetic
	Water waves
	String
	Spring (Slinky)
	Surface acoustic
	Bulk acoustic
Longitudinal	Sound in air
	Spring (Slinky)
	Bulk acoustic
Torsional	Spring (Slinky)
	Bulk acoustic

Maxwell's equations. They can be derived for sound waves, water waves, and earthquake waves. A very similar equation can even be derived for the matter waves used in quantum mechanics (1). This is why it is possible to use a Slinky to illustrate all the basic kinds of waves that are normally encountered.

Microwave engineers use other kinds of waves besides electromagnetic waves. They make use of *surface acoustic waves* in SAW devices. They also make use of bulk acoustic waves. The various kinds of waves and some of the media they are associated with are given in Table 5.1.

The wave equation can be expanded for electromagnetic waves to include losses in a transmission line. This form of the wave equation is called the *telegrapher's equation*. (Find the derivation of the telegrapher's equation in a text on electromagnetic theory or transmission lines (2). Justify each step. If you are really ambitious, try to derive the three-dimensional form of the telegrapher's equation. Warning: This is fairly difficult. You will be describing propagation of electromagnetic waves in a generalized lossy medium. Examples include propagation in human beings, buildings, and the Earth.)

5.2   FUNDAMENTAL TRANSMISSION LINE EQUATIONS

In ordinary circuit theory a voltage may be connected to an LC circuit, as shown in Fig. 5.6.

Notice that neither the inductor nor the capacitor is considered to be perfect. They have losses associated with them. Notice, also, that the effect of the voltage generator is felt instantaneously at every point in the circuit. The phase of the output voltage may differ from the phase of the input, but the effect appears instantly. The output phase difference depends on the

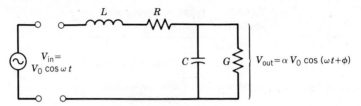

Figure 5.6. *Simple lossy LC circuit.*

components R, L, C, G and the frequency in accordance with the following equation:

$$V_{out} = \left(\frac{1/(jwC + G)}{R + jwL + [1/(jwC + G)]} \right) V_{in} \qquad (5.2)$$

For the more elaborate circuit of Fig. 5.7 with two identical lossy LC meshes in cascade, the output voltage will have a different phase, which again depends on the components.

It is evident that this process could be continued ad infinitem, adding more and more stages. In any real circuit, as opposed to the idealized circuit, the physical dimension of the circuit would steadily expand in the direction of the new stages. It is clear that each lossy inductor and capacitor would have associated with it a certain length, Δx. If the length is the same for both components, we can then associate a certain inductance with a certain length and a certain capacitance with the same length. Such a unit piece of circuit is shown in Fig. 5.8.

Notice that this elementary piece of circuit has associated with it a certain length, Δx, as well as L, C, R, and G. These are distributed parameters as opposed to lumped parameters.

Intuitively, it can be seen that there will be a certain time lag while an input change propagates through the length of the circuit and becomes a change at the output. Even in the case of continuous sinusoidal excitation the phase now has to be a function of length, since the lossy Ls and Cs have lengths associated with them. Thus length, as well as time and the velocity of signal propagation, becomes an important factor in circuit calculations.

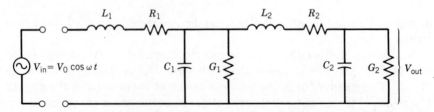

Figure 5.7. *Two mesh lossy LC circuit.*

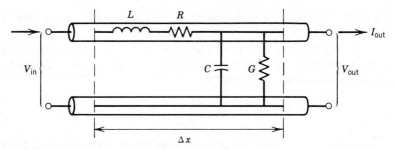

Figure 5.8. Unit circuit piece.

This length factor is what makes microwave circuits different from regular circuits. Everything must be specified in terms of distance as well as time.

In order to find the telegrapher's equation, shrink the above piece of circuit down to a differential element. Also, redefine L, C, R, and G to be inductance, capacitance, resistance, and conductance *per unit length*. They now become distributed values. Consider the differential changes in current and voltage from one side to the other. If you haven't already found the telegrapher's equation your answer should be

$$\frac{\partial^2 V(x, t)}{\partial x^2} = LC \frac{\partial^2 V(x, t)}{\partial t^2} + (RC + LG) \frac{\partial V(x, t)}{\partial t} + RGV(x, t) \qquad (5.3)$$

and

$$\frac{\partial^2 I(x, t)}{\partial x^2} = LC \frac{\partial^2 I(x, t)}{\partial t^2} + (RC + LG) \frac{\partial I(x, t)}{\partial t} + RGI(x, t) \qquad (5.4)$$

Notice the similarity to the wave equation. Notice, also, that there are equations for both voltage and current.

A solution for the voltage equation is a traveling wave:

$$V = V_0 \cos(\omega t \pm \beta x) \qquad (5.5)$$

where $\omega = 2\pi f$ and $\beta = 2\pi/\lambda$. Notice that the sign between the time and distance terms can be either negative or positive.

As an aid to understanding the above equation, consider what happens when the circular frequency, ω, is factored outside the parenthesis. The equation

$$V = V_0 \cos \omega \left(t + \frac{\beta}{\omega} x \right) \qquad (5.6)$$

results.

The ratio ω/β has the dimensions (distance/time), which is velocity. This is, in fact, the phase velocity of the wave.

One interesting property of traveling waves is that under certain conditions they can create a standing wave. The derivation uses the following trigonometric identity:

$$\cos(x \pm y) = \cos x \cos y \mp \sin x \sin y \tag{5.7}$$

Substituting Eq. 5.7 into the equations for waves traveling in both directions we get

$$V_+ = V_0(\cos \omega t \cos \beta x - \sin \omega t \sin \beta x)$$
$$V_- = V_0(\cos \omega t \cos \beta x + \sin \omega t \sin \beta x) \tag{5.8}$$

Adding these two equations together gives

$$V_{\text{total}} = V_+ + V_- = 2V_0 \cos \omega t \cos \beta x \tag{5.9}$$

This is no longer a traveling wave. The time variable and the space vari-

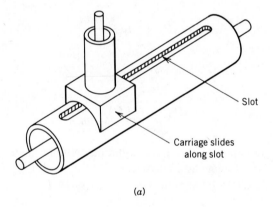

(a)

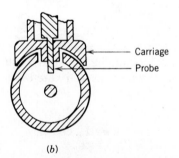

(b)

Figure 5.9. *Slotted line.* (a) *Isometric view;* (b) *cross section.*

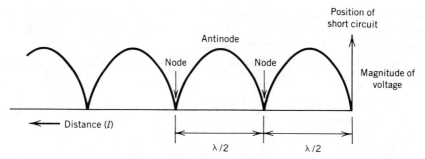

Figure 5.10. *Standing waves on a shorted transmission line*

able are now in arguments of different trigonometric functions. The effect of one cannot be canceled by the other. This is the equation for a *standing wave*. Thus a stationary nonmoving wave can be made from two moving waves. This frequently happens in microwave circuits.

A standing wave can be detected on a transmission line with a slotted line. This is a special piece of line that allows a detector probe to be introduced into a line. In the case of coaxial lines and waveguides there actually is a slot, as shown in Fig. 5.9.

The probe is a short monopole with a diode detector. The pattern of detected voltage it sees as a function of position (*l*) looks like Fig. 5.10 when the line is shorted. The points of zero voltage are nodes. The position of the short fixes the position of all the nodes, since they are exactly one-half wavelength apart. The short has to have a node positioned at it, too, since the voltage across a short must be zero.

The short is a discontinuity in the line. Discontinuities and nonuniformities of any kind cause reflections. For example, opens cause a pattern similar to a short but shifted one-quarter wavelength so that the maximum voltage, rather than zero voltage, occurs across an open as in Fig. 5.11.

Not all discontinuities need be shorts or opens. Shorts and opens reflect 100% of the incident wave. Many discontinuities reflect only part of it. In

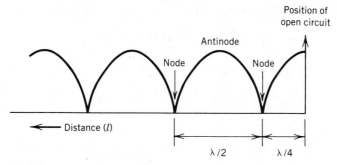

Figure 5.11. *Standing waves on an open transmission line.*

this case the pattern will look something like Fig. 5.12. The detected pattern along the slotted line does not go all the way to zero. Instead there is a maximum value and a minimum value.

The VSWR, or voltage standing wave ratio, is defined in terms of these voltages:

$$\text{VSWR} = \rho = \frac{V_{\max}}{V_{\min}} \tag{5.10}$$

It would also be possible to define a current standing wave ratio, but this is rarely done since diode detectors measure voltage.

VSWR is probably the most used parameter is describing microwave circuits. It can be related to the underlying properties of the circuit. There are, however, several other important concepts that must be discussed first. In order to develop them it will be necessary to use some slightly more sophisticated mathematical techniques. The voltages and currents will have to be represented as complex numbers:

$$v(x, t) = V(x)e^{j\omega t}, \qquad i(x, t) = I(x)e^{j\omega t} \tag{5.11}$$

Substituting these into the telegrapher's equation, it becomes

$$\frac{d^2 V(x)}{dx^2} - \gamma^2 V(x) = 0$$

and $\tag{5.12}$

$$\frac{d^2 I(x)}{dx^2} - \gamma I(x) = 0$$

where

$$\gamma = \alpha + j\beta = \sqrt{(R + j\omega L)(G + j\omega C)}$$

α = the attenuation coefficient in nepers per meter

β = the propagation constant in radians per meter $(2\pi/\lambda)$

The general solutions of the above equations are

$$V(x) = Ae^{-\gamma x} + Be^{\gamma x} \tag{5.13}$$

$$I(x) = \frac{A}{Z_0} e^{-\gamma x} - \frac{B}{Z_0} e^{\gamma x} \tag{5.14}$$

where

$$Z_0 = \sqrt{\frac{R + j\omega L}{G + j\omega C}} \tag{5.15}$$

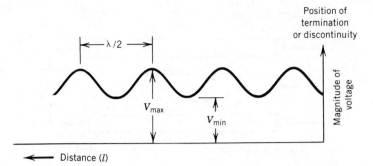

Figure 5.12. *Standing waves on a transmission line with arbitrary termination (or discontinuity).*

which is known as the *complex characteristic impedance* of the transmission line. (Verify that those equations are the general solutions, and formally derive Z_0.) These equations apply only to *uniform* transmission lines. That is, the distributed parameters of the line must be independent of position along the line.

The reflection coefficient for a transmission line terminated in a load Z_L is defined as

$$\Gamma(x) = \frac{Be^{\gamma x}}{Ae^{-\gamma x}} = \frac{B}{A} e^{2\gamma x} = \Gamma_0 e^{2\gamma x} \tag{5.16}$$

where

$$\Gamma_0 = \Gamma(0) = B/A$$

$$e^{-\gamma x} = e^{-\alpha x} e^{-j\beta x} \text{ (incident wave)}$$

$$e^{\gamma x} = e^{\alpha x} e^{j\beta x} \text{ (reflected wave)}$$

$$\beta x = \text{the electrical length of the line}$$

Here Γ_0 is the reflection coefficient at the load, and $x = 0$ in Fig. 5.13 at the load Z_L as shown.

Using these results, the reflected wave can be written as $A\Gamma_0 e^{\gamma x}$, and 5.13 and 5.14 can be rewritten as

$$V(x) = A(e^{-\gamma x} + \Gamma_0 e^{\gamma x}) \tag{5.17}$$

$$I(x) = \frac{A}{Z_0} (e^{-\gamma x} - \Gamma_0 e^{\gamma x}) \tag{5.18}$$

The input impedance of the line can then be written at any point x along

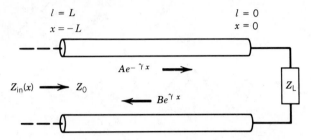

Figure 5.13. *Waves traveling on a terminated line.*

the line as

$$Z_{in}(x) = \frac{V(x)}{I(x)} = Z_0 \frac{e^{-\gamma x} + \Gamma_0 e^{\gamma x}}{e^{-\gamma x} - \Gamma_0 e^{\gamma x}} \tag{5.19}$$

where Γ_0 is evaluated at $x = 0$ and $Z_{IN}(0) = Z_L$. Setting $x = 0$ in Eq. 5.19,

$$Z_L = Z_0 \frac{1 + \Gamma_0}{1 - \Gamma_0} \tag{5.20}$$

and solving for Γ_0 yields

$$\Gamma_0 = \frac{Z_L - Z_0}{Z_L + Z_0} \tag{5.21}$$

It is easy to see from Eq. 5.21 that the reflection coefficient of the load must be zero ($\Gamma_0 = 0$) when the load impedance equals the characteristic impedance ($Z_L = Z_0$).

The input impedance at any point along the line is found by substituting Eq. 5.21 into Eq. 5.19 and letting $x = -l$ to give

$$Z_{in}(l) = Z_0 \frac{Z_L \cosh \gamma l + Z_0 \sinh \gamma l}{Z_0 \cosh \gamma l + Z_L \sinh \gamma l} \tag{5.22}$$

The substitution $x = -l$ allows measurement of positive distances from the load toward the source.

These equations are difficult to deal with because they have complex variables and parameters. In practical microwave circuits it is often possible to assume lossless transmission lines, so $R = 0$ and $G = 0$. In this case $\alpha = 0$ and $\gamma = j\beta$, where $\beta = \omega\sqrt{LC}$. The velocity is simply $v = \omega/\beta = 1/\sqrt{LC}$. The lossless traveling wave equations are

$$V(x) = Ae^{-j\beta x} + Be^{j\beta x} \tag{5.23}$$

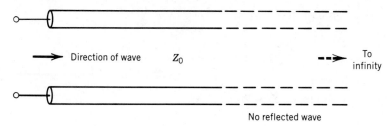

Figure 5.14. Infinite uniform transmission line.

and

$$I(x) = \frac{A}{Z_0} e^{-j\beta x} - \frac{B}{Z_0} e^{j\beta x} \tag{5.24}$$

And, most importantly, the equation for characteristic impedance reduces to

$$Z_0 = \sqrt{\frac{L}{C}} . \tag{5.25}$$

The input impedance for a lossless line then becomes

$$Z_{in}(l) = Z_0 \frac{Z_L \cosh \beta l + j Z_0 \sinh \beta l}{Z_0 \cosh \beta l + j Z_L \sinh \beta l} \tag{5.26}$$

The equation for characteristic impedance Z_0 on a lossless line (Eq. 5.25) is very interesting. The real characteristic impedance, a pure resistance, is a function of two distributed *reactive elements*, the inductance (L) and the capacitance (C). How can this be?

Consider a transmission line extending to infinity, as shown in Fig. 5.14. A wave launched into the line would have no reflection. The signal would simply continue on forever and never return. In other words power can continually be absorbed by the line.

Now suppose the transmission line is cut and terminated with different resistances until one is found that causes no reflection, as shown in Fig. 5.15. This value of resistance is the characteristic impedance of the line, Z_0,

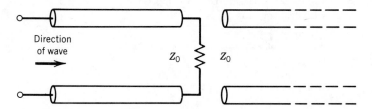

Figure 5.15. Interrupted transmission line terminated by Z_0.

in accordance with Eq. 5.21:

$$\Gamma_0 = \frac{Z_L - Z_0}{Z_0 + Z_L}\bigg|_{Z_L = Z_0} = 0 \qquad (5.27)$$

which says reflectance is zero when the load equals the characteristic impedance.

In the first case, the energy goes into the line and never comes back because it goes to infinity and is permanently stored in the line. In the second case, it is absorbed by the resistor, turns into heat, and never comes back. The source has no way of telling which mechanism is active. They both look the same. In either case, power disappears never to return. Therefore, the characteristic impedance of a lossless transmission line, with no resistances in it, is a *pure resistance*.

It makes no difference where the line is cut! The characteristic impedance is always the same.

The characteristic impedance cannot be measured using an ohm meter across the line unless it is pulsed like a time domain reflectometer (see Chapter 19).

There are really three different kinds of impedances in transmission lines:

1. The distributed parameters L, R, C, G
2. The characteristic impedance, Z_0
3. The actual impedance at a point in the line, Z_{IN}

Table 5.2 compares their characteristics. All of these impedances are represented in Eqs. 5.22 and 5.15.

The characteristic impedance is an example of a natural principle that includes not only transmission lines, but many phenomena. Consider radar waves propagating in space. In the empty sky the waves go on forever. Space can be thought of as a multidimensional infinite transmission line! If the waves strike an airplane, the plane is a discontinuity, and there is a reflected wave.

Space, in fact, has a characteristic impedance of about 377 Ohms/square. Cloth coated with resistive material of this impedance acts as a perfect ter-

Table 5.2   Characteristics of Impedances Associated with Transmission Lines

L, R, C, G	Z_0	Z_{in}
Directly measurable	Not directly measurable	Directly measurable
Not a function of position	*Not* a function of position	Function of position
Distributed	*Not* distributed	*Not* distributed

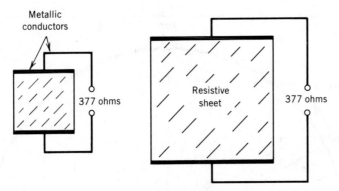

Figure 5.16. *Examples of square sheet resistances.*

mination, totally absorbing incident waves. This is one approach used in the stealth aircraft. The definition of a square is given in Fig. 5.16. Each square has the same resistance for cloth with the same surface resistivity, no matter what size it is. The resistance changes only when the geometry of the square is changed from having equal sides to being a rectangle.

In conclusion, VSWR, described earlier in terms of measurements, can be related to the preceding equations. In Eq. 5.9 two waves traveling in opposite directions on a transmission line produce a standing-wave pattern. From Eq. 5.9 the maximum value of the voltage is

$$| V(x) |_{max} = | A | (1 + | \Gamma_0 |) \qquad (5.28)$$

and the minimum value is

$$| V(x) |_{min} = | A | (1 - | \Gamma_0 |) \qquad (5.29)$$

These values substituted into Eq. 5.10 give the VSWR

$$\text{VSWR} = \rho = \frac{| V(x) |_{max}}{| V(x) |_{min}} = \frac{1 + | \Gamma_0 |}{1 - | \Gamma_0 |} \qquad (5.30)$$

or

$$| \Gamma_0 | = \frac{\text{VSWR} - 1}{\text{VSWR} + 1} = \frac{\rho - 1}{\rho + 1} \qquad (5.31)$$

5.3 SMITH CHART

The Smith Chart is a graphic tool used by engineers for designing microwave circuits (3,4). It provides a pictorial way of displaying the relationships oc-

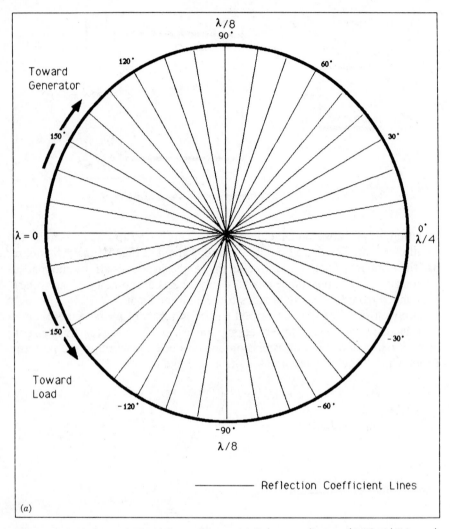

Figure 5.17. Polar and "Smith" coordinates. (a) Polar coordinates; (b) "Smith" impedance coordinates.

curring in transmission lines that often helps in the design process. The human brain is more effective at thinking pictorially than thinking in terms of equations. It is essential that all prospective microwave engineers become so familiar with the Smith chart that its use becomes second nature and intuitive.

The original Smith chart is a superposition on one chart of both the reflection coefficient (Γ) in polar coordinates and the input impedance

$$Z_{\text{in}} = Z_0 \frac{Z_{\text{L}} \cos \beta l + j Z_0 \sin \beta l}{Z_0 \cos \beta l + j Z_{\text{L}} \sin \beta l} \tag{5.32}$$

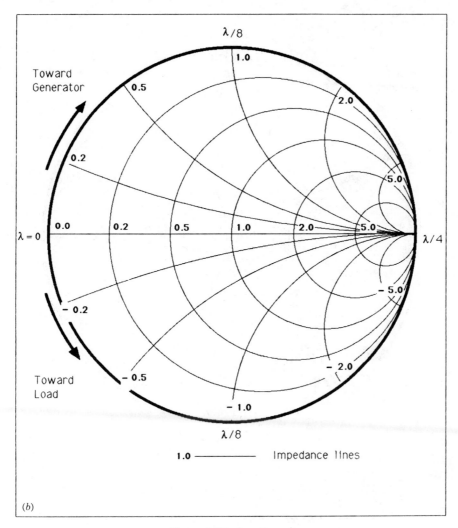

Figure 5.17. *(continued)*

These two coordinate systems are shown in Fig. 5.17 side by side for a standard single color Smith chart.

Examine the advanced Smith chart reproduced in Fig. 5.18. You will see the lines of the special coordinates for the impedance, Z_{in}, and the admittance, Y_{in}, but not the lines for the polar coordinates representing the reflection coefficient, Γ. They have been suppressed to avoid making the chart too busy. However, they are represented by angle graduations around the circumference of the chart (Fig. 5.17a). Notice that $0°$ is at the right hand side of the chart and $180°$ is at the left. The polar coordinates are easily generated whenever needed by laying a straightedge between the center of

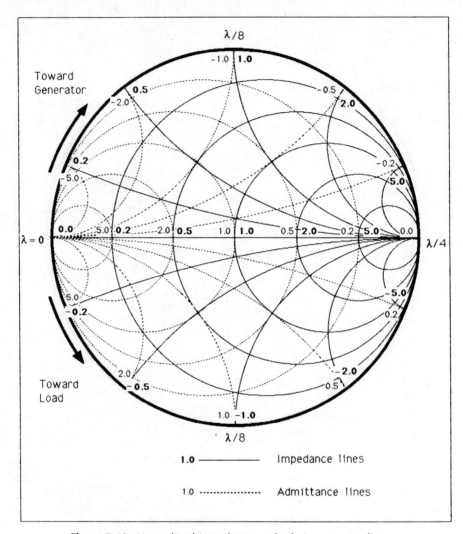

Figure 5.18. *Normalized impedance and admittance coordinates.*

the chart and the appropriate angle marking on the circumference. The limits of the polar coordinates go linearly from 0 at the center to 1 at the circumference. There is a linear scale on commercial charts that can be used with a pair of dividers for this purpose.

Whenever you look at a Smith chart always imagine the polar coordinate system for reflection coefficients as well as the explicitly drawn coordinate systems for impedance and admittance. On a Smith chart a point simultaneously represents three different things:

Reflection coefficient, Impedance, Admittance

depending on the coordinate system you use as a reference. Commercial Smith charts are available with impedance coordinates printed in red and admittance coordinates in green.

The special coordinate systems for impedance and admittance are derived by a transformation from the ordinary representation of complex values on the two-dimensional, linear orthogonal plane shown in Fig. 5.19.

The transformation simply uses the relationship between reflection coefficient (Γ) and impedance (Z) we derived in Eq. 5.21:

$$\Gamma = \frac{Z - Z_0}{Z + Z_0} \qquad (5.33)$$

Usually it is normalized by dividing through by Z_0 to give

$$\Gamma = \frac{z - 1}{z + 1} \qquad (5.34)$$

This type of transformation is called "conformal" because it preserves the angles between coordinates even though they are undergoing considerable distortion. In this case, the 90° angles between the real and imaginary coordinates are preserved, allowing them to serve their same function in the transformed system.

The actual mechanics of the transformation are accomplished as follows. Divide the equation for reflection coefficient into its real and imaginary parts

$$\Gamma = U + jV = \frac{(r - 1) + jx}{(r + 1) + jx} . \qquad (5.35)$$

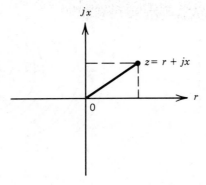

Figure 5.19. *Impedance coordinate system.*

Then separate the real and imaginary parts to obtain

$$U = \frac{r^2 - 1 + x^2}{(r + 1)^2 + x^2} \qquad (5.36)$$

$$V = \frac{2x}{(r + 1)^2 + x^2} \qquad (5.37)$$

Eliminating x from these equations yields

$$\left(U - \frac{r}{r + 1}\right)^2 + V^2 = \left(\frac{1}{r + 1}\right)^2 \qquad (5.38)$$

In the U–V, or Smith chart plane, the lines of constant resistance (r) are plotted in Fig. 5.20. The constant resistance lines become a family of circles centered at $U = r/(r + 1)$, $V = 0$ with radii $1/(r + 1)$.

Eliminating r from Eqs. 5.36 and 5.37 results in

$$(U - 1)^2 + \left(V - \frac{1}{x}\right)^2 = \left(\frac{1}{x}\right)^2 \qquad (5.39)$$

Plotting this equation in the U–V, or Smith chart, plane the constant reactance lines in Fig. 5.21 are obtained.

The constant reactance lines become a family of circles centered at $U = 1$, $V = 1/x$, with radii $1/x$.

The net effect of this transformation is equivalent to a giant mathematical ogre grasping the positive and negative reactance coordinates at infinity,

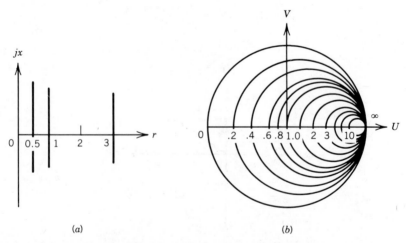

(a)   (b)

Figure 5.20. Resistance lines in the two planes. (a) Constant resistance lines in $z = r + jx$ plane; (b) constant resistance lines in $U + jV$ or Smith chart plane.

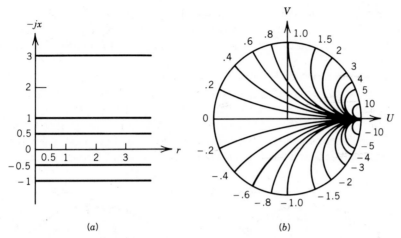

Figure 5.21. *Constant reactance lines in two planes. (a) Constant reactance lines in z = r + jx plane; (b) constant reactance lines in U + jV or Smith chart plane.*

pulling them around to the positive resistance coordinate at infinity, and then pushing them in toward zero, stopping at unit distance as shown in Fig. 5.22.* These coordinates are then superimposed on the polar coordinates of the reflection coefficient.

The above Smith chart coordinates are the impedance lines. The admittance lines are derived using the transformation

$$\Gamma = \frac{1 - y}{1 + y} \qquad (5.40)$$

* Analogy provided by Melvin Zisserson during class lecture.

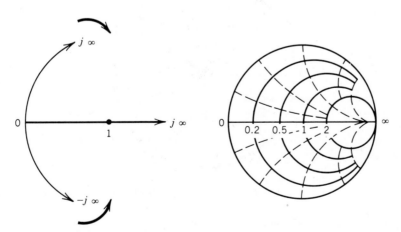

Figure 5.22. *Distortion of the impedance plane by the "Ogre."*

The impedance and admittance coordinate systems are shown side by side in Fig. 5.23 for your reference, along with the suppressed reflection coefficient system. Problem 6 should be worked now to confirm use of the Smith chart.

Outside the Smith chart the reflection coefficient is greater than 1, which means more power is reflected back than is incident! Notice that the resistances and conductances are negative. This region is often useful for representing circuits with amplification. The so called, "compressed" Smith chart, which maps part of this region, is shown in Fig. 5.24. Displays on network analyzers and various automated devices (see Chapter 20) can often

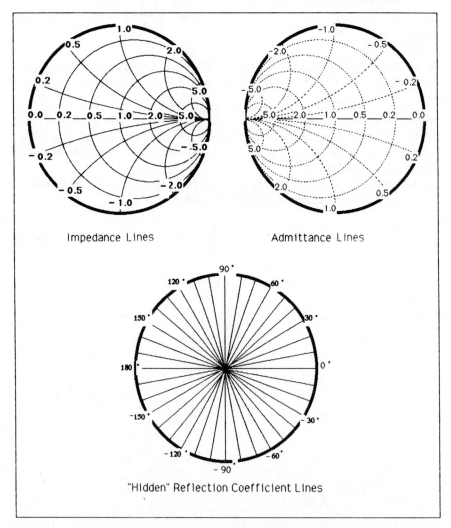

Figure 5.23. Impedance, admittance, and reflection coefficient lines in the unit circle.

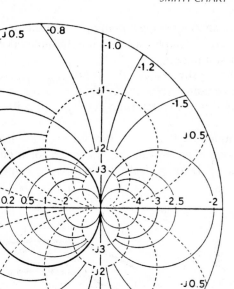

Figure 5.24. "Compressed" Smith chart.

expand or contract the Smith chart to the degree required to display the system under test.

Any lossless uniform transmission line can be plotted as a concentric circle about the center of the Smith chart. Only one point is needed to establish a circle. Thus, a single impedance or admittance value is enough to enter the chart. The phase relationship along a transmission line is represented by movement around the circle, and the electrical distance along the line is also directly related to the distance around the circle.

There is a wavelength scale around the circumference of the chart. It shows that a complete revolution of the chart is equivalent to half a wavelength. *Thus, 180 electrical degrees on the transmission line are represented by 360 physical degrees of revolution on the Smith chart. The degrees printed along the circumference of the chart are not electrical degrees.* They are the phase angle of the reflection coefficient. They show that, in a complete revolution of the chart, the reflection coefficient goes through a complete cycle of 180° positive and 180° negative. The reflection coefficient, admittance, and impedance at any point in the line all vary cyclically as the point of reference moves along a line terminated in anything other than its characteristic impedance.

These properties of the Smith chart can be used to determine the impedance or admittance at any point on a transmission line after the impedance or admittance is determined for any specific point.

Example 5.1. Point A in Fig. 5.25 represents the impedance $(0.1 + j0.5)$ of the load. The impedance is desired for a point 0.1 wavelength away from the load. Draw a circle through A and centered on the center of the chart.

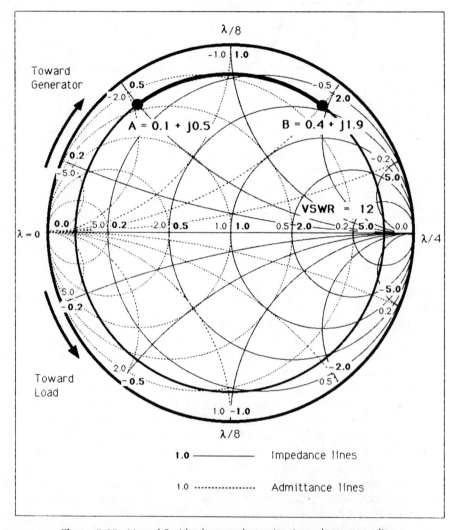

Figure 5.25. *Use of Smith chart to determine impedance on a line.*

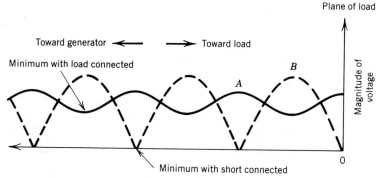

Figure 5.26. *Example of minimum shift for reactive loads.*

Move clockwise 0.1 wavelength toward the generator. This point can be read as $0.4 + j1.9$ on the impedance coordinates.

The example illustrates another fact. Since any uniform transmission line has a constant VSWR, the VSWR can be read directly off the chart from the point where the circle crosses the right half of the center line, giving a VSWR of about 12 in this case. The reciprocal of the VSWR can be read where the circle crosses the left half of the center line. The admittance can be readily found by moving λ/4 away, to a point directly opposite on the chart and the same distance from the center.

The Smith chart was originally developed for use with the slotted line. On the slotted line the actual impedance or admittance values are not known. The impedance (admittance) is measured in an indirect way by sampling the standing wave pattern. This sample provides enough information to calculate the VSWR of the line and therefore the radius of the circle.

Position on the line, and therefore impedance or admittance, is determined by replacing the load with a short. Figure 5.26 shows the standing wave patterns for the conditions with (a) normal load and (b) a short in place of the load. The voltage minimum point for the shorted condition has a different position from the minimum with load. This shift (from loaded to shorted line) contains the desired information.

The following rules are helpful in interpreting the meaning of the minimum shift from a loaded to a shorted line.

1. The shift in the minimum after the load has been shorted is never more than $+/-$ one-quarter wavelength.
2. If shorting the load causes the minimum to move toward the load, a capacitive component exists in the load.
3. If shorting the load causes the minimum to shift toward the generator, an inductive component exists in the load.
4. If shorting the load causes no shift in the minimum, a completely resistive load exists equal to $Z_0/$VSWR.

5. If shorting the load causes the minimum to shift exactly one-quarter wavelength, the load is completely resistive and has a value of (Z_0) (VSWR).

6. When the load is shorted, the minimum will always be a multiple of one-half wavelengths from the load.

Example 5.2. A coaxial air line is used in measuring an unknown load at some undetermined distance from the load. The VSWR is 3.3, and the source frequency is 3 GHz. Replacing the load with a short causes the voltage minimum (null) to move 1.0 cm toward the generator. What is the impedance of the load? For the solution, refer to Fig. 5.27.

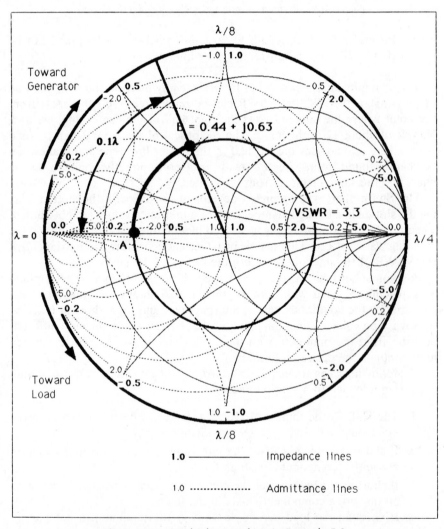

Figure 5.27. *Smith chart analysis in Example 5.2.*

1. Draw the VSWR circle (VSWR = 3.3).
2. Determine the number of fractional wavelengths represented by 1 cm. (Wavelength at 3000 MHz is 10 cm if coaxial line is used, so 1 cm = 0.1 wavelength.)
3. Draw a line from the chart out to the center scale representing 0.1 wavelength toward the generator.
4. Read the impedance of the load as $0.44 + j0.63$ on the impedance coordinate system.

Notice that the minimum associated with the load is always located at the point where the VSWR circle crosses the center line to the left of center, i.e., point A. Notice, also, that it is not necessary to know the distance to the load. In fact, in general this will not be known. All measurements are relative to an arbitrary minimum for the loaded case.

5.4 MATCHING

In most cases, loads and other terminations for transmission lines in practical circuits will not have impedances equal to the characteristic impedance of the line. One of the major problems is how to reduce, or eliminate, the resulting reflections and high VSWRs.

One method is to interpose an arrangement of transmission line sections or lumped components between the mismatched transmission line and its termination. This is called matching. The following are three examples.

Example 5.3. Consider the circuit in Fig. 5.28. The load is an antenna, which, because of design requirements, has a normalized impedance of $0.4 - j0.4$. We will match it to a normalized line using a stub. A stub in this case is a

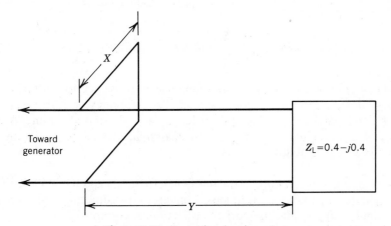

Figure 5.28. Example of stub tuning.

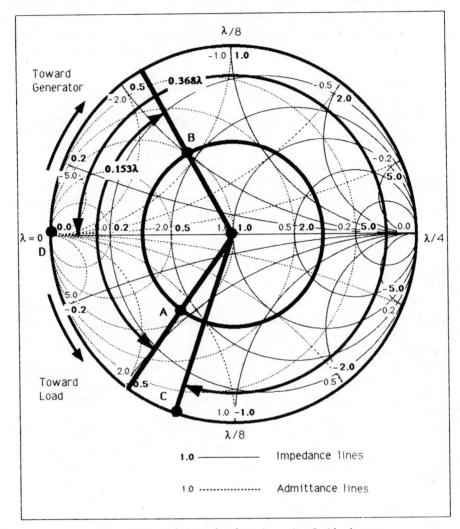

Figure 5.29. *Solution of stub tuning using Smith chart.*

second section of transmission line, identical in cross section, attached to the first line at a distance Y from the load. The frequency is 3 GHz, and the line is a coaxial air line. The stub of length X is shorted at its end. What values of X and Y will eliminate the mismatch at 3 GHz? The solution, as shown in Fig. 5.29 follows:

1. Plot the antenna impedance $0.4 - j0.4$ (point A) and the VSWR circle on the impedance coordinates. Convert to the admittance coordinates and read the admittance $1.25 + j1.25$ mhos (point A).

2. Move around the VSWR circle toward the generator to the constant conductance circle representing a normalized conductance of 1 (point *B*). At point *B* the admittance of the line is $1.0 - j1.15$. If the reactive component could be eliminated at this point the line would be matched. Adding a susceptance of $+j1.15$ in parallel would do this. This means the input admittance of stub *X* should look like $+j1.15$, capacitive susceptance.

3. The distance *Y*, which can be easily computed from the phase angle between points *A* and *B*, is found to be 0.153 wavelengths, or, 1.53 cm from the antenna to the matching stub.

4. To determine the length of stub *X*, enter the Smith chart with the desired capacitive susceptance of $+j1.15$ (point *C*). Note that this point is on an infinite VSWR circle precisely at the circumference of the chart because the matching stub is completely reactive and has no resistive component. From point *C*, move around the Smith chart toward the load until a shorted point is reached at point *D*. This movement requires 0.386 wavelength, or, at 3 GHz, 3.86 cm.

By attaching a parallel, shorted matching stub 3.86 cm long of the same characteristic impedance as the main line at a point 1.53 cm from the mismatched antenna, a perfect match results at 3 GHz. However, by itself, this method is useful only over a narrow frequency range.

Example 5.4. A second method uses quarter-wave sections of transmission line between mismatched transmission lines and their terminations. In this case the additional section of transmission line is added, in series, and does not have the same Z_0 as the line it is matching. Refer to Figs. 5.30 and 5.31. Several quarter-wave matching lines are drawn on it. Line *B*–*B'* goes from a normalized input impedance of 0.3 to 3.33. That is, a quarter-wave section matches an impedance at one end that is the inverse of the impedance at the other end. *A*–*A'* is a pathological case, since it matches 1 to 1 but has zero length! A quarter-wave line evidently provides a way of matching circuit

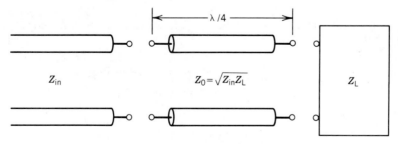

Figure 5.30. *Quarter-wave transformer impedance matching.*

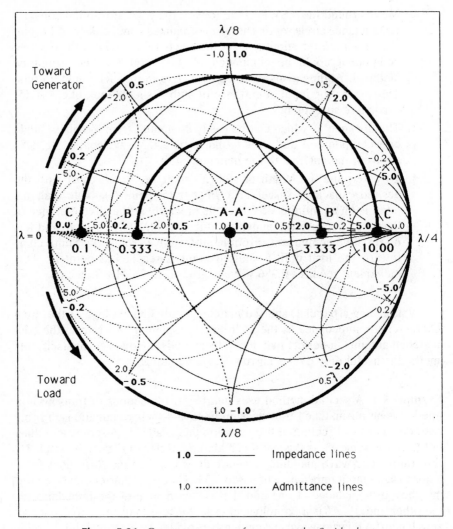

Figure 5.31. *Quarter-wave transformers on the Smith chart.*

elements with different impedances. But, what characteristic impedance should the line have? Recall Eq. 5.32.

$$Z_{in} = Z_0 \frac{Z_L \cos \beta l + jZ_0 \sin \beta l}{Z_0 \cos \beta l + jZ_L \sin \beta l} \tag{5.41}$$

where

$$\beta = \frac{2\pi}{\lambda} \tag{5.42}$$

If

$$l = \lambda/4, \; Z_{in} = Z_0 \frac{Z_0}{Z_L} \tag{5.43}$$

and

$$Z_0 = \sqrt{Z_{in}Z_L} \tag{5.44}$$

So for C–C' the normalized matching impedance is

$$Z_0 = \sqrt{0.1 \times 10} = 1 \tag{5.45}$$

Thus by interposing a quarter wavelength line with a characteristic impedance, which is the geometric mean of the impedances of the two mismatched circuit components, a perfect match is achieved. Again, this method is good only over a narrow frequency range. And, in the most straightforward applications, it requires a real impedance at either end.

Example 5.5. A third matching method uses lumped components to match a mismatched transmission line to its termination. Consider Fig. 5.32. Point B corresponds to an impedance of $1 + j$. It is clear that if a capacitor of reactance $-j$ is interposed between the line and its termination they will be matched. The circuit is shown in Fig. 5.33.

More elaborate combinations of lumped components can be used to match almost any two points on the Smith chart using the following rules:

1. Up is inductive impedance, capacitive admittance.
2. Down is capacitive impedance, inductive admittance.

Point C in Fig. 5.32 is reached by first going to B by adding an inductive susceptance. Movement from B to A is achieved by adding a capacitive reactance. It corresponds to a series capacitance and a parallel inductance, as in Fig. 5.34.

Always draw the matching components in the order you generate them while moving from the circuit element requiring matching to the element being matched. In general, it is good practice to use only reactive matching elements since they don't waste energy or add noise.

5.5 NON-TEM LINES

The transmission lines discussed above have two conductors. There are other types that use only one conductor, or even no conductor at all. Wave-

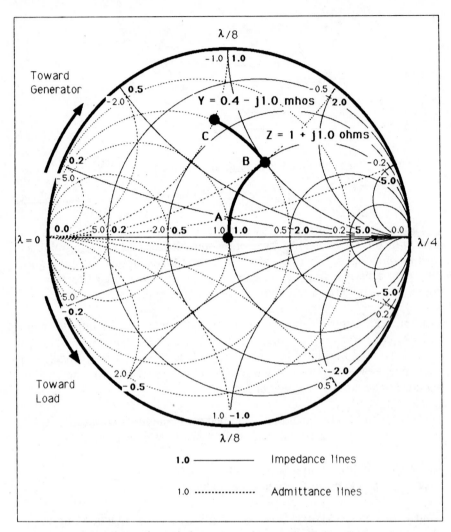

Figure 5.32. *Lumped element matching using the Smith chart.*

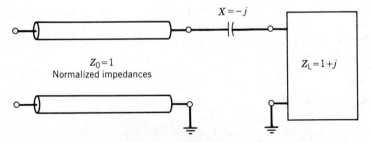

Figure 5.33. *Lumped element match to the load.*

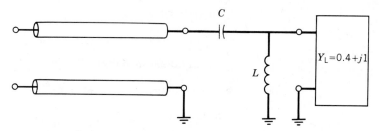

Figure 5.34. *Lumped element ell network to match the load.*

guides are an example. But even ordinary two-conductor lines may start acting like waveguides under certain conditions at microwave frequencies.

It is important to understand the differences between the ideal transmission lines described above and real transmission lines. One of the most important differences occurs between TEM and non-TEM transmission lines.

To study non-TEM lines it is necessary to introduce some additional mathematics. You may want to review vector algebra at this point (5).

Maxwell's equations are the starting point. They are

$$\nabla \times \mathbf{E} = -\frac{\partial \mathbf{B}}{\partial t} \tag{5.46a}$$

$$\nabla \cdot \mathbf{D} = \rho \tag{5.46b}$$

$$\nabla \times \mathbf{H} = \mathbf{J} + \frac{\partial \mathbf{D}}{\partial t} \tag{5.46c}$$

$$\nabla \cdot \mathbf{B} = 0 \tag{5.46d}$$

where $\mathbf{D} = \epsilon \mathbf{E}$ and $\mathbf{B} = \mu \mathbf{H}$.

The symbol $\nabla \times$ is the curl and $\nabla \cdot$ is the divergence. The symbol ∇ means

$$\nabla = \mathbf{x}\frac{\partial}{\partial x} + \mathbf{y}\frac{\partial}{\partial y} + \mathbf{z}\frac{\partial}{\partial z} \tag{5.47}$$

where $\mathbf{x}$, $\mathbf{y}$, and $\mathbf{z}$ are the unit vectors in the three directions.

The other symbols include

$$\mathbf{E} = \text{the electric field vector}$$

$$\mathbf{H} = \text{the magnetic field vector}$$

$$\mathbf{J} = \text{the current density}$$

ρ = the charge density

ϵ = the dielectric constant

μ = the magnetic permeability

$\times$ = the cross product

$\cdot$ = the dot product

The curl is simply a measure of how fast the field intensity changes as one moves across the vector field (see Fig. 5.35), and the divergence is a measure of how fast the field intensity changes along the field (see Fig. 5.36).

Water flowing in a channel exhibits curl. The water near the walls moves slower than the water in the middle, as shown in Fig. 5.37.

The velocity distribution is parabolic if there is no turbulence. A curl meter (see Fig. 5.38), which is like a child's windmill with the blades perpendicular to the flow, will twirl if placed near the walls, but not in the center (5). Thus the curl is high near the walls but low in the center.

Mathematically, curl is represented in a determinant format as

$$\nabla \times \mathbf{E} = \begin{vmatrix} \mathbf{x} & \mathbf{y} & \mathbf{z} \\ \dfrac{\partial}{\partial x} & \dfrac{\partial}{\partial y} & \dfrac{\partial}{\partial z} \\ E_x & E_y & E_z \end{vmatrix} \qquad (5.48)$$

and divergence is

$$\nabla \cdot \mathbf{E} = \frac{\partial E_x}{\partial x} + \frac{\partial E_y}{\partial y} + \frac{\partial E_z}{\partial z} \qquad (5.49)$$

There is nothing mysterious about Maxwell's equations. They relate changes in the intensity of the fields as one moves across or along them to changes in time or to quantities like current and charge. Again note the fundamental importance of space as well as time in microwave circuits as distinguished from lower frequency circuits.

Consider waves propagating in the space between and around conductors that are uniform in the z-direction as shown in Fig. 5.39. There may be two or three conductors, or only one conductor, which may be hollow. A common type of transmission line is a pair of parallel wires shown in Fig. 5.39*a*.

Sometimes the pair may be shielded as shown in Fig. 5.39*b*. Or, only one conductor may be shielded by the other, as in a coaxial line (Fig. 5.39*c*). The central conductor may be left out entirely as with a waveguide (Fig.

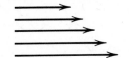

Figure 5.35. A field with curl.

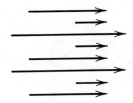

Figure 5.36. A field with divergence.

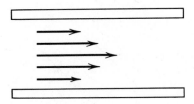

Figure 5.37. Water channel with curl.

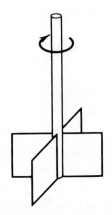

Figure 5.38. Curl meter.

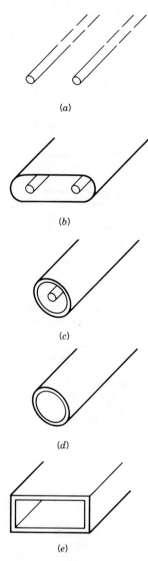

Figure 5.39. *Common transmission line configurations. (a) Two conductors; (b) two-conductor coax; (c) single-conductor coax; (d) circular waveguide; (e) rectangular waveguide.*

5.39*d*). The waveguide need not be round. Waveguides (Fig. 5.39*e*) are often rectangular. Modern microwave circuits often use microstrip, which is discussed in Chapter 8.

When Maxwell's equations are used only in the space around conductors, or enclosed by them, where there is no charge ($\sigma = 0$) or current ($\mathbf{J} = 0$), they can be simplified to

$$\nabla \times \mathbf{E} = -\frac{\partial \mathbf{B}}{\partial t} = -\mu\frac{\partial \mathbf{H}}{\partial t} \qquad (5.50a)$$

$$\nabla \cdot \mathbf{D} = 0 \qquad (5.50b)$$

$$\nabla \times \mathbf{H} = \frac{\partial \mathbf{D}}{\partial t} = \epsilon\frac{\partial \mathbf{E}}{\partial t} \qquad (5.50c)$$

$$\nabla \cdot \mathbf{B} = 0 \qquad (5.50d)$$

These equations can be combined as follows:

$$\nabla \times (\nabla \times \mathbf{E}) = \nabla(\nabla \cdot \mathbf{E}) - \nabla^2\mathbf{E} = \mu\nabla \times \frac{\partial \mathbf{H}}{\partial t} \qquad (5.51)$$

$$\frac{\partial}{\partial t}(\nabla \times \mathbf{H}) = \epsilon\frac{\partial^2 \mathbf{E}}{\partial t^2} \qquad (5.52)$$

to give

$$\nabla^2\mathbf{E} = \mu\epsilon\frac{\partial^2 \mathbf{E}}{\partial t^2} \qquad (5.53)$$

This is the wave equation. Eq. 5.53 states that the electric field propagates with a velocity

$$v = \frac{1}{\sqrt{\mu\epsilon}} \qquad (5.54)$$

and substituting the wave equation becomes

$$\nabla^2\mathbf{E} = \frac{1}{v^2}\frac{\partial^2 \mathbf{E}}{\partial t^2} \qquad (5.55)$$

An equation identical in form can be found for the magnetic field:

$$\nabla^2\mathbf{H} = \frac{1}{v^2}\frac{\partial^2 \mathbf{H}}{\partial t^2} \qquad (5.56)$$

If the waves are sinusoidal in time Maxwell's equations become

$$\nabla \times \mathbf{E} = j\omega\mathbf{B} \qquad (5.57a)$$

$$\nabla \times \mathbf{H} = -j\omega\mathbf{D} \qquad (5.57b)$$

and the wave equation becomes

$$(\nabla^2 + \mu\epsilon\omega^2)\mathbf{E} = 0 \qquad (5.58a)$$

$$(\nabla^2 + \mu\epsilon\omega^2)\mathbf{H} = 0 \qquad (5.58b)$$

With transmission lines it is often more convenient to assume that the waves are sinusoidal not only in time, but also along the line (z-direction). In this case, assume that the electric and magnetic fields have the following form:

$$\mathbf{E}(x, y, z, t) = \mathbf{E}(x, y)e^{\pm jkz - j\omega t}$$
$$\mathbf{H}(x, y, z, t) = \mathbf{H}(x, y)e^{\pm jkz - j\omega t} \qquad (5.59)$$

Here k is the wavenumber (the number of radians per unit length for lossless lines), which may be real or complex.

The wave equation reduces, upon substitution from Eqs. 5.50 and 5.57, to

$$[\nabla_t^2 + (\mu\epsilon\omega^2 - k^2)]\mathbf{E} = 0 \qquad (5.60)$$

or

$$[\nabla_t^2 + (\mu\epsilon\omega^2 - k^2)]\mathbf{H} = 0 \qquad (5.61)$$

where ∇_t^2 is the transverse part of the laplacian operator

$$\nabla_t^2 = \frac{\partial^2}{\partial x^2} + \frac{\partial^2}{\partial y^2} \qquad (5.62)$$

For the analysis of TEM lines it is useful to divide the electric and magnetic fields into parallel and transverse components, as below (1,7):

$$\mathbf{E} = \mathbf{E}_z + \mathbf{E}_t \qquad (5.63a)$$

$$\mathbf{H} = \mathbf{H}_z + \mathbf{H}_t \qquad (5.63b)$$

where

$$\mathbf{E}_z = \mathbf{z}E_z \quad (\mathbf{E}_z = \mathbf{E}_p) \qquad (5.64a)$$

$$\mathbf{E}_t = (\mathbf{z} \times \mathbf{E}) \times \mathbf{z} = \mathbf{x}E_x + \mathbf{y}E_y \qquad (5.64b)$$

and

$$\mathbf{H}_z = \mathbf{z}H_z \quad (\mathbf{H}_z = \mathbf{H}_p) \qquad (5.65a)$$

$$\mathbf{H}_t = (\mathbf{z} \times \mathbf{H}) \times \mathbf{z} = \mathbf{x}H_x + \mathbf{y}E_y \qquad (5.65b)$$

and $\mathbf{x}$, $\mathbf{y}$, and $\mathbf{z}$ are unit vectors in the x, y, and z directions, respectively, and the subscript p denotes the parallel field component.

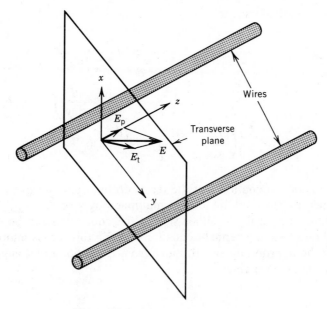

Figure 5.40. Transverse components.

It is also possible to define a transverse curl, divergence, and gradient, as below

$$\mathbf{\nabla}_t \times \mathbf{A}_t = \mathbf{z}\left(\frac{\partial A_y}{\partial x} - \frac{\partial A_x}{\partial y}\right) \tag{5.66a}$$

$$\mathbf{\nabla}_t \cdot \mathbf{A}_t = \frac{\partial A_x}{\partial x} + \frac{\partial A_y}{\partial y} \tag{5.66b}$$

$$\mathbf{\nabla}A_z = \left(\mathbf{x}\frac{\partial}{\partial x} + \mathbf{y}\frac{\partial}{\partial y}\right)A_z \tag{5.66c}$$

The parallel and transverse coordinate system is illustrated for a transmission system in Fig. 5.40.

Begin the detailed analysis of TEM and non-TEM lines by expanding the vector operators in Eq. 5.57 into their components and rearranging them from

$$\mathbf{x}\left(\frac{\partial E_z}{\partial y} - \frac{\partial E_y}{\partial z}\right) + \mathbf{y}\left(\frac{\partial E_x}{\partial z} - \frac{\partial E_z}{\partial x}\right) + \mathbf{z}\left(\frac{\partial E_y}{\partial x} - \frac{\partial E_x}{\partial y}\right)$$
$$= j\omega(\mathbf{x}B_x + \mathbf{y}B_y + \mathbf{z}B_z) \tag{5.67a}$$

$$\frac{\partial D_x}{\partial x} + \frac{\partial D_y}{\partial y} + \frac{\partial D_z}{\partial z} = 0 \tag{5.67b}$$

for the electric field case to

$$\left(\mathbf{x}\frac{\partial E_y}{\partial z} - \mathbf{y}\frac{\partial E_x}{\partial z}\right) + j\omega(\mathbf{x}B_x + \mathbf{y}B_y) + \left(\mathbf{x}\frac{\partial}{\partial y} - \mathbf{y}\frac{\partial}{\partial x}\right)E_z$$

$$= \mathbf{z}\left(\frac{\partial y}{\partial x} - \frac{\partial E_x}{\partial y}\right) - j\omega\mathbf{z}B_z \quad (5.68a)$$

$$\left(\frac{\partial D_x}{\partial x} + \frac{\partial D_y}{\partial y}\right) + \frac{\partial D_z}{\partial z} = 0 \quad (5.68b)$$

Note in the above equation that the terms on the right all have **z** as a multiplier, whereas those on the left are multiplied by **x** or **y**.

Therefore, the terms on either side of the equal sign can be physically separated into two independent equations. Doing this, cross multiplying by **z**, making the appropriate substitutions from Eq. 5.66, and doing the same for the magnetic case gives

$$\frac{\partial \mathbf{E}_t}{\partial z} + j\omega\mathbf{z} \times \mathbf{B}_t = \mathbf{\nabla}_t E_z \quad (5.69a)$$

$$\mathbf{z} \cdot (\mathbf{\nabla}_t \times \mathbf{E}_t) = j\omega B_z \quad (5.69b)$$

$$\frac{\partial \mathbf{H}_t}{\partial z} - j\omega\mathbf{z} \times \mathbf{D}_t = \mathbf{\nabla}_t H_z \quad (5.69c)$$

$$\mathbf{z} \cdot (\mathbf{\nabla}_t \times \mathbf{H}_t) = -j\omega D_z \quad (5.69d)$$

$$\mathbf{\nabla}_t \cdot \mathbf{E}_t = -\frac{\partial E_z}{\partial z} \quad (5.69e)$$

$$\mathbf{\nabla}_t \cdot \mathbf{H}_t = -\frac{\partial H_z}{\partial z} \quad (5.69f)$$

In a TEM wave, by definition, all components in the z-direction are zero. Setting all z-components to zero in the above equations makes *the right side disappear* in every case! Maxwell's equations in a TEM line thus become

$$\frac{\partial \mathbf{E}_{\text{TEM}}}{\partial z} + j\omega\mathbf{z} \times \mathbf{B}_{\text{TEM}} = 0 \quad (5.70a)$$

$$(\mathbf{\nabla}_t \times \mathbf{E}_{\text{TEM}}) = 0 \quad (5.70b)$$

$$\frac{\partial \mathbf{H}_{\text{TEM}}}{\partial z} - j\omega\mathbf{z} \times \mathbf{D}_{\text{TEM}} = 0 \quad (5.70c)$$

$$(\mathbf{\nabla}_t \times \mathbf{H}_{\text{TEM}}) = 0 \quad (5.70d)$$

$$\mathbf{\nabla}_t \cdot \mathbf{E}_{TEM} = 0 \qquad (5.70e)$$

$$\mathbf{\nabla}_t \cdot \mathbf{H}_{TEM} = 0 \qquad (5.70f)$$

From the above equations

$$\mathbf{\nabla}_t \times \mathbf{E}_{TEM} = 0 \qquad (5.71)$$

$$\mathbf{\nabla}_t \cdot \mathbf{E}_{TEM} = 0 \qquad (5.72)$$

This means that $\mathbf{E}_{TEM}$ is a solution of an *electrostatic* problem in two dimensions. A consequence of this is that the TEM mode cannot exist inside a single, hollow, cylindrical conductor of perfect conductivity. The surface is an equipotential; *the electric field therefore vanishes inside*. It is necessary to have two or more separate conductors to support the TEM mode. The parallel wire and coaxial cable transmission lines are common TEM lines; stripline (see Chapter 8) is another.

There are two consequences that can be deduced from this solution of the equations. First, the wave number of Eq. 5.59 is given by

$$k = \omega\sqrt{\mu\epsilon} \qquad (5.73)$$

Combining Eqs. 5.54 and 5.73 gives

$$v = \frac{1}{\sqrt{\mu\epsilon}} = \frac{\omega}{k} \qquad (5.74)$$

and Eq. 5.59 becomes

$$\mathbf{E} = E(x, y) \exp\left[-j\omega\left(t \pm \frac{z}{v}\right)\right] \qquad (5.75)$$

Thus a TEM wave exists down to zero frequency and has a constant velocity related to μ and ϵ. DC circuits are the limiting case at zero frequency.

Second, it can be deduced from the above equations that

$$\mathbf{H}_{TEM} = \pm \sqrt{\frac{\epsilon}{\mu}}\, \mathbf{z} \times \mathbf{E}_{TEM} \qquad (5.76)$$

where the relationship between $\mathbf{H}_{TEM}$ and $\mathbf{E}_{TEM}$ is just like that in free space.

In hollow cylinders (and on transmission lines at high frequencies) two other kinds of field configurations can also propagate. Their existence can be demonstrated using Eqs. 5.60 and 5.61, reproduced below:

$$[\mathbf{\nabla}_t^2 + (\mu\epsilon\omega^2 - k^2)]\mathbf{E} = 0 \qquad (5.77)$$

$$[\mathbf{\nabla}_t^2 + (\mu\epsilon\omega^2 - k^2)]\mathbf{H} = 0 \qquad (5.78)$$

and the boundary conditions at the surface of a conductor

$$\mathbf{n} \times \mathbf{E} = 0 \tag{5.79a}$$

$$\mathbf{n} \cdot \mu\mathbf{H} = 0 \tag{5.79b}$$

where $\mathbf{n}$ is a unit normal vector. These two modes are as follows:

Transverse magnetic (TM) wave

$$H_z = 0 \text{ everywhere } (E_z \text{ at surface } = 0)$$

Transverse electric (TE) wave

$$E_z = 0 \text{ everywhere } \left(\frac{\partial H_z}{\partial n} \text{ at surface } = 0 \right)$$

Also, it can be shown that for both TE and TM waves,

$$\mathbf{H}_t = \pm \frac{1}{Z} \mathbf{z} \times \mathbf{E}_t \tag{5.80}$$

where Z is called the wave impedance and is given by

$$Z = \frac{k}{k_0} \sqrt{\frac{\mu}{\epsilon}} \quad \text{for} \quad \text{TM} \tag{5.81}$$

$$Z = \frac{k_0}{k} \sqrt{\frac{\mu}{\epsilon}} \quad \text{for} \quad \text{TE} \tag{5.82}$$

where for free space

$$k_0 = \omega\sqrt{\mu\epsilon} \tag{5.83}$$

as noted previously.

The transverse fields are related to the longitudinal fields as follows:

$$\mathbf{E}_t = \pm \frac{jk}{\gamma^2} \mathbf{\nabla}_t E_z \quad \text{for} \quad \text{TM} \tag{5.84}$$

$$\mathbf{H}_t = \pm \frac{jk}{\gamma^2} \mathbf{\nabla}_t H_z \quad \text{for} \quad \text{TE} \tag{5.85}$$

and

$$\gamma^2 = \mu\epsilon\omega^2 - k^2 \tag{5.86}$$

H_z and E_z satisfy the equations

$$(\nabla_t^2 + \gamma^2)E_z = 0 \qquad \text{for} \quad \text{TM} \tag{5.87}$$

and

$$(\nabla_t^2 + \gamma^2)H_z = 0 \qquad \text{for} \quad \text{TE} \tag{5.88}$$

These equations, when solved together with the boundary conditions, lead to an eigenvalue problem. There is a spectrum of eigenvalues γ_λ^2 and corresponding solutions ψ_λ, $\lambda = 1, 2, 3, \ldots$, which form an orthogonal set. These different solutions are called *modes of the guide*.

For a given frequency ω, the wave number k is determined for each value of λ via Eq. 5.86:

$$k_\lambda^2 = \mu\epsilon\omega^2 - \gamma_\lambda^2 \tag{5.89}$$

Define a *cutoff frequency* ω_λ,

$$\omega_\lambda = \frac{\gamma_\lambda}{\sqrt{\mu\epsilon}} \tag{5.90}$$

The wave number can be written

$$k_\lambda = \sqrt{\mu\epsilon}\,\sqrt{\omega^2 - \omega_\lambda^2} \tag{5.91}$$

so that for $\omega > \omega_\lambda$ the wave number k_λ is real; waves of the λ mode can propagate in the guide. For frequencies less than the cutoff frequency, k_λ is imaginary; such modes cannot propagate and are called *cutoff modes* or *evanescent modes*.

The behavior of the modes is shown in Fig. 5.41. We see that at any given frequency only a finite number of modes can propagate. It is often convenient

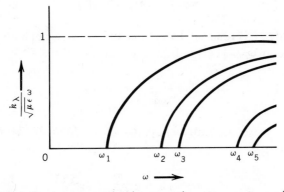

Figure 5.41. *Normalized wave number k_λ versus frequency w. λ, mode number; ω_λ, cutoff frequency; below ω_2 only ω_1 propagates.*

to choose the dimensions of the guide so that at the operating frequency only the lowest, or dominant, mode can occur.

The concept of an evanescent wave has wide application. Even though it cannot propagate, the evanescent wave can penetrate into a section of cutoff waveguide. Its amplitude decays exponentially with distance. For example, a section of waveguide below cutoff can be used as calibrated attenuator. If a dry drinking glass is held in the hands loosely, the ridges on the fingertips are not visible. When the grip is tightened, they become visible because the evanescent wave propagates only a few wavelengths beyond the glass and can interact with the surfaces of the fingers only when they are tightly pressed against the glass. A similar concept in quantum mechanics leads to the tunneling phenomenon that is important in semiconductors. Thus the wave may be evanescent, but it exists.

The most common kind of waveguide is rectangular and normally uses only TE waves. Since it is TE, the only parallel component is H_z. This leads to the equation

$$\left(\frac{\partial^2}{\partial x^2} + \frac{\partial^2}{\partial y^2} + \gamma_\lambda^2\right)H_z = 0 \tag{5.92}$$

with boundary conditions

$$\frac{\partial H_z}{\partial n} = 0 \quad \text{at } x = 0, a \quad \text{and} \quad y = 0, b \tag{5.93}$$

The solution is

$$H_{z,m,n}(x, y) = H_0 \cos\left(\frac{m\pi x}{a}\right) \cos\left(\frac{n\pi y}{b}\right) \tag{5.94}$$

where

$$\gamma_{m,n}^2 = \pi^2\left(\frac{m^2}{a^2} + \frac{n^2}{b^2}\right) \tag{5.95}$$

The single index λ is replaced by two, m and n. They can both be zero. The cutoff frequency $\omega_{m,n}$ is given by

$$\omega_{m,n} = \frac{\pi}{\sqrt{\mu\epsilon}}\left(\frac{m^2}{a^2} + \frac{n^2}{b^2}\right)^{1/2} \tag{5.96}$$

If $a > b$, the lowest cutoff frequency, that of the dominant TE mode, occurs for $m = 1$, $n = 0$:

$$\omega_{1,0} = \frac{\pi}{\sqrt{\mu\epsilon}\, a} \tag{5.97}$$

This corresponds to one-half of a free-space wavelength across the guide, or

$$\frac{\lambda}{2} = a \tag{5.98}$$

The explicit fields for this mode, denoted $TE_{1,0}$, are

$$H_z = H_0 \cos\left(\frac{\pi x}{a}\right) e^{jkz - j\omega t} \tag{5.99}$$

$$H_x = -\frac{jka}{\pi} H_0 \sin\left(\frac{\pi x}{a}\right) e^{jkz - j\omega t} \tag{5.100}$$

$$E_y = j\frac{\omega a \mu}{\pi} H_0 \sin\left(\frac{\pi x}{a}\right) e^{jkz - j\omega t} \tag{5.101}$$

where $k_\lambda = k_{1,0}$ is given by

$$k_\lambda = \sqrt{\mu\epsilon} \sqrt{\omega^2 - \omega_\lambda^2} \tag{5.102}$$

with

$$\omega_\lambda = \omega_{1,0} \tag{5.103}$$

The presence of a factor j means there is both a spatial and a temporal phase difference between H_x (and E_y) and H_z in the propagation direction. It happens that the $TE_{1,0}$ mode has the lowest cutoff frequency of both TE and TM modes and is the one used in most practical situations. For a typical choice, $a = 2b$ and the ratio of cutoff frequencies $\omega_{m,n}$ for the next few modes to $\omega_{1,0}$ are given in Table 5.3.

Table 5.3   Ratio of Rectangular Waveguide Cutoff Frequencies[a]

		$n \rightarrow$			
		0	1	2	3
	0		2.00	4.00	6.00
	1	1.00	2.24	5.13	
m	2	2.00	2.84	4.48	
$\downarrow$	3	3.00	3.61	5.00	
	4	4.00	4.48	5.66	
	5	5.00	5.39		
	6	6.00			

[a] Assumes $a = 2b$.

Circular and elliptic waveguides are also used. A variant of the rectangular waveguide has ridges down the middle of both broad sides. It has a greater bandwidth than the simple rectangular form. The relationship between the electric and magnetic fields in a rectangular waveguide is illustrated in Fig. 5.42 for the $TE_{1,0}$ mode. There are many advanced books covering waveguides (2,5,8,11,12,14), as more elementary introductions (9,10).

TE and TM waves are not restricted to waveguides. They can exist on nominally TEM lines at high enough frequencies, such as coaxial and two wires. Microstrip lines, which you will learn about in more detail in a later chapter, have some non-TEM components at their operating frequencies, although they can be thought of as TEM for practical purposes.

The way to avoid non-TEM modes on a nominally TEM line, such as a coaxial cable, is to operate below the cutoff frequencies for TE and TM waves. For TE waves in coaxial line the cutoff wavelength is approximately

$$\lambda_c \cong \frac{2\pi}{n} \frac{r_1 + r_0}{2}, \qquad n = 1, 2, 3 \qquad (5.104)$$

as illustrated in Fig. 5.43. The cutoff wavelength is close to the arithmetic mean of the circumferences of the inner and outer conductor surfaces, with a radius r_c. For TM waves on coaxial lines the cutoff wavelength is given

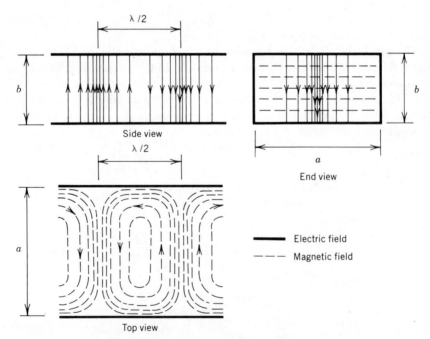

Figure 5.42. *Field configuration of the dominant or TE_{10} mode in a rectangular waveguide.*

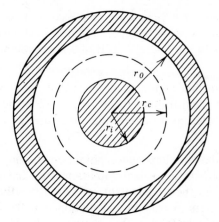

Figure 5.43. *Cutoff wavelength radius r_c for a coaxial transmission line.*

by

$$\lambda_c \approx \frac{2}{P}(r_0 - r_i) \qquad P = 1, 2, 3 \ldots \qquad (5.105)$$

PROBLEMS

1. Every kind of wave demonstrated here is used in some way in microwave systems. If you have any problem visualizing the differences among traveling waves, standing waves, transverse waves, polarized waves, longitudinal waves, or torsional waves get hold of a Slinky and experiment with it until you do understand the differences, not in terms of mathematics but intuitively. See if you can set up standing longitudinal waves and standing torsional waves. Instead of having the far end rigidly fixed, hold the Slinky vertically with the bottom end free. This corresponds to having an "open" termination as opposed to a "short" on the end of a transmission line. Attach the end of the Slinky to a small block of wood and float it on water while holding it vertically. This corresponds to having a "lossy" termination. See what kind of standing waves you get under each of these conditions. See what happens to a traveling wave when it hits each kind of termination.

2. Prove that Eq. 5.5 is a solution of the telegrapher's equation. Also determine which direction the wave moves when the sign is positive and the direction when it is negative.

3. What happens when two waves with different frequencies interact with each other?

4. Find the reflection coefficient for VSWRs of 1, 3, and 10. What percent of the incident power is being reflected in each case? From this you should be able to see that a VSWR of 3 is acceptable, but one of 10 is not.

5. Derive the constant conductance lines and the constant susceptance lines and plot them in both the $y = g + jb$ plane and the $U + jV$, Smith chart planes, as we did for the impedance case. What differences do you see?

6. Locate the following impedance points on the impedance coordinate system. Read their equivalent admittances on the admittance coordinate system and their equivalent reflection coefficients on the polar coordinate system. Verify by calculation that these admittances and reflection coefficients are the correct ones for each normalized impedance.

$$z = 1 + j1$$
$$= 1 + j0$$
$$= 3 - j3$$
$$= 0$$
$$= 0.5 + j0.5$$

7. a. What is meant by normalized impedance?
   b. If a 100-ohm resistive load is connected to a transmission line with a characteristic impedance of 50 ohms, what is the normalized impedance of the load?
   c. If the 100-ohm load is connected to a 75-ohm transsission line, what is the normalized impedance of the load?
   d. If the impedance of the load is $100 + j200$, what is the normalized impedance of the load?

8. a. If the load on a 50-ohm line is a 100-ohm resistor, what is the VSWR?
   b. Would the voltage at the load in (a) be a high voltage or a low voltage with respect to the voltage at the load of a matched line?
   c. If the load on a 50-ohm transmission line were a 25-ohm resistor, what would the VSWR be?
   d. Would the voltage at the load in (c) be a high voltage or a low voltage with respect to the voltage at the load of a matched line?

9. a. If the load on a 50-ohm transmission line is 100-ohms resistive, what is the impedance at a point one-quarter wavelength toward the generator? At one-half wavelength?
   b. If the load on a 50-ohm line is 25-ohms resistive, what is the imped-

ance at a point one-quarter wavelength toward the generator and at one-half wavelength?

  c. From (a) and (b) does a quarter wavelength line repeat or transform the load impedance?

  d. From (a) and (b) does a half wavelength line repeat or transform the load impedance?

  e. What kind of a matching transformer will an eighth-wave transmission line make?

10. **a.** What is the impedance of a quarter-wave shorted stub (a quarter wavelength of transmission line shorted at the load end)? Does this represent a series or parallel resonant circuit?

  b. What is the impedance of a quarter-wave open stub? Is it a series or parallel resonant circuit?

  c. What is the impedance of a one-eighth wave shorted stub?

  d. What is the impedance of a three-eighths wave shorted stub?

  e. What kind of circuits do the one-eighth wave and three-eighth wave stubs represent?

11. **a.** A 1 GHz frequency is used. In an air dielectric transmission line, what is the number of centimeters in a wavelength? In a half wavelength? A quarter wavelength?

  b. If the load causes a VSWR of 3.0 and a minimum voltage point 0.1 wavelengths from the load, what is the normalized load impedance and the actual load impedance, assuming Z_0 equals 50 ohms?

12. **a.** On a 50-ohm line what normalized and actual admittance does an impedance of $200 + j300$ ohms correspond to? Calculate first using the *impedance lines on the Smith chart only*. Then verify using *impedance and admittance lines*.

  b. When would you use normalized impedance values and normalized admittance values for computations using the Smith chart?

  c. Why is it easier to use admittances when computations are made using parallel matching stubs?

13. A space ship has an external antenna that has recently been slightly damaged by a micrometeoroid strike, changing the VSWR. The antenna is connected to the transmitter by a coaxial cable. The captain elects to rematch the antenna using a parallel coaxial stub attached to the cable inside the space ship in an arrangement similar to the one in Fig. 5.28. He does not know the exact frequency of the transmitter since the operating manual has been lost (a bad move). Also he does not know the exact length of the coaxial cable as it goes through the hull of the space ship.

  With the antenna in place, a slotted line shows that the first minimum voltage inside the hull is at 11 cm relative to the inside surface of the hull. The next minimum is at 22 cm. The captain then sends a spaceman outside the ship to short the antenna. The first voltage minimum moves to 15 cm. The VSWR is determined to be 5.1. Find the length of the stub and its location relative to the first minimum. Do this for both shorted and open stubs. Which would be better?

14. Redo Problem 13 using a lumped parallel reactance to ground to compensate for the reactance of the damaged antenna exactly. Then use a quarter wave matching section to transform the remaining resistive part of the load to 50 ohms. (A procedure like this is often used to match transistors in microstrip circuits. Why is it a less practical solution when using coaxial cables?)

15. Redo Problem 13 using a single section of transmission line that acts as a matching transformer that has both resistive and reactive transformation properties. (Why is this approach usually less practical?)

16. In what direction do the electric and magnetic fields point in a TEM line? In a TE line? In a TM line?

17. Why will TEM waves not propagate on a single conductor line or a hollow tube?

18. In what way does the field in a DC line resemble that in a TEM line? What is the lowest frequency that can be passed on a TEM line?

19. What is the lowest frequency that can be passed on waveguide? What kind of a filter does a waveguide make?

20. Find an ordinary piece of coaxial cable, such as might be found in any laboratory. Measure the radii of the inner and outer conductors. Calculate the frequency above which TE waves will begin propagating; also TM waves. (Remember the wavelength in nonair dielectric is divided by the square root of the relative dielectric constant.) Why is this undesirable? What kind of coaxial cable can be used to suppress propagation of TE and TM waves at higher frequencies?

21. The broad dimension of a waveguide is 1.0 in. and the narrow is 0.5 in. Calculate the cutoff frequency and wavelength of the dominant mode.

22. A high-power waveguide has a hole in it. The hole is too small for energy to radiate. Nevertheless, in close proximity to the hole electric and magnetic fields can be detected. Explain.

REFERENCES

1. P. Morse and H. Feshbach, *Methods of Theoretical Physics,* McGraw-Hill, New York, 1953.

2. Phillip C. Magnusson, *Transmission Lines and Wave Propagation,* Allyn and Bacon, Boston, 1965.

3. Phillip H. Smith, "Transmission Line Calculator," *Electronics,* Vol. 12, 1939, p. 31.

4. Phillip H. Smith, *"An Improved Transmission Line Calculator,"* *Electronics,* Vol. 17, p. 130, January 1944.

5. John D. Kraus, *Electromagnetics,* McGraw-Hill, New York, 1953.

6. J. D. Jackson, *Classical Electrodynamics,* Wiley, New York, 1975.

7. Whinnery Ramo and Van Duzer, *Fields and Waves in Communication Electronics,* Wiley, New York, 1965.

8. Robert E. Collin, *Field Theory of Guided Waves,* McGraw-Hill, New York, 1960.

9. A. L. Lance, *Introduction to Microwave Theory and Measurements,* McGraw-Hill, New York, 1965.

10. Albert Camps and Joseph A. Markum, *Microwave Primer,* Sams, Indianapolis, 1965.

11. John C. Slater, *Microwave Electronics,* Dover, New York, 1969.

12. Samuel Y. Liao, *Microwave Devices and Circuits,* Prentice-Hall, Englewood Cliffs, NJ, 1980.

13. Phillip H. Smith, *Electronic Applications of the Smith Chart,* Krieger, 1983 (1969).

14. Guillarmo Gonzales, *Microwave Transistor Amplifiers: Analysis and Design,* Prentice-Hall, Englewood Cliffs, NJ, 1984.

15. Irving L. Kosow (Ed.), Hewlett Packard Co. Engineering Staff, *Microwave Theory and Measurements,* Prentice-Hall, Englewood Cliffs, NJ, 1962.

6 | Computational Methods for Transmission Line Problems

The first level of computational methods for transmission line problems includes the Smith chart and equations used with it. These methods do not permit calculations of phase shift and attenuation of multiple element circuits.

The second level of computational methods makes use of matrices. *ABCD* matrices are used for representing and combining circuit elements, and *S*-parameter matrices are used to calculate familiar measurable parameters; input reflection coefficient (Smith chart) and transmission attenuation and phase shift. Matrices quickly lead to algebraic chaos. With the aid of the computer, however, problems can be easily solved numerically. The matrices can be reduced to a fairly elegent set of FORTRAN IV subroutines, which become a very powerful tool for predicting the responses of circuits. These second-level computational methods will be fully disclosed below.

Higher-level computational methods are needed for combining *S*-parameter matrices directly and for calculating parallel path problems. Microwave transistors are characterized by their *S*-parameters, and these higher-level methods are needed to design circuits using them. The higher-level methods are included in conventional microwave software programs such as TOUCH-STONE® and SUPER-COMPACT®. These software programs also include optimizing design subroutines.

6.1 REVIEW OF LEVEL I COMPUTATIONAL METHODS

Any input impedance and reflection coefficient can be calculated with the transmission line equations below:

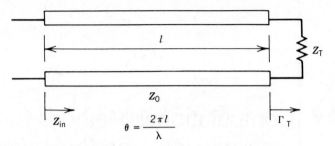

Figure 6.1. Transmission line with terminations.

$$Z_{in} = Z_0 \frac{Z_T \cos \theta + jZ_0 \sin \theta}{Z_0 \cos \theta + jZ_T \sin \theta} \tag{6.1}$$

$$\Gamma_T = \frac{Z_T - Z_0}{Z_T + Z_0} \tag{6.2}$$

$$Z_T = Z_0 \frac{1 + \Gamma_T}{1 - \Gamma_T} \tag{6.3}$$

Working from right to left, elements can be combined by using conventional series and shunt additions and conversions. The transmission line equation is used when transmission line elements are encountered. In order to work with admittance, $1/Y_x$ is substituted for Z_x in all places in Eqs. 6.1–6.3 and Fig. 6.1.

The attenuation of a single discrete series or shunt element embedded in a bilaterally matched transmission line (see Fig. 6.2) can be calculated with the diode attenuation equation:

$$\alpha = 10 \log \left| \left(\frac{R}{2Z_0} + 1 \right)^2 + \left(\frac{X}{2Z_0} \right)^2 \right| \tag{6.4}$$

For shunt circuits, all impedance elements are replaced by their admittance equivalents (R to G, X to B, and Z to Y). This equation is useful for determining the performance of simple diode switches over very wide bandwidths.

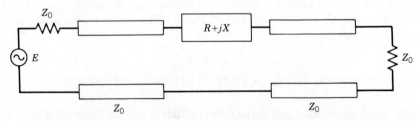

Figure 6.2. Series element in a transmission line circuit.

6.2 REVIEW OF LOOP ANALYSIS USING MATRIX METHODS

A matrix is a convenient notation for keeping track of coefficients in simultaneous equations such as those shown below, which represent the loops in Fig. 6.3.

$$E = I_1 R_1 + (I_1 - I_2)R_2 = (R_1 + R_2)I_1 + (-R_2)I_2 \tag{6.5a}$$

$$0 = (I_2 - I_1)R_2 + I_2(R_3 + R_4) = (-R_2)I_1 + (R_2 + R_3 + R_4)I_2 \tag{6.5b}$$

The same equations can be represented in matrix form as shown in Eq. 6.6:

$$\begin{bmatrix} E \\ O \end{bmatrix} = \begin{bmatrix} R_1 + R_2 & -R_2 \\ -R_2 & R_2 + R_3 + R_4 \end{bmatrix} \cdot \begin{bmatrix} I_1 \\ I_2 \end{bmatrix} \tag{6.6}$$

The matrix can be inverted (solving for I_1 and I_2) by forming a fraction in which the denominator is the determinant and the numerator is the determinant with the left-hand (voltage source column) matrix substituted for the appropriate column in the determinant as shown in Eq. 6.7:

$$I_2 = \frac{\begin{bmatrix} R_1 + R_2 & E \\ -R_2 & O \end{bmatrix}}{\begin{bmatrix} R_1 + R_2 & -R_2 \\ -R_2 & R_2 + R_3 + R_4 \end{bmatrix}} \tag{6.7}$$

Conventional rules for evaluating determinants break the fraction out into a conventional algebraic expression:

$$I_2 = \frac{(R_1 + R_2)(0) - (E)(-R_2)}{(R_1 + R_2)(R_2 + R_3 + R_4) - (-R_2)(-R_2)} \tag{6.8}$$

$$= \frac{ER_2}{R_1(R_2 + R_3 + R_4) + R_2(R_1 + R_3 + R_4) - R_2^2}$$

These methods are used for conventional loop circuit analysis and are similar to those needed for solving transmission line problems.

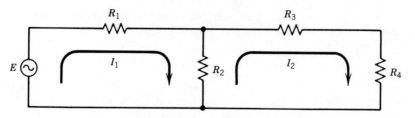

Figure 6.3. Loop analysis circuit.

6.3   REVIEW OF MATRIX MULTIPLICATION

Matrix multiplication of two matrices is shown below:

$$A \cdot B = C = \begin{bmatrix} a_{11} & a_{12} \\ a_{21} & a_{22} \end{bmatrix} \cdot \begin{bmatrix} b_{11} & b_{12} \\ b_{21} & b_{22} \end{bmatrix} = \begin{bmatrix} c_{11} & c_{12} \\ c_{21} & c_{22} \end{bmatrix} \tag{6.9}$$

$$\left. \begin{array}{l} c_{11} = a_{11}b_{11} + a_{12}b_{21} \\ c_{12} = a_{11}b_{12} + a_{12}b_{22} \\ c_{21} = a_{21}b_{11} + a_{22}b_{21} \\ c_{22} = a_{21}b_{12} + a_{22}b_{22} \end{array} \right\} \; c_{mn} = a_{m1}b_{1n} + a_{m2}b_{2n} \tag{6.10}$$

The elements of the row of matrix A are progressively multiplied by the elements of the column of matrix B and summed to give that element of matrix C corresponding to the intersection of the row and column. This systematic multiplication produces the general rule shown in c_{mn}, which becomes a simple FORTRAN subroutine:

```
      SUBROUTINE CASCAD
      COMMON/MATRIX/A(2,2),B(2,2)
      COMPLEX A,B,C(2,2)
      DO 10 M=1,2
      DO 10 N=1,2
   10 C(M,N) = A(M,1)*B(1,N)+A(M,2)*B(2,N)
      DO 20 M=1,2
      DO 20 N=1,2
   20 B(M,N)=C(M,N)
      RETURN
      END
```

In using the subroutine, A is the new element to be added, and B is the old element (everything to the right). The intermediate product is C, which is converted to the old element B to be ready for the next calculation. (*Note:* This is for working from right to left. For working left to right, let 20 A(M,N)=C(M,N) and put new element in as B.)

6.4   *ABCD* MATRIX

The *ABCD* matrix is defined so that pairs can be cascaded by simple multiplication as shown in Fig. 6.4 and Eqs. 6.11 and 6.12:

$$v_1 = Av_2 + Bi_2 \tag{6.11a}$$

$$i_1 = Cv_2 + Di_2 \tag{6.11b}$$

$$\begin{bmatrix} v_1 \\ i_1 \end{bmatrix} = \begin{bmatrix} A & B \\ C & D \end{bmatrix} \cdot \begin{bmatrix} v_2 \\ i_2 \end{bmatrix} \tag{6.12}$$

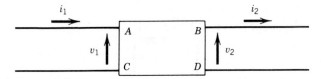

Figure 6.4. *Positive current and voltages for a two-port network.*

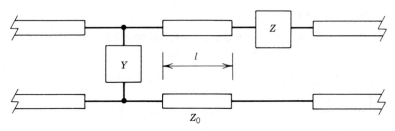

Figure 6.5. *Lumped elements in a transmission line circuit.*

When the elements in the lumped element/transmission line circuit (see Fig. 6.5) are combined using *ABCD* matrices, each element is represented by an *ABCD* matrix and the elements are cascaded by multiplication working from right to left. Subroutines can be written for each type of element such as SER(Z), TLINE(THETA,ZO), and SHUNT(Y). At the end of the calculation, the entire circuit is represented by one *ABCD* matrix, which can be converted to measurable *S*-parameters using relationships that are derived later under "*S*-Parameters" (Section 6.6).

6.5   *ABCD* MATRIX REPRESENTATION OF ELEMENTS

The *ABCD* matrices for common elements are now derived. From these elements, more complex circuits are developed by cascading two-port networks.

6.5.1   Series Impedance

The positive voltages and currents for a series impedance are shown in Fig. 6.6. The equations and matrix representation are shown below:

$$v_1 = i_2 Z + v_2 = v_2 + Z i_2 \qquad \longrightarrow \begin{bmatrix} 1 & Z \\ 0 & 1 \end{bmatrix}$$

$$i_1 = i_2 = 0 + i_2 \qquad\qquad\qquad\qquad\qquad \tag{6.13}$$

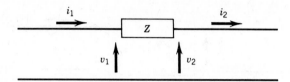

Figure 6.6. *Positive currents and voltages for a series impedance.*

6.5.2   Shunt Admittance

Figure 6.7 defines the positive voltages and currents for a shunt admittance. The equations and matrix follow:

$$v_1 = v_2 = v_2 + 0$$

$$i_1 = i_2 + v_2 Y = Y v_2 + i_2 \qquad \longrightarrow \begin{bmatrix} 1 & 0 \\ Y & 1 \end{bmatrix} \qquad (6.14)$$

6.5.3   Transformer

The ideal transformer shown in Fig. 6.8 is represented by the equations below:

$$v_1 = n v_2 = n v_2 + 0$$

$$i_1 = \left(\frac{1}{n}\right) i_2 = 0 + \left(\frac{i}{n}\right) i_2 \qquad \longrightarrow \begin{bmatrix} n & 0 \\ 0 & 1/n \end{bmatrix} \qquad (6.15)$$

From inspecting the above three derivations it can be seen that the B term (see Eq. 6.12) is for series impedance, the C term is for shunt admittance, and the A and D terms are for transforming ratios.

6.5.4   Lossless Transmission Line

Most microwave problems will require analysis of transmission lines. For short lines, the lossless approximation is usually valid unless high accuracy is required. The voltage and current definitions for a lossless TEM line are

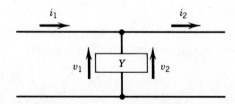

Figure 6.7. *Positive currents and voltages for a shunt admittance.*

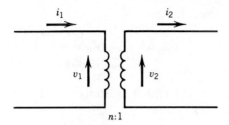

Figure 6.8. *Positive currents and voltages for a transformer.*

shown in Fig. 6.9 and the equations follow:

$$v_1 = v_2 e^{j\theta} = v_2 \cos \theta + jv_2 \sin \theta = v_2 \cos \theta + ji_2 Z_0 \sin \theta$$

$$i_1 = i_2 e^{j\theta} = i_2 \cos \theta + ji_2 \sin \theta = jY_0 v_2 \sin \theta + i_2 \cos \theta \qquad (6.16)$$

$$\longrightarrow \begin{bmatrix} \cos \theta & jZ_0 \sin \theta \\ jY_0 \sin \theta & \cos \theta \end{bmatrix}$$

The evolution of Eq. 6.16 makes use of several required relationships. The definition of the transmission line Z_0 is satisfied only when $v_1/i_1 = Z_0 = v_2/i_2$. The transmission line is a symmetrical network requiring that $A = D$. It is reciprocal that requires that $AD - BC = 1$. It is also lossless, which requires that A and D be real and B and C be imaginary.

6.5.5 Attenuator

When the transmission line equation was derived, the general terms with attenuation were hyperbolic functions. These were simplified to cosine and j sine by assuming no losses. The more general term is given by the following matrix. Taking α' as the attenuation in nepers and assuming no phase shift,

$$\begin{bmatrix} \cosh \gamma l & Z_0 \sinh \gamma l \\ Y_0 \sinh \gamma l & \cosh \gamma l \end{bmatrix} \qquad (6.17)$$

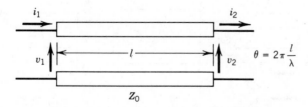

Figure 6.9. *Positive currents and voltages on a lossless transmission line.*

an attenuator is represented by

$$\begin{bmatrix} \cosh \alpha' & Z_0 \sinh \alpha' \\ Y_0 \sinh \alpha' & \cosh \alpha' \end{bmatrix} \qquad (6.18)$$

Expressing α more conventionally in decibels gives

$$\begin{bmatrix} \frac{1}{2}(10^{\alpha/20} + 10^{-\alpha/20}) & \frac{Z_0}{2}(10^{\alpha/20} - 10^{-\alpha/20}) \\ \frac{Y_0}{2}(10^{\alpha/20} - 10^{-\alpha/20}) & \frac{1}{2}(10^{\alpha/20} + 10^{-\alpha/20}) \end{bmatrix} \qquad (6.19)$$

The attenuation of slightly lossy transmission lines is usually modeled by two small attenuators, one at each end of the line, each having half of the line attenuation.

6.6 S-PARAMETERS

S-parameters are frequently encountered in microwave analyses. Network analyzers measure the magnitude and relative phase of the traveling waves (see Fig. 6.10) on the lines that are used to define the S-parameters.

V_I represents the voltage of the incident (forward) traveling wave on the transmission line. V_R represents the voltage of the reflected wave, traveling back toward the generator. V_T represents the voltage of the wave transmitted through the device. The voltage reflection coefficient is defined as Γ:

$$\Gamma = \frac{V_R}{V_I} \qquad (6.20)$$

S-parameters are defined as voltage ratios in which the first subscript indicates the port at which the output voltage is being measured (numerator of the ratio) and the second subscript indicates the port at which the input is taken (denominator of the ratio). Both input and output must be connected

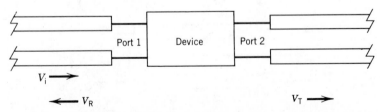

Figure 6.10. *Traveling waves for a device on a matched transmission line.*

to matched transmission lines, and the characteristic impedance must be specified with the S-parameters. Any good transmission line system will be well matched in both directions so S-parameters are meaningful in microwave measurements. Most TEM transmission line systems have a 50-ohm characteristic impedance so that when the characteristic impedance is not specified, 50 ohms is assumed.

The term S_{11} represents the ratio of voltage out port 1 (V_R) to voltage incident to port 1 (V_I):

$$S_{11} = \frac{V_R}{V_I} \tag{6.21}$$

which is equal to Γ, the input reflection coefficient. The term S_{21} represents the ratio of voltage out port 2 (V_T) to voltage incident to port 1 (V_I):

$$S_{21} = \frac{V_T}{V_I} \tag{6.22}$$

This is similar to attenuation defined earlier in Eq. 6.4:

$$\alpha = 20 \log\left(\frac{V_I}{V_T}\right) = 20 \log\left(\frac{1}{S_{21}}\right) \tag{6.23}$$

6.6.1   Derivation of S_{11}

S_{11} in terms of the $ABCD$-parameters is derived using a two-step process. First, Z_{in} is derived using Eq. 6.11 and the definition of the $ABCD$ matrix (Fig. 6.4). Second, the reflection coefficient is equated to the impedances Z_{in} and Z_0. These steps are carried out in Eqs. 6.24 and 6.25:

$$Z_{in} = \frac{v_1}{i_1} = \frac{Av_2 + Bi_2}{Cv_2 + Di_2} = \frac{AZ_0i_2 + Bi_2}{CZ_0i_2 + Di_2} = \frac{AZ_0 + B}{CZ_0 + D} \tag{6.24}$$

$$S_{11} = \Gamma = \frac{Z_{in} - Z_0}{Z_{in} + Z_0} = \frac{AZ_0 + B - CZ_0^2 - DZ_0}{AZ_0 + B + CZ_0^2 + DZ_0}$$

$$= \frac{A + BY_0 - CZ_0 - D}{A + BY_0 + CZ_0 + D} \tag{6.25}$$

Since the device must see Z_0 in both directions, $v_2/i_2 = Z_0$ was used.

6.6.2   Derivation of S_{21}

The derivation of S_{21} is similar to that used later (Chapter 12) in deriving the diode attenuation equation. The source impedance (see Fig. 6.11) must be

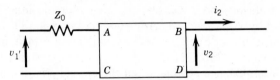

Figure 6.11. Circuit for derivation of S_{21}.

included in the calculation to separate incident and reflected waves. The incident wave V_I is defined by letting the matrix represent a straight-through circuit. Here $v_1 = v_2$, $i_1 = i_2$, $A = 1 = D$, and $B = 0 = C$.

Since the source impedance is a network in series with the circuit under test, it is represented in Eq. 6.26 as a matrix similar to Eq. 6.13:

$$\begin{bmatrix} 1 & Z_0 \\ 0 & 1 \end{bmatrix} \cdot \begin{bmatrix} A & B \\ C & D \end{bmatrix} = \begin{bmatrix} A + CZ_0 & B + DZ_0 \\ C & D \end{bmatrix} = \begin{bmatrix} A' & B' \\ C' & D' \end{bmatrix} \tag{6.26}$$

Equations 6.27 and 6.28 are the derivation of the transmitted wave $V_T (= V_2)$, which then combine to give S_{21}:

$$v_1' = A'v_2 + B'i_2 = A'v_2 + BY_0v_2 \tag{6.27a}$$

$$v_2 = \frac{v_I'}{A' + B'Y_0} = \frac{v_I'}{A + CZ_0 + BY_0 + D} \tag{6.27b}$$

$$V_I = v_2(A = D = 1, B = C = 0) = \frac{v_1'}{2} \tag{6.27c}$$

$$S_{21} = \frac{v_2}{V_I} = \frac{2}{A + BY_0 + CZ_0 + D} \tag{6.28}$$

6.7 APPLICATION OF S-PARAMETERS

Suppose a resistor and matched load terminate a transmission line. The VSWR should be equal to $(R + Z_0)/Z_0$. The derivation follows. For the above circuit (Fig. 6.12), the S_{11} is derived using Eqs. 6.13 and 6.25, yielding the result in Eq. 6.29:

$$S_{11} = \frac{A + BY_0 - CZ_0 - D}{A + BY_0 + CZ_0 + D} = \frac{RY_0}{2 + RY_0} \tag{6.29}$$

The VSWR, defined as $(1 + |S_{11}|)/(1 - |S_{11}|)$, yields the desired relation as shown in Eq. 6.30:

$$\rho = \frac{1 + |S_{11}|}{1 - |S_{11}|} = \frac{2 + RY_0 + RY_0}{2 + RY_0 - RY_0} = 1 + RY_0 = \frac{R + Z_0}{Z_0} \tag{6.30}$$

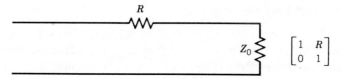

Figure 6.12. Matched transmission line with resistor.

As a second example of the application of S-parameters, what does the diode attenuation equation look like when derived from $ABCD$ matrices? Consider the diode as a series impedance Z having the matrix shown in Eq. 6.31:

$$\begin{bmatrix} 1 & Z \\ 0 & 1 \end{bmatrix} \tag{6.31}$$

Using Eq. 6.28, the S_{21} for the series diode is derived as shown in Eq. 6.32:

$$S_{21} = \frac{2}{A + BY_0 + CZ_0 + D} = \frac{2}{2 + ZY_0} \tag{6.32}$$

Finally, using Eq. 6.23, the attenuation is derived in Eq. 6.33:

$$\alpha = 20 \log \left| \frac{1}{S_{21}} \right| = 20 \log \left| 1 + \frac{Z}{2Z_0} \right|$$

$$= 10 \log \left| \left(\frac{R}{2Z_0} + 1 \right)^2 + \left(\frac{X}{2Z_0} \right)^2 \right| \tag{6.33}$$

This yields the result quoted earlier in Eq. 6.4.

The final example of the application of the S-parameters is the phase shift from an $ABCD$ matrix. Again, use the S_{21} relation and group the real and imaginary terms in the denominator as shown in Eq. 6.34:

$$S_{21} = \frac{2}{A + BY_0 + CZ_0 + D} = \frac{2}{\Sigma \text{Re} + j\Sigma \text{Im}} = \frac{2(\Sigma \text{Re} - j\Sigma \text{Im})}{(\Sigma \text{Re})^2 + (\Sigma \text{Im})^2} \tag{6.34}$$

Completing the square the denominator yields the real and imaginary terms over a common real term. The phase angle is simply the inverse tangent of the ratio of these terms as shown in Eq. 6.35:

$$\phi = \tan^{-1} \left[\frac{\text{Im}(S_{21})}{\text{Re}(S_{21})} \right] = \tan^{-1} \left[-\frac{\Sigma \text{Im}}{\Sigma \text{Re}} \right] \tag{6.35}$$

6.7.1   Loaded Line Phase Shifter

In this section a more realistic example of the utility of *ABCD* matrices to the analysis of microwave problems is given. The example selected is a shunt-loaded phase shifter, as shown in Fig. 6.13. By inspection the *ABCD* matrices for the three two-port subelements (two shunt elements at the ends of a lossless transmission line) are developed, as shown in Eq. 6.36a; Eqs. 6.36b–6.36d show the steps to derive the overall *ABCD* matrix for the two port:

$$\begin{bmatrix} 1 & 0 \\ jB & 1 \end{bmatrix} \begin{bmatrix} \cos\theta & jZ_0\sin\theta \\ jY_0\sin\theta & \cos\theta \end{bmatrix} \begin{bmatrix} 1 & 0 \\ jB & 1 \end{bmatrix} \tag{6.36a}$$

$$\begin{bmatrix} \cos\theta & jZ_0\sin\theta \\ jB\cos\theta + jY_0\sin\theta & -BZ_0\sin\theta + \cos\theta \end{bmatrix} \begin{bmatrix} 1 & 0 \\ jB & 1 \end{bmatrix} \tag{6.36b}$$

$$\begin{bmatrix} \cos\theta - BZ_0\sin\theta & jZ_0\sin\theta \\ jB\cos\theta + jY_0\sin\theta & -BZ_0\sin\theta + \cos\theta \\ -jB^2Z_0\sin\theta + jB\cos\theta & \end{bmatrix} \tag{6.36c}$$

$$\begin{bmatrix} \cos\theta - BZ_0\sin\theta & jZ_0\sin\theta \\ j[2B\cos\theta + (Y_0 - B^2Z_0)\sin\theta] & \cos\theta - BZ_0\sin\theta \end{bmatrix} \tag{6.36d}$$

In Eq. 6.37, the phase shift derived using the technique presented in Eq. 6.35 yields the final result in Eq. 6.37c:

$$\phi = \tan^{-1}\left[-\frac{\Sigma\,\mathrm{Im}}{\Sigma\,\mathrm{Re}} \right] = \tan^{-1}\left[-\frac{\sin\theta + \dfrac{B}{Y_0}\cos\theta - \dfrac{1}{2}\dfrac{B^2}{Y_0^2}\sin\theta}{\cos\theta - BZ_0\sin\theta} \right] \tag{6.37a}$$

$$= \tan^{-1}\left[\frac{-\dfrac{B}{Y_0}\cos\theta + \left(1 - \dfrac{1}{2}\dfrac{B^2}{Y_0^2}\right)\sin\theta}{\cos\theta - \dfrac{B}{Y_0}\sin\theta} \right] \tag{6.37b}$$

$$= \tan^{-1}\left[\frac{-\dfrac{B}{Y_0} + \left(1 - \dfrac{1}{2}\dfrac{B^2}{Y_0^2}\right)\tan\theta}{1 - \dfrac{B}{Y_0}\tan\theta} \right] \tag{6.37c}$$

6.7.2   Reactive Shunt Double Diode Switch

Referring again to Fig. 6.13 and considering the diodes as switches, the attenuation of the switch is derived. The results are given in Eq. 6.38:

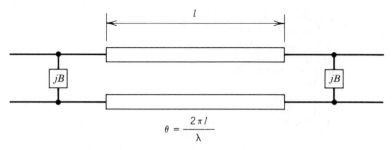

Figure 6.13. Loaded line phase shifter.

$$\alpha = 20 \log\left[\frac{1}{S_{21}}\right] = 10 \log \frac{1}{4}[(\Sigma\mathrm{Re})^2 + (\Sigma\mathrm{Im})^2] \qquad (6.38a)$$

$$\alpha = 10 \log\left\{\left(\cos\theta - \frac{B}{Y_0}\sin\theta\right)^2 + \left[\frac{B}{Y_0}\cos\theta + \left(1 - \frac{1}{2}\frac{B^2}{Y_0^2}\right)\sin\theta\right]^2\right\}$$

$$(6.38b)$$

If both diodes were in the same plane, $\theta = 0$, the attenuation is

$$\alpha = 10 \log\left[1 + \left(\frac{B}{Y_0}\right)^2\right] \qquad (6.39)$$

6.8 FORTRAN SUBROUTINES

The analysis of microwave circuits using the *ABCD* matrices uses several basic modules. These modules are readily computerized for added speed. The FORTRAN listings for the series impedance, shunt admittance, ideal transformer, and lossless transmission line follow. The last two subroutines calculate the attenuation and *S*-parameters given the *ABCD* matrix representing the cascaded network:

```
SUBROUTINE SER(Z)
COMMON/MATRIX/A(2,2),B(2,2)
COMPLEX A,B,C(2,2),Z
A(1,1) = CMPLX(1.,0.)
A(1,2) = Z
A(2,1) = CMPLX(0.,0.)
A(2,2) = CMPLX(1.,0.)
CALL CASCAD
RETURN
END
```

```
SUBROUTINE SHUNT(Y)
COMMON/MATRIX/A(2,2),B(2,2)
COMPLEX A,B,C(2,2),Y
A(1,1) = CMPLX(1.,0.)
A(1,2) = CMPLX(0.,0.)
A(2,1) = Y
A(2,2) = CMPLX(1.,0.)
CALL CASCAD
RETURN
END


SUBROUTINE TFORMR(EN)
COMMON/MATRIX/A(2,2),B(2,2)
COMPLEX A,B,C(2,2)
A(1,1) = CMPLX(EN,0.)
A(1,2) = CMPLX(0.,0.)
A(2,1) = CMPLX(0.,0.)
A(2,2) = CMPLX(1./EN,0.)
CALL CASCAD
RETURN
END


SUBROUTINE TLINE(THETA,Z0)
COMMON/MATRIX/A(2,2),B(2,2)
COMPLEX A,B,C(2,2)
A(1,1) = CMPLX(COS(THETA),0.)
A(1,2) = CMPLX(0.,Z0*SIN(THETA))
A(2,1) = CMPLX(0.,1/Z0*SIN(THETA))
A(2,2) = A(1,1)
CALL CASCAD
RETURN
END


SUBROUTINE ATTEN(ATT,Z0)
COMMON/MATRIX/A(2,2),B(2,2)
COMPLEX A,B,C(2,2)
BD = 10.**(ATT/Z0.)
A(1,1) = CMPLX(0.5*(BD + 1./BD),0.)
A(1,2) = CMPLX(Z0/2.*(BD - 1./BD,0.)
A(2,1) = 1./Z0**2*A(1,2)
A(2,2) = A(1,1)
CALL CASCAD
RETURN
END


SUBROUTINE CALCS(Z0,S21,S11)
COMMON/MATRIX/A(2,2),B(2,2)
COMPLEX B,S21,S11
S21 = 2./(B(1,1)+B(1,2)/Z0+B(2,1)*Z0+B(2,2))
S11 = S21/2.*(B(1,1)+B(1,2)/Z0-B(2,1)*Z0-B(2,2))
RETURN
END
```

6.9 DEFINITION OF *S*-PARAMETER MATRIX

The two-port network in Fig. 6.14 will be used to define the *S*-parameters. a_n is the voltage magnitude of waves incident on the network characterized by the *S*-parameter matrix, and b_n is the voltage magnitude of waves leaving the network. The resulting equations are

$$b_1 = S_{11}a_1 + S_{12}a_2 \tag{6.40a}$$

$$b_2 = S_{21}a_1 + S_{22}a_2 \tag{6.40b}$$

For a perfect (matched) load, $a_2 = 0$, yielding

$$b_1 = S_{11}a_1 \tag{6.41a}$$

$$S_{11} = \frac{b_1}{a_1} \tag{6.41b}$$

$$b_2 = S_{21}a_1 \tag{6.41c}$$

$$S_{21} = \frac{b_2}{a_1} \tag{6.41d}$$

The *S*-parameters in terms of *ABCD* parameters are presented in Eq. 6.42 (1):

$$\begin{bmatrix} S_{11} & S_{12} \\ S_{21} & S_{22} \end{bmatrix} = \frac{\begin{bmatrix} A + BY_0 - CZ_0 - D & 2(AD - BC) \\ 2 & -A + BY_0 - CZ_0 + D \end{bmatrix}}{A + BY_0 + CZ_0 + D}$$

$$\tag{6.42}$$

6.9.1 *ABCD*-Parameters from *S*-Parameters

Most microwave devices are characterized by their *S*-parameters. Catalog components are specified by their input VSWR, isolation, or directivity. These are easily converted to magnitudes of *S*-parameters. Microwave transistors are sold with their *S*-parameter specifications. Components measured with a slotted line, bridge, or network analyzer are all either in *S*-parameters or easily converted to *S*-parameters. Many theories for designing and char-

Figure 6.14. *Circuit for S-parameter matrix.*

acterizing components such as switches, circulators, filters, isolators, and directional couplers give the component performance parameters in the *S*-format. In order to facilitate calculations including these components, the *S*-parameters must be converted to *ABCD* parameters. The equivalence is given in Eq. 6.43 followed by the FORTRAN listing to generate the elements of the *ABCD* matrix:

$$\begin{bmatrix} A & B \\ C & D \end{bmatrix} = \frac{1}{2S_{21}} \times$$

$$\begin{bmatrix} [(1 + S_{11})(1 - S_{22}) + S_{12}S_{21}] & Z_0[(1 + S_{11})(1 + S_{22}) - S_{12}S_{21}] \\ Y_0[(1 - S_{11})(1 - S_{22}) - S_{12}S_{21}] & [(1 - S_{11})(1 + S_{22}) + S_{12}S_{21}] \end{bmatrix}$$

$$(6.43)$$

```
SUBROUTINE S2ABCD(Z0)
COMMON/MATRIX/S(2,2),ABCD(2,2)
COMPLEX S,ABCD
ABCD(1,1) = ((1.+S(1,1))*(1.-S(2,2))+S(1,2)*S(2,1))/2./S(2,1)
ABCD(1,2) = Z0*((1.+S(1,1))*(1.+S(2,2))-S(1,2)*S(2,1))/2./S(2,1)
ABCD(2,1) = ((1.-S(1,1))*(1.-S(2,2))-S(1,2)*S(2,1))/Z0/2/S(2,1)
ABCD(2,2) = ((1.-S(2,1))*(1.+S(2,2))+S(1,2)*S(2,1))/2./S(2,1)
RETURN
END
```

6.9.2 Side Arm Admittance

Frequently, when calculating the response of a transmission line circuit a side arm is encountered. Examples are a bias tee, an off arm of a double throw switch, or an out-of-band arm of a channel dropping filter (frequency division multiplexer). The side arm shown in Fig. 6.15*a* is accommodated as a shunt admittance, as shown in Fig. 6.15*b*. The equations for deriving *Y* follow:

$$v_1 = Av_2 + Bi_2 \qquad (6.44a)$$

$$i_1 = Cv_2 + Di_2 \qquad (6.44b)$$

$$Y = \frac{i_1}{v_1} = \frac{Cv_2 + Di_2}{Av_2 + Bi_2} = \frac{CR_T + D}{AR_T + B} \qquad (6.45)$$

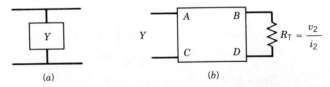

(a)   *(b)*

Figure 6.15. (a) Side arm admittance and (b) its ABCD representation.

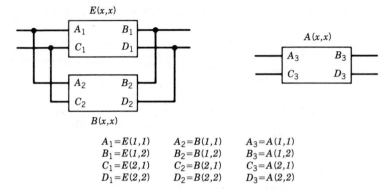

$$A_1=E(1,1) \qquad A_2=B(1,1) \qquad A_3=A(1,1)$$
$$B_1=E(1,2) \qquad B_2=B(1,2) \qquad B_3=A(1,2)$$
$$C_1=E(2,1) \qquad C_2=B(2,1) \qquad C_3=A(2,1)$$
$$D_1=E(2,2) \qquad D_2=B(2,2) \qquad D_3=A(2,2)$$

Figure 6.16. *Circuit for parallel transmission lines.*

The *ABCD* matrix of the side arm is calculated in the conventional manner using cascading matrices. Note that it should be calculated first and saved using notations other than *A*, *B*, and *C* to avoid confusion with the main path calculation. The shunt side arm FORTRAN statement is

$$\text{YSA} = (B(2,1)*RT + B(2,2))/(B(1,1)*RT + B(1,2)) \tag{6.46}$$

In using the equation, B is the working *ABCD* matrix and RT is defined using the statement above (RT will often be 50 ohms). If several side arms are encountered, they should be appropriately indexed YSA1, YSA2, etc. When the side arm is encountered in the main calculation, Y = YSA is stated before calling subroutine SHUNT(Y).

6.9.3   Parallel Transmission Line Paths

In the design of amplifiers with feedback and multiple path diode phase shifters, the transmission line is split into two paths and then rejoined. The circuits for the parallel transmission line paths are shown in Fig. 6.16 and the equations (matrices) are given in Eq. 6.47:

$$
\begin{bmatrix} A_3 & B_3 \\ C_3 & D_3 \end{bmatrix} =
\begin{bmatrix} \dfrac{A_1B_2 + A_2B_1}{B_1 + B_2} & \dfrac{B_1B_2}{B_1 + B_2} \\[2em] (C_1 + C_2) + \dfrac{(A_2 - A_1)(D_1 - D_2)}{B_1 + B_2} & \dfrac{D_1B_2 + D_2B_1}{B_1 + B_2} \end{bmatrix}
\tag{6.47}
$$

The FORTRAN statements to compute to $ABCD_3$ matrix follow:

```
SUBROUTINE PARLIN
COMMON/MATRIX/A(2,2),B(2,2),D(2,2),E(2,2)
COMPLEX A,B,C,(2,2),D,E
A(1,1) = (E(1,1)*B(1,2)+B(1,1)*E(1,2))/(E(1,2)+B(1,2))
A(1,2) = E(1,2)*B(1,2)/(E(1,2)+B(1,2))
A(2,1) = E(2,1)+B(2,1)+(B(1,1)-E(1,1))*(E(2,2)-B(2,2))/(E(1,2)+B(1,2))
A(2,2) = (E(2,2)*B(1,2)+B(2,2)*E(1,2))/(E(1,2)+B(1,2))
B(1,1) = D(1,1)
B(1,2) = D(1,2)
B(2,1) = D(2,1)
B(2,2) = D(2,2)
CALL CASCAD
RETURN
END
```

The working matrix B must be converted to a D matrix to be saved before calling the subroutine. The top matrix ($ABCD_1$) is then derived starting with initializing and then converted to E from B to be saved. Finally the bottom matrix ($ABCD_2$) is calculated in the same way but leaving it in B for calling PARLIN. The accumulated matrix after PARLIN is in B as always after calling CASCAD and includes the parallel lines and everything to their right.

6.9.4   Initializing the Main Calculation

The CASCAD subroutine called as each element is added makes use of a B matrix, which is the accumulated ABCD matrix working away from the load (on the right). In order to begin a calculation, a B matrix must be defined either by using statements for the first element, which are like the subroutine for that element but with B for A and not calling CASCAD, or by using an initializing subroutine corresponding to a transformer ratio of 1 with no call for CASCAD:

```
SUBROUTINE EINIT
COMMON/MATRIX/A(2,2),B(2,2)
COMPLEX B
B(1,1) = CMPLX(1.,0.)
B(1,2) = CMPLX(0.,0.)
B(2,1) = CMPLX(0.,0.)
B(2,2) = CMPLX(1.,0.)
RETURN
END
```

6.9.5   Functions

The FORTRAN statements for several functions frequently required are given below:

```
FUNCTION TPHASE(S21)
COMPLEX S21
TPHASE = ATAN(AIMAG(S21)/REAL(S21))
RETURN
END


FUNCTION TLOSS(S21)
COMPLEX S21
TLOSS = -10.*ALOG10(REAL(S21)**2+AIMAG(S21)**2)
RETURN
END


FUNCTION GAMMAG(S11)
COMPLEX S11
GAMMAG = SQRT(REAL(S11)**2+AIMAG(S11)**2)
RETURN
END


FUNCTION VSWR(GAMMAG)
VSWR = (1.+GAMMAG)/(1.-GAMMAG)
RETURN
END


COMPLEX FUNCTION YSA(B,RT)
COMPLEX B,YSA
YSA = (B(2,1)*RT+B(2,2))/(B(1,1)*RT+B(1,2))
RETURN
END
```

6.9.6 Summary of Tools

The following subroutines and functions have been developed in this chapter:

SUBROUTINES	FUNCTIONS
CASCAD	TPHASE(S21)
SER(Z)	TLOSS(S21)
SHUNT(Y)	GAMMAG(S11)
TFORMR(EN)	VSWR(GAMMAG)
TLINE(THETA,Z0)	COMPLEX YSA(B,RT)
ATTEN(ATT,Z0)	
CALCS(Z0,S21,S11)	
S2ABCD(Z0)	
PARLIN	
EINIT	

6.10 HIGHER-LEVEL *S*-PARAMETER COMPUTATIONAL METHODS

The computational methods described so far are limited to two-port devices. Frequently in measurements, directional couplers are used, which are three-port devices, and hybrid junctions and magic T's are used, which are four-port devices. Signal flow graphs provide a convenient method for accommodating the many interactive terms in making computations on circuits with more than two ports.

6.10.1   Signal Flow Graphs Applied to Microwave Networks

Signal flow graphs are representative of the scattering matrix for n-port networks and allow generation of network equations by the use of general rules. The graph is generated by representing the incident a_i and reflected b_i waves as nodes and the scattering parameters S_{ij} by directed line segments. The waves shown in Fig. 6.14 and represented by Eq. 6.40 can be represented by the signal flow graph given in Fig. 6.17. The arrows pointing to the b_i nodes from the a_i nodes generate the S-parameter equations; e.g., $b_2 = S_{21}a_1 + S_{22}a_2$. The signal flow graphs for a load, generator, video detector, lossless transmission line, shunt admittance, and series impedance (2, 3) are shown in Figs. 6.18a through f. In Fig. 6.18c, the video detector, the calibration of M is assumed to include the effects of the detector's response to the total voltage or the power absorbed. In the generator case, Fig. 6.18b, the generator represents a constant source of outgoing waves b with a reflection coefficient Γ_g looking back into the generator (2).

The networks can be cascaded by connecting the outgoing wave from one network to the input waves of the next network. Each total network flow graph contains paths (a series of directed lines followed sequentially so that no node is touched twice) whose total value is the product P_i of the coefficients (S-parameters) encountered while traversing the path. Each network is made up of "loops" of various orders. For example,

Order of Loop	Definition
First	A series of directed lines (S-parameters) coming to closure and followed sequentially without touching a node (wave) more than once
Second	Product of two first-order loops that do not touch at any point
Third	Product of three first-order loops that do not touch at any point

Figure 6.17. Signal flow graph for a two-port network.

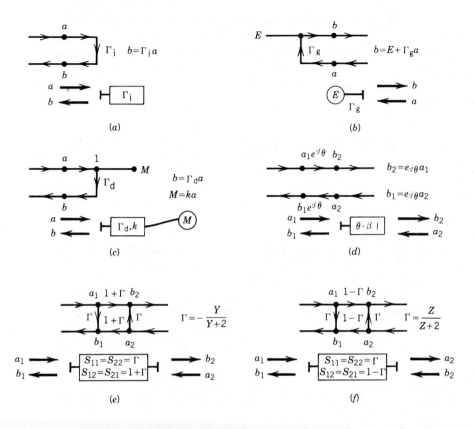

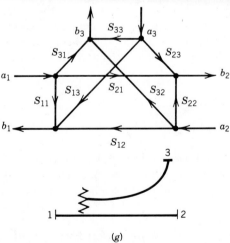

Figure 6.18. Signal flow graph of common microwave components. (a) Load; (b) generator; (c) video detector; (d) lossless line; (e) shunt admittance; (f) series impedance; (g) three-port directional coupler.

Usually these higher-order loops will not actually be evaluated because they tend to be the product of small terms. The solution of the flow graph is accomplished by application of Mason's "nontouching loop" rule (3):

$$\frac{\text{Dependent variable}}{\text{Independent variable}}$$

$$= \frac{\begin{aligned}&P_1(1 - \sum L(1)^{(1)} + \sum L(2)^{(1)} - \sum L(3)^{(1)} + \cdots) \\ &+ P_2(1 - \sum L(1)^{(2)} + \sum L(2)^{(2)} \cdots) + P_3(1 - \cdots) + \cdots\end{aligned}}{1 - \sum L(1) + \sum L(2) - \sum L(3) + \cdots}$$

(6.48)

where

P_i = values of scattering parameters around the ith path from independent variable to dependent variable

$\sum L(1)$ = sum of all first-order loops

$\sum L(2)$ = sum of all second-order non-touching loops

$\sum L(1)^{(1)}$ = sum of all first-order loops that do not touch path P_1 at any point

and so on

As an example of the use of the above equation, consider a two-port network driven by a generator of reflection coefficient Γ_g and terminated by a load Γ_l as shown in Fig. 6.19. The transmission of the network is b_2/E_g, and the reflection coefficient is b_1/a_1. For this example, b_2 and b_1 are the dependent variables, whereas E_g and a_1 are the independent variables. Working out the reflection coefficient, there are two paths from a_1 to b_1, as shown in Fig. 6.19b. There are no second-order loops and only one first-order loop that does not touch P_1, namely S_{22l}. Thus applying Eq. 6.48, the reflection coefficient is

$$\Gamma_{\text{in}} = \frac{b_1}{a_1} = \frac{P_1(1 - \sum L(1)^{(1)}) + P_2}{1 - \sum L(1)} = \frac{S_{11}(1 - S_{22}\Gamma_l) + S_{21}\Gamma_l S_{12}}{1 - S_{22}\Gamma_l} \quad (6.49)$$

which is a logical result, since if $\Gamma_l = 0$, $\Gamma_{\text{in}} = S_{11}$. The generator flow graph was not used because no path exists from a_1 to b_1 through the generator. Γ_g appears in the expression for the transmission coefficient b_2/E_g.

6.10.2   Worst-Case Limits from Signal Flow Graph Relations

The quantities in Eq. 6.48 are complex numbers. Since their magnitude and phase can vary rapidly with frequency, it is instructive to compute both an upper and lower bound. This is accomplished by making all the terms in the

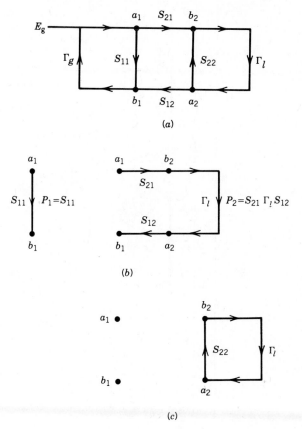

Figure 6.19. *Signal flow graph and loops for a two-port network with generator and load. (a) Signal flow graph for a generator, two ports and a load; (b) first-order loop* $L(1)^{(1)}$ *does not touch* P_1; *(c) reflection coefficient paths.*

numerator positive (negative) and all the terms in the denominator negative (positive). The logarithmic limits are

$$20 \log_{10} \left| \frac{\text{(Positive terms)}}{\text{(Negative terms)}} \right| \quad \text{upper limit} \qquad (6.50)$$

and

$$20 \log_{10} \left| \frac{\text{(Negative terms)}}{\text{(Positive terms)}} \right| \quad \text{lower limit} \qquad (6.51)$$

The development of the relations for the ratio of dependent to independent variables (Eq. 6.48) is obviously complex. Kuhn (3) manipulates the flow graph and reduces the complexity by using four rules to avoid use of the

Table 6.1 Manipulation Rules for Flow Graphs

Statement	Graph

RULE I: Two branches, whose common node has only one incoming and one outgoing branch (branches in series), may be combined to form a single branch whose coefficient is the product of the coefficients of the original branches. Thus the common node is eliminated.

$$E_1 \;\xrightarrow{S_{21}}\; E_2 \;\xrightarrow{S_{32}}\; E_3 \;=\; E_1 \;\longrightarrow\; E_3, \quad S_{31} = S_{21}S_{32}$$

RULE II: Two branches pointing from a common node to another common node (branches in parallel) may be combined into a single branch whose coefficient is the sum of the coefficients of the original branches.

$$= \quad E_1 \;\longrightarrow\; E_2, \quad S_{21} = S_A + S_B$$

RULE III: When node n possesses a self loop (a branch which begins and ends at n) of coefficient S_{nn} the self loop may be eliminated by dividing the coefficient of every other branch entering node n by $1 - S_{nn}$.

$$= \quad E_1 \;\xrightarrow{\dfrac{S_{21}}{1 - S_{22}}}\; E_2 \;\xrightarrow{S_{32}}\; E_3$$

RULE IV: A node may be duplicated, i.e., split into two nodes which may be subsequently treated as two separate nodes, so long as the resulting signal flow graph contains, once and only once, each combination of separate (not a branch which forms a self loop) input and output branches which connected to the original node. Any self loop attached to the original node must also be attached to each node resulting from duplication.

Node duplication rule: (a) node with single input branch; (b) node with single output branch; (c) node with neither single input nor single output branches.

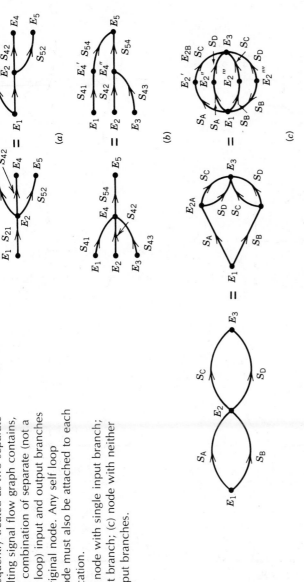

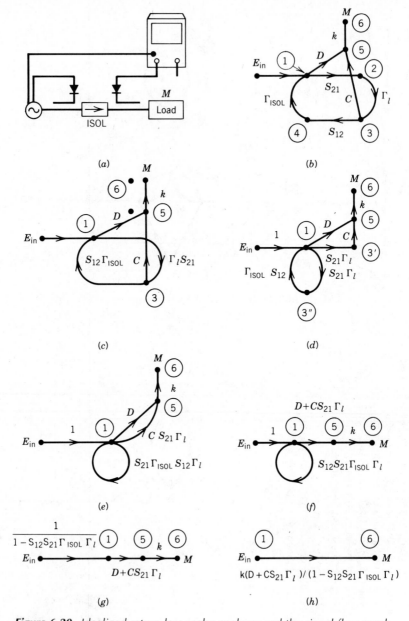

Figure 6.20. *Idealized return loss scalar analyzer and the signal flow graph.*

nontouching loop relation (Eq. 6.48). This method is recommended for complex networks where missing a path is likely to occur.

6.10.3 Manipulation of Flow Graphs

The following manipulations reduce the flow graph to simpler forms to find the value at a node. The manipulations are based on the four rules (2) shown in Table 6.1. Each of these rules may be verified analytically (see Problems). A particularly interesting analysis for rule 3 follows:

$$E_2 = S_{21}E_1 + S_{22}E_2 \quad \text{and} \quad E_3 = S_{32}E_2 \qquad (6.52)$$

Solving for E_2 and E_3 in terms of the independent (input) voltage E_1 yields

$$E_2 = \frac{S_{21}}{1 - S_{22}} E_1 \quad \text{and} \quad E_3 = \frac{S_{32}S_{21}}{1 - S_{22}} E_1 \qquad (6.53)$$

which appears similar to the $1 - \mu\beta$ feedback term in that discipline. It is interesting to note that if S_B is reversed in rule 2, $S_{21} \neq S_A - S_B$.

As an example of the use of this technique, consider the reflectometer in Fig. 6.20a, with a perfect isolator ($|S_{21}| = 1, |S_{11}| = |S_{22}| = 0$) and no DUT. A simplified flow diagram is shown in Fig. 6.20b, where k is the detector constant, $C = |S_{53}|$ is the coupling constant, and $D = |S_{51}|$ is the coupler's directivity. The other paths have not been shown because they are usually small effects and add little to the understanding of the impact of the coupler's directivity.

The analysis of this example follows these steps:

1. Combine S_{21} and Γ_l, and S_{12} and Γ_{isol} using rule 1 (see Fig. 6.20c).
2. Duplicate node 3 using rule 4 to create nodes 3′ and 3″ (see Fig. 6.20d).
3. Eliminate node 3′ and 3″ by rule 1 (see Fig. 6.20e).
4. Combine the two parallel branches between nodes 1 and 5 using rule 2 (see Fig. 6.20f).
5. Remove the loop at node 1 using rule 3 (see Fig. 6.20g).
6. Eliminate nodes 1 and 5 by successive application's of rule 1 (see Fig. 6.20h).

The resulting relation was derived without use of the nontouching loop rule. For a perfect isolator ($\Gamma_{isol} = 0$), $M/E_{in} = k(D + CS_{21}\Gamma_l)$. If $\Gamma_l = 0$, $M/E_{in} = kD$, which is the limiting condition for direct measurement of the coupler's directivity.

The complete analysis of the directional coupler (including all the terms in Fig. 6.18g) with a detector reflection coefficient Γ_{det} has been analyzed by the flow graph technique (4). A computer is used to generate all the paths and loops yielding the following:

paths from node 1 to node 6:

$$P_1 = kD = kS_{31}, \ P_2 = kS_{21}\Gamma_i C = kS_{21}\Gamma_i S_{32};$$

first-order loops, L(1):

$$S_{21}\Gamma_i S_{12}\Gamma_{iso l}, \ D\Gamma_{det}S_{23}\Gamma_i S_{12}\Gamma_{iso l}, \ D\Gamma_{det}S_{13}\Gamma_{iso l}, \ S_{21}\Gamma_i C\Gamma_{det}S_{13}\Gamma_{iso l},$$
$$S_{11}\Gamma_{iso l}, \ S_{33}\Gamma_{det}, \ \Gamma_{det}S_{23}\Gamma_i C, \ S_{22}\Gamma_i;$$

second-order loops, L(2), "non-touching," no common nodes:

$$(D\Gamma_{det}S_{13}\Gamma_{iso l})(S_{22}\Gamma_i), \ (S_{21}\Gamma_i S_{12}\Gamma_{iso l})(\Gamma_{det}S_{33}), \ (S_{11}\Gamma_{iso l})(S_{33}\Gamma_{det}),$$
$$(S_{11}\Gamma_{iso l})(\Gamma_{det}S_{23}\Gamma_i C), \ (S_{11}\Gamma_{iso l})(\Gamma_i S_{22}), \ (S_{33}\Gamma_{det})(\Gamma_i S_{22});$$

third-order loops, L(3), "non-touching," all other loops touch:

$$(S_{11}\Gamma_{iso l})(S_{33}\Gamma_{det})(S_{22}\Gamma_i);$$

first-order loops not touching P_1, L(1)[1]:

$$S_{22}\Gamma_i;$$

first-order loops not touching P_2, L(1)[2]:

0.

The total relation for the transmission from the generator to the detector is found by substituting these terms into Eq. 6.48.

PROBLEMS

1. For the circuit shown in Fig. P1, write the matrix equation using only a 2 × 2 matrix. Find the currents i_2 and i_3. *Hint:* To find i_3 use the relations $i_2 = i_L + i_R$ and $i_3 = i_R$.

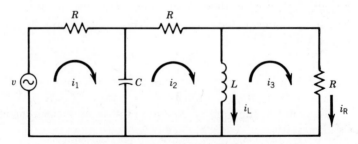

Figure P1. *Three-loop lumped-element network.*

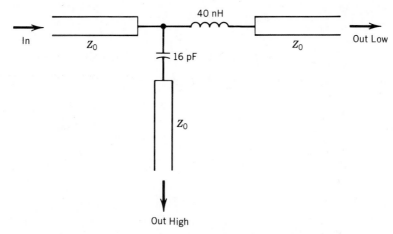

Figure P2. *Lumped-element diplexer. All $Z_0 = 50$ ohms.*

2. The diplexer in Fig. P2 passes low frequencies to the OUT Low port and high frequencies to the OUT High port. Write a FORTRAN program to compute the input VSWR, S_{21} to Low output and S_{21} to High output over the frequency range 0–1 GHz.

3. Derive the relations for rules 1 and 2 in Table 6.1.

REFERENCES

1. R. W. Beatty and D. M. Kerns, "Relationship Between Different Kinds of Network Parameters, Not Assuming Reciprocity or Equality of the Waveguide or Transmission Line Characteristic Impedances," *Proc. IEEE,* Vol. 52, No. 1, January 1964, p. 84.
2. J. K. Hunton, "Analysis of Microwave Measurement Techniques by Means of Signal Flow Graphs," *IRE Trans. Micro. Th. Tech.,* Vol. MTT-8, No. 2, March 1960, p. 206.
3. N. Kuhn, "Simplified Signal Flow Graph Analysis," *Microwave Journal,* Vol. VI, No. 11, November 1963, p. 59.
4. P. I. Somlo, "Signal Flow Graph Reduction," *Microwave Journal,* Vol. 30, No. 1, January 1987, p. 163.

7 Computer-Aided Design of Microwave Circuits

7.1 MOTIVATION FOR COMPUTER-AIDED DESIGN

Microwave engineers are always under pressure to develop new products that are faster, cheaper, and at or beyond the state-of-the-art. With the average engineer costing the company over twice the salary each year for equipment and support, management is hesitant to hire more engineers. Computers can be powerful tools for the microwave engineer in performing arduous and error-prone calculations. These computer tools have allowed engineers to try many circuits and components faster and cheaper than the trial-and-error methods of the past. Computers cannot replace the engineer, and a good knowledge of the basic principles of microwave design is still an absolute necessity.

This chapter discusses the need for computers in the design process. It includes a short history of microwave design and a discussion of conventional and computer design methods. The application of computer techniques to the design of a complete microwave amplifier is presented in Chapter 15.

7.1.1 History

Early engineers investigating microwaves tuned waveguides by screws and irises (and even by denting!) to achieve the desired results. This trial-and-error method was a practical tool for microwave engineers for many years. Many discoveries by early engineers, including the principle of reflections and components—such as traveling wave detectors and wavemeters—were made using these methods. Some of the more obvious disadvantages were the lack of repeatability, little mathematical basis for observed results, and test equipment that was unreliable, if it existed at all.

The development of the Smith chart was perhaps the greatest single contribution to the engineering analysis of microwave circuits. Developed by Phillip H. Smith (1), it provided a graphical tool for solving complex transmission line problems. It reduced the number of laborious mathematical calculations, thus reducing the possibility of error and allowing the engineer to visualize the step-by-step process of problem solving. This, along with the advent of the scattering matrix based on the concept of transmitted and reflected waves to describe network performance, allowed simpler representation of multiport microwave networks. These developments lead to rapid developments in the microwave field during and after the World War II.

In the early 1950s a microwave transmission media called stripline was developed, which consists of a metal conductor sandwiched between two dielectric plates that are metalized on the outside (see Chapter 8). Early stripline work used razor blades and glue to cut the thin strips and paste them on dielectric sheets (a somewhat unpredictable art). With the availability of copper-clad laminates (dielectrics), stripline became a predictable technology. The advantage of stripline is that the impedance can be controlled by the width of the center strip, which is fabricated by photoetching a copper-clad dielectric substrate. Many components can be connected without the need to break the outer shielding. Also, striplines are very convenient in constructing parallel line couplers due to the natural coupling of two strips placed close together.

The next advance in microwave transmission media was microstrip line, but it wasn't until the late 1960s that it became widely used as it wasn't until then that low-loss dielectric materials and methods for deposition of metallic films became widely available. Like the printed circuit board used at RF and VHF, microstrip has the advantage of allowing lumped elements, both active and passive, to be easily mounted directly on the substrate. This hybrid technology is a large part of the microwave integrated circuit (MIC) work that is done today.

The next generation of MICs was the monolithic microwave integrated circuit on a semiconductor substrate. The semiconductor materials used include silicon and gallium arsenide (GaAs). Recent trends indicate that GaAs technology may hold the key to MICs and also to gigahertz bandwidth analog amplifiers and gigabit-rate digital integrated circuits.

Both hybrid and monolithic MICs exhibit advantages similar to lower frequency and digital integrated circuits. These are improved system reliability, reduced weight and volume, and eventual cost reduction due to standardization and automated processing methods. As with low-frequency integrated circuits, the MICs are responsible for both the expansion of present markets and the opening of many new applications, including a host of new nonmilitary uses.

There are some disadvantages to MICs. Before MICs became popular, the microwave engineer had the flexibility (and the necessity) of having tuners and adjustment screws in circuits in order to optimize the performance of the circuit after fabrication. Today's MICs, especially if they have to meet high reliability standards, normally do not have these trimming arrangements. Consequently, the devices used have to be characterized precisely and the circuits need to be designed with little room for error. Computer-aided design, simulation, and optimization techniques have, therefore, become a necessity.

7.1.2 Computer-Aided Design Savings Analysis

Before the advent of microwave CAD tools, the engineer had to design new products by trial and error, which was time-consuming and expensive. With the advent of microwave programs such as COMPACT®, SUPER-COMPACT®, and TOUCHSTONE®, the engineer can rapidly analyze and optimize circuits without the need to build a breadboard. Further development of microwave tools has allowed the analysis of monolithic structures and discontinuities and the rapid configuration of these structures. Figure 7.1 is a cost-savings analysis comparing various CAD software programs with manual design methods.

The significant reduction in circuit design time from 3 months to 2 weeks with COMPACT® was due to the fact that many initial designs could be tried without the need of constructing and bench testing breadboarded circuits. SUPER-COMPACT® reduced that time to 2 days because of the accurate modeling of discontinuities and other parasitic effects. Accurate modeling allows the designer to build the circuit analytically without the need for breadboards. The design time reduction is limited by component deliveries and delays in drafting.

Computers have allowed many engineering departments to increase their productivity substantially while easing the burden on engineers.

7.2 MICROWAVE DESIGN PROCEDURES

7.2.1 Classical Procedure for Microwave Design

The classical approach for the design of microwave circuits is outlined in Fig. 7.2. The first step is to determine the requirements for the circuit, such as gain, noise figure, and stability. The next step involves combining these requirements with design data to arrive at an initial circuit configuration. Design includes data from manufacturers' catalogs for active devices that may meet the design criteria, analysis and synthesis procedures used in deciding values of various parameters, and a good dose of experience. A

TOUCHSTONE is a trademark of EEsof, Westlake Village, CA

Computer-Aided-Design Savings Analysis

One Project	Manual	Time and Cost			Comparative Savings
		COMPACT®	SUPER-COMPACT®	SUPER-COMPACT® and AUTOART®	
Circuit design time	3 months	2 weeks	3 days	3 days	COMPACT® to SUPER COMPACT® with AUTOART®
Total design effort (person months)	6 person months	4 person months	2 person months	1 person month	
Product design cycle (calendar month)	10 months	5 months	3 months	1 month	4 months
Total design cost (at $6,000/ person month)	$36,000	$24,000	$12,000	$6,000	$18,000
Annual 30 projects	$1,080,000	$720,000	$360,000	$180,000	$540,000

® Registered trademarks of Compact Software, Paterson, NJ

Figure 7.1. Computer-aided design cost analysis.

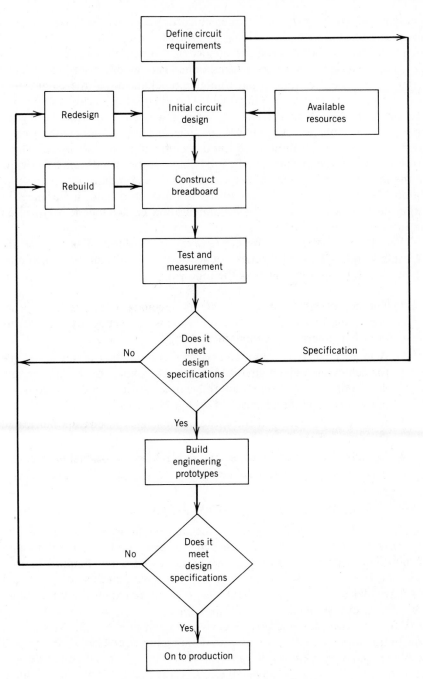

Figure 7.2. Classical design procedure.

laboratory model (breadboard) is then constructed from this initial data, and measurements are made to evaluate its characteristics. This performance is then compared with the initial circuit specifications; if they are not met, the circuit is modified to meet these conditions. This modification may simply require adjustment of tuning and trimming mechanisms in the circuit or it may require completely rethinking and rebuilding all or part of the circuit. Measurements are then carried out again and compared with the design criteria. This sequence of modifications, measurements, and comparison is carried out iteratively until the desired specifications are achieved. There are times when the circuit specifications have to be compromised in view of the feasible performance of the circuit. This may require making up the difference in other stages or going back to the customer for a variance. In either case, when the final circuit configuration is developed, it is passed on for prototype fabrication.

This classical procedure for microwave circuit design has become increasingly difficult to use because of its iterative and empirical methods. There are considerations such as the following:

1. The complexity of modern systems demands more precise and accurate design of circuits and subsystems; consequently, the effects of circuit tolerance become increasingly important.
2. A large variety of active and passive components is now available for achieving a given circuit function. The choice of the appropriate device or transmission structure becomes difficult and time-consuming when using the empirical design approach.
3. It is very difficult to incorporate any changes in circuits fabricated by MIC technology.
4. Breadboarded circuits, especially in MIC, are notoriously inaccurate in predicting the performance of the final design.

7.2.2 CAD Procedure for Microwave Design

Computer-aided design (CAD) techniques are methods of dealing with the problems of developing state-of-the-art microwave circuits. Computer-aided design can be interpreted as any design process where the computer is used as a tool. However, CAD has come to imply that without this computer tool the particular design process would have been more expensive, more time-consuming, or less reliable, if not impossible.

A flow diagram for microwave CAD is shown in Fig. 7.3. As with the conventional procedure, it begins with defining the circuit requirements and combines them with known design data to produce an initial circuit design. The amount of design data available in CAD is greatly expanded, though. SUPER-COMPACT® and TOUCHSTONE®, the programs used as examples in this text, contain large data banks of active devices including S-parameters and noise figures, passive devices such as transmission media,

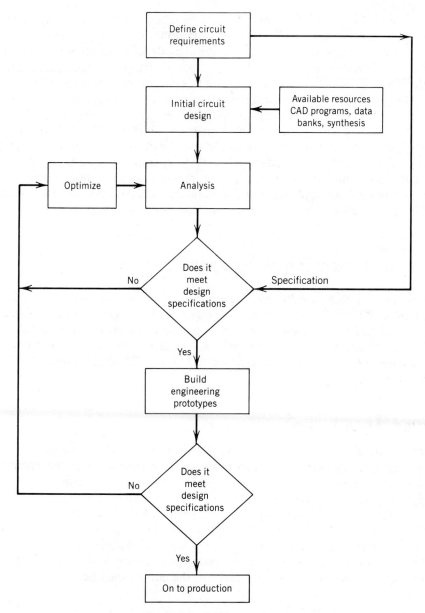

Figure 7.3. *CAD procedure.*

and any existing circuits that may have been designed previously. One of the fastest ways to design a circuit is to use one that already exists. Computer analysis allows rapid evaluation of many existing circuits, which can save the engineer much time and money especially if high reliability standards must be met.

The performance of the initial circuit is evaluated by a computer-aided circuit analysis package. This set of subroutines converts the components as described in the input file into numerical models for computer evaluation. Circuit characteristics obtained as a result of analysis are compared with the given specifications. If the results of this comparison are not satisfactory, certain parameters can be changed within the circuit in a systematic manner. This constitutes the key step in optimization. This sequence of analysis, comparison, and modification is performed iteratively until the specifications are met or the optimum performance of the circuit (within given constraints) is achieved. The circuit is fabricated, and experimental measurements are made. Some modifications may still be required, but it is hoped that with accurate modeling and analysis, these changes will be small, and the aim of CAD is to reduce these postdesign modifications to a minimum.

7.3   AN INTRODUCTION TO SUPER-COMPACT®

This section presents the information needed to operate the Version 1.7 SUPER-COMPACT® (S-C) software program. This program is not presented to avoid hand calculation but rather to be used as a check on hand calculations. If this or another program is not available, the material in the remainder of this text does not require CAD. TOUCHTONE® is described in Section 7.4.

7.3.1   Inputs to SUPER-COMPACT®

The structure of the inputs to S-C is shown in Fig. 7.4. The inputs are arranged in blocks. The analysis portion of the program uses the computational methods described in Chapter 6. For example, when using the LADder or cascaded circuit block, the S-C uses *ABCD*-matrix multiplication. This allows cascaded circuits to be analyzed without feedback elements in the circuit. The LAD circuits only operate on one- or two-port circuits. The second-to-last entry in the circuit description block defines the circuit name followed by a colon and one or two node number entries (n^1 [n^2]). If a one-port circuit named A is being described, the entry would be

    A:1POR n1 0

The 1POR means a one port, and n1 would be the port terminal. The 0 (zero) is optional for a one port and if not used implies the second terminal is ground. Node 0 is reserved for use as the ground terminal. A two-port circuit named B is described

    B:2POR 1 5

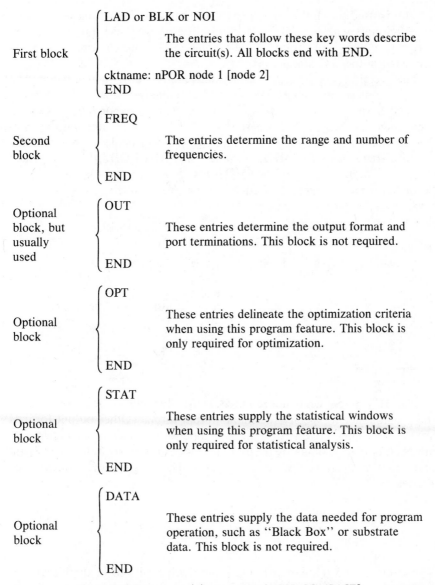

First block
- LAD or BLK or NOI
  - The entries that follow these key words describe the circuit(s). All blocks end with END.
- cktname: nPOR node 1 [node 2]
- END

Second block
- FREQ
  - The entries determine the range and number of frequencies.
- END

Optional block, but usually used
- OUT
  - These entries determine the output format and port terminations. This block is not required.
- END

Optional block
- OPT
  - These entries delineate the optimization criteria when using this program feature. This block is only required for optimization.
- END

Optional block
- STAT
  - These entries supply the statistical windows when using this program feature. This block is only required for statistical analysis.
- END

Optional block
- DATA
  - These entries supply the data needed for program operation, such as "Black Box" or substrate data. This block is not required.
- END

Figure 7.4. *Structure of the inputs to SUPER-COMPACT®.*

where node 1 is the input and node 5 is the output. The input signal is applied between nodes 1 and 0, while the output appears between nodes 5 and 0.

The BLK circuit description technique uses a Y-matrix nodal analysis, which is more computer time-consuming than the $ABCD$-matrix multiplication. However, the BLK does allow almost any interconnection of nodes for the analysis of complex circuitry (2). The BLK technique is used for circuits with feedback.

NOI is used to tell the computer to perform noise figure analysis on cascaded circuits with limited feedback (3). The use of these three circuit description options is illustrated in the examples to follow. The circuit elements are listed in the S-C Manual (4).

The second required block of the input file is the frequency block. The three options are to do the analyses at (1) discrete frequencies, (2) j linearly stepped frequencies ($f_j = f_{initial} + jf_{step}$), and (3) exponentially stepped frequencies [$f_j = (f_{initial})r^j$, where $r = e^{\ln[(f_{final}/f_{initial})/(n-1)]}$ and n is the number of frequencies between $f_{initial}$ and f_{final}]. An example of exponentially stepped frequencies is 1 MHz, 10 MHz, 100 MHz, 1 GHz, 10 GHz, where $f_{initial} = 1$ MHz, $f_{final} = 10{,}000$ MHz, $n = 5$, $r = e^{2.3} = 10$, $j = 1$–5. The formats for requesting these frequencies are, respectively,

FORMAT				EXAMPLE			
1. FREQ				FREQ			
1st Freq	2nd Freq . . . nth Freq			1 MHz	2.5 MHz	1 GHz	
END				END			
2. FREQ				FREQ			
STEP	$f_{initial}$	f_{final}	f_{step}	STEP	1 GHz	10 GHz	1 GHz
END				END			
3. FREQ				FREQ			
ESTP	f_{inital}	f_{final}	n	ESTP	1 Hz	1 GHz	10
END				END			

The OUT block, while not required, is usually included because it allows the programmer to specify the initial output format. Other output formats are available interactively. In addition, if input source and output load impedances other than 50 ohms are desired, this block must be included. The PRInt and PLOt options are the key statements. The output block contains the entries

```
OUT
PRI     cktname     [impedance]     keyword
END
```

The circuit name (cktname) is the same as that labeled in the circuit description block. If several circuits are described in the circuit description block, this selects the one whose printed output is desired. The impedances are needed (50 ohms, if not specified) if the S-parameters or the voltage gain output values are desired. Both real and complex impedances are possible, which are independent or dependent on frequency. They are formatted by port number. R1 = 30 is a 30-ohm source resistance on port 1, and Z2 = $20 - 30$ is a 20-j30-ohm load impedance on port 2 (node 5). The keyword requests S-parameters, Z or Y matrix, or voltage gain. Continuing the example started earlier, the output block might be

```
OUT
PLO B S
END
```

which gives a plot of the *S*-parameters of the two-port circuit B with 50-ohm source and load impedances. If other than 50 ohms is desired, the format is

```
OUT
PLO B     R1 = 30     Z2 = 20 -30 S
END
```

The impedances default to ohms if not labeled. The frequency defaults to hertz, so MHz and GHz are almost always needed for microwave circuit designs.

7.3.2   Using SUPER-COMPACT® for Analysis

With this short introduction and reference to the Manual (4) the reader can begin to use S-C. The S-C input for the simple circuit of Fig. 6.2 appears as shown in Fig. 7.5. S-C allows the operator to create a NEW program named SERIESRL (line 13) from the CMD > level after the log on procedure (lines 1 to 12). Line 14 uses a / to indicate that the new file is completed. Lines 15–17 are comment lines the author uses to identify the file name and its purpose. Each line is terminated with a carriage return (CR). Comment lines start with an *, and continuation lines (up to 160 total characters) start with a + . For this simple circuit, the LADder analysis (line 18) is used. Line 19 is a TRansmission Line connected from nodes 1 to 2 of 50-ohm charac-

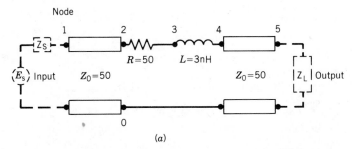

(a)

Figure 7.5. SUPER-COMPACT® program for an impedance in a 50-ohm transmission line (see Fig. 6.2). (a) Series R-L circuit in two TEM lines; (b) S-C program listing to analyze series R-L (includes an error); (c) analysis results of selected frequencies; (d) magnitude and phase of S_{21} for the two port in (a); (e) S_{11} over a limited frequency range plotted on a Smith chart.

```
 1  AATTDDTT44  11  66  33
 2  DIAL TONE
 3  4 1 6 3
 4  CONNECT 1200
 5  select a service: tso wylbur cms pci
 6  cms users depress break at logon prompt
 7  first request: CMS
 8  enter logon
 9  LOGON A 5
10  LOGON AT 10:24:16 GMT FRIDAY 01/03/86
11  SUPER-COMPACT VERSION 1.7 + 001 04/09/84
12  CMD >
13  .NEW SERIESRL
14  ADD LINES - '/' TO TERMINATE
15  .* FILE NAME:SERIESRL
16  .* THIS S-C PROGRAM COMPUTES THE S-PARAMETERS
17  .+ FOR THE CIRCUIT IN FIG. 6.2.
18  .LAD
19  .TRL 1 2 Z=50 P=100CM A=0.03/M
20  .RES 2 3 R=0.05KOH
21  .IND  3 4 L=3NH
22  .TRL 4 5 Z=50 E=90 F=4GHZ
23  .C:2POR 1 5
24  .END
25  .*******
26  .FREQ
27  .STEP 2GHZ 6E9 200MHZ
28  .END
29  .*******
30  .OUT
31  .PRI C S
32  .PRI PLO C Z
33  .END
34  ./
35  EDIT >
36  .SAVE
37  FILE SAVED - SERIESRL
38  EDIT >
39  .ANA
40  TRL 1 2 Z=50 P=100CM A=0.03/M
41  REQUIRED KEYWORD IS MISSING
42  C:2POR 1 5
43  THE ABOVE LADDER IS INCOMPLETE
44  END
45  LAST LINE OF BLOCK MUST BE A PORT DEFINITION
46  CMD >
47  .EDIT
48  EDIT >
49  .F/TRL 1/
50  TRL1 2 Z=50 P=100CM A=0.03/M
51  EDIT >
52  .C/A=/V=0.8 A=/
53  TRL 1 2 Z=50 P=100CM V=0.8 A=0.03/M
54  EDIT >
55  .ANA
```

(b)

.ANA

CIRCUIT: C
S-MATRIX, ZS = 50.0+J 0.0 ZL = 50.0+J 0.0

| Freq | S11 | | S21 | | S12 | | S22 | | S21 |
GHZ	Mag	Ang	Mag	Ang	Mag	Ang	Mag	Ang	dB
3.80000	0.522	141.7	0.600	−55.0	0.600	−55.0	0.526	−141.4	−4.44
4.00000	0.535	−98.5	0.594	−0.8	0.594	−0.8	0.539	−150.2	−4.53
4.20000	0.548	21.2	0.587	53.3	0.587	53.3	0.552	−159.1	−4.62

CIRCUIT: C

Z-Matrix (Ohms)

| Freq | Z11/Z21 | | Z12/Z22 | |
GHZ	Re	Im	Re	Im
3.8000	14.38	4.821	11.81	−23.97
	11.81	−23.97	9.878	−23.86
4.0000	30.47	−54.84	27.18	−27.61
	27.18	−27.61	24.42	−24.89
4.2000	65.96	92.93	5.358	56.86
	5.358	56.86	.6155	8.256

Press return to continue to PLOT (<CR>):

(*c*)

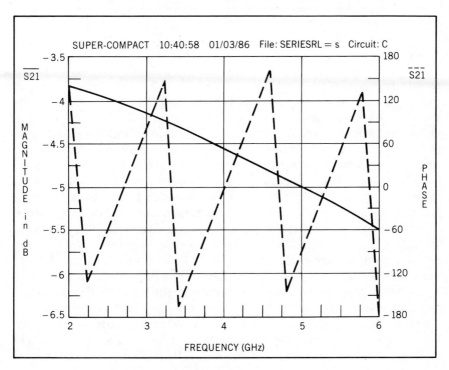

(*d*)

Figure 7.5. (*continued*)

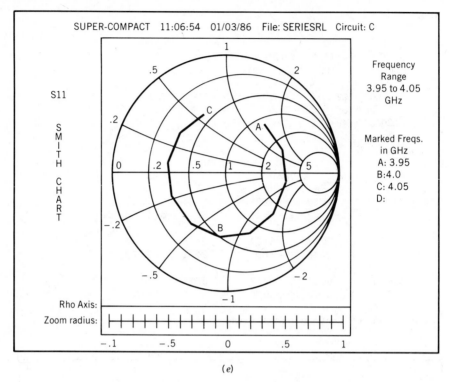

(e)

Figure 7.5. *(continued)*

teristic impedance, Physical length 100 cm and 0.03 dB Attenuation per meter. This line contains an error, which appears in lines 40–45 when ANAlysis (line 39) is attempted. The corrected line 19 appears in line 53 and was corrected using the EDITor mode available from the CMD level. The error in line 19 was the failure to give the relative Velocity = 0.8 before the Attenuation. A RESistor between nodes 2 and 3 of 0.05 kilohms = 50 ohms and an INDuctor between nodes 3 and 4 completes the lumped elements in series with the distributed lines. The TRansmission Line between nodes 4 and 5 is specified using another method, namely characteristic impedance of 50 ohms and 90° (quarter-wavelength) long at 4 GHz. Line 23 is used to label the circuit C in the LAD as being a 2 port with the input port between nodes 1 and 0 and the output port between nodes 5 and 0. The circuit description ends with END in Line 24. Lines 25 and 29 are comment lines to break the program into its major blocks. The frequencies that are analyzed are specified in line 27 using the GHz, exponential, and MHz formats. The output formats are given in lines 31 and 32. When the program is run, the first PRIntout of circuit C is the *S*-parameters assuming 50-ohm source and load impedances. The second PRIntout of C (Line 32) is the *Z*-parameters (see Fig. 7.5c), and then a PLOt interactive format appears. In this graphic interactive mode, the Z- or S-parameters can be requested and are automatically formatted, if desired. Typical results for the transfer function S_{21} magnitude and phase are shown in Fig. 7.5d (computed with 50-MHz steps rather than 200 MHz as in Fig. 7.5b). Because of the rapid phase variation of the reflection coefficient S_{11}, only a 100-MHz range of frequencies was analyzed looking into the source port. When plotted on the Smith chart, the input impedance can be estimated. As expected, the VSWR is $3:1$ since $|R + jX| = 100$ ohms in series with 50 ohms. A, B, and C on the Smith chart indicate the start, center, and stop frequencies. The frequency step for this analysis was 10 MHz (see Fig 7.5e).

The procedure for correcting mistakes identified when the ANA (line 39) is attempted is now summarized. S-C lists three errors (lines 40 to 45) and then returns to the CMD mode (Line 46). Edit the first error first by calling the EDIT mode (line 47) and the program response (line 48). To find the offending line in the program rapidly, type line 49. If TRL1 appears more than once, the editor finds the first occurrence and displays it as line 50 (a repeat of line 19) and retypes line 51. To add in the velocity, type Change as shown on Line 52. S-C responds with lines 53 and 54. The program is now ready to attempt ANAlysis. Since the program runs properly, the other errors noted earlier were generated by the lack of entries in line 19.

The program should also be saved (line 36) again with the correction. If the user logs off, returns, and recalls the file SERIESRL, the erroneous file is recalled.

The example in Fig. 7.5 clearly shows the ease with which a circuit is described and analyzed using S-C compared to the techniques in Chapter 6. The *ABCD*-matrix format in Chapter 6 was used by S-C to perform this analysis. If a more complex circuit is analyzed, the ease of inputting to S-C becomes more evident since it is a logical format similar to the physical layout.

7.3.3   Using SUPER-COMPACT® for Optimization

While the analysis feature of S-C is frequently used, the optimization feature is invaluable to the designer. In the example shown in Fig. 7.6, a microwave circuit involving both waveguide and coax is optimized for a given gain over a frequency band. This circuit is similar to the Kenyon (5) circuit for matching a two-terminal microwave diode (see Chapter 13), which exhibits a negative resistance to a waveguide at microwave frequencies. Two analyses are performed. The first analysis shows the gain of the amplifier with no additional tuning, whereas the second shows the response when a reactive element is added to the waveguide transmission line (6). The optimization done before the second analysis was used to determine the location and value of the capacitor to be added across the waveguide.

```
NEW KENYON
ADD LINES - '/' TO TERMINATE
.* FILE NAME IS KENYON
.* THIS PROGRAM MODELS THE KENYON CIRCUIT AT 8GHZ
*****
.
.LAD
.RWG   1 2 A=1.122IN B=0.5IN WAVI
.IWG:  1POR 1 0
.END
*
.
.BLK
.C1: ?0.0PF 0.0PF 0.1PF?
.P1: ?0.5IN  4.0IN  4.0IN?
.CAP   1 0 C=C1
.RWG   1 2 A=1.122IN B=0.5IN P=P1 WAV2
.TRF   2 3 R1=437.0 R2=30.0
.TRL   3 4 Z=4.46 P=0.0899IN K=1.0
.TRL   4 5 Z=22.42 P=0.0899IN K=1.0
.SRX   5 0 R=-1.5 L=0.86NH C=0.44PF
.KC:   1POR 1 0
.END
*****
.
.FREQ
.STEP 7GHZ 10GHZ 0.1GHZ
.STEP 8GHZ 8.4GHZ 0.05GHZ
.END
*****
.
.OUT
.PLO KC Z1=IWG S
.END
*****
.
.OPT
.KC Z1=IWG F=8GHZ 8.4GHZ MS11 6DB 6.5B
8GHZ
.END
*****
.
.DATA
.WAV1:RWG ER=1.0
.WAV2:RWG ER=1.0
.END
./
EDIT >
.SAVE
FILE SAVED - KENYON
EDIT >
```

(a)

Figure 7.6. Example of optimization (broadbanding) of a Kenyon-like amplifier. (a) Input file for analysis and optimization; (b) initial amplifier performance with C1 = 0; (c) optimization via interactive display; (d) optimized (broadband) response C1 = 0.0126 pF, P1 = 4.0337 in.

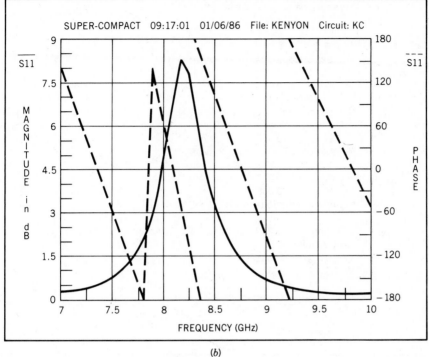

SUPER-COMPACT   09:17:01   01/06/86   File: KENYON   Circuit: KC

(b)

```
CMD >
.OPT
Gradient, Random or Quit? (G/R/Q):
.G
<OPTIM>    4
Gradient optimization does not allow negative or zero element values.
Use Random optimization.

CMD >
.R
WHAT ?
CMD >
.OPT
Gradient, Random or Quit? (G/R/Q):
.R
Err. F. =       2.263
Number of trials? (n/<0>):
.10
Randomize random number generator? (Y/<N>):

.
Record of improvement
    Trial       Err. F.
       3        1.661
       9        1.021
      10        0.935
```
(c)

Figure 7.6. *(continued)*

Number of additional trials? (n/<0>):
.10
    15   0.385
    19   0.244
Number of additional trials? (n/<0>):
.10
    23   0.148
    25   0.110
Number of additional trials? (n/<0>):
.10
    34   0.067
Number of additional trials? (n/<0>):
.10
    41   0.067
    45   0.066
Number of additional trials? (n/<0>):

CPU time =   6.37 Secs.   101 Function evaluations
Best Values:
Var(1) =   .12633E-01PF
Var(2) =   4.0337   IN

Press return to continue to PLOT (<CR>):

(C) CONT.

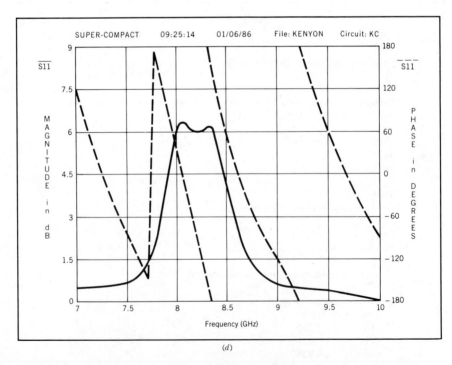

(d)

Figure 7.6. (continued)

7.4   AN INTRODUCTION TO TOUCHSTONE®

This section presents information needed to operate the TOUCHSTONE® (TS) microwave CAD program, Version 1.45. This CAD program is distributed by EEsof, Westlake Village, CA 92362, and is popular because it operates on suitably sized personal computers (PCs).

7.4.1   Inputs to TOUCHSTONE®

The structure of the inputs for TS is similar to S-C and is shown in the example in Fig. 7.7 (same problem solved earlier using S-C). Three blocks are required to operate TS. These are the circuit (CKT, line 14), OUTput (line 22), and FREQuency (line 27) blocks. The other blocks—DIMension (line 7), VARiable (line 13), TERMination (line 20), PROcess (line 21), GRID (line 29), OPTimization (line 33), and TOLerance (line 34)—are optional depending on the type of analysis being performed. Details of all these can be found in the TS Manual tutorial section. Only the commonly used blocks will be described below.

The comment lines in TS start with an !. Lines 5 and 6 are helpful to remind the user of the purpose of the program at a later date. The circuit being analyzed is shown in Figs. 6.2 and 7.5*a*.

The DIMension statements in lines 8 through 12 are the same (except for LeNGth) as the default values and are included here to assure the programmer of this fact. Common abbreviations are used except for OHm.

The CKT is described in lines 15 through 19. The circuit descriptions are similar, but not identical, to S-C. TLINP is a physical transmission line connected between nodes 1 and 2. TS uses only a nodal format for inputs. The impedance Z is 50 ohms and the physical length is 100 cm. The effective dielectric constant K of the line is 1.5625, which corresponds to line 53, Fig. 7.5*b* where the velocity on the line is 8/10 the speed of light. The attenuation on the line is 0.0003 dB per unit length, where the unit length is defined as centimeter (line 12). The F = 0 notation is used to indicate that the attenuation is independent of frequency. A nonzero value for F indicates what frequency the attenuation varies around via the relation

$$\alpha(\text{dB/unit length, } f) = \alpha(\text{dB/unit length, } f_x)[f/f_x]^{1/2} \qquad (7.1)$$

The two lumped elements, RESistor and INDuctor, can be represented individually or as a series RL element in TS. Here the 50-ohm resistor is connected between nodes 2 and 3, while the 3 nH inductor is connected between 3 and 4.

The transmission LINe between nodes 4 and 5 is described by its impedance, Z = 50 ohms, and its electrical length of 90° at F = 4 GHz. This is a lossless line.

```
 1  Touchstone (TM) -Ver(1.45-Lot 100)-Ser(16299-2275- 10000
 2                  SERIESRL.CKT     01/16/87 - 14:52:00

 3  !  File Name:SERIESRL

 4  !  Date: 15 JAN 1987
 5  !  Circuit: THIS TOUCHSTONE PROGRAM COMPUTES THE S-PARAMETERS
 6  !          FOR THE CIRCUIT IN FIGURE 6.2.
 7  DIM      !  OPTIONAL INPUT
 8        FREQ GHZ
 9        RES   OH
10        IND   NH
11        ANG   DEG
12        LNG   CM
13  VAR      !  OPTIONAL INPUT
14  CKT      !  REQUIRED INPUT
15        TLINP 1 2 Z = 50 L = 100 K = 1.5625 A = 0.0003 F = 0
16        RES 2 3 R = 50
17        IND  3 4 L = 3
18        TLIN 4 5 Z = 50 E = 90 F = 4
19        DEF2P 1 5 C
20  TERM     !  OPTIONAL INPUT, DEFAULT VALUES ARE 50-0HMS
21  PROC     !  OPTIONAL INPUT
22  OUT      !  REQUIRED INPUT
23        C SPAR
24        C S11
25        C DB[S21] GR1
26        C ANG[S21] GR1A
27  FREQ     !  REQUIRED INPUT
28        SWEEP 2 6 0.2
29  GRID      !  OPTIONAL INPUT, USE AUTORANGING
30        RANGE 2 6 0.2
31        GR1  −6.5  −3.5 0.25
32        GR1A  − 180 180 30
33  OPT      !  OPTIONAL INPUT
34  TOL      !  OPTIONAL INPUT
```

Figure 7.7. TOUCHSTONE® program for an impedance in a 50-ohm transmission line (see Figs. 6.2 and 7.5).

The CKT block is ended by DEFining a two-port (2P) between nodes 1 and 5, named C as shown in line 19.

If terminations for computing the S-parameters other than 50 ohms are desired, these are entered into the TERMination block (line 20). For this analysis, the 50-ohm default is used.

The OUTput block beginning with line 22 tells TS what parameters are needed for the various output formats. SPARameters (line 23) computes the S_{ij} shown in Fig. 7.8 for circuit C and will list these on the SCreeN of the

Output Format menu. This menu will appear at the bottom of the screen and allows us to manipulate the various outputs, reenter the EDITor, etc. C S11 (line 24) computes the S11 for circuit C to allow presentation on the regular Smith chart (perimeter $\Gamma = 1$) SC2 of the input reflection coefficient S11. Line 25 (C DB[S21] GR1) plots the magnitude of S21 in decibels on the GRid number 1 to be defined in line 31. This yields the output shown in Fig. 7.9a, which is equivalent to Fig. 7.5d for S-C. Like S-C, TS can put two plots on the same graph, as shown in line 26. Here the angle of S21 is plotted on GRid 1A as defined in line 32. The A indicates that two ordinates will appear on the output plot.

The FREQuency block, lines 27 and 28, is required. This block tells TS what frequency range to compute over. Up to 100 frequency steps may be included. For this example, the computations SWEEP from 2 to 6 GHz with 200 MHz (0.2 GHz) steps. Remember that all frequencies must be given in gigahertz as defined in line 8.

The GRID block is an optional block and is used for this example so that the TS output formats will be the same as the S-C formats in Fig. 7.5. TS does have autoranging that will set up the ordinate values based on the output. Within this block, the frequency range (abscissa) of the plots is called out again. If plots over a frequency range less than the range in the FREQ block are desired, the plot range is defined on line 30. For this example, the values in line 28 are repeated. GRid 1 is defined as -6.5 to -3.5 dB, with 0.25 dB steps for the left-hand ordinate in Fig. 7.9a. The right-hand ordinate, GR1A, extends from $-180°$ to $180°$ in $30°$ steps.

```
Touchstone (TM) -Ver(1.45-Lot 100)-Ser(16299-2275- 1000)
          SERIESRL.CKT   01/16/87 - 14:52:45
```

FREQ-GHZ	MAG[S11] C	ANG[S11] C	DB[S21] C	ANG[S21] C	MAG[S12] C	ANG[S12] C	MAG[S22] C	ANG[S22] C
2.00000	0.402	138.824	-3.818	178.846	0.644	178.846	0.405	-67.093
2.20000	0.415	-100.278	-3.872	-127.204	0.640	-127.204	0.417	-74.785
2.40000	0.427	20.447	-3.930	-73.237	0.636	-73.237	0.430	-82.645
2.60000	0.441	141.020	-3.992	-19.253	0.632	-19.253	0.444	-90.667
2.80000	0.454	-98.561	-4.059	34.751	0.627	34.751	0.457	-98.838
3.00000	0.468	21.726	-4.129	88.777	0.622	88.777	0.471	-107.139
3.20000	0.481	141.897	-4.203	142.821	0.616	142.821	0.485	-115.562
3.40000	0.495	-98.047	-4.280	-163.112	0.611	-163.112	0.498	-124.096
3.60000	0.508	21.912	-4.360	-109.023	0.605	-109.023	0.512	-132.725
3.80000	0.522	141.789	-4.444	-54.912	0.600	-54.912	0.526	-141.442
4.00000	0.535	-98.415	-4.530	-0.777	0.594	-0.777	0.539	-150.237
4.20000	0.548	21.312	-4.619	53.381	0.588	53.381	0.552	-159.100
4.40000	0.561	140.980	-4.710	107.561	0.581	107.561	0.565	-168.023
4.60000	0.574	-99.406	-4.804	161.766	0.575	161.766	0.578	-177.000
4.80000	0.586	20.159	-4.899	-144.006	0.569	-144.006	0.590	173.976
5.00000	0.598	139.688	-4.997	-89.755	0.563	-89.755	0.602	164.912
5.20000	0.610	-100.823	-5.096	-35.480	0.556	-35.480	0.614	155.810
5.40000	0.621	18.634	-5.197	18.818	0.550	18.818	0.626	146.679
5.60000	0.632	138.067	-5.299	73.138	0.543	73.138	0.637	137.520
5.80000	0.643	-102.524	-5.402	127.482	0.537	127.482	0.648	128.337
6.00000	0.654	16.864	-5.507	-178.150	0.530	-178.150	0.658	119.134

Figure 7.8. Analysis results for the TOUCHSTONE® program in Fig. 7.7.

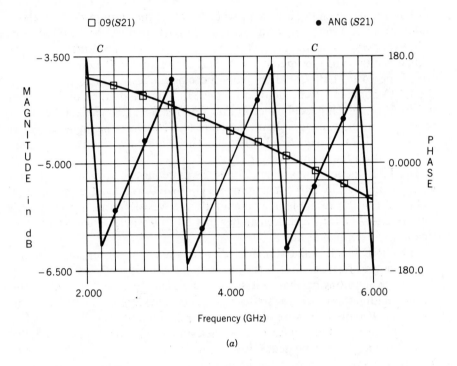

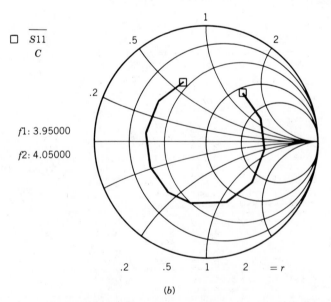

Figure 7.9. TOUCHSTONE® results for circuit in Fig. 6.2. (a) Magnitude and phase of S_{21}; (b) S_{11} over a limited frequency range.

The OPTimization block tells TS what to optimize, as described in Section 7.3.3 for S-C. The TOLerance block is used when statistical analyses similar to those described in Chapter 15 are desired. Refer to the TS Manual for more details on these inputs.

Once all the inputs for the TS input file are provided, the computer is taken out of the editor mode with EXIT in the menu at the bottom of the screen. This sets up the circuit file, and when SWEEP is commanded, the TS analysis is performed with the required outputs being generated. If the SCreeN output format is selected, the printout in Fig. 7.8 appears on the screen (to the extent that the screen can accommodate all the data). If a hard copy is desired, Print File is selected from the menu. Print Window will print the items on the screen without outputting the entire file. By comparing the results at 3.8, 4.0, and 4.2 GHz in Fig. 7.8 and the S-C results in Fig. 7.5c, the two CADs yield the same S-parameters. The Z-parameters are also available in TS.

The Smith chart presentation in Fig. 7.9b (compare with Fig. 7.5e) was obtained by changing the SWEEP range from 3.95 to 4.05 GHz in 0.01 GHz steps. Normalized values of resistance and reactance are labeled on the SC2 chart. SC1 is a compressed ($\Gamma_{max} = 3$) Smith chart, while SC3 is an expanded ($\Gamma_{max} = 0.33$) Smith chart available by toggling the output format menu.

7.5 OTHER FEATURES AVAILABLE WITH CAD

These examples have demonstrated the power of CAD techniques in microwave engineering. Other features in these programs are available to assist further in the design. For example, in Chapter 15 the use of the network synthesis capability for selecting distributed element input and output matching networks is demonstrated. Also a statistical variation of the element values is performed, which determines which values are critical to the performance of the circuit.

PROBLEMS

1. A large microwave component manufacturer expects to design 30 projects in the first 12 months but only 15 projects in the following 12 months. Compute the total costs (use Fig. 7.1) assuming: (1) manual design and (2) using SUPER-COMPACT® if engineers can*not* be reassigned to other projects during the second year. Each CAD system costs $30,000 and is depreciated linearly to zero in 2 years. Redo this problem if engineers are reassigned when not required.

2. Write a S-C program to compute E/I_1 for the circuit in Fig. 6.3 at 1 GHz.

3. Redo Problem 2 assuming some of the resistors are transmission lines as follows:

    $R_1 \Rightarrow Z = 50$, effective dielectric constant $= 1$, physical length $= 30$ cm

    $R_2 \Rightarrow Z = 100$, physical length $= 11.8$ in., relative phase velocity $= 1.0$

    $R_3 \Rightarrow Z = 75$, electrical length $= 180°$ at 1 GHz

    $R_4 \Rightarrow$ two 50-ohm resistors in parallel.
    Use transmission line theory to estimate the input impedance at 1 GHz.

4. Write an S-C program to compute the phase shift for the circuit in Fig. 6.13 at 1 GHz steps between 1 and 20 GHz. Assume $B = \omega C$, $C = 1$ pF, the line length is $45°$ at 10 GHz and has a characteristic impedance of 100 ohms. The source impedance is 50 ohms and the load impedance is $20 + j10$ ohms.

5. Can the circuit in Fig. 6.16 be modeled in a LAD block? Write the circuit model block for this circuit.

6. Redo Problem 2 using TOUCHSTONE® (TS).

7. Redo Problem 3 using TS.

8. Redo Problem 4 using TS.

REFERENCES

1. P. H. Smith, "Transmission Line Calculator," *Electronics,* Vol. 12, No. 1, January 1939, p. 29; "An Improved Transmission Line Calculator," *Electronics,* Vol. 17, No. 1, January 1944, p. 130; and *Electronic Applications of the Smith Chart,* McGraw-Hill, New York, 1969.

2. W. Chen, *Active Network and Feedback Amplifier Theory,* McGraw-Hill, New York, 1980, Chap. 2.

3. H. Hillbrand and P. H. Russer, "An Efficient Method for Computer Aided Noise Analysis of Linear Amplifier Networks," *IEEE Trans. Circ. Syst.,* Vol. CAS-23, No. 4, April 1976, p. 235.

4. *SUPER-COMPACT® User Manual,* Comsat General Integrated Systems, Version 1.7 (now owned by Communications Consulting Corp., Paterson, NJ).

5. N. D. Kenyon, "A Circuit Design for MM-Wave IMPATT Oscillators," *Digest Tech. Papers, G-MTT Intl. Micro. Symp.,* Newport Beach, CA, May 11–14, 1970, IEEE Cat. No. 70C10-MTT, p. 300.

6. K. Garbrecht and W. Heinlein, "A Simple Broad-Banding Technique for Microwave Reflection-Type Amplifiers," *Microwave Journal,* Vol. 13, February, 1970, p. 77.

7. *EEsof TOUCHSTONE® User's Manual,* Pub. 1.4-01186, January 1986.

8
Planar Transmission Lines and Components

8.1 INTRODUCTION

8.1.1 Motivation for Planar Configurations

Planar microwave transmission lines allow mounting of semiconductor devices with minimal parasitic effects compared to coaxial or waveguide configurations. Planar microwave designs have begun to dominate low-power applications for both commercial and military environments because of their performance, ease of fabrication, and resulting low cost. Planar designs using special substrates have been applied through 100 GHz. These planar configurations have had as big an effect on microwave circuit design as printed circuit boards at VHF had over wired chassis radio construction. A historical review of the development of the principal planar configuration is given in Ref. 1.

8.1.2 Configurations

A listing of many planar transmission line configurations is shown in Table 8.1 (2). The four common configurations along with their electric field profiles and launching techniques from coax are shown in Fig. 8.1. Today's most common configuration is microstrip (one-half of stripline).

8.1.2.1 Stripline. Stripline appears as a flattened coaxial transmission line (see Fig. 8.2). Assuming a uniform dielectric media, the fields are transverse electric and magnetic (TEM), and the line shows low dispersion (the wavelength is nearly inversely proportional to frequency). For the solid dielectric

Table 8.1 Characteristics of Planar Transmission Lines

Type of Line	Configuration	Transition to Waveguide	Fabrication	Compatibility with Active Devices	Compatibility with Passive Devices	Isolation	Propagation Loss	Radiation Loss	Compatibility with Radiators
Dielectric surface waveguide		Moderate	Injection molding, machining	Difficult	Requires develop.	Poor to moderate at length	Low	High	Yes dielectric antennas
Dielectric filled waveguide		Easy	Injection molding, machining	Difficult	Requires develop.	Excellent	Low	High	Yes dielectric antennas
Image guide		Moderate	Injection molding, machining	Difficult	Requires develop.	Good at 3"–35 GHz 1/4"–70 GHz	Low	High	Yes dielectric antennas
Insulated image guide		Moderate	Injection molding, machining	Difficult	Requires develop.	Poor to moderate at length	Low	High	Yes dielectric antennas
Inverted strip dielectric guide		Moderate	Machining	Difficult	Requires develop.	–10 db @ 1 wavelength array	Low	High	Yes dielectric antennas
M-guide			Machining molding	Difficult	Difficult	Poor to moderate	Low	Moderate to high	Yes dielectric antennas
Groove guide			Machining	Difficult		Poor to moderate	Low	Moderate to high	Yes dielectric antennas
Finline		Moderate	Etching	Yes	Yes	Moderate to good	High	High	Need transition to waveguide

Slot line	Moderate	Etching	Yes	Yes	Moderate to poor	Same or > tran μ-strip	Moderate	Horns or slot antennas
Coplanar waveguide	Moderate	Etching	Yes	Yes	Poor @ 1/2 wavelength	> tran μ-strip	Moderate	Coplanar antennas
Strip line	Easy	Etching	Yes, but not easy	Yes	Moderate	Moderate	More with side shield	Yes at low frequencies
Microstrip	Moderate	Etching	Yes	Yes	Moderate	High	Moderate	μ-Strip antennas or transition to waveguide worn
Microguide	Difficult	Etching	Yes	Yes		Moderate		
Inverted microstrip		Etching	Yes	Yes	Moderate to good	75% less than μ-strip	Low	μ-Strip or transition to waveguide worn
Trapped inverted microstrip	Easy	Etching	Yes	Yes	Moderate to good	70% less than μ-strip	Low	
Strip slot line	Possibly easy	Etching	Yes	Yes	Moderate to good	May be controlled	Low	Good

(From J. Rivera and T. Itoh, University of Texas Microwave Lab. Report 79–1, August 1979.)

form of stripline, the center conductor is photolithographically etched on one side of both halves and sandwiched together with their ground-plane sides on the outside. Alternatively, the center conductor is etched on only one-half and the metallization is completely removed from one side of the other half. Since the dielectric material is usually soft, the air gap formed by the thickness of the metallization is removed when the two halves are compressed. Figure 8.1a is not to scale since the thickness of the metallization is typically 1/50 of the dielectric thickness.

The stripline configuration is used to build circulators and couplers. The isolation between adjacent circuits is only fair, so grounding pins between the ground planes are periodically placed as a Faraday shield as shown.

Other configurations of stripline include dielectric posts periodically placed to support a metallic conductor between metal plates for high-power applications and suspended stripline where the center conductor and dielectric are suspended in a waveguide to form a coaxial configuration. The latter configuration is widely used at millimeter wavelengths since the E-fields are mostly contained in lowloss air.

Launching a wave into stripline is done by connecting the coax's center conductor to the center strip and the coax shield to the top and bottom plates.

8.1.2.2 Microstrip.
Microstrip is configured with the conductor parallel to the ground plane as shown in Fig. 8.1b. In this configuration more E-field

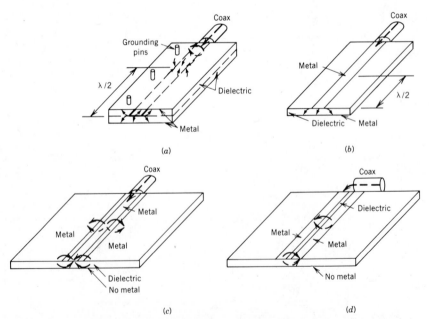

Figure 8.1. *Popular planar transmission lines.* (a) *Stripline;* (b) *microstrip;* (c) *coplanar waveguide;* (d) *slotline. Arrows indicate E-field.*

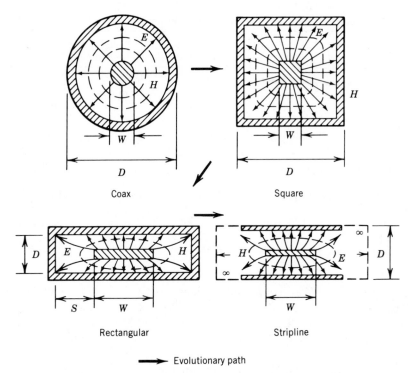

Coax

Square

Rectangular

Stripline

➡️ Evolutionary path

Figure 8.2. *Evolution of stripline from Coax. (From Ref. 1, © 1984 IEEE.)*

is located in air than in stripline, making the dispersion larger. The propagation is termed quasi-TEM. The isolation between circuits on the same substrate is not good, so the adjacent circuits are mounted in channelized compartments having absorber on some walls to prevent propagation of power radiated from the microstrip. Another technique for minimization of propagation between circuits is to make the channel width small so the energy will not propagate (similar to a cutoff waveguide).

Microstrip is usually the lowest-cost media because all the circuitry is photolithographically etched on one side of the substrate, and it is easy to mount devices requiring connection between two surface conductors. For devices generating heat, a hole is drilled through the substrate to allow connection to the ground-supporting material. Frequently the microstrip is epoxied to an aluminum housing, which conducts the heat away. Because of its popularity, design equations are presented later in this chapter.

Launching energy from coax to microstrip is done by connecting the coax center conductor to the circuit conductor and the shield to the ground plane. These launchers are available commercially through 18 GHz with VSWR less than 1.2:1.

Table 8.2   Relative Characteristics of Planar Transmission Lines

Characteristic	Stripline	Microstrip	Coplanar Waveguide	Slotline
Attenuation	Low	Low	Medium	High
Dispersion	Low	Low	Medium	High
Range of Z_0 (ohms)	10–100	15–110	25–125	50–300
Integrate shunt elements	Difficult	Difficult	Easy	Easy
Integrate series elements	Moderate	Easy	Easy	Difficult
Ease of fabrication	Moderate	Easy	Easy	Easy
Ease of launching	Easy with coax	Easy with coax	Easy with coax	Easy with waveguide

8.1.2.3   Coplanar Waveguide. In this planar configuration the ground planes are positioned on either side of the center conductor in a planar coax format as shown in Fig. 8.1*c*. The propagation is quasi-TEM and exhibits moderate dispersion and loss.

8.1.2.4   Slotline. Extending the width of the center conductor of the co-planar waveguide and removing one ground plane yields the slotline configuration shown in Fig. 8.1*d*. Propagation is non-TEM and shows high dispersion and loss.

8.1.2.5   Comparison of Popular Planar Transmission Lines. Table 8.2 compares some of the parameters of the four planar transmission lines described above. The more the field configurations differ from coax, the poorer the performance. Because of the ease of production and integration with other low-power devices, their faults are tolerated and they are used extensively.

8.2   SUBSTRATE MATERIALS

Two basic types of substrates are used—hard and soft. The hard substrates are fabricated from ceramic, crystalline materials, or semiconductors, whereas the soft substrates use ceramic or glass-filled Teflon. The hard substrates usually have a different thermal coefficient of expansion than the circuit fixture so that special metals and adhesives are needed to match between the substrate and the circuit fixture. Also, hard substrate circuits are limited to small areas (<1 cm squares) to prevent excessive stresses. These substrates are preferred where fine line resolution on smooth (polished) surfaces and low loss are required (e.g., at millimeter waves). Hard substrate circuits are more expensive to fabricate because they are brittle and difficult to drill and cut to size for device and fixture mounting.

The soft substrates are available in a range of dielectric constants, depending on material. Low dielectric constant substrates use glass microfibre reinforced polytetrafluorethylene (PTFE Teflon) and Teflon-impregnated woven glass laminates. These materials have become known by their trade names, such as Duroid® (manufactured by Rogers Corp., Chandler, AZ) and Di-Clad® (Keene Corp., Bear, DE). The Teflon dielectric is clad with various thicknesses of copper. The thickness is specified by the weight of Cu per square foot of laminate. Typical values are 1/2, 1, and 2 oz (corresponding to 0.7, 1.4, and 2.8 mils thickness, respectively). The rolled copper has a smoother (7 microinch) surface than the electrodeposited (95 microinch) copper. Because the soft substrates can follow the temperature-induced dimensional changes of the fixture, large circuits with a minimum of interconnections can be built.

Tables 8.3 and 8.4 provide the key design data needed for common substrate materials.

Table 8.3 Properties of Microwave Dielectric Substrates

Material	Relative Dielectric Constant	Loss Tangent at 10 GHz (tan δ)	Thermal Conductivity K (W/cm/°C)	Dielectric Strength (kV/cm)
Sapphire	11.7	10^{-4}	0.4	4×10^3
Alumina	9.7	2×10^{-4}	0.3	4×10^3
Quartz (fused)	3.8	10^{-4}	0.01	10×10^3
Polystyrene	2.53	4.7×10^{-4}	0.0015	280
Beryllium oxide (BeO)	6.6	10^{-4}	2.5	—
GaAs ($\rho = 10^7$ ohm-cm)	12.3	16×10^{-4}	0.3	350
Si ($\rho = 10^3$ ohm-cm)	11.7	50×10^{-4}	0.9	300
3M 250 type GX	2.5	19×10^{-4}	0.0026	200
Keene DI-clad 527	2.5	19×10^{-4}	0.0026	200
Rogers 5870	2.35	12×10^{-4}	0.0026	200
3M Cu-clad 233	2.33	12×10^{-4}	0.0026	200
Keene DI-clad 870	2.33	12×10^{-4}	0.0026	200
Rogers 5880	2.20	9×10^{-4}	0.0026	200
3M Cu-clad 217	2.17	9×10^{-4}	0.0026	200
Keene DI-clad 880	2.20	9×10^{-4}	0.0026	200
Rogers 6010	10.5	15×10^{-4}	0.004	160
3M epsilam 10	10.2	15×10^{-4}	0.004	160
Keene DI-clad 810	10.2	15×10^{-4}	0.004	160
Air	1.0	0	0.00024	30

(E. C. Niehenke, "Microstrip Circuits," WMEC Class Notes, November, 1981.)

Table 8.4 Typical Properties of RT-duroid® Soft Substrates

Product Number, Type	5880 Random Fiber			5870 Random Fiber			6006 Ceramic Filled			6010.5 Ceramic Filled		
Dielectric constant, Z-direction, ASTM D3380, 10 GHz	2.20			2.33			6.00			10.50		
Standard tolerance	±.02			±.02			±.15			±.25		
Isotropy Ratio X, Y/Z	<1.04			1.04			—			1.03		
Dissipation factor, ASTM D3380	.0009			.0012			.0019			.0023		
Specific gravity	2.2			2.2			2.7			2.9		
Specific heat (J/g/K)	.96			.96			.97			1.00		
Thermal conductivity, Z-direction, 23–100°C Rogers T.R. #2721 (W/m/K)	.26			.26			.37			.41		
Direction of test	X	Y	Z	X	Y	Z	X	Y	Z	X	Y	Z
Tensile, ASTM D638												
Modulus (MPa)	1070	860	—	1300	1280	—	510	627	—	931	559	—
Ultimate stress (MPa)	29	27	—	50	42	—	20	17	—	17	13	—
Ultimate strain (%)	6.0	4.9	—	9.8	9.8	—	12.5	5	—	12	10	—
Compression, ASTM D695												
Modulus (MPa)	710	710	940	1210	1360	830	—	—	1069	2144	—	—
Ultimate stress (MPa)	27	28	52	30	37	54	—	—	54	47	—	—
Ultimate strain (%)	8.5	7.7	12.5	4.0	3.3	8.7	—	—	33	25	—	—
Linear thermal expansion, ASTM D3386 mm/m total change from 35°C at −50°C	−6.8	−8.0	−18	−2.6	−5.0	−19	—	—	—	−2.0	−2.1	−2.6
0°C	−2.8	−3.8	−12	−0.9	−2.2	−11	—	—	—	−1.0	−1.0	−1.4
150°C	2.1	3.8	29	1.6	3.9	29	—	—	—	2.2	2.2	1.7
250°C	4.6	8.0	80	2.6	6.7	76	—	—	—	3.7	3.8	4.3

Note: 1 MPa (mega Pascal) = 145 psi, ASTM, American Society for Testing Materials, Philadelphia, PA. RT-duroid is a registered trademark of Rogers Corporation. (Courtesy Rogers Corp., Chandler, AZ.)

8.3 PARAMETERS OF PLANAR TRANSMISSION LINES

Microstrip is the most common circuit design used in microwaves. For this reason it is presented first in some detail, followed by design techniques for stripline, coplanar waveguide, and slotline.

8.3.1 Microstrip

8.3.1.1 Effective Dielectric Constant. Some of the electric field on a microstrip line passes through air (see Fig. 8.1*b*) so that the effective dielectric constant (ϵ_{eff}) is less than the bulk relative dielectric constant (ϵ_r) of the substrate. The wavelength on the transmission line is

$$\lambda = \frac{c}{(f\sqrt{\epsilon_{eff}})} \tag{8.1}$$

where c is the velocity of light in $\epsilon_r = 1$, and f is the frequency. The convenient closed form solution for ϵ_{eff} when the line is not in an enclosure is (3)

$$\epsilon_{eff} = \left(\frac{(\epsilon_r + 1)}{2}\right) + \left(\frac{(\epsilon_r - 1)}{2}\right)\left[\left(1 + \frac{12h}{W}\right)^{-1/2} + 0.04\left(1 - \frac{W}{h}\right)^2\right] \tag{8.2}$$

for $W/h \leq 1$, and

$$\epsilon_{eff} = \left(\frac{(\epsilon_r + 1)}{2}\right) + \left(\frac{(\epsilon_r - 1)}{2}\right)\left(1 + \frac{12h}{W}\right)^{-1/2} \tag{8.3}$$

for $W/h > 1$. The dimensional terms are defined in Fig. 8.3.

8.3.1.2 Characteristic Impedance. An approximation to the characteristic impedance is found using the following general relations for TEM lines. Assuming that the primary mode of propagation is TEM or quasi-TEM, the phase velocity is

$$v_{phase} = \frac{c}{(\epsilon_{eff})^{-1/2}} \tag{8.4}$$

and the characteristic impedance Z_0 is (see Chapter 5)

$$Z_0 = (v_{phase}C_u)^{-1} \tag{8.5}$$

where C_u is the capacitance of the transmission line per unit length. For the case where the electric field lines along the microstrip are small compared to the static transverse fields, a measurement of C_u, given ϵ_{eff}, approximates the Z_0. For more accurate results, conformal mapping, Green's functions, and variational and moment methods are used, but for simplicity these are not included here. To compute the capacitance of an isolated microstrip line, the fields are computed for a given charge on the microstrip conductor.

Two modes (shown in Fig. 8.4) are used to compute the even and odd mode impedances for coupled lines in Section 8.4. As indicated above, this has been done several ways. A convenient formulation is found to be (3)

$$Z_0 = [60(\epsilon_{eff})^{-1/2}] \ln\left[\left(\frac{8h}{W}\right) + \left(\frac{0.25W}{h}\right)\right] \text{ ohms} \qquad (8.6)$$

for $W/h < 1$, and

$$Z_0 = \frac{[120\pi(\epsilon_{eff})^{-1/2}]}{[(W/h) + 1.393 + 0.667 \ln(1.444 + W/h)]} \qquad (8.7)$$

for $W/h \geq 1$.

8.3.1.3   Effect of Nonzero Conductor Thickness. Equations 8.1 through 8.4 assume the thickness t of the microstrip line is zero. For $t > 0$, the electric fields from the edge of the line make the line width W appear larger to an

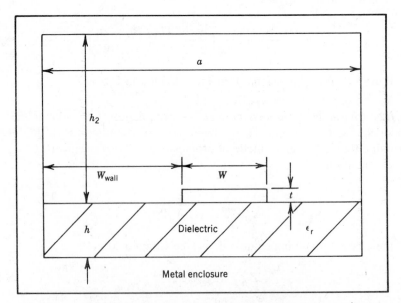

Figure 8.3. *Parameters of a microstrip transmission line in an enclosure.*

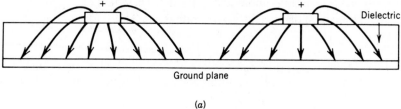

(a)

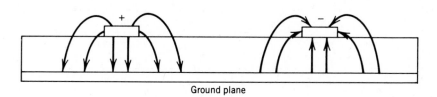

(b)

Figure 8.4. Electric field diagrams for even and off mode coupled microstrip transmission lines. (a) Even mode; (b) odd mode.

effective width W_{eff}. This two-dimensional case has been derived (4)

$$W_{\text{eff}} = W + \left(\frac{t}{\pi}\right)\left[1 + \ln\left(\frac{2h}{t}\right)\right] \tag{8.8}$$

for $W/h > 1/2\pi$, and

$$W_{\text{eff}} = W + \left(\frac{t}{\pi}\right)\left[1 + \ln\left(\frac{4\pi W}{t}\right)\right] \tag{8.9}$$

for $W/h \leq 1/(2\pi)$. Both relations can be used for the common configuration where $t \leq h$ and $t \leq W/2$. This correction is applied in the formulas for ϵ_{eff} and Z_0' by substituting W_{eff} for W in all cases.

8.3.1.4 Enclosure Effects. The sides and lid of the circuit enclosure add capacitance which tends to lower ϵ_{eff} and Z_0. These effects can be neglected (5) when $W_{\text{wall}} > 5W$ and $h_2 > 5h$ (see Fig. 8.3). Because these equations are lengthy, consult the reference if this case is encountered.

8.3.1.5 Coupling Between Microstrip Circuits. To maintain high isolation between microstrip circuits, they are mounted in a cutoff waveguide type enclosure (see Fig. 8.3) where a is less than the dimension required for cutoff ($f_{\text{cutoff}} = c/2a$) and the operating frequency $f_{\text{op}} < f_{\text{cutoff}}$. This design technique (6) yields isolations approaching 100 dB. The effect on f_{cutoff} of loading the waveguide with the dielectric substrate is small (7%) for normal substrate designs.

8.3.1.6   Tables of ϵ_{eff} and Z_0.   Table 8.5 gives ϵ_{eff} and Z_0 as a function of W/h for common substrate materials for ease in design. The table also in-

Table 8.5   Effective Dielectric Constant and Characteristic Impedance for Common Substrates

W/h Ratio	Substrate Dielectric Constant, ϵ_r							
	$\epsilon_r = 2.2$		2.35		2.5		3.8	
	ϵ_{eff}	$Z_0{}^a$	ϵ_{eff}	Z_0	ϵ_{eff}	Z_0	ϵ_{eff}	Z_0
0.1	1.68	203	1.77	198	1.85	194	2.56	164
0.2	1.70	170	1.79	166	1.87	162	2.62	137
0.3	1.71	151	1.80	148	1.89	144	2.64	122
0.5	1.73	127	1.82	124	1.91	121	2.70	102
0.7	1.75	112	1.84	109	1.94	106	2.73	89
1.0	1.77	96	1.87	93	1.96	91	2.79	76
1.2	1.79	87	1.89	85	1.98	83	2.82	70
1.5	1.81	78	1.91	76	2.01	74	2.87	62
2.0	1.84	66	1.94	64	2.04	63	2.93	52
3.0	1.88	51	1.99	50	2.10	49	3.04	40
4.0	1.91	42	2.03	41	2.14	40	3.12	33
5.0	1.94	36	2.05	35	2.17	34	3.18	28
7.0	1.99	27	2.11	27	2.23	26	3.30	21
v at 50 ohms	2.18×10^6 cm/s		2.13×10^6		2.07×10^6		1.75×10^6	

W/h Ratio	Substrate Dielectric Constant, ϵ_r							
	$\epsilon_r = 6.6$		9.7		10.2		10.5	
	ϵ_{eff}	Z_0	ϵ_{eff}	Z_0	ϵ_{eff}	Z_0	ϵ_{eff}	Z_0
0.1	4.14	130	5.78	109	6.15	106	6.32	105
0.2	4.21	108	5.98	91	6.27	89	6.44	88
0.3	4.27	96	6.07	80	6.36	78	6.53	77
0.5	4.36	80	6.21	67	6.51	66	6.69	65
0.7	4.45	70	6.34	59	6.64	57	6.82	57
1.0	4.55	60	6.50	50	6.82	49	7.01	48
1.2	4.72	54	6.61	46	6.93	44	7.12	44
1.5	4.71	48	6.75	40	7.07	39	7.27	39
2.0	4.84	41	6.95	34	7.29	33	7.50	33
3.0	5.05	31	7.28	26	7.64	25	7.85	25
4.0	5.21	26	7.53	21	7.90	21	8.12	20
5.0	5.34	22	7.72	18	8.11	18	8.34	17
v at 50 ohms	1.38×10^6 cm/s		1.18×10^6		1.15×10^6		1.14×10^6	

^{a} Z_0 in ohms.

dicates the practical design limits for W/h. Other estimates of ϵ_{eff} and Z_0 have been derived and are compared in Ref. 7.

8.3.1.7 Dispersion in Microstrip. Above the critical frequency (8)

$$f'_c = 0.3\left(\frac{Z_0}{(h\sqrt{\epsilon_r} - 1)}\right) \text{ GHz} \tag{8.10}$$

where h is in centimeters, ϵ_{eff} begins to vary with frequency because hybrid modes can propagate. The variation in ϵ_{eff} is (9)

$$\epsilon_{\text{eff}} = \epsilon_r - \frac{[\epsilon_r - \epsilon_{\text{eff}}]}{[(0.6 + 0.009Z_0)(8\pi hf/Z_0)^2]} \tag{8.11}$$

where f is in gigahertz and h is in centimeters. The increase in the characteristic impedance can be found from Eq. 8.6 or 8.7, but more detailed calculations are found in Ref. 10. Z_0 increases 2–4% for 0.025-in. thick alumina substrate between 0 and 10 GHz.

8.3.1.8 Loss in Microstrip. The loss in microstrip circuits arises from conductor loss due to resistive losses in the strip, dielectric losses due to polarization heating by the time-varying fields in the substrate, and radiation due to the antenna action of the microstrip.

Conductor Loss. The conductor loss is the primary loss mechanism at low frequencies. For nonuniform current distributions, W/h values near 1, $\epsilon_{\text{eff}} = 10$, and $Z_0 = 50$ ohms, the conductor attenuation per unit length is (11)

$$\alpha_c = \frac{8.68R_s}{2\pi Z_0 h_{\text{cm}}}\left(1 - \frac{W_{\text{eff}}^2}{16h^2}\right)\left\{1 + \frac{h}{W_{\text{eff}}} + \frac{h}{\pi W_{\text{eff}}}\left[\ln\left(\frac{2h}{t}\right) - \frac{t}{h}\right]\right\} \text{ dB/cm} \tag{8.12}$$

The current flows in the surface skin depth $\delta_s = (\pi f \mu \sigma)^{-1/2}$, where μ is the permeability and σ is the metal conductivity. For the common metals (Au, Cu, Al) the skin depth is about 0.8 μm at 10 GHz. As the surface roughness increases, the current path increases and the conductor loss increases. Measured loss per inch data is given in Table 8.6 (12). For this reason smooth substrates are essential for long circuits at high microwave and millimeter wave frequencies. Surface roughnesses less than 0.25 μm root mean square (rms) (includes rolled Cu on soft substrates) should be usable to 18 GHz.

Dielectric Loss. The dielectric loss arises because of polarization reversal losses in the substrate. A measure of this loss mechanism is the loss tangent

Table 8.6 Loss per Centimeter versus Surface Roughness[a]

Surface Roughness (rms)	Frequency		
	3 GHz	6 GHz	9 GHz
0.6 μm	0.065 dB/cm	0.087	0.131
0.25	0.055	0.073	0.096
0.05	0.043	0.055	0.070

[a] Material: 99.5% alumina, 0.64 mm(25 mil) thick, chrome/copper metallization.

(tan δ, see Table 8.3). The specific dielectric loss is

$$\alpha_{\text{diel}} = 27.3 \, \frac{\epsilon_r^2}{(\epsilon_{\text{eff}})^{3/2}} \left(\frac{\tan \delta}{\lambda_0} \right) \text{ dB/cm} \tag{8.13}$$

where $\lambda_0 = c/f$ = free space wavelength (13). Generally the dielectric loss is less than the conductor loss except at high frequencies.

Radiation Loss and Q. Open stubs, bends, and discontinuities on transmission lines excite higher-order modes, which can radiate energy. The ratio of the reactive energy to radiated energy is called the Q associated with the radiation, Q_{rad}. For example, for the open stub, the following expressions have been derived (14):

$$Q_{\text{rad}} = \frac{Z_0 \lambda_0^2}{480 \pi h^2 F} \tag{8.14}$$

where

$$F = \frac{\epsilon_{\text{eff}}(f) + 1}{\epsilon_{\text{eff}}(f)} - \frac{[\epsilon_{\text{eff}}(f) - 1]^2}{2[\epsilon_{\text{eff}}(f)]^{2/3}} \cdot \ln \left\{ \frac{1 + [\epsilon_{\text{eff}}(f)]^{1/2}}{-1 + [\epsilon_{\text{eff}}(f)]^{1/2}} \right\} \tag{8.15}$$

Clearly the expressions are complex and generally increase rapidly with frequency because the stub becomes a significant percentage of a wavelength (an efficient antenna).

The total Q of the transmission line is a measure of the loss of the line while propagating energy. Assigning a Q to each of the three loss mechanisms, the total Q_t is related by

$$\frac{1}{Q_t} = \frac{1}{Q_{\text{cond}}} + \frac{1}{Q_{\text{diel}}} + \frac{1}{Q_{\text{rad}}} = \frac{1}{Q_o} + \frac{1}{Q_{\text{rad}}} \tag{8.16}$$

where Q_{cond} relates to conductor loss, Q_{diel} relates to dielectric loss, and Q_o is related to the dielectric and conductor losses. Figure 8.5 shows the Q

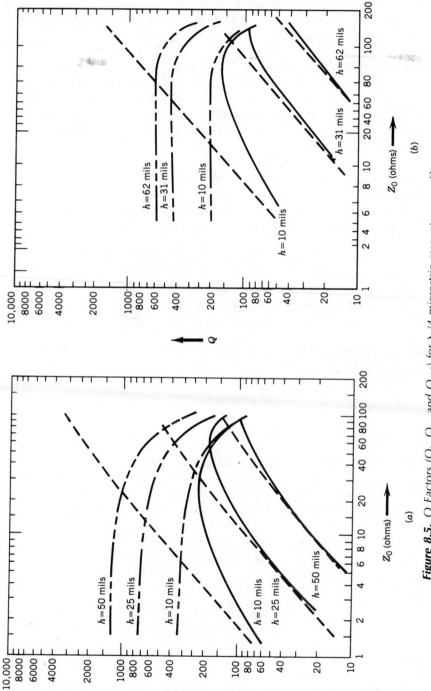

Figure 8.5. Q Factors (Q_l, Q_o, and Q_{rad}) for $\lambda_g/4$ microstrip resonators. ---, Q_l; ----, Q_{o}; ---, Q_l. (a) Alumina substrate; (b) Duroid[TM] ($\epsilon_r = 2.35$) substrate. $f = 8$ GHz. (From Ref. 39, © 1975 IEEE.)

factors for quarter-wave resonators fabricated on alumina and Duroid® substrates at 8 GHz. This data will be found useful in the design of filters since it relates to the insertion loss.

8.3.1.9 Microstrip Discontinuities. The design of microstrip circuits frequently involves connecting two lines of differing impedances and bends in the line to minimize the circuit real estate. The effects of the parasitics associated with these discontinuities has been estimated both analytically and experimentally. Refs. 15 and 16 are general references for more details of these discontinuities.

Step-Impedance Discontinuities. Step-impedance discontinuities (changes in microstrip or stripline linewidth) are encountered in the design of quarterwave transformers, couplers, and filters. Reference 17 presents formulas for both symmetric and assymmetric steps.

T- and Cross-Junctions. T-Junctions are encountered in the fabrication of branch line couplers, switches, and tuning stubs. Reference 18 describes a compensated *T*-junction used on couplers through 17 GHz. Reference 19 presents equivalent circuits for cross-junctions encountered with symmetric stubs used to match transmission lines to reactive elements.

Bends. Abrupt bends in microstrip transmission lines usually provide better performance than gradual bends. A compensation technique involving the chamfering of the outside of the bend to remove some capacitance has been developed for 30°, 60°, 90°, and 120° bends (20).

8.3.2 Stripline

8.3.2.1 Characteristic Impedance. Referring to the parameters in Fig. 8.6 for symmetric stripline ($b' = b''$ and $\epsilon_r' = \epsilon_r'' = \epsilon_r$), the characteristic impedance can be computed in closed form in the common range $0.05 < W/(b - t) < 0.35$ and $t/b < 0.25$ to 1% accuracy using (Refs. 14 and 21):

$$\frac{W_{\text{eff}}}{b} = \frac{W}{b} - \left(\frac{(0.35 - W/b)^2}{(1 + 12t/b)} \right) \tag{8.17}$$

and

$$Z_0 = \left(\frac{30\pi}{(\epsilon_r)^{1/2}} \right) \frac{(1 - t/b)}{[(W_{\text{eff}}/b) + (C_f/\pi)]} \tag{8.18}$$

where

$$C_f = 2 \ln\left(\frac{1}{1 - (t/b)} + 1 \right) - \frac{t}{b} \ln\left\{ \frac{1}{[1 - (t/b)]^2} - 1 \right\} \tag{8.19}$$

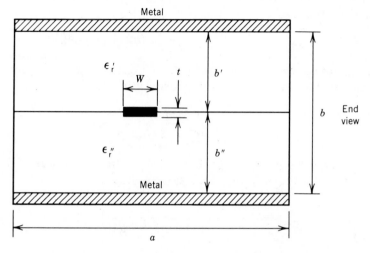

Figure 8.6. *Dimensions of stripline.*

For stripline, $\epsilon_{\text{eff}} = \epsilon_r$ and the propagation is considered to be TEM. Since the stripline is usually fabricated in soft substrate material, the center conductor is compressed into the dielectric leaving no air gap between dielectrics.

8.3.2.2 Loss. Because stripline uses two ground planes, radiation loss can be neglected unless $a < 10 W$. The conductor loss is

$$\alpha_{\text{cond}} = \frac{0.0231 R_s \epsilon_r}{30\pi(b-t)} Z_0 \left[1 + \frac{2W}{b-t} + \frac{(b+t)}{\pi(b-t)} \ln\left(\frac{2b-t}{t}\right) + B \right]$$

(8.20)

in decibels per unit length, where the sheet resistivity

$$R_s = \left(\frac{\pi f \mu_0}{\sigma}\right)^{1/2} \text{ ohms/square}$$

(8.21)

and

$$B = \frac{[0.35 - (W/b)]}{[b-t][1 + 12(t/b)]^2} \left[\left(\frac{t}{b}\right)(17.45b + 35W) \right.$$

$$\left. - 9W + 5.85 - 32.4\left(\frac{t^2}{b}\right) \right]$$

(8.22)

for $Z_0' \geq 120$ ohms, where $Z_0' = Z_0(\epsilon_r)^{1/2}(1 + 2.3t/b)$. $B = 0$ for $Z_0' < 120$ ohms.

The dielectric loss is

$$\alpha_{diel} = 27.3(\epsilon_r)^{1/2}\,\lambda_0^{-1}\,\tan\,\delta \quad \text{(dB/unit length)} \tag{8.23}$$

where the loss tangent was defined earlier (Eq. 8.13). At high frequencies the dielectric loss can exceed the conductor loss. The Q of stripline is

$$Q = \frac{8.68\pi(\epsilon_r)^{1/2}}{[\lambda_0(\alpha_{cond} + \alpha_{diel})]} \tag{8.24}$$

8.3.2.3 Upper Frequency Limit. For stripline, the lowest-order TE mode will be excited if the operating frequency exceeds

$$f_{cutoff} = \left(\frac{15}{b(\epsilon_r)^{1/2}}\right)\left(\frac{1}{(W/b) + (\pi/4)}\right) \quad \text{GHz} \tag{8.25}$$

where W and b are in centimeters (22). If b is increased, the Q increases but the f_{cutoff} decreases. Thus there is a range of values of b for a given value of ϵ_r and Q_{min} versus frequency. Reference 14 gives the values, which are given here in Table 8.7.

8.3.2.4 Other Considerations. Reference 14 also presents data on the effects of fabrication tolerances on characteristic impedance (VSWR), the equivalent circuits for discontinuities in stripline, and the power-handling limitations.

8.3.3 Coplanar Waveguide

Coplanar waveguide is a surface-strip transmission line configuration where the center conductor is located between two symmetrically positioned coplanar ground planes. It can be considered a planar stripline configuration as shown in Fig. 8.1c. This transmission line is used to mount nonreciprocal magnetic components (isolators, etc) because both axial and transverse magnetic fields exist near the center conductor, which readily couples into the

Table 8.7 Range of Stripline Thicknesses versus Frequency

Frequency (GHz)	b_{min}(cm)[a]	b_{max}(cm)[b]
1	0.7	5
10	0.22	0.62
20	0.18	0.3

[a] Determined by minimum Q of 100 and ϵ_r = 2.53.
[b] Determined by f_{cutoff} > operating frequency and ϵ_r = 2.53.
(From Ref. 14, Courtesy Microwaves and RF, Jan 1978.)

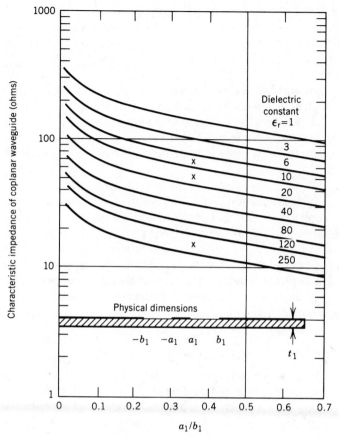

Figure 8.7. *Characteristic impedance of coplanar waveguide. (From Ref. 23, © 1969 IEEE.)*

magnetic material (23). This configuration readily accommodates mounting beam lead devices across a transmission line as in a balanced mixer.

8.3.3.1 Characteristic Impedance. The characteristic impedance of coplanar waveguide as a function of line dimensions and substrate dielectric constant has been calculated by estimating the capacitance per unit length and applying the relation (22)

$$Z_0 = \frac{1}{C v_{ph}} = \frac{1}{\text{(Capacitance per unit length)(phase velocity)}} \quad (8.26)$$

The capacitance is computed via a conformal mapping (22), and the phase velocity is assumed to be equal to a transmission line imbedded in a media with one-half the dielectric constant of the substrate. The resulting Z_0 values are shown in Fig. 8.7. The three experimental points for $\epsilon_r = 9.5$, 16, and 130 with $a_1/b_1 = 0.33$ confirm the accuracy of these approximations.

The calculations assume an infinitely thick substrate, but estimates indicate Z_0 increases only 10% when the substrate thickness is decreased to the slot width. The measured effective dielectric constant has been found to be within 80% of $(\epsilon_r + 1)/2$ for $t_1/b_1 > 0.5$ and $\epsilon_r = 16$ (24).

8.3.3.2 Loss and Dispersion. Experimental measurements of the unloaded Q (see Chapter 20) and resonant frequency of coplanar waveguide resonators (23) have shown that losses per unit length are comparable to microstrip (both configurations have significant current crowding at the edges of the center conductor) and show little dispersion over the 1–10-GHz frequency range (24).

8.3.3.3 Upper Frequency Limit. The width of the ground planes between adjacent coplanar circuits can be decreased if a cover is positioned over the circuit as shown in Fig. 8.8. The cover increases the isolation between circuits by preventing intercircuit radiation coupling. The distance $2b_1$ (see Fig. 8.8) should be less than a half-wavelength in the dielectric to prevent the line from radiating as a dipole. This ultimately sets an upper frequency limit on the use of the line for a given ϵ_r and suggests the use of lower ϵ_r for high-frequency applications.

8.3.4 Slotline

Slotline, like coplanar waveguide, uses the dielectric substrate to support metallization on one side and confines the fields. Like coplanar waveguide it allows components (such as beam lead diodes that do not dissipate significant heat) to be mounted in shunt across the line. A variation of the

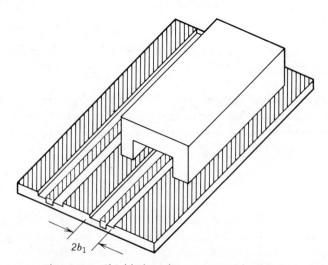

Figure 8.8. *Shielded coplanar waveguide circuits.*

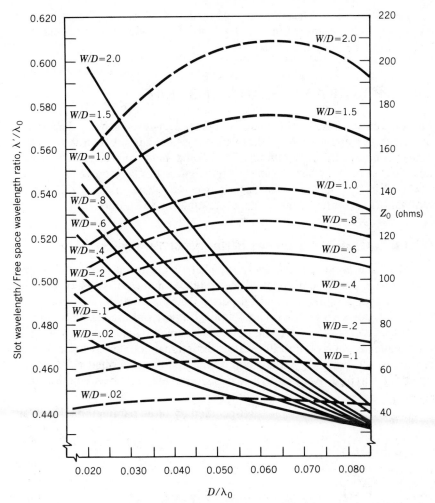

Figure 8.9. Dispersion and characteristic impedance of slotline on $\epsilon_r = 9.6$ substrate. —, λ'/λ_0; ---, Z_0; W = slot width. D = dielectric thickness. (From Ref. 27, © 1969 IEEE.)

standard one-sided dielectric slotline is sandwich slotline using a second dielectric sandwiching the slotline conductors (25).

8.3.4.1   Characteristic Impedance. The analysis of slotline characteristic impedance (Z_0 is used even though the line is non-TEM) is complex (26) and does not yield results as accurate as those for other line configurations. Slotline impedances are generally higher than those associated with microstrip with similar dimensional ratios as shown in Fig. 8.9 for $\epsilon_r = 9.6$ (27). For 50-ohm lines on 10-mil Duroid®, this result yields a slot width of about 13 μm. Experimental results show that W/D ratios are actually larger than predictions, which helps reduce the effects of fabrication tolerances.

8.3.4.2   *Dispersion.* The frequency dependence of the wavelength also varies significantly, as shown in Fig. 8.9. This effect must be considered when fabricating tuned circuits in slotline.

8.4   APPLICATIONS OF PLANAR TECHNOLOGY

Earlier sections showed how to design planar transmission lines. This section describes the application of these transmission lines to common passive planar components. A commonly used component, microstrip couplers, is described in detail, whereas other components are only referenced. Couplers are used in traveling wave bridges, balanced mixers, balanced amplifiers, modulators, discriminators, attenuators, and phase shifters.

8.4.1   Types of Couplers and Performance Parameters

Couplers are either conductively coupled or electromagnetically coupled transmission line circuits with three, four, or more ports. They have the property that a signal incident on port 1 (see Fig. 8.10) will be coupled to the output ports 3 and 4, with no coupling to port 2 at the design frequency. For the special case where the power is coupled equally to ports 3 and 4, the coupler is called a hybrid or 3-dB directional coupler. If the phase difference is 90° between ports 3 and 4, the coupler is termed a 90° hybrid, quadrature hybrid. Another hybrid design provides 180° between the output ports.

The performance of couplers is specified by three parameters. These are

$$C = \text{coupling factor in dB} = 10 \log_{10}(P_1/P_3) \qquad (8.27)$$

$$D = \text{directivity in dB} = 10 \log_{10}(P_3/P_2) \qquad (8.28)$$

$$I = \text{isolation in dB} = 10 \log_{10}(P_1/P_2) \qquad (8.29)$$

where P_i means the power at port i. The ratios are inverted so that the numbers are positive. For a lossless coupler, these three parameters are related (see Problem 9).

Conductively coupled circuits include the branch line, rat race, and Wilkinson couplers. The backward wave and Lange couplers are in the electromagnetically coupled group.

8.4.2   Conductively Coupled Couplers

8.4.2.1   *Wilkinson Power Divider.* This transmission line coupler style allows N-way equal amplitude and phase division of a signal. Since the network is reciprocal, it can combine N in-phase sources into a single output

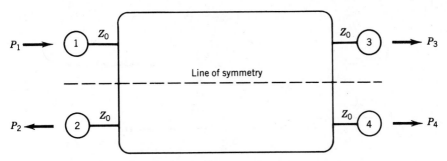

Figure 8.10. *Schematic of a four-port coupler.*

(termed a combiner). This latter mode is useful for combining the output of N medium-power solid-state devices for high-power requirements.

The Wilkinson divider (28) uses quarter-wavelength transmission lines emanating from a common input port to N outputs as shown in Fig. 8.11. The terminating resistors R_x carry no current if all the port voltages are of equal amplitude and phase since their common node is not grounded. As will be shown, these resistors perform the function of providing isolation between output ports for signals reflected from unmatched loads. This is accomplished by canceling the reflected wave, which has been delayed π radians through the two quarter-wavelength transmission lines, with another equal amplitude wave through the resistors delayed 0 radians (assuming resistors of zero length). The Wilkinson divider was envisioned in coax but can be applied to other transmission media.

In the following discussion, the simple ($\lambda/4$, two-way) Wilkinson will be derived. References to modifications are given later. The derivation will be done in two stages at the design frequency. Its performance off the design frequency is best analyzed using computer-aided design techniques (see Chapter 7). The first step is to derive the characteristic impedance of the N branch lines Z_w terminated in Z_0 and driven from $Z_0 = Z_{in}$. The Z_{in} of the ith quarter-wavelength branch is $Z_{in,i} = Z_w^2/Z_0$, a relation derived in lossless transmission line theory. For N branches in parallel, $Z_{in} = Z_{in,i}/N$, so $Z_{in} = Z_w^2/NZ_0$. If Z_{in} is to be matched to Z_0, then $Z_w = (N)^{1/2}Z_0$. Thus $Z_w > Z_0$, so consideration must be given to linewidths and losses in microstrip when N is large.

The second step is to determine the value of R_x to maximize the isolation between output ports. This case arises when the jth load is not matched. For simplicity of notation, the following derivation considers only a two-way divider. The reflected wave is modeled as a voltage source V with internal resistance $R_0 = Z_0$, as shown in Fig. 8.11b. The I_{ia} are branch currents traveling on the lossless transmission lines, whereas I_{ib} are the currents in the resistors. For the currents and voltages on the branch lines, the voltage and current at a given point θ degrees along the line of char-

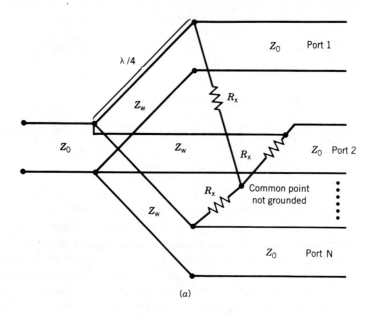

(a)

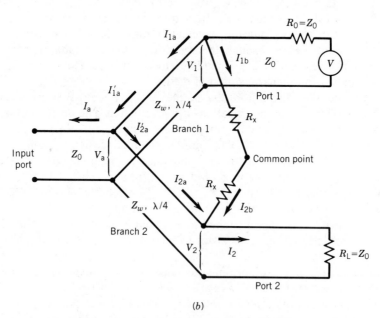

(b)

Figure 8.11. The Wilkinson power divider. (a) Schematic of the N way; (b) current and voltage definitions for a two way.

acteristic impedance Z_0 from the load L is (29)

$$V_\theta = V_L \cos \theta + jI_L Z_0 \sin \theta \qquad (8.30)$$

$$I_\theta = I_L \cos \theta + j\left(\frac{V_L}{Z_0}\right) \sin \theta \qquad (8.31)$$

For branch 1

$$V_1 = jI'_{1a} Z_w \qquad (8.32)$$

$$I_{1a} = \frac{jV_a}{Z_w} \qquad (8.33)$$

and for branch 2,

$$V_a = jI_{2a} Z_w \qquad (8.34)$$

$$I'_{2a} = \frac{jV_2}{Z_w} \qquad (8.35)$$

Applying Kirchoff's laws at the nodes yields the relations

$$I'_{1a} = \frac{V_a}{Z_0} + I'_{a2} \quad \text{at the input port} \qquad (8.36)$$

$$I_{2a} = \frac{V_2}{Z_0} - I_{2b} = \frac{V_2}{Z_0} - I_{1b} \quad \text{at port 2} \qquad (8.37)$$

$$V - V_1 = (I_{1a} + I_{1b})R_0 \quad \text{at port 1} \qquad (8.38)$$

$$I_{1b} = \frac{(V_1 - V_2)}{2R_x} \quad \text{at port 1} \qquad (8.39)$$

A simultaneous solution for the unknown voltage V_1, V_2, and V_a is found. Using Eqs. 8.34, 8.37, and 8.39, yields

$$V_a - jV_2\left[\left(\frac{Z_w}{Z_0}\right) + \left(\frac{Z_w}{2R_x}\right)\right] + j\left(\frac{V_1 Z_w}{2R_x}\right) = 0 \qquad (8.40)$$

Using Eqs. 8.32, 8.36, and 8.35 yields

$$j\left(\frac{V_a Z_w}{Z_0}\right) - V_2 - V_1 = 0 \qquad (8.41)$$

Using Eqs. 8.38, 8.33, and 8.39 yields

$$V = \left(\frac{jR_0}{Z_w}\right)V_a - \left(\frac{R_0}{2R_x}\right)V_2 + \left[\left(\frac{R_0}{2R_x}\right) + 1\right]V_1 \qquad (8.42)$$

For perfect isolation $V_2 = 0$ so Eqs. 8.40 and 8.41 yield

$$R_x = \frac{Z_w^2}{2Z_0} \qquad (8.43)$$

Using Eqs. 8.42 and 8.40 provides a relation between V_1 and V of

$$\frac{V}{V_1} = \frac{(R_0 + R_x)}{R_x} \qquad (8.44)$$

If the output at port 1 is to match to R_0, then using Eq. 8.44,

$$\frac{V_1}{(I_{1a} + I_{1b})} = \frac{V_1 R_0}{(V - V_1)} = R_0 \quad \text{or} \quad R_x = R_0 \qquad (8.45)$$

Substituting Eq. 8.45 into Eq. 8.43 yields

$$Z_w = (2)^{1/2}Z_0 \qquad (8.46)$$

which confirms the result calculated earlier for $N = 2$. Reference 28 shows that $R_x = R_0$ independent of N.

For $N = 2$ the VSWR at the octave bandwidth limits is 1.42:1 at the input and less than 1.1:1 at the two output ports. The isolation is reduced to 14 dB at this frequency limit. For greater bandwidths, more quarter-wavelength sections are cascaded (30). The divider can also be used for unequal power splits (31). For cases where quarter-wavelength lines are physically too large, a shortened design technique has been derived (32).

8.4.2.2   Branch Line. The branch line coupler uses branching transmission lines to couple 3–9 dB from the input line. A branch line configuration shown in Fig. 8.12 matches at all ports to Z_0 (typically 50 ohms). If port 1 is the input, the coupled power appears at port 4 and almost no power appears at port 2. The remainder of the input power exists at port 3 (except for internal losses). The operation of the branch line coupler is derived below. A straightforward application of the Smith chart is not possible because several traveling waves exist in the network (simultaneous solutions are required) (38).

Figure 8.12*a* shows the microstrip pattern for a two-section branch line coupler and shows the symmetry. If ports 1 and 2 are considered inputs and excited with sinusoids of $E_1 = E_2 = (1/2)E\underline{/0°}$ (even-mode excitation), a

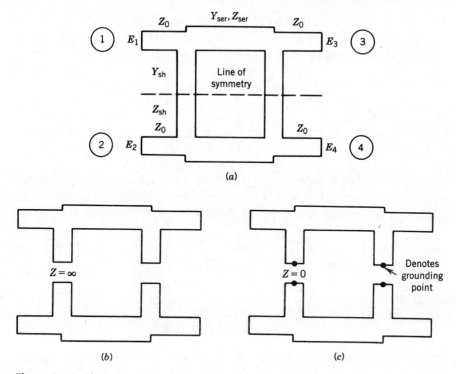

Figure 8.12. The two-section branch line coupler. (a) Parameters of a two-section branch line coupler; (b) equivalent circuit for even-mode excitation; (c) equivalent circuit for odd-mode excitation.

voltage maximum occurs at the line of symmetry and zero net current flows across this line. This condition is equivalent to an infinite load ($Z = \infty$, $Y = 0$) so the circuit can be split open as shown in Fig. 8.12b without any change in its operation. If ports 1 and 2 are excited so $E_1 = (1/2)E\underline{/0°}$ and $E_2 = (1/2)E\underline{/180°}$ (odd-mode excitation), a voltage minimum ($Z = 0$, $Y = \infty$) occurs at the line of symmetry yielding the circuit in Fig. 8.12c. By super-position (the network is linear), the excitations sum to $E_1 = E\underline{/0°}$ and $E_2 = 0$, so if the resultant signals at ports 3 and 4 are derived for each excitation mode they can be summed to find the resultant voltage.

Using the *ABCD* matrix notation (see Chapter 6), the voltages emerging from the four ports are

$$E_1 = (\tfrac{1}{2})(\Gamma_{\text{even}} + \Gamma_{\text{odd}})E \tag{8.47}$$

$$E_2 = (\tfrac{1}{2})(\Gamma_{\text{even}} - \Gamma_{\text{odd}})E \tag{8.48}$$

$$E_3 = (\tfrac{1}{2})(T_{\text{even}} + T_{\text{odd}})E \tag{8.49}$$

$$E_4 = (\tfrac{1}{2})(T_{\text{even}} - T_{\text{odd}})E \tag{8.50}$$

where Γ_{even}, Γ_{odd} are reflection coefficients with matched loads and T_{even}, T_{odd} are transmission coefficients for matched generators and loads. These coefficients are related to the *ABCD*-matrix coefficients as

$$T = \frac{2}{(A + B + C + D)} \qquad (8.51)$$

$$\Gamma = \frac{(A + B - C - D)}{(A + B + C + D)}. \qquad (8.52)$$

The problem is now reduced to finding the *ABCD* matrix for each of the equivalent circuits in Fig. 8.12. For the even mode, the matrix elements are (presented for brevity in normalized admittance units) as follows:

$$\begin{aligned}
\begin{matrix} \text{Even-mode} \\ ABCD \\ \text{matrix} \end{matrix} &= \begin{bmatrix} \lambda/8\,\text{open-} \\ \text{circuited} \\ \text{stub} \end{bmatrix} \begin{bmatrix} \lambda/4\,\text{lossless} \\ \text{transmission} \\ \text{line} \end{bmatrix} \begin{bmatrix} \lambda/8\,\text{open-} \\ \text{circuited} \\ \text{stub} \end{bmatrix} \\[2mm]
&= \begin{bmatrix} 1 & 0 \\ jY_{sh} & 1 \end{bmatrix} \begin{bmatrix} 0 & j/Y_{ser} \\ jY_{ser} & 0 \end{bmatrix} \begin{bmatrix} 1 & 0 \\ jY_{sh} & 1 \end{bmatrix} \\[2mm]
&= \begin{bmatrix} -a/b & j/b \\ j(b^2 - a^2)/b & -a/b \end{bmatrix} = \begin{bmatrix} A & B \\ C & D \end{bmatrix}_{even} \qquad \begin{aligned} a &= Y_{sh} \\ b &= Y_{ser} \end{aligned}
\end{aligned}$$

$$(8.53)$$

and a similar matrix exists for the odd mode. As a result

$$\Gamma_{even} = \frac{(j/b) - j(b^2 - a^2)/b}{-2(a/b) + j[(1/b) + b - (a^2/b)]} \qquad (8.54)$$

and a similar relation exists for Γ_{odd}. The ports will be matched if $\Gamma_{even} = \Gamma_{odd} = 0$. This occurs when the numerators are zero, namely,

$$\frac{1}{b} = \frac{(b^2 - a^2)}{b} \qquad \text{or} \qquad 1 + a^2 = b^2 \qquad (8.55)$$

For this case, the emerging waves E_1 and E_2 are zero. The coupler is also perfectly directive, namely, $E_2 = 0$.

The coupling into port 4 is found using the relations

$$T_{even} = \frac{b}{(j - a)} \qquad (8.56)$$

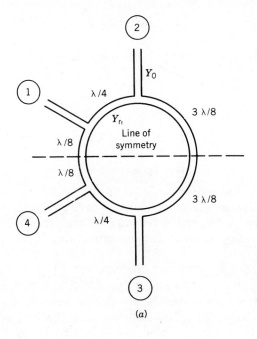

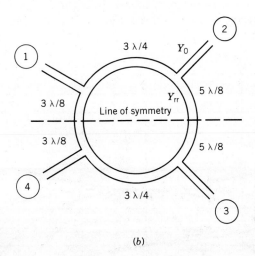

Figure 8.13. Two types of rat race couplers. (a) 3λ/2 design; (b) 7λ/2 design.

and

$$T'_{\text{odd}} = \frac{b}{(a + j)} \qquad (8.57)$$

The resulting voltages at ports 3 and 4 are

$$\frac{E_3}{E} = \frac{-j}{b} = \frac{-j}{Y_{\text{ser}}} \tag{8.58}$$

and

$$\frac{E_4}{E} = \frac{-a}{b} = \frac{-Y_{\text{sh}}}{Y_{\text{ser}}} \tag{8.59}$$

Note the 90° phase difference between the output ports. For additional bandwidth, more series and shunt quarter-wavelength transmission lines can be added to make an *N*-section branch line coupler. The analysis follows directly by expanding on the multiplications in Eq. 8.53.

8.4.2.3   Rat Race. The rat race coupler shown in Fig. 8.13 also uses transmission lines to split the input power and isolate one port. Like the branch line coupler, the rat race is also DC coupled between ports. The bandwidth of the 3λ/2 rat race (Fig. 8.13*a*) is better than the 7λ/2 unit (Fig. 8.13*b*) because of reduced phase imbalance between ports when off the design frequency since ports are λ/2 closer together for the 3λ/2 design.

The analysis for these units follows the even-and-odd-mode approach

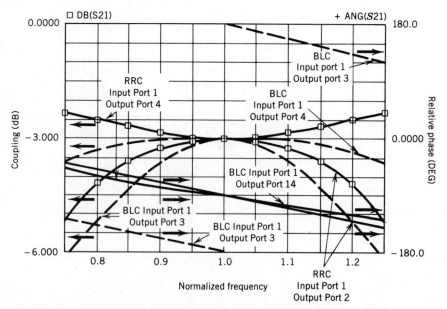

Figure 8.14. *Performance of ideal 3λ/2 rat race and branch line couplers. RRC, rat race coupler; BLC, branch line coupler.*

developed above because of the symmetry planes shown in Fig. 8.13. The results are (38)

$$
\begin{array}{c}
\text{Even-mode} \\
ABCD \\
\text{matrix}
\end{array}
=
\begin{bmatrix}
1 & j/Y_{rr} \\
j/Y_{rr} & -1
\end{bmatrix}
\tag{8.60}
$$

$$
\begin{array}{c}
\text{Odd-mode} \\
ABCD \\
\text{matrix}
\end{array}
=
\begin{bmatrix}
-1 & j/Y_{rr} \\
j/Y_{rr} & 1
\end{bmatrix}
\tag{8.61}
$$

yielding

$$
\Gamma_{even} = \frac{-jY_{rr}}{Y_0} \tag{8.62a}
$$

$$
\Gamma_{odd} = \frac{jY_{rr}}{Y_0} \tag{8.62b}
$$

$$
T_{even} = \frac{-jY_{rr}}{Y_0} \tag{8.62c}
$$

$$
T_{odd} = \frac{-jY_{rr}}{Y_0} \tag{8.62d}
$$

For power fed into port 1 ($E_1 = 0$), the power divides equally (if $Y_{rr} = 0.707$ Y_0) in phase at ports 2 and 4 ($E_2 = E_4 = -jY_{rr}/Y_0$), and none appears at port 3 ($E_3 = 0$). Doing a similar analysis (see Problem 12), power fed into port 2 divides and appears out of phase at ports 1 and 3 with no output at port 4. Figure 8.14 shows the performance of a $3\lambda/2$ rat race coupler. In general the rat race has more bandwidth than the branch line coupler, but the choice between couplers is usually made based on the phase difference needed for a particular application.

8.4.3 Electromagnetically Coupled Couplers

8.4.3.1 *Backward Wave Coupler.* Two transmission lines spaced less than the substrate thickness apart are coupled as shown in Fig. 8.15. As a result, energy is coupled from one line to another. Consider the two lines of electrical length θ and the defining currents and voltages in Fig. 8.15a. As shown earlier, the voltage source $2E$ is equivalent to the two excitations in Figs. 8.15b and c for the even and odd mode, respectively. Each transmission line has a given characteristic impedance depending on the mode, namely, Z_{oe} for the even mode and Z_{oo} for the odd mode. Bryant and Weiss (33) have computed these values for various configurations on microstrip (see Fig.

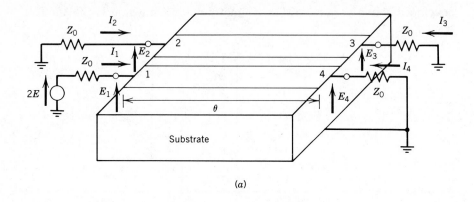

(a)

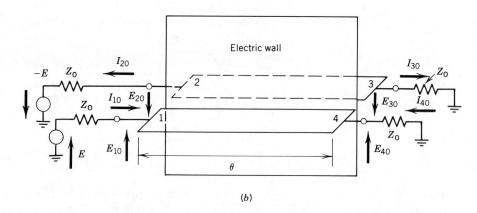

(b)

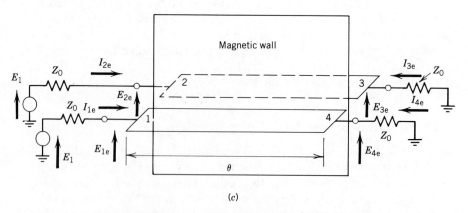

(c)

Figure 8.15. Coupled-strip directional coupler. (a) *Definition of voltages and currents;* (b) *odd-mode excitation;* (c) *even-mode excitation. Substrate omitted in (b) and (c) for clarity.*

8.16). The input impedance Z_{in} of the coupler in Fig. 8.15 at port 1 is

$$Z_{in} = \frac{(E_{1odd} + E_{1even})}{(I_{1odd} + I_{1even})} \qquad (8.63)$$

where

$$E_{1odd} = Z_{in,odd} I_{1odd} \qquad (8.64)$$

$$Z_{in,odd} = Z_{oo} \left[\frac{(Z_0 + jZ_{oo} \tan \theta)}{(Z_{oo} + jZ_0 \tan \theta)} \right] \qquad (8.65)$$

Similar relations exist for the even mode when terminated in Z_0. Referring to Fig. 8.15b, the current $I_{1odd} = E/(Z_0 + Z_{in,odd})$ and so forth for I_{1even},

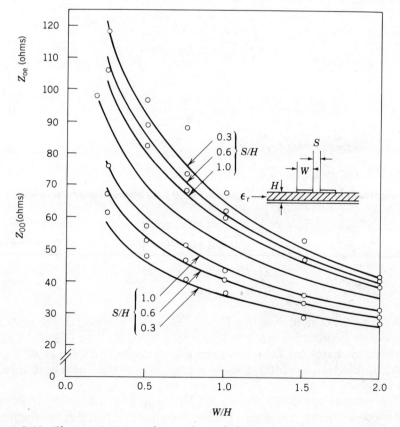

Figure 8.16. Characteristic impedance of coupled-microstrip transmission lines, $\epsilon_r = 9.0$. Z_{0e}, even mode; Z_{00}, odd mode. (From Ref. 33, © 1968 IEEE.)

$E_{1\text{odd}}$, and $E_{1\text{even}}$, yielding

$$Z_{\text{in}} = \frac{E \left| \dfrac{Z_{\text{in,odd}}}{Z_0 + Z_{\text{in,odd}}} + \dfrac{Z_{\text{in,even}}}{Z_0 + Z_{\text{in,even}}} \right|}{E \left| \dfrac{1}{Z_0 + Z_{\text{in,odd}}} + \dfrac{1}{Z_0 + Z_{\text{in,even}}} \right|} \qquad (8.66)$$

If Eqs. 8.64 and 8.65 are substituted in Eq. 8.66, the $Z_{\text{in}} = Z_0$ for all θ if $Z_0 = (Z_{\text{oo}} Z_{\text{oe}})^{1/2}$.

For this $Z_{\text{in}} = Z_0$ case, the voltages at the other ports can be determined for each node. Using Eqs. 8.30 and 8.31 for the even-mode relations, $E_{1e} = E_{2e}$ and $E_{3e} = E_{4e}$, whereas the characteristic impedance Z_{oe} is terminated in $(Z_{\text{oe}} Z_{\text{oo}})^{1/2}$. For the odd mode $E_{2o} = -E_{1o}$ and $E_{3o} = -E_{4o}$. The total voltage appearing at the ports is $E_i = E_{io} + E_{ie}$, where $i = 2, 3, 4$. Carrying through all the algebra yields

$$E_4 = \frac{E(1 - C_v^2)^{1/2}}{[(1 - C_v^2)^{1/2} \cos\theta + j\sin\theta]} \qquad (8.67)$$

$$E_3 = 0 \qquad (8.68)$$

$$E_2 = \frac{jEC_v \sin}{[(1 - C_v)^{1/2} \cos\theta + j\sin\theta]} \qquad (8.69)$$

$$C_v = \frac{(Z_{\text{oe}} - Z_{\text{oo}})}{(Z_{\text{oe}} + Z_{\text{oo}})} = \text{voltage coupling ratio} \qquad (8.70)$$

Using the relation $Z_0^2 = Z_{\text{oe}} Z_{\text{oo}}$, Eq. 8.70 can be rearranged to yield

$$Z_{\text{oo}} = Z_0 \left[\frac{(1 - C_v)}{(1 + C_v)} \right]^{1/2} \qquad (8.71)$$

$$Z_{\text{oe}} = Z_0 \left[\frac{(1 + C_v)}{(1 - C_v)} \right]^{1/2} \qquad (8.72)$$

Thus, given a desired coupling $C = -20 \log_{10} C_v$ to port 2, the odd- and even-mode impedances are determined for a given Z_0. Use of Ref. 33 yields the dimensions of the lines and line spacing. Note that the coupling is to port 2 rather than to port 3, thus appearing to propagate backward through the coupler.

The outputs are in quadrature for all frequencies where the even- and odd-mode velocities are equal, but the coupling is a function of frequency. In microstrip, the even-mode fields are more confined to the dielectric, so the velocities are not equal and the above relations are only approximate. A technique for reducing this effect proposed a dielectric overlay (34). For

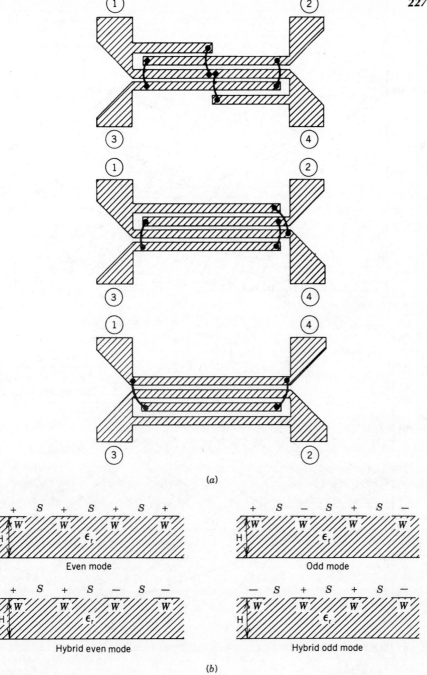

(a)

(b)

Figure 8.17. Configuration and modes of an interdigitated coupler. (a) Interdigitated configurations and bonding straps; ports: (1), input; (2), isolated; (3), coupled; (4), direct; (b) nondegenerate modes. (From Ref. 36, Courtesy Microwaves and RF, May 1976.)

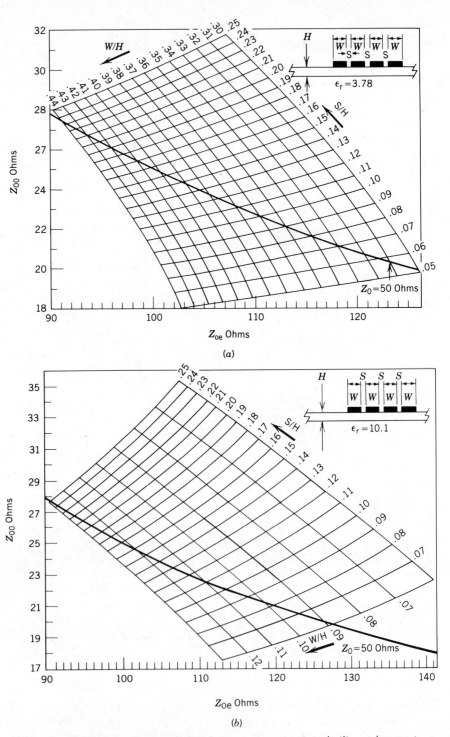

Figure 8.18. *Four-strip Lange coupler design curves. (a) Fused silica substrate (ϵ_r = 3.78); (b) alumina substrate (ϵ_r = 10.1). (From Ref. 36, Courtesy Microwaves and RF, May 1976.)*

a 3-dB coupler, $C_v = 0.707$, $Z_{oo} = 20.7$, and $Z_{oe} = 121$ ohms, yielding unrealizable S/H ratios for most etching techniques (see Fig. 8.16). As a result the Lange interdigital coupler was developed. Backward wave couplers work over the 6–20 dB range in microstrip.

8.4.3.2   Lange Coupler. The interdigitated Lange coupler (35 and 36) cannot be designed using the two-coupled line technique above because the four-coupled line charge distributions are not the same as those in Ref. 33. With four lines as shown in Fig. 8.17a there are four nondegenerate modes shown in Fig. 8.17b. The two modes with two equal adjacent potentials are a hybrid even and odd mode. To suppress these two modes the bond wires in Fig. 8.17a are used. The Lange coupler has adjacent alternating potentials in the odd mode.

A Green's function analysis (36) has been used and experimentally verified yielding the design curves in Figs. 8.18a and b for $\epsilon_r = 10.1$ and 3.78, respectively. The four-strip design yields coupling values of 2–6 dB with realizable line spacings and widths. The other configurations of Lange coupler in Fig. 18.17a are derivable from the design first proposed by Lange (37). These designs yield 3-dB coupling over an octave bandwidth with nearly constant $\pi/2$ phase shift between the output ports.

8.4.4   Other Applications

Besides couplers, planar circuitry is used for patch antennas (see Chapter 11), for diode switches and phase shifters (Chapter 12), as the media to launch into ferrite devices (Chapter 10), for planar filters (Chapter 9). In fact, for nearly every device and circuit to be described in the remainder of this book, a planar format should be considered.

PROBLEMS

1. For 0.635 mm (25 mil) thick Rogers 5880 substrate use Eqs. 8.6 and 8.7 to compute Z_0 for $W/h = 0.1$, 1.0, and 5 assuming $t = 0$. Compare with Table 8.5.

2. Redo Problem 1 for 57-g (2-oz) Cu conductor.

3. What is the physical length of a quarter-wavelength transmission line in Problem 2 at 9 GHz neglecting end effects.

4. Estimate the line dimensions for $Z_0 = 50$ ohms on 0.254-mm (10-mil) thick alumina.

5. Compute the conductivity loss for 28-g (1-oz) rolled Cu conductor 50-ohm microstrip at 9 GHz and compare with the value in Table 8.6. Use Eq. 8.21 for R_s, with the conductivity of Cu = $(1.7 \times 10^{-6} \text{ ohm-cm})^{-1}$ and the permeability of free space is $4\pi \times 10^{-7}$ H/m.

6. Compute the dielectric loss for the microstrip in Problem 4 on alumina and Rogers 5880, both 0.254 mm (10 mils) thick at 8 GHz.

7. Compute the impedance of two 100-ohm, 14-g (1/2-oz) Cu microstrip lines configured as a stripline on 0.254 mm (10-mil) thick Rogers 6010.

8. For $0.05 < w/(b - t) < 0.35$ and $t/b < 0.25$, derive W/b in stripline as a function of Z_0 and t/b. Select the proper root of the solution based on engineering judgment.

9. Derive the relation $I(\text{dB}) = C(\text{dB}) + D(\text{dB})$ for a lossless coupler. Compute the insertion loss of the matched lossless coupler versus the coupling and isolation factors.

10. Derive the values for R_x and Z_w for an N-way Wilkinson divider using the approach in Section 8.4.2.1.

11. Compute the odd-mode $ABCD$ matrix and Γ_{odd} for a two-section branch line coupler.

12. Derive the power and phase distribution for the $3\lambda/2$ rat race coupler when excited at port 2.

13. Show that $Y_{rr} = 0.707\, Y_0$ for equal power split in a rat race coupler.

14. Calculate the microstrip line dimensions for a 10-dB backward wave coupler using $\epsilon_r = 9.0$ substrate material. Select a reasonable h. The port impedances are to be 50 ohms.

15. Design a 3-dB Lange coupler on alumina with $Z_0 = 50$ ohms for $h = 0.254$ mm (10 mils).

REFERENCES

1. R. M. Barrett, "Microwave Printed Circuits—The Early Years," *IEEE Trans. Micro. Th. Tech.,* Vol. MTT-32, No. 9, September, 1984, p. 983.

2. T. Itoh, IEEE Washington MTTS Course, November 13, 1984, and R. Mittra and T. Itoh, "Analysis of Microstrip Transmission Lines," in *Advances in Microwaves,* Vol. 8, L. Young and H. Sobol, eds., Academic Press, New York, 1974, p. 67.

3. E. O. Hammerstad, "Equations for Microstrip Circuit Design," *Proc. European Micro. Conf.,* Hamburg, W. Germany, September, 1975, p. 268.

4. M. A. R. Gunston and J. R. Weale, "Variation of Microstrip Impedance with Strip Thickness," *Electronics Letters,* Vol. 5, December 27, 1969, p. 697.

5. S. March, "Microstrip Packaging: Watch The Last Step," *MicroWaves,* Vol. 20, No. 13, December, 1981, p. 83, and *MicroWaves,* Vol. 21, No. 2, February, 1982, p. 9.

6. K. W. Craft, "Microstrip in Instrumentation," *Micro. Jrnl.,* Vol. 16, No. 4, April, 1973, p. 9.

7. J. J. Lev, "Synthesize and Analyze Microstrip Lines," *Microwaves and RF,* Vol. 24, January, 1985, p. 111.

8. W. J. Chudobiak, O. P. Jain, and V. Makios, "Dispersion in Microstrip," *IEEE Trans. Micro. Th. Tech.,* Vol. MTT-19, September, 1971, p. 783.

9. W. J. Getsinger, "Microstrip Dispersion Model," *IEEE Trans. Micro. Th. Tech.,* Vol. MTT-21, January, 1973, p. 34.

10. R. P. Owens, "Predicted Frequency Dependence of Microstrip Characteristic Impedance Using the Planar-Waveguide Model," *Electronics Letters,* Vol. 12, May 27, 1976, p. 269.

11. R. A. Pucel, D. J. Masse, and C. P. Hartwing, "Losses in Microstrip," *IEEE Trans. Micro. Th. Tech.,* Vol. MTT-16, June, 1968, p. 342, and "Corrections to Losses in Microstrip," ibid., Vol. MTT-16, December 1968, p. 1046.

12. W. Schilling, "The Real World of Micromin Substrates—Part I," *MicroWaves,* Vol. 7, December, 1968, p. 52.

13. M. V. Schneider, "Dielectric Loss in Integrated Microwave Circuits," *Bell Syst. Tech. Jrnl.,* Vol. 48, No. 7, September, 1969, p. 2325.

14. I. J. Bahl and R. Garg, "A Designer's Guide to Stripline Circuits," *MicroWaves,* Vol. 17, January, 1978, p. 89.

15. T. C. Edwards, *Foundations for Microstrip Circuit Design,* Wiley, New York, 1981, Chap. 5.

16. E. O. Hammerstad and F. Bekkadal, "Microstrip Handbook," Electronics Res. Lab., Univ. of Trondheim, Norwegian Inst. Tech., Rpt. No. STF44A74169, February 1975.

17. P. Benedek and P. Silvester, "Equivalent Capacitances for Microstrip Gaps and Steps," *IEEE Trans. Micro Th. Tech,* Vol. MTT-20, No. 11, November 1972, p. 729.

18. P. Silvester and P. Benedek, "Microstrip Discontinuity Capacitances for Right-Angle Bends, T-Junction and Crossing's," *IEEE Trans. Micro. Th. Tech.,* Vol. MTT-21, No. 5, May 1973, p. 341.

19. M. Dydyk, "Master the T-junction and Sharpen Your MIC Designs," *MicroWaves,* Vol. 16, No. 5, May 1977, p. 84.

20. P. Anders and F. Arndt, "Moment Method of Designing Matched Microstrip Bands," Dig. Papers, *1979 European Micro. Conf.,* Brighton, England, p. 430.

21. S. B. Cohn, "Problems in Strip Transmission Lines," *IRE Trans. Micro. Th. Tech.,* Vol. MTT-3, March, 1955, p. 119.

22. G. D. Vendelin, "Limitations on Stripline Q," *Micro. Jrnl.,* Vol. 13, No. 5, May, 1970, p. 63.

23. C. P. Wen, "Coplanar Waveguide: A Surface Strip Transmission Line Suitable for Nonreciprocal Gyromagnetic Device Applications," *IEEE Trans. Micro. Th. Tech.,* Vol. MTT-17, No. 12, December, 1969, p. 1087.

24. C. P. Wen, "Attenuation Characteristics of Coplanar Waveguides," *Proc. IEEE,* Vol. 58, No. 1, January, 1970, p. 141.

25. S. B. Cohn, "Sandwich Slot Line," *IEEE Trans. Micro. Th. Tech.,* Vol. MTT-19, No. 9, September, 1971, p. 773.

26. S. B. Cohn, "Slot Line on a Dielectric Substrate," *IEEE Trans. Micro. Th. Tech.,* Vol. MTT-17, No. 10, October, 1969, p. 768.

27. E. A. Mariani, C. P. Heinzman, J. P. Agrios, and S. B. Cohn, "Slot Line Characteristics," *IEEE Micro. Th. Tech.*, Vol. MTT-17, No. 12, December, 1969, p. 1091.

28. E. J. Wilkinson, "An N-Way Hybrid Power Divider," *IRE Trans. Micro. Th. Tech.*, Vol. MTT-8, January, 1960, p. 116.

29. T. Moreno, *Microwave Transmission Design Data*, Dover, New York, 1948, Chap. 1.

30. S. B Cohn, "A Class of Broadband Three-Port TEM-Mode Hybrids," *IEEE Trans. Micro. Th. Tech.*, Vol. MTT-16, No. 2, February, 1968, p. 110.

31. L. I. Parad and R. L. Moynihan, "Split-Tee Power Divider," *IEEE Trans. Micro. Th. Tech.*, Vol. MTT-13, January, 1965, p. 91.

32. R. C. Webb, "Power Divider/Combiners: Small Size, Big Specs," *MicroWaves*, Vol. 20, November, 1981, p. 67.

33. T. G. Bryant and J. A. Weiss, "Parameters of Microstrip Transmission Lines and of Coupled Pairs of Microstrip Lines," *IEEE Trans. Micro. Th. Tech.*, Vol. MTT-16, December, 1968, p. 1021.

34. B. Sheleg and B. E. Spielman, "Broadband (7-18GHz) 10 dB Overlay Coupler for MIC Application," *Electronics Letters*, Vol. 11, No. 8, April 17, 1975, p. 175.

35. J. Lange, "Interdigitated Stripline Quadrature Hybrid," *IEEE Trans. Micro. Th. Tech.*, Vol. MTT-17, No. 12, December, 1969, p. 1150.

36. D. D. Paolino, "Design More Accurate Interdigitated Couplers," *MicroWaves*, May, 1976, p. 34.

37. R. Waugh and D. LaCombe, "Unfolding The Lange Coupler," *IEEE Trans. Micro. Th. Tech.*, Vol. MTT-20, November 1972, p. 777.

38. J. Reed and G. J. Wheeler, "A Method of Analysis of Symmetrical Four-Port Networks," IRE Trans. *Micro. Th. Tech.*, Vol. MTT-4, No. 10, October 1956, p. 246.

39. E. Belohoubek and E. Denlinger, "Loss Considerations for Microstrip Resonators," *IEEE Trans. Micro. Th. Tech.*, Vol. MTT-23, June, 1975, p. 522.

9 Microwave Filter Theory

Microwave filter theory plays an important role in the design of microwave systems. Filters are used for rejecting harmonics and excluding out-of-band signals. Filter theory is used for designing frequency division multiplexers and for broadband matching into discrete discontinuities such as the input capacitance of a microwave transistor. The parasitics of diodes can be incorporated into filter structures to obtain broadband switches and limiters. A multiplexer permits signals at different frequencies to be combined on a single transmission line. Separating the bias and RF ports on a diode switch is a form of multiplexing.

9.1 BASIC FILTER THEORY

The lumped element equivalent circuit of a transmission line is a low-pass filter as shown in Fig. 9.1. At low frequencies C is a high impedance and L is a low impedance so all available power is delivered to the matched load. At high frequencies, L blocks incident power and C shunts it to ground so that incident power is reflected.

The attenuation of this simple circuit can be calculated using $ABCD$ matrices. Figure 9.2 shows this derivation. The attenuation for various values of k is shown in Fig. 9.3. x_1 is the normalized frequency at which the attenuation increases to the ripple maximum.

Given a shunt diode of 0.5 pF and a 0.5 dB maximum insertion loss up to the highest possible frequency, the L and highest frequency (assume Z_0 = 50 ohms) are determined as follows:

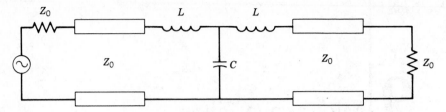

Figure 9.1. Lumped element equivalent circuit.

From Fig. 9.3 for 0.5 dB insertion loss

$$k = 1.71 \tag{9.1a}$$

$$Z_T = kZ_0 = (1.71)(50) = 85.5 \text{ ohms} \tag{9.1b}$$

$$L = (\tfrac{1}{2})Z_T^2 C = (\tfrac{1}{2})(85.5)^2(0.5 \times 10^{-12}) = 1.83 \times 10^{-9} \tag{9.1c}$$

$$L = 1.83 \text{ nH} \tag{9.1d}$$

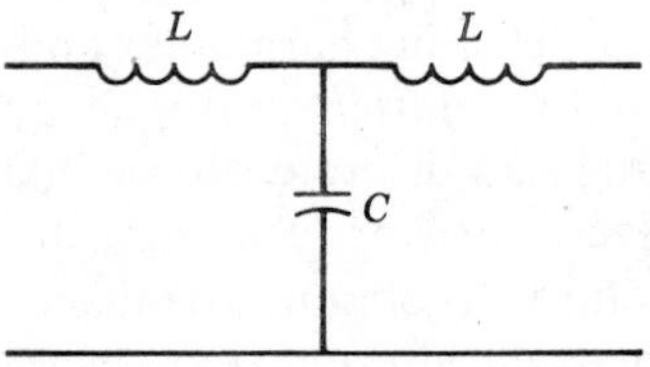

$$\begin{vmatrix} 1 & j\omega L \\ 0 & 1 \end{vmatrix} \cdot \begin{vmatrix} 1 & 0 \\ j\omega C & 1 \end{vmatrix} \cdot \begin{vmatrix} 1 & j\omega L \\ 0 & 1 \end{vmatrix} = \begin{vmatrix} 1 - \omega^2 LC & j\omega L \\ j\omega C & 1 \end{vmatrix} \cdot \begin{vmatrix} 1 & j\omega L \\ 0 & 1 \end{vmatrix}$$

$$= \begin{vmatrix} 1 - \omega^2 LC & j\omega L + j\omega L(1 - \omega^2 LC) \\ j\omega C & 1 - \omega^2 LC \end{vmatrix}$$

$$\alpha = 10 \log \frac{1}{4} [\text{Re}^2 + \text{Im}^2]_N^*$$

$$= 10 \log \frac{1}{4} \left[4(1 - \omega^2 LC)^2 + \left(\frac{\omega C}{Y_0} + \frac{\omega L}{Z_0}(2 - \omega^2 LC) \right)^2 \right]$$

Define $Z_T = \sqrt{\dfrac{2L}{C}}$, $\quad \omega_0 = \sqrt{\dfrac{1}{LC}}$, $\quad k = \dfrac{Z_T}{Z_0}$, $\quad x = \dfrac{\omega}{\omega_0}$

$$\alpha = 10 \log \left\{ (1 - x^2)^2 + \frac{1}{2} \left[\frac{x}{k} + kx \left(\frac{1 - x^2}{2} \right) \right]^2 \right\}$$

* Matrix B and C terms normalized by Y_0 and Z_0 respectively (see Eq. 6.34. 6.38a)

Figure 9.2. Matrix attentuation calculation.

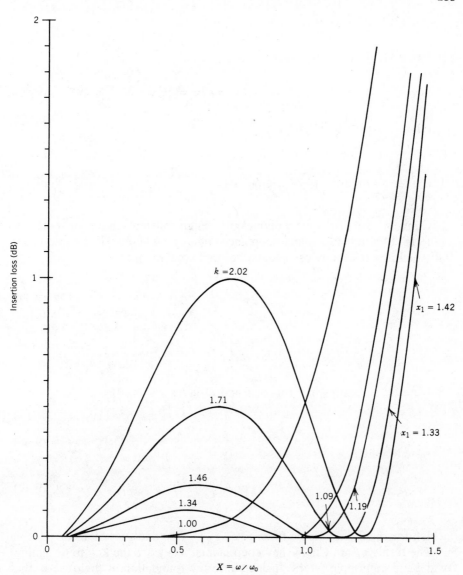

Figure 9.3. *Insertion loss.*

From Fig. 9.3, for $k = 1.71$

$$x_1 = 1.33 \tag{9.2a}$$

$$\omega_0 = \frac{1}{\sqrt{LC}} = \frac{1}{\sqrt{(1.83 \times 10^{-12})(0.5 \times 10^{-1/2})}} = 3.31 \times 10^{10}, \tag{9.2b}$$

$$\omega_1 = x_1\omega_0 = (1.33)(3.31 \times 10^{10}) = 4.40 \times 10^{10} \tag{9.2c}$$

and

$$f = \frac{1}{2\pi}(4.40 \times 10^{10}) = 7.00 \text{ GHz} \tag{9.2d}$$

Conventional filter theory defines the design constants more conveniently in terms of reactances and susceptances of ω_1 (1) where ω_1 is the frequency at which the insertion loss equals the ripple maximum.

$$g_1 = \frac{\omega_1 L}{Z_0} \tag{9.3a}$$

$$g_2 = \frac{\omega_1 C}{Y_0} \tag{9.3b}$$

Solving for g_1 and g_2 in terms of k and x_1 gives

$$g_1 = \frac{kx_1}{\sqrt{2}} \tag{9.4a}$$

$$g_2 = \frac{\sqrt{2}\, x_1}{k} \tag{9.4b}$$

which for the 0.5-dB ripple filter gives $g_1 = 1.60$ and $g_2 = 1.10$.

The terms g_n are Chebyshev coefficients. They are the key to designing wideband equiripple filters. Insights into the foundation of their use in the design of filters and the broad possibilities of their application are given by Riblet (2).

The coefficients are given in Table 9.1 and Fig. 9.4 for various values of maximum insertion loss δ_m and maximum VSWR ρ_m, where n is the number of elements in the filters. Note that the filter derived above using *ABCD* matrices had three elements and also that the filters are symmetrical ($g_1 = g_3$) and tapered at the ends (the middle elements of long filters are alike).

In general, for a low-pass filter $g_k = \omega_1 L_k / Z_0$ and $g_k = \omega_1 C_k / Y_0$. The filter must have an odd number of components and can be started with a series inductor or a shunt capacitor.

The coefficients can be derived using the following equations:

$$g_1 = \frac{2a_1}{\gamma} \tag{9.5a}$$

$$g_k = \frac{4a_{k-1}a_k}{(b_{k-1}g_{k-1})} \tag{9.5b}$$

$$a_k = \sin \pi \frac{(2_{k-1})}{(2n)} \tag{9.5c}$$

$$b_k = \gamma^2 + \sin^2 \pi \frac{k}{n} \tag{9.5d}$$

$$\gamma = \sinh \left[\frac{1}{2n} \ln \left(\coth \frac{\delta_m}{17.37} \right) \right] \tag{9.5e}$$

$$= 20 \log \left(\frac{\rho_m + 1}{2\sqrt{\rho_m}} \right) \tag{9.5f}$$

n is the total number of elements in the low-pass filter prototype and k is the number of specific elements.

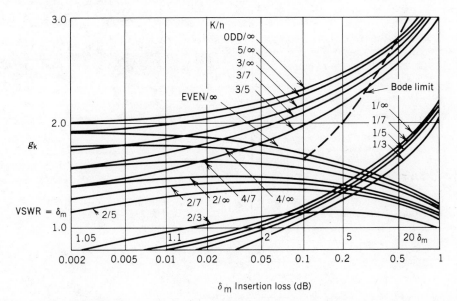

Figure 9.4. g Coefficients versus maximum insertion loss.

Table 9.1 Chebyshev Coefficients

VSWR ρ_m	δ_m (dB)	n	g_1	g_2	g_3	g_4	g_5	(g_{13}) g_6	(g_{13}) g_7
1.05	.0026	(25)	(.7088)	(1.4098)	(1.7621)	(1.7894)	(1.9356)	(1.9177)	(2.0141)
		7	.6624	1.3097	1.6143	1.5924	1.6143	1.3097	.6624
		5	.6181	1.2026	1.4130	1.2026	.6181		
		3	.4861	.8259	.4861				
1.1	.01	(25)	(.8370)	(1.4712)	(1.8663)	(1.7845)	(2.0086)	(1.8818)	(2.0704)
		7	.7953	1.3917	1.7466	1.6329	1.7466	1.3917	.7953
		5	.7547	1.3040	1.5754	1.3040	.7547		
		3	.6274	.9688	.6274				
1.2	.04	(25)	(1.0138)	(1.4950)	(2.0071)	(1.7381)	(2.1219)	(1.8081)	(2.1701)
		7	.9765	1.4353	1.9116	1.6277	1.9116	1.4353	.9765
		5	.9359	1.3677	1.7693	1.3677	.9395		
		3	.8187	1.0896	.8187				
1.36	.1	(25)	(1.2153)	(1.4676)	(2.1761)	(1.6544)	(2.2722)	(1.7053)	(2.3116)
		7	1.1811	1.4228	2.0966	1.5733	2.0966	1.4228	1.1811
		5	1.1468	1.3712	1.9750	1.3712	1.1468		
		3	1.0315	1.1474	1.0315				

1.54	.2	(25)	(1.4047)	(1.4136)	(2.3460)	(1.5632)	(2.4316)	(1.6027)	(2.4661)
		7	1.3722	1.3781	2.2756	1.5001	2.2756	1.3781	1.3722
		5	1.3394	1.3370	2.1660	1.3370	1.3394		
		3	1.2275	1.1525	1.2275				
2.0	.5	(25)	(1.7682)	(1.2826)	(2.6997)	(1.3869)	(2.7748)	(1.4134)	(2.8046)
		7	1.7372	1.2583	2.6381	1.3444	2.6381	1.2583	1.7372
		5	1.7058	1.2296	2.5408	1.2296	1.7058		
		3	1.5963	1.0967	1.5963				
2.66	1	(25)	(2.1975)	(1.1287)	(3.1515)	(1.2033)	(3.2223)	(1.2218)	(3.2501)
		7	2.1664	1.1116	3.0934	1.1736	3.0934	1.1116	2.1664
		5	2.1349	1.0911	3.0009	1.0911	2.1349		
		3	2.0236	0.9941	2.0236				
4.1	2	(25)	(2.8980)	(.9234)	(3.9360)	(.9732)	(4.0079)	(.9853)	(4.0358)
		7	2.8655	0.9119	3.8780	0.9535	3.8780	0.9119	2.8655
		5	2.8310	0.8985	3.7827	0.8985	2.8310		
		3	2.7107	0.8327	2.7107				
Maximally flat = 3 dB at ω_i'		7	0.4450	1.247	1.802	2.000	1.802	1.247	0.4450
		5	0.6180	1.618	2.000	1.618	0.6180		
		3	1.000	2.000	1.000				

The insertion loss can be calculated directly from the following equations:

$$\delta = 10 \log\left\{1 + \epsilon \cos^2\left[n \cos^{-1}\left(\frac{\omega}{\omega_1}\right)\right]\right\} \tag{9.6a}$$

$$\epsilon = 10^{\delta m/10} - 1 = \frac{(\rho_m^{-1})^2}{(4\rho_m)} \tag{9.6b}$$

The isolation (attenuation above ω_1) is calculated using

$$\eta = 10 \log\left\{1 + \epsilon \cosh^2\left[n \cosh^{-1}\left(\frac{\omega}{\omega_1}\right)\right]\right\} \tag{9.7}$$

For isolation greater than 10 dB this is approximated by

$$\eta = 6(n - 2) - 10 \log\left(\frac{1}{\delta_m}\right) + 20n \log\left(\frac{\omega}{\omega_1}\right) \tag{9.8}$$

A low-pass filter is specified by having an insertion loss less than some

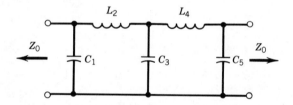

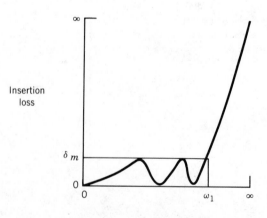

Figure 9.5. Low-pass filter prototype.

maximum up to a certain frequency and an isolation greater than some minimum above another higher frequency. A typical low-pass filter is shown in Fig. 9.5.

9.2 HIGH-PASS FILTER

In order to make a high-pass filter, series capacitors and shunt inductors are used as shown in Fig. 9.6. The elements are defined by

$$\frac{\omega_1 L_k}{Z_0} = \frac{1}{g_k} \tag{9.9a}$$

$$\frac{\omega_1 C_k}{Y_0} = \frac{1}{g_k} \tag{9.9b}$$

and ω_1/ω is used for ω/ω_1 in the response equations.

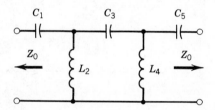

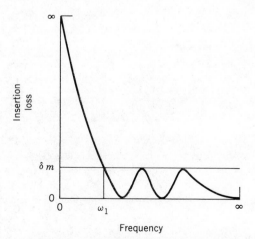

Figure 9.6. *High-pass filter prototype.*

9.3   BANDPASS FILTER

By series resonating shunt elements and parallel resonating series elements, the low-pass filter is made into a bandpass filter as shown in Fig. 9.7. The passband bandwidth remains constant and is simply translated to the higher frequency. The elements must be resonated at ω_0, where ω_0 is defined by

$$\omega_0 = \sqrt{\omega_u \omega_L} \tag{9.10}$$

ω_u and ω_L are, respectively, the upper frequency and lower angular frequency of the passband.

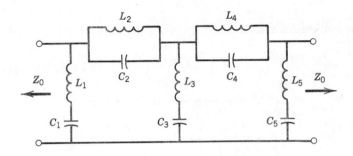

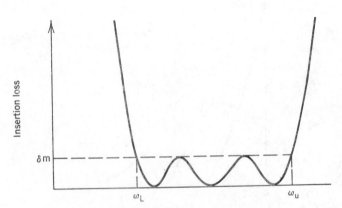

Figure 9.7. *Bandpass filter prototype.*

9.4   TRANSMISSION LINE TRANSFORMATION

There are two relationships that aid in the design of filters using transmission line elements—Richards equation and Kuroda's identity.

9.4.1   Richards Equation

Richards equation (3) allows transmission line elements to be used in filter structures in place of lumped elements, making the realization of filters at microwave frequencies possible. A short-circuited length of transmission line is used for an inductor, and an open-circuited length of transmission line is used for a capacitor. The reactance of an indicator is $X_L = \omega L$, whereas the reactance of a shorted length of transmission line of characteristic impedance Z_{0L} as shown in Fig. 9.8 is

$$X_L = Z_{0L} \tan 2\pi\frac{l}{\lambda} = Z_{0L} \tan 2\pi\frac{lf}{v_g} = Z_{0L} \tan \omega\left(\frac{l}{v_g}\right) \qquad (9.11a)$$

where

$$v_g = \frac{c}{\sqrt{\epsilon_r}} \qquad (9.11b)$$

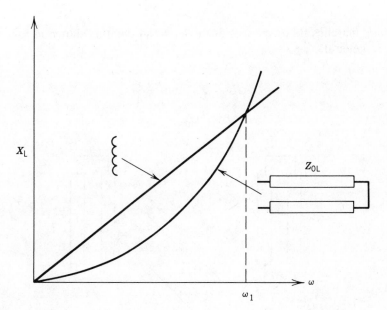

Figure 9.8. Reactance of a lumped inductor and shorted transmission line versus frequency.

Similarly the susceptance of a capacitor and an open circuited line can be compared.

The circuit of Fig. 9.2 can therefore be changed, as represented by Fig. 9.9, when transmission line elements are substituted. For the purpose of filter design, the following equivalence between inductive reactance and reactance of a shorted transmission line element is used:

$$X_L = \omega L = Z_{0L} \tan 2\pi \frac{l}{\lambda} = Z_{0L} \tan\left(\frac{\pi}{2} \frac{\omega'}{\omega'_Q}\right) \tag{9.12}$$

where angular frequency with transmission line elements is designated at ω', and ω'_Q is the angular frequency at which the line length is $\lambda/4$. Similarly, capacitive susceptance is represented by

$$B_c = \omega C = Y_{0c} \tan\left(\frac{\pi}{2} \frac{\omega'}{\omega'_Q}\right) \tag{9.13}$$

The response of a lumped element low-pass filter is shown in Fig. 9.10, and the response of a transmission line element filter is shown in Fig. 9.11. The parameters of the transmission line filter are selected to satisfy

$$\omega_1 L = Z_{0L} \tan\left(\frac{\pi}{2} \frac{\omega'_1}{\omega'_Q}\right) \tag{9.14}$$

which matches the responses at the $\omega_1 = \omega'_1$ point as shown in Fig. 9.8.

In general,

$$\omega L = Z_{0L} \tan\left(\frac{\pi}{2} \frac{\omega'}{\omega'_Q}\right) \tag{9.15}$$

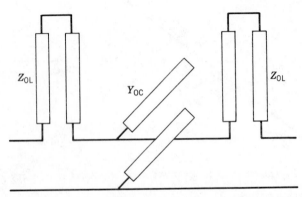

Figure 9.9. *Transmission line representation of a low-pass filter. All lines of length l.*

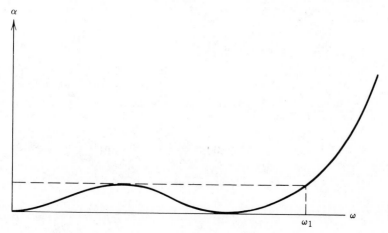

Figure 9.10. *Frequency response of a lumped element low-pass filter.*

The ratio of these two equations defines the transformation of the frequency variable:

$$\frac{\omega}{\omega_1} = \frac{\tan\left(\dfrac{\pi}{2}\dfrac{\omega'}{\omega'_Q}\right)}{\tan\left(\dfrac{\pi}{2}\dfrac{\omega'_1}{\omega'_Q}\right)} \tag{9.16}$$

When selecting transmission elements, the length l is selected to give the desired rejection at a selected frequency. Then Z_{0L} is selected to match the reactance $\omega_1 L$.

For example, to design a low-pass filter that passes up to 1 GHz with 0.5 dB maximum insertion loss and stops 2 GHz, the lines are a quarter-wave-

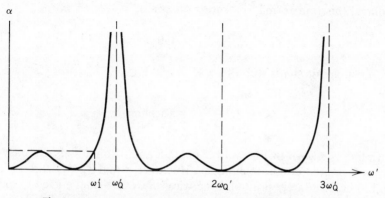

Figure 9.11. *Response of a transmission line low-pass filter.*

length long at 2 GHz:

$$\omega'_1 = 2\pi \times 10^9 \tag{9.17a}$$

$$\omega'_Q = 2\pi \times (2 \times 10^9) \tag{9.17b}$$

$$\tan\left(\frac{\pi}{2}\frac{\omega'_1}{\omega'_Q}\right) = \tan\frac{\pi}{4} = 1 \tag{9.17c}$$

$$g_1 = 1.60 = \frac{\omega_1 L}{Z_0} = \frac{Z_{0L}}{Z_0}\tan\left(\frac{\pi}{4}\right) \tag{9.17d}$$

$$Z_{0L} = g_1 Z_0 = 80 \text{ ohms} \tag{9.17e}$$

$$g_2 = 1.10 = \frac{\omega_1 C}{Y_0} = \frac{Y_{0C}}{Y_0}\tan\left(\frac{\pi}{4}\right) \tag{9.17f}$$

$$Z_{0C} = \frac{Z_0}{g_2} = 45.5 \text{ ohms} \tag{9.17g}$$

This structure might be realized in coaxial transmission line where the series inductors are very small coaxial transmission lines inside the center conductor of the main 50-ohm transmission line. The open circuit shunt line is easy to realize with a shunt T. The circuit would be much easier to make in coax and stripline if all of the lines were shunt Ts. Kuroda's identity permits the filter structure to be so configured.

9.4.2 Kuroda's Identity

Kuroda's identity (4) is the equivalence between two different transmission line structures, as shown in Fig. 9.12. All lines are a quarter-wavelength long. The series inductance stubs of 80 ohms (Z_1) from above are transformed to shunt capacitive stubs using the upper part of Fig. 9.12. Since $Z_0 = 50$ ohms, this transforms to a series section of

$$Z'_0 = Z_0 + Z_1 = 130 \text{ ohms} \tag{9.18}$$

and an open circuit shunt stub of

$$Z'_1 = Z_0 + \frac{Z_0^2}{Z_1} = 81 \text{ ohms} \tag{9.19}$$

The stripline is shown in Fig. 9.13. The 45.5 and 81-ohm lines are easy to realize with printed circuit techniques. However, when $Z_0 > 100$, lines are too thin to be accurately printed. Fine wire, however, can be used to make the higher Z_0 structures. The characteristic impedance of slab line is given

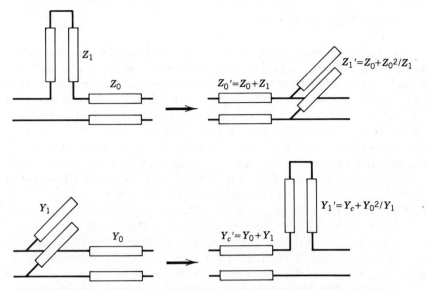

Figure 9.12. Kuroda's identities for two transmission line structures.

by

$$Z_0 = \frac{60}{\sqrt{\epsilon_r}} \ln\left(\frac{4}{\pi} \cdot \frac{D}{d}\right) \qquad (9.20)$$

where d is the diameter of the wire centered between the slabs, D is the distance between the slabs, and ϵ_r is the relative dielectric constant of the stripline board. Conventional stripline board is available with $\epsilon_r = 2.65$ and thickness 0.0625 in. (giving $D = 0.125$ in.). A wire 0.0048 in. in diameter gives the desired 130 ohm Z_0. Note that the length of the wire in the gap should be $\lambda/4$ (i.e., the width of the 45.5-ohm line should not be allowed to shorten these lines). This filter will work at any frequency by appropriately selecting the quarter wavelength.

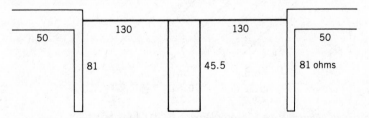

Figure 9.13. Stripline implementation of a low-pass filter.

9.5   FILTER DESIGN EXAMPLE

A 1–2-GHz sweeper has second and third harmonics, which contribute errors to the measurements made with it. A filter is required that gives a 1/2-dB ripple over the band and suppresses the second and the third harmonics by at least 20 dB for most of the band (1.1–1.9 GHz). The harmonics will appear at 2–4 GHz and 3–6 GHz; therefore, it is required to stop 2–6 GHz with 20 dB of attenuation between 2.2 and 5.7 GHz. The center of the stop band is at 4 GHz. When the filter is made up of quarter-wavelength series and short stubs as in Fig. 9.9, the stubs must be a quarter wavelength at $\omega_0 = 2\pi \times 4 \times 10^9$, and they must provide a 1/2-dB equal ripple cutoff at $\omega_1' = 2\pi \times 2 \times 10^9$. The filter must provide 20 dB at $\omega_\eta = 2\pi \times 2.2 \times 10^9$. Equation 9.16 yields $\omega_\eta/\omega_1 = 1.17085$. For $\sigma = 1/2$ dB, $\eta = 20$ dB, and ω_η/ω_1 as above, Eqs. 9.7 and 9.6b yield $n = 7$. That is, a filter is required that has seven elements. Direct realization with $\lambda/4$ sections would be as in Fig. 9.14 by use of Eqs. 9.17d and 9.17f to calculate the characteristic impedances. Note that all of the series and short stubs are in the same electrical plane as in Fig. 9.9. The spaces between Z_{0L1}, Z_{0L3}, etc., in Fig. 9.14 are spaced apart in the figure to carry forward the spatial representation of series inductors and shunt capacitors. The series stubs are not readily constructed in stripline, so Kuroda's identity is repeatedly applied until all the stubs are strung out as shown in Fig. 9.15. The identity is first applied to Z_{0L7} three times, providing the 178-ohms open circuit stub and three quarter-wavelength sections of transmission line with $Z_0 \neq 50$ ohms between the stub and the rest of the filter structure. Next, we apply it to Y_{0c6} two times, being

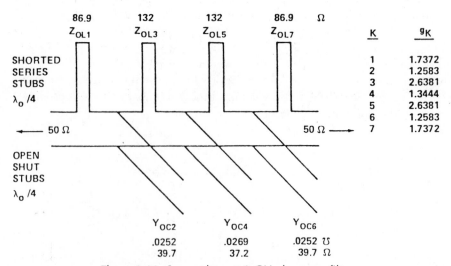

Figure 9.14. *Seven-element 2-GHz low-pass filter.*

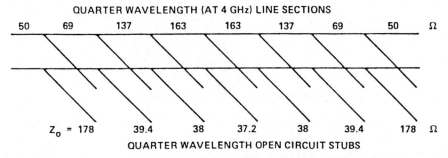

QUARTER WAVELENGTH (AT 4 GHz) LINE SECTIONS

| 50 | 69 | 137 | 163 | 163 | 137 | 69 | 50 | Ω |

| Z_0 = 178 | 39.4 | 38 | 37.2 | 38 | 39.4 | 178 | Ω |

QUARTER WAVELENGTH OPEN CIRCUIT STUBS

Figure 9.15. *2-GHz low-pass filter transformed using Kuroda's identity.*

careful to use the $Z_0 \neq 50$ ohm characteristic impedances. Similarly, Z_{OL5} is done once. Since the filter is symmetrical, the transforming algebra is complete. This filter was derived to demonstrate the application of Kuroda's identity. It is impractical because different $Z_0 > 100$ ohms are hard to realize and the filter takes up too much room.

A more practical design of these frequencies might use a basic capacitive input prototype, as in Fig. 9.5. The shunt capacitance is made up of pairs of open 30-ohm stubs, and the series inductors are made up of 161-ohm series line sections. The element values are derived by Eqs. 9.3a and 9.3b and are given in Fig. 9.16. The inductance per unit length of a line section is given by $L = Z_0/v_g$, where $v_g = 1.843 \times 10^{10}$ cm/sec, the speed of light in the stripline dielectric of 2.65. The capacitance per unit length is given by $C = 1/Z_0 v_g$. The capacitance associated with an inductance line section is $C = L/Z_0^2$ and should be equally placed, half at each end of the line section, as shown in Fig. 9.16. As shown in Fig. 9.17, this filter is 3.5 cm long compared to 6.9 cm for the transformed filter shown in Fig. 9.15. The small filter should not have resonances until the sections are a half-wavelength long—about 15 GHz. The 30-ohm stubs can be etched on the board and the 44 AWG wire soldered on to create the 180-ohm slab line.

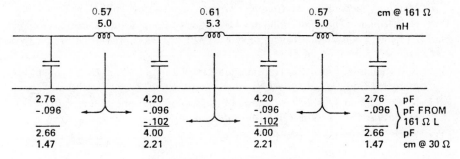

| 0.57 | | 0.61 | | 0.57 | cm @ 161 Ω |
| 5.0 | | 5.3 | | 5.0 | nH |

2.76		4.20		4.20		2.76	pF
-.096		-.096		-.096		-.096	pF FROM 161 Ω L
		-.102		-.102			
2.66		4.00		4.00		2.66	pF
1.47		2.21		2.21		1.47	cm @ 30 Ω

Figure 9.16. *2-GHz low-pass filter.*

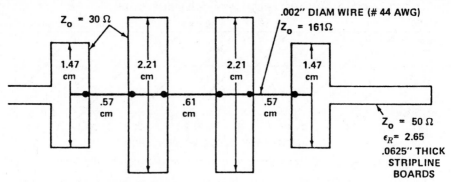

Figure 9.17. *Stripline pattern for a 2-GHz low-pass filter.*

PROBLEMS

1. How high in frequency could you go with 0.5 pF and no filter structure and still get 0.5 dB maximum?

2. Solve Eq. 9.8 for *n*. How many elements are required for a low-pass filter that must provide less than 0.5 dB insertion loss up to 2 GHz and greater than 30 dB isolation above 3 GHz? 2.2 GHz?

3. Design a lumped circuit baseband limiter to 12.4 GHz (0–12.4 GHz) using two shunt limiter chip diodes having 0.33 pF capacitance at zero bias. The diodes are the shunt capacitances of a low-pass filter and the filter is to work in a Z_0 = 50-ohm system. What is the maximum insertion loss of the filter? (Hint: A lower insertion loss results for a five-element filter.)

REFERENCES

1. G. L. Matthaei, L. Young, and E. M. T. Jones, *Microwave Filters, Impedance Matching Networks, and Coupling Structures*, McGraw-Hill, New York, 1964.
2. H. J. Riblet, "The Application of a New Class of Equal-Ripple Functions to Some Familiar Transmission-Line Problems," *IEEE Trans. Micro. Th. Tech.*, Vol. MTT-12, July 1964, p. 415.
3. P. I. Richards, "Resistor-Transmission-Line Circuits," *Proc. IRE,* Vol. 36, February 1948, p. 217.
4. N. Ozaki, and J. Ishii, "Synthesis of a Class of Strip-line Filters," *IRE Trans. on Circuit Theory*, Vol. CT-5, June 1958, p. 104.

10 | Ferrites

Ferrite devices are widely used in microwave systems. The most common ferrite device is the circulator that allows transmitter and receiver to share one antenna. An example of this application is shown in Fig. 10.1a. This is a prototype X-band (8–12.4 GHz) solid-state phased-array module containing GaAs monolithic power amplifiers, low-noise amplifiers, and phase shifters. Two ferrite circulators are seen to the right of the module. A schematic diagram of the module is shown in Fig. 10.1b. In addition to allowing the transmit and receive channels to share the same antenna, the circulator enables the output power amplifiers to see an impedance that is constant and independent of the beam pointing angle of the phased array. Placing a matched termination on one of the three ports of the circulator allows it to function as an isolator to protect oscillators and amplifiers from load variations. Other common ferrite devices include phase shifters, which electronically control the beam direction of a phased array antenna, and yttrium iron garnet (YIG) resonators, which are used in microwave sweep oscillators and tunable filters. An understanding of ferrite materials is necessary to see how these devices work (1,2,3).

10.1 FERRITE MATERIALS

All materials can be grouped into one of two magnetic classes. Some behave linearly. When a magnetic field, H, is applied, the materials show a linear change in flux density, B, as shown in Fig. 10.2a. A nonmagnetic material such as free space (i.e., a vacuum) is represented by the diagonal line. B in

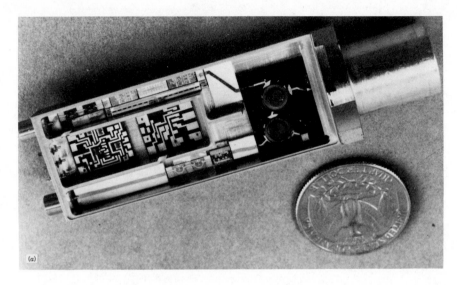

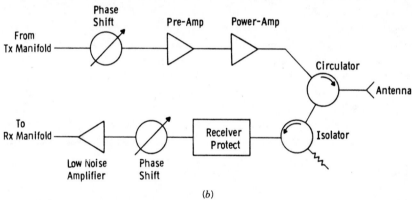

Figure 10.1. X-band solid state phased array transmit-receive module; (a) Prototype; (b) schematic.

a vacuum is proportional to H by the constant μ_0 $(4\pi \times 10^{-7}$ H/m). Relative permeability μ_r is used for convenience to describe the characteristics of magnetic materials so that

$$B = \mu_0\mu_r H \qquad (10.1)$$

The relative permeability of free space is 1, and linear magnetic materials generally have a relative permeability slightly greater or less than 1, corresponding to paramagnetic or diamagnetic material, respectively. For these weakly magnetized materials, the small difference of μ_r from unity is often expressed as a magnetic susceptibility, χ_m, so that

$$\mu_r = 1 + \chi_m \tag{10.2}$$

χ_m ranges from -10^{-5} for diamagnetic to $+10^{-5}$ for paramagnetic materials.

Magnetic properties are frequently specified in CGS units instead of the MKS units used in Fig. 10.2. A flux density of 10^{-4} Wb/m (MKS) equals a

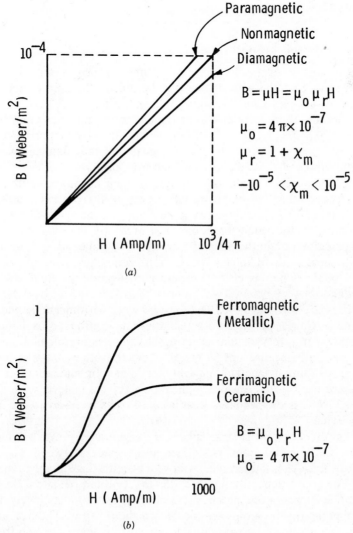

Figure 10.2. Type of magnetic materials. (a) Linear; (b) nonlinear.

gauss (CGS), whereas a magnetic field of 79.6 A/m ($1000/4\pi$) equals 1 Oe. The Earth's magnetic field is about 40 A/m (1/2 Oe).

Ferrites are nonlinear magnetic ceramic materials with B–H characteristics shown in Fig. 10.2b, which saturate at 0.5 Wb/m^2 or less. Metallic materials such as iron and nickel are ferromagnetic and saturate at about 2 Wb/m^2 and 0.5 Wb/m^2, respectively. Ferri- and ferromagnetic materials have a relative permeability μ_r, which depends on the magnetic field, H, and can have values in the range 1 to >10,000. In order to understand these characteristics, we must look more closely at the behavior of ferri- and ferromagnets.

10.1.1 Material Structure

Magnetism results from the spin of electrons. In most atoms the electrons are paired so that the spin of one is balanced by the spin of another and there is no net spin. But in atoms of magnetic materials like iron, nickel, and chromium, there are single unpaired electrons and a net spin. This spinning electron is like a current in a very small wire loop and produces a field at its center that is equivalent to a very small magnetic dipole. One of these magnetic dipoles is associated with each iron atom.

Free magnets tend to align themselves so their north and south poles are touching each other. In solid materials, however, there are very strong crystal (exchange) forces that hold magnetic dipoles of the atoms aligned so they are pointing in the same direction. In a magnetic material, the exchange forces can either align the magnetic dipoles all parallel giving a ferromagnet or align them antiparallel giving an antiferromagnet. Iron and nickel are ferromagnetic, whereas chromium and nickel oxide are antiferromagnetic (nonmagnetic).

Ferrites are ferrimagnetic. They are between ferromagnetic and antiferromagnetic. Some sites in the crystal lattices have stronger magnetic dipoles than others in a very regular pattern. The dipoles are aligned in opposite directions but, because one is greater than the other, there is a net, but smaller, resulting magnetization. Only ferrimagnetic materials formed from oxides are used in microwave applications; typical materials are shown in Fig. 10.3. YIG is of special interest because of its widespread use in microwave devices.

In the YIG crystal the large spherical oxygen atoms form a cubic framework, with the metal atoms positioned in the spaces between. The arrangement of the oxygen atoms results in spaces with three distinct shapes and sizes. For convenience these are called (a), (c), and (d) sites. Because the size of the yttrium atom matches that of the (c) sites most closely, all of the nonmagnetic yttrium atoms are located on the (c) sites. The iron atoms are distributed between the (a) and (d) sites so that there are two iron atoms on the (a) sites and three iron atoms on the (d) sites. The exchange forces make all of the dipoles on the (a) sites line up together and those on the (d)

- Oxides (Ferrites)

- $NiFe_2O_4$, $MgFe_2O_4$, $LiFe_5O_8$, $Y_3Fe_5O_{12}$

- $Y_3Fe_5O_{12}$ — Yttrium Iron Garnet (YIG)

$$Y_3 \quad Fe_2 \quad Fe_3 \quad O_{12}$$

(c)   (a)   (d)
Site   Site   Site

Exchange Force Makes

— All Dipoles on (a) Sites Parallel

— All Dipoles on (d) Sites Parallel

— Dipoles on (a) Antiparallel to (d)

(d) Site ——— ⟶
         } ⟶ Total
(a) Site ◀———

Figure 10.3. *Ferrimagnetic materials.*

sites line up together, but the dipoles in (*a*) and (*d*) are in opposite directions. The (*d*) site dipoles are stronger because there are three iron atoms as opposed to two at the (*a*) site. There is thus a total magnetization corresponding to a single uncompensated iron atom.

Most microwave ferrite materials have polycrystalline structures made by sintering together powders of different oxides. Figure 10.4 shows the solid-state chemical reaction for nickel ferrite and yttrium iron garnet. Other common materials are magnesium manganese ferrite and lithium ferrite. The fabrication process is as follows. The prereacted metal oxide materials are mixed into a slurry and ground to obtain microparticles of a controlled size. They are then pressed into the correct shape and dimension as with a ceramic pottery piece. Next they are fired at a very high temperature (1500°K) causing the particles to fuse together into a polycrystalline structure. The dimensions of the original microparticles are retained in the fusing. The fired ferrite is then ground to its final dimensions.

10.1.2 Direct Current and Low-Frequency Effects

As mentioned earlier, the exchange forces cause the magnetic dipoles in a ferrimagnetic material to be aligned parallel, thus producing a net magnetic moment. As the volume of the material is increased, however, it eventually becomes favorable for the magnetic moment in some areas to change direction so as to minimize the magnetic field energy outside the material.

Nickel Ferrite

$$NiO + Fe_2O_3 \longrightarrow NiFe_2O_4$$
Solid
State
Reaction

Yttrium Iron Garnet

$$1.5\, Y_2O_3 + 2.5\, Fe_2O_3 \longrightarrow Y_3Fe_5O_{12}$$
Solid
State
Reaction

Figure 10.4. Chemistry of ferrite formation.

These areas of differently aligned magnetic moment are called magnetic domains. Figure 10.5 shows an example of domains in the absence of an applied magnetic field so that there is zero total magnetic moment. When an external magnetic field is applied, the walls of the magnetic domains move. Those domains pointing in the direction of the applied field grow in size, whereas those pointing in other directions diminish. When the external field is strong enough, the dipoles all line up with it producing a saturated magnetization in the ferrite.

When an alternating magnetic field is applied at low frequencies, the domains respond to it giving a high permeability. This effect is useful up to about 200 MHz in ferrites that are insulators, but for conducting metals (ferromagnets) this is possible only up to a few kilohertz. At higher frequencies, the domain walls do not have time to move and the permeability approaches that of other nonmagnetic materials—relative permeability of one. Permeability is normally complex. The real part corresponds to stored energy or inductance, and the imaginary part corresponds to loss or resistance. Figure 10.6 shows the frequency dependence of the initial permeability (μ_i) of a typical ferrite material. Initial permeability is measured close to zero applied magnetic field where the domain walls are just beginning to move from their equilibrium state. At low frequencies, the permeability is high because the domains can follow the applied field. As frequency increases, domains fall behind and encounter much friction in trying to follow, contributing to loss. At very high frequencies, the domains cannot follow and there is no loss from trying to follow.

Because of the high permeability of ferrites at low frequencies, they are very useful in making miniature transformers for general electronic equipment and inductors (antennas) in AM radios.

10.1.3 Microwave Effects

In order to understand the behavior of ferrite materials at microwave frequencies, we must look more closely at the effect of a magnetic bias field on a spinning electron.

When a DC magnetic field is applied to a spinning electron its magnetic moment precesses like a gyroscope in a gravitational field as shown in Fig. 10.7. Also (like the gyroscope), it precesses at a rotational frequency depending on the strength of the magnetic field. This coupling between the gyroscopic character of a ferrite magnetic dipole and applied magnetic field is called gyromagnetic coupling, and the relationship between the microwave precessional frequency (ω) and DC magnetic field (H) is termed the gyromagnetic ratio, typically 35 kHz/A/m (2.8 MHz/Oe); that is, $f = \gamma H$. The

● Portions of Material with Their Net Dipole Moments Aligned Parallel Are Domains

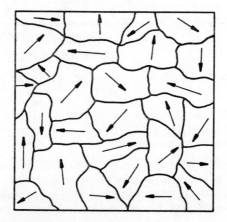

● No External Applied Field
 — Domains Have Random Orientation
 — No Net Magnetic Moment
 — Not True for Permanent Magnets

● Applied Static External Field
 — Aligns Domains
 — Growth of Favorably Oriented Domains

● Alternating External Field (Low Frequency)
 — Effects Due to Domain Wall Motion and Domain Rotation
 — Useful Up to ~ 200 MHz in Ferrites
 — Limited by Induced Currents in Metals to Few kilohertz

Figure 10.5. *Magnetic domains and the influence of external fields.*

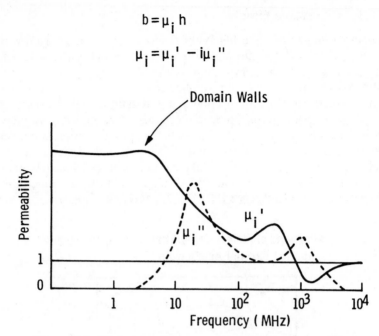

Figure 10.6. Typical frequency spectrum of initial permeability.

direction of electron precession is determined by the direction of the DC magnetic field and when the DC magnetic field strength is changed the precession frequency changes.

Useful microwave devices require a means to couple the microwave energy into the spins to produce an interaction (Fig. 10.8). The best coupling is obtained when the microwave magnetic field is perpendicular to the applied DC magnetic field and the microwave frequency is at the precession frequency. It is like pushing a child on a swing. Pushing at the right frequency and time gives the biggest swing. When the microwave frequency is at the precession frequency, the precession is in resonance.

The gyromagnetic resonance causes the permeability to become a much more complicated factor. It is characterized by a three-by-three tensor (a form of matrix) as shown in Fig. 10.8. In this tensor, $\mu = \mu_0(1 + X_{xx})$, which gives the microwave flux density in the X (or Y) direction produced by a microwave magnetic field in the X (or Y) direction. $jK = \mu_0 X_{xy}$ and relates the microwave flux density in the Y (or X) direction to the microwave field in the X (or Y) direction:

$$\chi_{xx} = \chi_{yy} = \frac{\gamma^2 HF}{(\gamma H)^2 - f^2} \tag{10.3}$$

$$\chi_{xy} = \chi_{yx} = \frac{jf\gamma M}{(\gamma H)^2 - f^2} \tag{10.4}$$

Microwave Effects

- Smaller Than Low Frequency Effects
- Due to Precession of Electron Spins

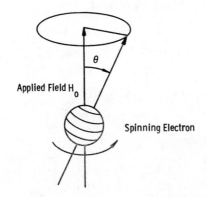

- Electron Spin System in a Magnetic Field is Analogous to a Precessing Gyroscope in the Earth's Gravity Field

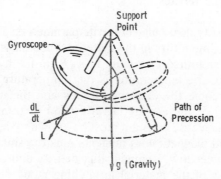

Figure 10.7. *Magnetic moment precession.*

where f is the frequency, M is the saturation magnetization, γ is the gyromagnetic ratio, and H is the applied DC field. The variation of the real and imaginary parts of the susceptibilities χ_{xx} and χ_{xy} are shown in Fig. 10.9. The frequency was fixed at 2.8 GHz, and the tensor susceptibility was found as the field was varied. Resonance linewidths (ΔH) of 16,000 and 40,000 (A/m) are shown. ΔH is a measure of the resonance Q and thus the losses in the ferrite material. Resonance linewidths of less than 800 A/m are

- Microwave Magnetic Field Perpendicular to Static Bias Field

- Coupling Maximum When Microwave Frequency is Same as Precession Frequency of Spin

- Resonance — F $=$ γ H

  Microwave   Gyromagnetic   Static Bias
  Frequency   Ratio   Field

- Gyromagnetic Ratio = 2.8 MHz/Oe

  or   = 35.18 kHz/A/m

- Tensor Permeability

$$
\begin{bmatrix} b_x \\ b_y \\ b_z \end{bmatrix} = \begin{bmatrix} \mu & -jK & 0 \\ jK & \mu & 0 \\ 0 & 0 & 1 \end{bmatrix} \begin{bmatrix} h_x \\ h_y \\ h_z \end{bmatrix}
$$

Figure 10.8. *Coupling of microwaves to spins.*

possible in polycrystalline ferrites and less than 40 A/m in single crystal YIG.

Figure 10.10 shows the typical range of ferrite parameters. The saturation magnetization is the maximum flux in the material when all of the domains are aligned. This is the saturated maximum flux (B) in the hysteresis loop of the B–H curve. The Curie temperature is the temperature at which the thermal vibrations overcome the exchange forces and the material is no longer ferrimagnetic. Since these materials are used in transmission line structures, the relative dielectric constant must be known to design matched devices. The electric and magnetic loss tangents must be small to keep the insertion loss of the devices low. Remnant flux density and coercive force determine the usefulness of the material in latching mode devices, which will be discussed later.

10.1.4   Circular Polarization and Faraday Rotation

The resonance responses shown in Fig. 10.9 were obtained by placing the ferrite in a linearly polarized RF magnetic field perpendicular to the DC applied magnetic field. Since the interaction of the microwave magnetic field with the ferrite is strongly affected by the polarization, this will be discussed here.

When the RF magnetic field is on a line, sinusiodally varying between plus and minus, it is linearly polarized as shown in Fig. 10.11a. When it appears as a vector in space of constant amplitude pivoting around its base point, it is circularly polarized. It can circle to the left (negatively polarized) or to the right (positively polarized) as shown in Fig. 10.11b. Circular polarization has a strong interaction only when the rotation of the vector is in the same direction as the spin of the magnetic dipole. When the vector is rotating in the opposite direction, there is practically no interaction. The variations of susceptibility with magnetic bias field for positively and negatively polarized waves are shown in Fig. 10.12. Note that linear polarization can be represented by the sum of two equal amplitude but oppositely rotating circular polarizations.

When a linearly polarized wave, for example, a TEM wave, travels through a ferrite material in the same direction as the magnetic bias field,

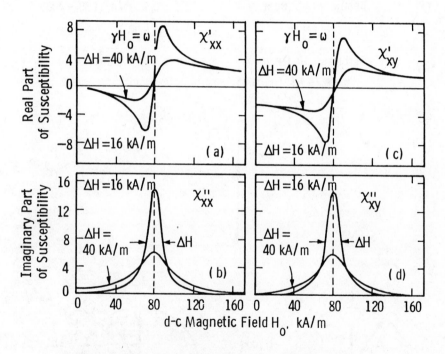

Real and imaginary parts of the susceptibility for a ferrite as a function of the magnetic bias field and two different resonance linewidths (ΔH)

— Microwave Frequency 2.8 GHz
— Saturation Magnetization (240 kA/m)

Figure 10.9. *Complex susceptibility for a ferrite.*

Symbol	Property	Range
$4\pi M_s$	Saturation Magnetization	12-400 kA/m (150 − 5000 Gauss)
T_c	Curie Temperature	$100 - 640\ °C$
ΔH	Resonance Linewidth	800 − 80, 000 A/M (10 − 1000 Oe)
ΔH_k	Spinwave Linewidth	80 − 1600 A/M (1 − 20 Oe)
ϵ'	Dielectric Constant	$12 - 18$
$\tan\delta_\epsilon$	Dielectric Loss Tangent	.0001 − .002
μ'	Permeability	$4\pi \times 10^{-7}$ Henry/m (App. Unity)
$\tan\delta_\mu$	Magnetic Loss Tangent	.0001 − 1
Br	Remnant Flux Density	.005 − 0.4 Wb/m^2 (50 − 4000 Gauss)
H_c	Coercive Force	16 − 240 A/M (0.2 − 3 Oe)
γ_{eff}	Gyromagnetic Ratio	35 kHz A/M (2.8 MHz/Oe)

Figure 10.10. *Ferrite characteristics used for devices design.*

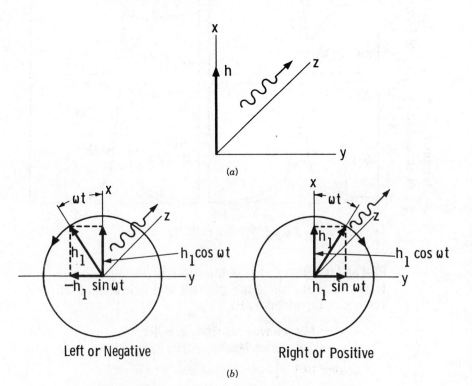

Figure 10.11. *(a) Linear and (b) circular polarization.*

the polarization of the wave is changed. This is called Faraday rotation and can be understood by referring to Fig. 10.13. A linearly polarized TEM wave enters the ferrite of length d at $z = 0$. The microwave magnetic field (h_{xo}) at $z = 0$ is in the x-direction and can be considered as composed of two oppositely rotating circularly polarized waves; that is, at $z = 0$,

$$h_{xo} = h_{ro} + h_{lo} \tag{10.5}$$

and

$$h_{yo} = 0 \tag{10.6}$$

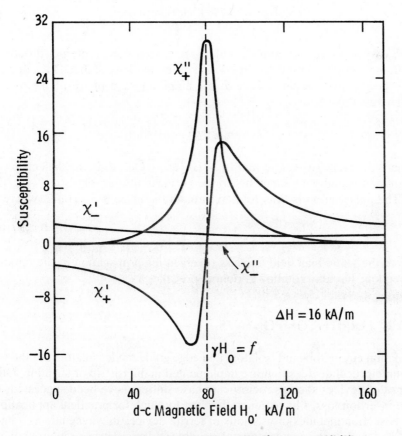

Real and imaginary parts of the effective susceptibility
for circularly polarized waves in a ferrite

- Large Effect for Positive Polarization

- Negligable Effect for Negative Polarization

Figure 10.12. *Complex susceptibility for circularly polarized waves.*

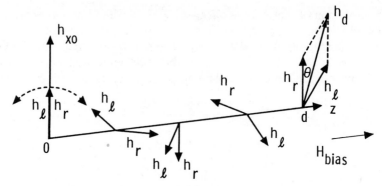

Figure 10.13. Faraday rotation.

The two oppositely rotating circularly polarized waves interact differently with the ferrite so that they experience different phase shifts. Thus, on leaving the ferrite material at $z = d$, the amplitude and phase of the linearly polarized wave h_d is

$$h_d = h_{ro}\underline{/\beta_r d} + h_{lo}\underline{/\beta_l d} \qquad (10.7)$$

where $\beta_r = \omega\sqrt{(\mu + k)\epsilon_r}$ and $\beta_l = \omega\sqrt{(\mu - k)\epsilon_r}$. ϵ_r is the dielectric constant and μ and k are elements of the permeability tensor.

The polarization is thus rotated through the angle $\theta = \Delta\beta d/2$, where

$$\Delta\beta = \beta_r - \beta_l \qquad (10.8)$$

Since the static bias field appears reversed for propagation in the opposite direction, Faraday rotation is nonreciprocal.

10.2 FERRITE COMPONENTS

Junction circulators and isolators and edge guide mode isolators are the only nonreciprocal devices that are implemented in microstrip or stripline. Other types of devices such as reciprocal phase shifters have been suggested, but the insertion losses have generally been too high for practical applications. This section includes discussions of ferrite devices in waveguide as a background leading to consideration of stripline and microstrip circulators, which are of special interest here.

10.2.1 Isolators

One of the earliest applications of gyromagnetic interaction in ferrites was the Faraday rotation isolator shown in Fig. 10.14. Here the input rectangular

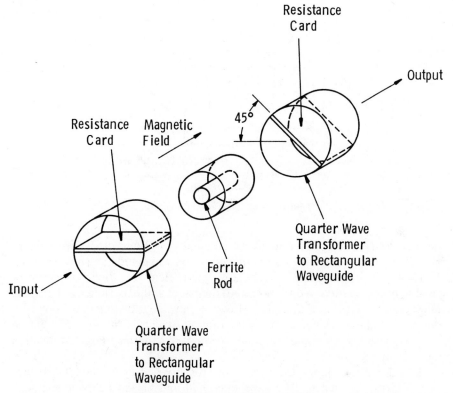

Figure 10.14. *Faraday rotation isolator.*

waveguide has a transition to a circular waveguide so that the electric field of the linearly polarized wave is perpendicular to the resistive card and is thus unattenuated. As the wave travels through the ferrite, its polarization is rotated through 45° in a clockwise direction by the Faraday effect. Thus the electric field of the linearly polarized wave is again perpendicular to the resistive card and is not attenuated. However, a wave reflected back into the output port by a mismatched load will experience a rotation in the ferrite of 45° in a counterclockwise direction, and thus the electric field of the reflected output will be parallel to the resistor card and will be highly attenuated.

Another example of an isolator is the resonance isolator shown in Fig. 10.15. Operation of this device can be understood by considering the microwave magnetic field of the TE_{01} mode in an empty waveguide of width w. In the middle of the waveguide, the magnetic field is linearly polarized in the x-direction and alternates in sense as the wave travels along. However, at approximately $x = w/3$ and $x = 2w/3$ the magnetic fields appear to be circularly polarized. Looking from below in Fig. 10.15 (i.e., along the positive z-direction) and with wave propagation along the positive y-direction,

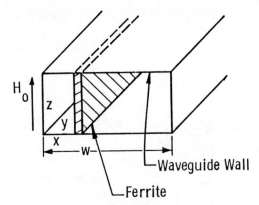

Figure 10.15. *Resonance isolator.*

the polarization is counterclockwise when $x = w/3$ and clockwise when $x = 2w/3$. Thus, if a ferrite slab is placed in the waveguide as shown in Fig. 10.15 at $x = w/3$ and is biased by a magnetic field H_o, then a wave propagating in the positive y-direction (into the page) will experience a negligible interaction (see χ''_- in Fig. 10.12). However, a wave traveling in the opposite direction will experience a strong interaction, which, if the ferrite is biased near its gyromagnetic resonance frequency, will result in a large attenuation (see χ''_+ in Fig. 10.12).

A modification to this structure, with a resistive card placed on the surface of the ferrite closest to the waveguide wall and biased below resonance, results in a field displacement isolator that generally has a wider bandwidth than resonance mode devices.

10.2.2 Phase Shifters

Without the resistance card the field displacement device acts as a differential phase shifter. The wave in the direction of high interaction sees a high susceptibility and experiences more phase delay than the wave traveling in the other direction. For high-power applications, the ferrite can be placed on the bottom waveguide wall giving less interaction but dissipating heat from the ferrite losses much more readily. The phase shifter can be made reciprocal by use of two symmetrically placed ferrite slabs in the waveguide.

10.2.3 Circulators

Two of these nonreciprocal differential phase shifters can be combined with 90° and 180° 3-dB couplers to make up a four-port circulator as shown in Fig. 10.16. Power entering port 1 is split by the magic T into two equal parts entering the two waveguide differential phase shifters in phase. The right-hand one is delayed 90° upon entering the sidewall coupler. The wave en-

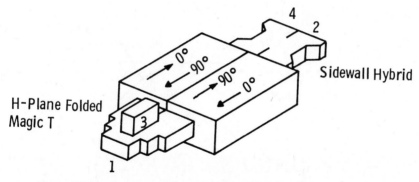

Figure 10.16. *Four-port differential phase shift waveguide circulator.*

tering the left side of the sidewall coupler is delayed 90° in moving over to the right side, making it in phase with the wave that came through the right waveguide section. Therefore, the waves add to emerge from port 2. Power entering from port 2 splits, with the left side (coming toward the observer) having an additional 90° delay because of crossing over. After the delay differential through the waveguides, the left side has 180° delay compared to the right side. Out-of-phase waves into the magic T add and emerge from port 3. A continuation of the logic shows that power entering port 3 emerges from port 4 and power entering port 4 emerges from port 1.

A three-port junction circulator of the type most commonly used in modern low- and medium-power applications is shown schematically in Fig. 10.17. Power entering port 1 emerges from port 2. Power entering port 2 comes out port 3, and power entering port 3 comes out port 1. When the transmitter is used to share an antenna, it is connected to port 1 and the antenna to port 2. The transmitter power thus is conveyed from the transmitter to the antenna. Received power from the antenna enters the circulator and is conveyed to port 3 where the receiver is placed. In order to use the circulator as an isolator, the amplifier output is connected to port 1 and the power conveyed to the load on port 2 with very little loss. Reflections from the load reenter the circulator at port 2 and are conveyed to port 3 where a matched load is placed. The reflections are absorbed by the matched load and no power goes back into the amplifier output on port 1. A four-port circulator as described above has an additional port. The additional port can be useful when the transmitter and receiver are sharing the antenna. By having the fourth port between the receiver and the transmitter and having a matched load on it an extra amount of isolation is available.

Examples of circulators are shown in Fig. 10.18. In the waveguide and stripline cases, *a* and *b*, the ferrite cylinder or disk acts as a dielectric resonator, which is strongly coupled to the input waveguide or stripline and thus has a broad bandwidth (low *Q*). The lowest dielectric resonant fre-

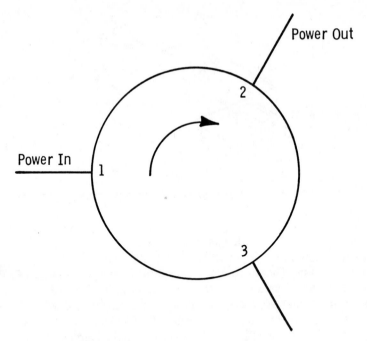

Figure 10.17. *Junction circulator.*

quency of the magnetized ferrite disk is approximately

$$\beta_r = 1.84 \qquad\qquad (10.9)$$

where $\beta^2 = \omega^2 \epsilon_0 \epsilon_r \mu_0 (\mu^2 - k^2)/\mu$.

This is a dipolar mode in which the electric fields are perpendicular to the plane of the disk and the microwave magnetic fields are in the plane of the disk, as shown in Fig. 10.19a. Thus, ports 2 and 3 will see voltages that are 180° out of phase with the input voltage and approximately half the amplitude of the input voltage at port 1.

The (linearly polarized) resonance of Fig. 10.19 can be considered to be made up of two circularly polarized waves of opposite sense. When a magnetic field is applied normal to the ferrite disk, one of the circularly polarized waves rotates in the same direction as the gyromagnetic dipoles (+), and the other is in the opposite direction (−). Thus, the field patterns of both waves are rotated in opposite directions. These two waves then add up to give the rotated field distribution shown in Fig. 10.19b so that the E-field vector at port 3 is very small (ideally zero) and therefore the port is isolated. The voltage at port 2 is almost equal to the input voltage at port 1, therefore, giving a low-insertion loss.

The center frequency of the circulator is given by Eq. 10.9, and the max-

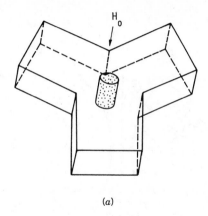

(a)

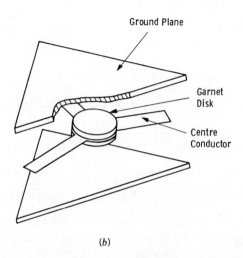

(b)

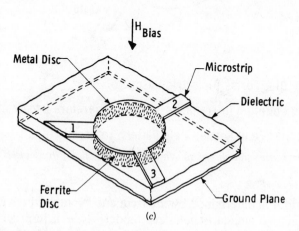

(c)

Figure 10.18. *Circulator structures. (a) Waveguide version; (b) stripline version; (c) microstrip version.*

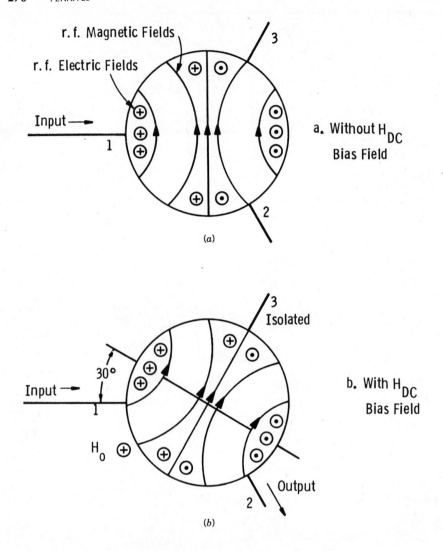

- Ferrite Disc Forms Dielectric Resonator
- Strong Coupling to Transmission Line Therefore Broadband
- Static Bias Field Rotates Field Pattern in Ferrite Disc

Figure 10.19. *Ferrite calculator. (a) Without H_{DC} bias field; (b) with H_{DC} bias field.*

imum bandwidth is determined by the frequency splitting between the res-onances of the two circularly polarized modes $(+, -)$. In order to use this bandwidth fully, it is necessary to match the input and output impedance of the circulator to 50 ohms to within some specified VSWR over the band-width. The isolation, η, between ports 1 and 3 and the insertion loss, δ, between ports 1 and 2 of a lossless circulator are given by

$$\eta = 10 \log\left(\frac{VSWR + 1}{VSWR - 1}\right)^2 \tag{10.10}$$

$$\delta = 10 \log\left(\frac{(VSWR + 1)^2}{4 (VSWR)}\right) \tag{10.11}$$

Thus a VSWR of less than 1.22 is required over the operating bandwidth in order to achieve an isolation of greater than 20 dB. The required matching is achieved by control of the resonator metalization geometry in a stripline circulator and use of multisection matching transformers on each port. Resonator metallization geometries other than disks, for example, triangles, are often used to prevent coupling to higher-order modes in broadband devices.

In practice, waveguide circulators commonly cover the waveguide bandwidth, for example, 2–4 GHz or 8.2–12.4 GHz. However, multioctave bandwidths can be achieved with stripline devices. For example, devices covering the 6–18-GHz frequency range with insertion losses of less than 1 dB and isolations greater than 20 dB are available commercially.

Microstrip circulators function in a manner similar to stripline circulators and can be physically equivalent to one-half of a stripline device, as shown in Fig. 10.18c. In this structure, a ferrite and garnet disk is inserted into a hole in the microstrip dielectric and the resonator metallization on the ferrite disk is bonded to the microstrip lines.

A more easily fabricated structure is obtained if the resonator metallization and the matching transformers are defined on the ferrite substrate. Then the center frequency is defined by the dimensions of the resonator metallization and the geometry of the permanent bias magnet, which produces a localized field at the junction. A more convenient way of obtaining microstrip compatible circulators or isolators is through use of plug-in stripline circulators, which have a built-in bias magnet and are bonded into the microstrip circuit.

Polycrystalline YIG ($4\pi M_s$ = 1760 Oe) is most often used in circulators at microwave frequencies below 20 GHz, although square loop ferrites are used when a latching or switching operation is required. In the low millimeter wave range, below 50 GHz, higher $4\pi M_s$ ferrites such as NiZn ferrites with a $4\pi M_s$ of 5000 Oe are suitable. However, optimum operation at higher millimeter wave frequencies requires a material with a $4\pi M_s$ higher than 5000 Oe, which is not available at present. An additional challenge to the ferrite device designer is compatibility with the ever-increasing bandwidth of GaAs monolithic integrated circuits.

10.2.4 Ferrite Switches and Latching Devices

The circulator can be made into a switch by reversing the direction of the DC magnetic field. When this is done, rotation goes in the opposite direction and an isolated port becomes an output port.

Latching ferrite devices are made with closed magnetic paths. When the magnetic path is a closed loop as shown in Fig. 10.20, the $B–H$ curve of certain ferrites demonstrates a strong hysteresis. When the applied DC magnetic field is removed, the flux remains at some high value, B_R, nearly as high as the saturated flux, and thus the ferrite device is self-biasing. In practice, the A/l ratio should be high, then the current driver can be a single wire run down the center of the hollow ferrite slab. The wire can be driven by a current pulse since it does not have to be retained at high current for a long time. If the hollow ferrite bar shown in Fig. 10.20 were the inside of a waveguide phase shifter, the phase shift would be switched in a latching mode by current pulses on the wire passing through the center of the ferrite.

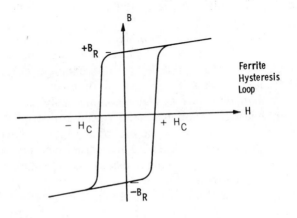

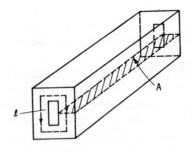

- Fast Switching
- Phase Shifters
- Circulators

Figure 10.20. Latching ferrite switch.

10.2.5 Filters

Another very useful ferrite device is the YIG resonator, which is normally a single crystal sphere, used because its gyromagnetic resonance has a very high Q. Single crystals of YIG are made by dissolving Y_2O_3 and Fe_2O_3, which form YIG, in a solvent comprising a mixture of PbO, PbF, and B_2O_3 at a temperature of 1200°C or higher. Once the Y_2O_3 and Fe_2O_3 are dissolved, the temperature of the solution is slowly reduced by a few degrees a day. Crystals nucleate on the walls of the container and grow larger as the temperature is reduced. Crystals of 2.5 cm diameter or greater have been produced using this technique.

The grown crystal is then cut to the approximate size desired and put into a tumbler for semiprecious stones with an abrasive powder. The tumbling action causes the YIG to become a perfect sphere. Successively finer abrasives are used until the sphere has a very smooth surface. Because the YIG is a single crystal, the gyromagnetic resonance has a very high Q. When the sphere is put into a DC magnetic field with an AC magnetic field perpendicular to it, as shown in Fig. 10.21, the AC energy is tightly coupled to the gyromagnetic resonance and is strongly absorbed. The structure is a high Q absorption or band stop filter, and since the frequency of resonance is tuned by the magnetic field as dictated by the gyromagnetic ratio, it can be tuned over a very broad (decade) band.

A bandpass structure using a YIG resonant sphere is shown in Fig. 10.22. The input and output loops are orthogonal to each other so that away from gyromagnetic resonance there is no coupling between them. When the DC magnetic field is applied to the YIG there is gryomagnetic coupling between them in the narrow frequency range of the gyromagnetic resonance. This bandpass filter is again tunable over a wide frequency range by the magnetic bias field. The variable bias field is supplied by a compact electromagnet that surrounds the microwave circuitry. YIG resonators are used to tune oscillators giving a very linear relationship between tuning (DC) current and frequency at frequencies up to 40 GHz. YIG-tuned bandpass filters are used as preselectors in spectrum analyzers, and both bandpass and bandstop filters find wide application in radar and EW systems.

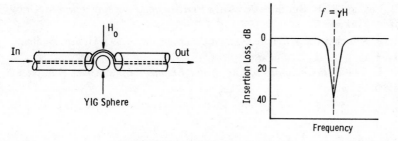

Figure 10.21. *YIG bandstop filter.*

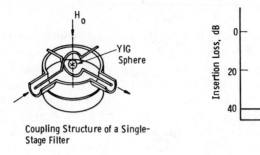

Coupling Structure of a Single-
Stage Filter

Figure 10.22. YIG bandpass filter.

PROBLEMS

1. What is the precession frequency of a ferrite material with a magnetic bias field of 142.8 kA/m?

2. Draw the microwave magnetic field lines looking down on the broad face of a TE_{01} mode waveguide of width w and show that a circularly polarized magnetic field is obtained at a spacing of $w/3$ from both side walls.

3. A Faraday rotation isolator containing a ferrite with $\epsilon = 13\epsilon_0$, $M = 160$ kA/m, and $\gamma = 35$ kHz/A/m is being designed to operate at 3 GHz with a magnetic bias field of 40 kA/m. Calculate the length of ferrite required to give 45° rotation.

4. **a.** Calculate the magnetic bias field change required to tune a YIG filter from 8 to 12 GHz.

   b. If the iron used in an electromagnet saturates when supplied a bias field of 1.5 MA/m, calculate the maximum frequency to which a YIG oscillator can be tuned.

REFERENCES

1. B. Lax and K. B. Button, *Microwave Ferrites and Ferrimagnetics,* McGraw-Hill, New York, 1962.
2. L. R. Whicker (ed.), *Ferrite Control Components,* Artech House, Dedham, MA, 1974.
3. R. E. Collin, *Foundations of Microwave Engineering,* McGraw-Hill, New York, 1966.

11 | Microwave Antennas

11.1 INTRODUCTION

The analysis and design of microwave antennas is a field nearly as large as the remainder of microwave engineering. This chapter presents definitions, analysis techniques, and typical results in a handbook format to help the microwave system and component engineer interact with microwave antenna designers.

Antennas are used to launch electromagnetic energy from a transmission line into space (an unbounded transmission line) and vice versa. Most antennas are reciprocal and have the same properties on transmit and receive. Microwave antennas are physically large compared to the operating wavelength. Many of the concepts applied to antennas are independent of their operating frequency if their mechanical dimensions are scaled with inverse frequency. Antennas for lower frequencies (AM, FM, and TV broadcasts) would be significantly larger if they were designed to have properties equal to microwave antennas. The general properties of antennas presented below can be applied to antennas throughout the spectrum. Microwave antennas are highly directive so their radiated fields do not interact strongly with nearby objects that are not in the direction of transmission. Thus ground effects, which play a major role in the design of megahertz-region antennas, are generally not important above 1 GHz. References 1, 2, and 3 are excellent for general antenna theory, and Refs. 4, 5, and Vol. 1 of 3 deal primarily with microwave antennas.

Typical microwave antenna types are shown in Fig. 11.1. The most common is the parabolic reflector used at many satellite Earth stations and in radar systems. More exotic antennas such as dielectric rods and lenses are used above 10 GHz. Microwave antennas can be steered by mechanically pointing them in the desired direction. Electronically steered phased array antennas are used for multiple beam and low-momentum steering applications.

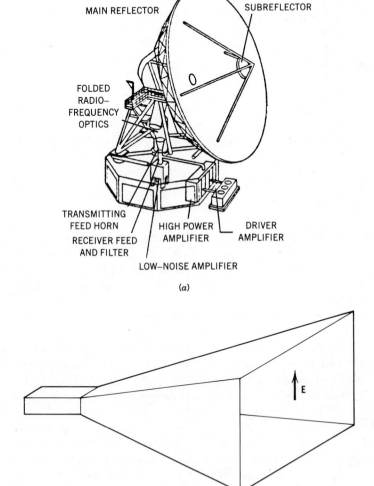

(a)

(b)

Figure 11.1. *Examples of microwave antennas. (a) Parabolic reflector; (b) pyramidal horn; (c) spiral; (d) patch (microstrip); (e) slotted waveguide array. ((a) is reprinted with permission of the publisher from* The Handbook of Digital Communications, 1979. *All rights reserved.)*

(c)

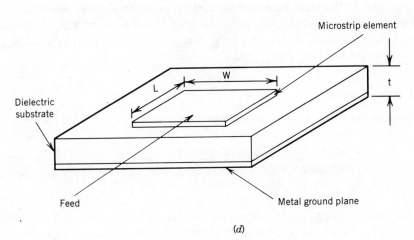

(d)

Figure 11.1. *(continued)*

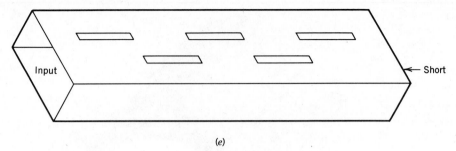

(e)

Figure 11.1. *(continued)*

11.2   ANTENNA PROPERTIES

Properties used to quantify antenna performance include pattern, gain, polarization, efficiency, impedance bandwidth, radiation resistance, reciprocity, and effective area. The importance of these properties for various systems has been described in Chapters 2–4.

11.2.1   Radiation Pattern

The radiation pattern is usually defined in a spherical coordinate system (Fig. 11.2). The pattern is the absolute or relative effective field strength at an angular direction θ, ϕ from the antenna axis of symmetry, denoted $| E(\theta, \phi) |$. The measurement of $| E |$ is made at a considerable distance from the antenna (called the far field) so that on a constant angle from the antenna the $| E |$ varies as the inverse radius.

The directivity can be derived from the radiation pattern. Directivity, D, is the ratio of the maximum radiation intensity to the average intensity. The directivity is a minimum and equal to 1 for an isotropic antenna (radiation intensity equal in all directions, a hypothetical case). Directivity is large for most microwave antennas, typically 20–50 dB.

11.2.2   Gain

The gain G of the antenna is the ratio

$$G = \frac{\text{Maximum radiated power density with given input power}}{\substack{\text{Maximum power density from a lossless isotropic} \\ \text{antenna with the same given input power}}} \quad (11.1)$$

Antenna gain is usually measured by comparing the radiated power density to that of a calibrated standard gain antenna. The efficiency, η, of an antenna is a measure of its ability to radiate power without resistive losses. Effi-

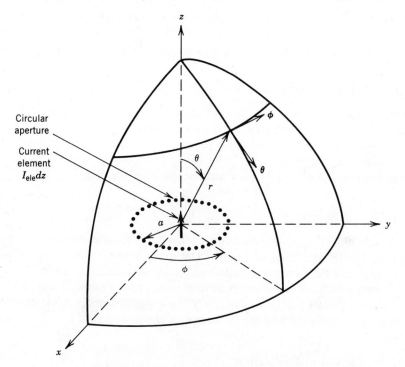

Figure 11.2. *Spherical coordinate system.*

ciency, gain, and directivity are related by

$$G = \eta D \qquad (11.2)$$

11.2.3 Polarization

The polarization of an electromagnetic wave is defined as the locus of the tip of the electric field vector. If the time-varying electric field vector remains parallel (normal) to the Earth's surface, the polarization is linear and horizontal (vertical). If the electric field rotates around the direction of propagation, it is right-hand or left-hand elliptically polarized. The electric field vector for a right-hand polarized wave rotates around the direction of propagation **k** in the direction the fingers on the right-hand point when the thumb points along **k**. For a circularly polarized wave, $|E|$ remains constant as the vector rotates around **k**.

11.2.4 Radiation Efficiency

The radiation efficiency ρ is the percentage of the total input power radiated by the antenna. The total losses associated with the antenna can include ohmic losses in the feed and reflector (if present) and spill over, i.e., energy not directed into the main beam but lost in the sidelobes of the antenna. The effect of the radiation efficiency is to reduce the power gain of the antenna.

11.2.5 Input Impedance

Antenna input impedance is the impedance measured at the antenna terminals. At resonance the antenna reactance is zero.

The impedance of the antenna is the same for receiving and transmitting (see reciprocity below). The current distribution on the surface of the antenna need not be the same for transmission as reception.

11.2.6 Bandwidth

The bandwidth (range of frequencies over which the antenna operates properly) can be defined in several ways. The four common bandwidth-defining parameters are the range over which the gain, impedance (VSWR or return loss), main lobe beamwidth, or polarization is within limits. The application and antenna design determines which of these parameters is more important or frequency sensitive and thus determines the bandwidth.

11.2.7 Radiation Resistance

The resistive portion of the antenna input impedance includes the resistance due to energy dissipated within the antenna and the resistance due to radiated energy. The resistance due to the energy radiated is the radiation resistance.

11.2.8 Reciprocity

Most antennas are composed of parts that are bidirectional. These antennas have the same properties when the electromagnetic wave is outgoing (transmitted) and incoming (received). The pattern, polarization, gain, and impedance are the same for transmission and reception. This property of an antenna is called reciprocity.

Reciprocity, when applied to electrical networks, states that if a voltage source v excites a current i in the load (see Fig. 11.3a), the reversal of the location of the load and voltage source excites the same current i as shown in Fig. 11.3b. To derive the reciprocity between transmission and reception, consider two arbitrary bidirectional antennas as shown in Fig. 11.3c. If a voltage v_1 at frequency f in antenna 1 excites a current i_2 in antenna 2, a transfer impedance Z_{12} can be defined similar to the mutual impedance between magnetically coupled circuits. Similarly, the transfer impedance Z_{21} can be defined for a voltage v_2 in antenna 2 exciting a current i_1 in antenna 1. Since the media between the antennas is linear and isotropic, the coupling between the two fixed antennas must be independent of transmission direc-

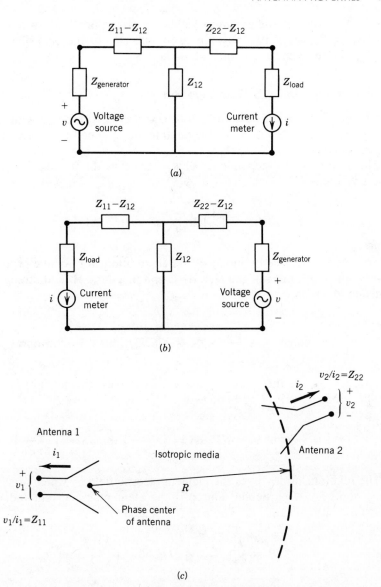

Figure 11.3. *Networks and antennas configured to demonstrate reciprocity. (a) Two-port network with mutual coupling; (b) two-port network with source and load reversed; (c) two antennas configured for reciprocity.*

tion. As a result, $Z_{12} = Z_{21}$ and there is no difference between the coupling for antenna 1 transmitting and antenna 2 transmitting.

In the above discussion, the two antennas were at fixed locations. If antenna 1 is given various orientations with respect to antenna 2, the current induced in 2 by 1 will be representative of the radiation pattern of 1. Reversing the roles of the antennas yields the absorption pattern of 1 due to currents in 2. These two patterns are equal (1), so the emission and absorption patterns of an antenna are equal.

11.2.9   Effective Area-Gain Relation

In Chapter 2 the useful relation $A_{eff} = G\lambda^2/4\pi$ was presented without derivation. The derivation follows for an antenna matched to its transmission line and load.

The study of time-dependent fields states that the average power flow into a volume V is

$$|P| = \left(\frac{1}{2}\right) \text{Re} \oint_S \mathbf{E} \times \mathbf{H}^* \cdot \mathbf{dS} \tag{11.3}$$

where $\mathbf{S}$ is the surface bounding V (6). In addition, for a wave propagating in free space the ratio of the electric $\mathbf{E}$ and magnetic $\mathbf{H}$ field strengths is a constant given as,

$$\frac{|\mathbf{E}|}{|\mathbf{H}|} = \text{const} = \sqrt{\frac{\mu_0}{\epsilon_0}} = Z_0 = 376.731 \text{ ohms} \approx 120\pi \text{ ohms} \tag{11.4}$$

Z_0, μ_0, and ϵ_0 are the wave impedance, free space permeability, and permitivity, respectively. Combining Eqs. 11.3 and 11.4 gives

$$|P| = \frac{1}{2Z_0} \oint_S \mathbf{E} \times \mathbf{E}^* \cdot \mathbf{dS} = \left(\frac{|\mathbf{E}|^2}{2Z_0}\right) \oint_S dS = \left(\frac{E^2}{8\pi Z_0}\right) \tag{11.5}$$

The relationship involves the incident electric field $\mathbf{E}$, which arises from many sources. A particular source is the elementary current radiator of length Δl with current I_{ele}. The power radiated is

$$P = \tfrac{1}{2}I_{ele}^2 R_{rad} = \tfrac{1}{2}I_{ele}^2 \, 80\pi^2 \left(\frac{\Delta l}{\lambda_0}\right)^2 \tag{11.6}$$

where R_{rad} is the radiation resistance. The power gain of the current element is $G_{ele} = 1.5 \sin^2 \theta$. Using the power density relation from Chapter 2

$$\frac{E_i^2}{2Z_0} = \frac{P_r G_{ele}}{4\pi R^2} \tag{11.7}$$

or

$$E_i = \frac{60\pi I_{ele} \, \Delta l \sin \theta}{R\lambda_0} \tag{11.8}$$

Now consider the power radiated by the antenna due to the current I_0 across the matched transmission line (6). One-half I_0 goes into the antenna and the remainder passes into the load, so

$$P_{\text{transmit}} = \frac{1}{2}\left(\frac{I_0}{2}\right)^2 R_{\text{rad}} \tag{11.9}$$

Again using the relation from Chapter 2,

$$\frac{(1/2)E_{\text{trans}}^2}{Z_0} = \frac{[P_{\text{trans}}G(\theta, \phi)]}{4\pi R^2} \tag{11.10}$$

where $G(\theta, \phi)$ is the power gain in the direction of the elementary radiator. The field radiated by the current I_0 at the elementary radiator location is

$$E_{\text{trans}} = [30\, G(\theta, \phi)\, R_{\text{rad}}]^{1/2}\, \frac{I_0}{2R} \tag{11.11}$$

Finally, the received signal voltage V_{rec} times current I_0 equals $E_{\text{trans}} \cdot I_{\text{ele}} \cdot \Delta l \cdot \cos \psi \cdot \sin \theta$ gives

$$V_{\text{rec}} = \frac{[E_{\text{trans}}I_{\text{ele}} \,\Delta l \cos \psi \sin \theta]}{I_0}$$

$$= \frac{\lambda_0}{2\pi}\left\{\frac{R_{\text{rad}}G(\theta, \phi)}{120}\right\}^{1/2} E_i \cos \psi \tag{11.12}$$

The power absorbed in the load is

$$P_{\text{load}} = \frac{1}{2}\left(\frac{V_{\text{rec}}^2}{R_{\text{rad}}}\right) = \frac{1}{8}\frac{\lambda_0^2}{\pi^2}\frac{E_i^2}{120}\,G(\theta, \phi) = A_{\text{eff}}\left(\frac{1}{2}\cdot\frac{E_i^2}{120\pi}\right) \tag{11.13}$$

Solving for A_{eff} yields the desired result:

$$A_{\text{eff}} = G(\theta, \phi)\,\frac{\lambda_0^2}{4\pi} \tag{11.14}$$

It has been shown (3) that this is the smallest effective area to yield a gain $G(\theta, \phi)$, namely,

$$A_{\text{eff}} \geq G(\theta, \phi)\,\frac{\lambda_0^2}{4\pi} \tag{11.15}$$

11.3   ANALYSIS OF BASIC ANTENNAS

Two antenna analysis methods assume the radiation is attributable to either current elements or apertures, respectively. Current elements are often used to analyze low-frequency antennas such as dipoles and wires. The simplest current element to analyze is the differential (elementary) current element $I_{ele}dz$ shown in Fig. 11.2.

Microwave antennas are usually analyzed using aperture techniques. Examples are reflector, lens, horn, and array antennas where the radiation is considered to emanate from an opening or aperture.

11.3.1   Linear Current Element

The radiation from a linear current element has been derived in numerous texts (2,6,7). The results are presented here and form the basis for definition of the zones (near and far field) around antennas.

The electric and magnetic components of the radiated field are derived using the vector potential to yield

$$
E = \frac{Z_0 k}{2\pi} (I_{ele}dz)e^{-jkr}\left\{ \left[\cos\theta \left(\frac{k}{r^2} - \frac{j}{r^3} \right) \right] \mathbf{a}_r \right.
$$

$$
\left. + j\left[\frac{\sin\theta}{2} \left(\frac{k^2}{r} - j\frac{k}{r^2} - \frac{1}{r^3} \right) \right] \mathbf{a}_\theta + 0\mathbf{a}_\phi \right\} \tag{11.16}
$$

$$
H = \frac{ke^{-jkr}}{4\pi} (I_{ele}dz)\left\{ \theta\mathbf{a}_r + 0\mathbf{a}_\theta + \left[\sin\theta \left(\frac{jk}{r} + \frac{1}{r^2} \right) \right] \mathbf{a}_\phi \right\}
$$

where the wave number $k = 2\pi/\lambda = 2\pi f/c = \omega/\sqrt{\mu_0 \epsilon_0}$ in free space.

The r^{-2} and r^{-3} terms decay rapidly with distance and are not considered to radiate energy from an antenna. Their magnitude is equal to the magnitude of the r^{-1} term for $r = k^{-1}$ (for other source distributions $r < k$). This region in the immediate vicinity of the current element is called the reactive near-field region.

The radiated field (r^{-1} dependence) region is subdivided into the near-field and far-field regions. In the near-field (Fresnel) region, the distance from different parts of the antenna to the observation point varies more than some arbitrary quantity. The far-field region extends from the boundary of the near-field region to infinity. The definition of this boundary is best derived from aperture dimensions of the antenna.

For longer antennas, the elemental current $I_{ele}dz = I_{ele}(z)dz$ is integrated over the length of the antenna. For example, for a half-wave dipole $I_{ele}(z) \approx I_0 \cos kz$ and using the approximation $r(z) = r - z\cos\theta$ yields

$$
E = j60I_0 \left(\frac{e^{-jkr}}{r} \right) \left(\frac{\cos[(\pi/2)\cos\theta]}{\sin\theta} \right) \tag{11.17}
$$

Clearly, the number of current distributions that can be solved in closed form is limited. Numerical techniques are used extensively for complex distributions. Many computer programs have been developed and are available for these calculations via time-share services.

11.3.2   Aperture

The aperture analysis technique derives the far-field radiated fields from the fields at the aperture surface. These aperture fields are derived from the sources in the antenna, or suitable approximations are made if exact fields cannot be derived. The radiated field is the Fourier transformation of the aperture field (8). The relation is

$$E_{ff}(\theta, \phi) = \cos^2\left(\frac{\theta}{2}\right) \left[1 - \tan^2\left(\frac{\theta}{2}\right) \cos 2\phi \right]$$

$$\int_{-\infty}^{\infty} \int_{-\infty}^{\infty} E_a(x, y) \exp[jk(x \sin \theta \cos \phi \qquad (11.18)$$

$$+ y \sin \theta \sin \phi)]dxdy$$

where E_{ff} is the electric field in the far field and E_a is the aperture field. A circular aperture of radius a is shown in Fig. 11.2, which might represent the plane immediately in front of a circular parabolic antenna.

This aperture formulation has been applied to the common configurations of a line, rectangular, and circular aperture (see Fig. 11.2). Usually for the rectangular aperture (applied to a horn) the $E_a(x, y) = E_a(x)E_a(y)$ so that a double application of the line source results can be used.

11.3.2.1   Line Source Distribution. Consider a line source along the z-axis extending from $-a/2$ to $a/2$. The resulting evaluation of Eq. (11.18), after a change of variables where $p = 2\pi z/a$, $u = (a \sin \theta)/\lambda$ and the integration over p is from $-\pi$ to π, yields for a $E_a(z) = E_z(p) = 1$, $E_{ff} \propto [\sin \pi u]/\pi u$. This is the familiar $\sin z/z$ distribution found so often in antenna theory and shown in Fig. 11.4a. Note that the first sidelobe is 13.2 dB down from the main beam peak, which is frequently not adequate, and the 3-dB beamwidth is $0.88\lambda/a$. The high sidelobe level can be reduced significantly if $E_a(z)$ is tapered to a low level at the ends of the line source. For example, if $E_z(p) = \cos(p/2)$, the first sidelobe level drops another 10 dB compared to the $E_z(p) = 1$ distribution. Unfortunately, the 3-dB beamwidth increases by 36%. A listing of the typical distributions and their beam characteristics is shown in the top part of Table 11.1. The Taylor (9) distribution (not involving a Taylor series) yields an optimized distribution where sidelobe levels are uniform and low while the mainbeam is only moderately spread. Tables for the Taylor design are given in Ref. 10.

Table 11.1 Line Source Pattern Parameters for Various Aperture Distributions

Distribution	Aperture Field	3-dB Beamwidth	Angular Position of 1st Zero	Level of 1st Sidelobe	Relative Peak Gain
Uniform	←2a→	$0.88\dfrac{\lambda}{a}$	$\dfrac{\lambda}{a}$	−13.2 dB	1
Cosine $\cos\left(\dfrac{\pi x}{2a}\right)$		$1.2\dfrac{\lambda}{a}$	$1.5\dfrac{\lambda}{a}$	−23.0 dB	0.810
Cosine square $\cos^2\left(\dfrac{\pi x}{2a}\right)$		$1.45\dfrac{\lambda}{a}$	$2.0\dfrac{\lambda}{a}$	−32.0 dB	0.667
Pedestal $1-(1-0.5)(x/a)^2$		$0.97\dfrac{\lambda}{a}$	$1.14\dfrac{\lambda}{a}$	−17.1 dB	0.970
Taylor $n=3$ −9 dB edge		$1.07\dfrac{\lambda}{a}$		−25.0 dB	

Circular Aperture Pattern Parameters for Various Distributions

Distribution	Aperture Field	3-dB Beamwidth	Angular Position of 1st Zero	Level of 1st Sidelobe	Relative Peak Gain
$0 \leq r \leq 1$ Uniform		$1.02\dfrac{\lambda}{D}$	$1.22\dfrac{\lambda}{D}$	-17.6 dB	$\left(\dfrac{\pi D}{\lambda}\right)^2 = 1$
Tapered to zero at edge $(1 - r^2)$		$1.27\dfrac{\lambda}{D}$	$1.63\dfrac{\lambda}{D}$	-24.6 dB	0.75
Tapered to zero at edge $(1 - r^2)^2$		$1.47\dfrac{\lambda}{D}$	$2.03\dfrac{\lambda}{D}$	-30.6 dB	0.56
Tapered to 0.5 at edge $0.5 + (1 - r^2)^2$		$1.16\dfrac{\lambda}{D}$	$1.51\dfrac{\lambda}{D}$	-26.5 dB	
Taylor distribution		$1.31\dfrac{\lambda}{D}$		-40.0 dB	

(Reproduced by permission of Peter Peregrinns Ltd. on behalf of the Institute of Electrical Engineers, UK, London and New York.)

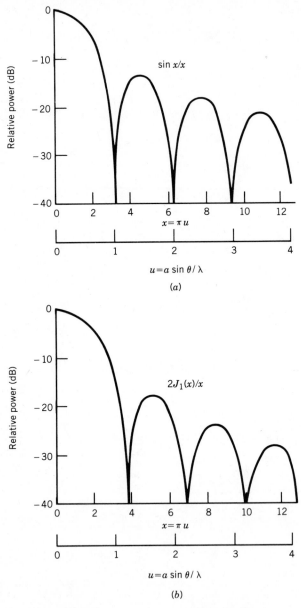

Figure 11.4. *Antenna patterns for two types of uniform illumination. (a) Line source; (b) circular aperture. (Reproduced by permission of Peter Peregrinns Ltd. on behalf of the Institute of Electrical Engineers, UK, London and New York.)*

11.3.2.2 Circular Aperture Distribution. Circular parabolic reflector antennas form the largest single group of microwave antennas. These antennas
are usually axially symmetric (no ϕ variation) in pattern (depends on the
type of feed). The analysis using Eq. 11.18 for common aperture distributions
yields the results in Fig. 11.4b and the bottom part of Table 11.1. As shown
in the figure, the results usually involve Bessel functions of low order. Assuming the aperture diameter $D = 2a$ for a line source, the 3-dB beamwidths
of the circular aperture are larger than the line source.

Because of its axial symmetry, the circular aperture is ideal for formulation of the transition between the near-field (Fresnel) region and the far-
field (Fraunhofer) region. In both these regions the radiating r^{-1} term dominates. As stated earlier, the transition between regions has been set (ar-
bitrarily) to be that distance where the difference in distance from the center
of the aperture plane to the edge of the aperture is $\lambda/16$ (22.5°). This distance
is $R_{ff} > 2D^2/\lambda$. As shown in Fig. 11.5, this $\lambda/16$ phase difference defines a
cone whose apex is at the center of the aperture plane. Within the range
$R_{pb} = D^2/2\lambda$, the beam can be envisioned as a cylinder of diameter D with
parallel rays.

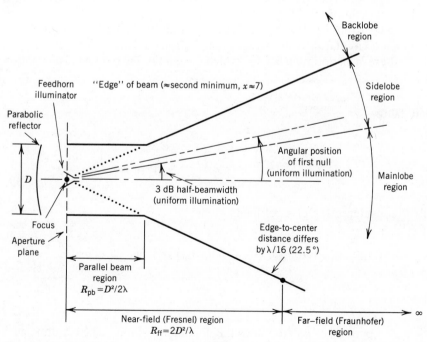

Figure 11.5. Parallel beam, near- and far-field regions for a circular aperture antenna.

11.4   MICROWAVE ANTENNA PERFORMANCE

This section presents the design parameters and performance of typical microwave antennas shown in Fig. 11.1. These results, although not exhaustive, fulfill the goal of this chapter to provide an introduction to microwave antennas.

11.4.1   Parabolic Reflector

The parabolic reflector antenna is generated using the geometry shown in Fig. 11.6, where all the path lengths from the focal point to the aperture plane are equal. In some cases the reflector is configured to locate the focal point in other positions, as shown in Fig. 11.7. The selection of the design depends on mechanical considerations and feed blockage. In the offset parabolic configuration, the blockage from the feed is eliminated.

The feed should be a point source with the proper pattern (primary pattern) to yield the required aperture distribution. Typically the feed is a horn antenna whose phase fronts are independent of angle. Circular horns give more uniform illumination of the reflector (11). The input impedance of the horn positioned in front of the reflector is affected because of the standing wave in this region (not present in the offset configurations). The magnitude of the reflection coefficient is $| \Gamma | = \lambda G_{\text{feed}}/4\pi$ (focal length), where G_{feed} is the peak gain of the feed (12).

The gain from a paraboloid is less than that available from a uniformly illuminated circular aperture with the same projected area. For a given G_{feed} (θ, ϕ), there is an optimum θ_{opt} for which the gain is a maximum. The relation is

$$\text{Gain factor} = \frac{G_{\text{paraboloid}}}{G_{\text{unif. illum.}}}$$

$$= (G_{\text{feed,max}}) \cot^2\left(\frac{\theta_{\text{opt}}}{2}\right) \left| \int_0^{2\pi} \int_0^{\theta_{\text{opt}}} \left(\frac{G_{\text{feed}}(\theta, \phi)}{G_{\text{feed,max}}}\right)^{1/2} \tan\frac{\theta}{2} \, d\theta d\phi \right|^2 \tag{11.19}$$

Silver considered the class of feeds

$$(G_{\text{feed,max}}) (G_{\text{feed}}(\theta, \phi)) = 2(n + 1) \cos^n \theta \quad \phi < \theta < \frac{\pi}{2} \tag{11.20}$$

$$= 0 \qquad\qquad \theta > \frac{\pi}{2}$$

where the parameter n determines the illumination taper. The resulting gain factor versus θ_{opt} is shown in Fig. 11.8. The gain factor is typically between 0.6 and 0.75, with edge illuminations 10 dB below the value at the center.

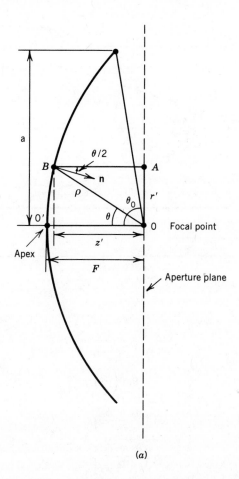

(a)

$$OB + BA = \rho + \rho \cos \theta' = \rho(1 + \cos \theta') = 2F$$

$$F = \frac{D}{4} \cot \left(\frac{\theta_0}{2} \right)$$

$$r' = \rho \sin \theta' = \frac{2F \sin \theta'}{1 + \cos \theta'} = 2F \tan \left(\frac{\theta'}{2} \right)$$

(b)

Figure 11.6. Geometry of a parabolic reflector antenna. (a) Geometry; (b) geometric relations.

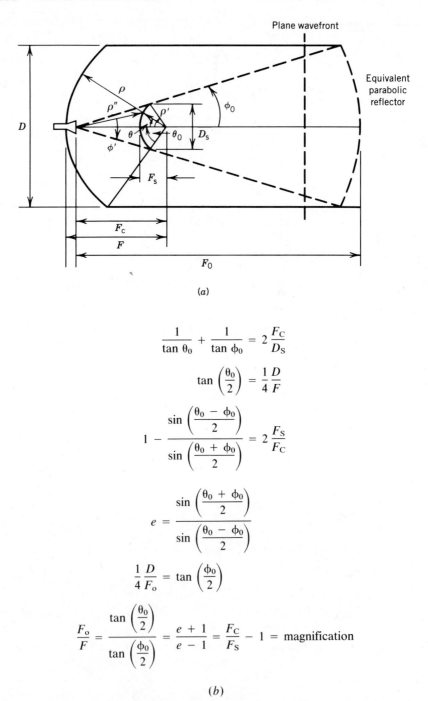

$$\frac{1}{\tan\theta_0} + \frac{1}{\tan\phi_0} = 2\frac{F_C}{D_S}$$

$$\tan\left(\frac{\theta_0}{2}\right) = \frac{1}{4}\frac{D}{F}$$

$$1 - \frac{\sin\left(\dfrac{\theta_0 - \phi_0}{2}\right)}{\sin\left(\dfrac{\theta_0 + \phi_0}{2}\right)} = 2\frac{F_S}{F_C}$$

$$e = \frac{\sin\left(\dfrac{\theta_0 + \phi_0}{2}\right)}{\sin\left(\dfrac{\theta_0 - \phi_0}{2}\right)}$$

$$\frac{1}{4}\frac{D}{F_o} = \tan\left(\frac{\phi_0}{2}\right)$$

$$\frac{F_o}{F} = \frac{\tan\left(\dfrac{\theta_0}{2}\right)}{\tan\left(\dfrac{\phi_0}{2}\right)} = \frac{e+1}{e-1} = \frac{F_C}{F_S} - 1 = \text{magnification}$$

(b)

Figure 11.7. *Geometry of parabolic reflector antennas. (a) Cassegrain geometry; (b) geometric relations; (c) offset reflector antenna configurations.*

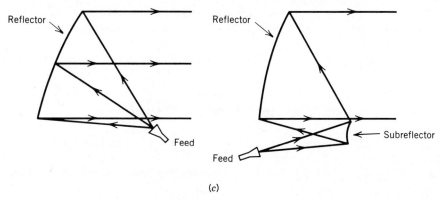

(c)

Figure 11.7. (continued)

The gain is also reduced due to mechanical tolerances because the reflector is not a perfect paraboloid. The loss in gain due to small phase errors (arising from mechanical imperfections) is (13)

$$\frac{G_{\text{phase error}}}{G_{\text{no phase error}}} = 1 - \bar{\delta}^2 \qquad (11.21)$$

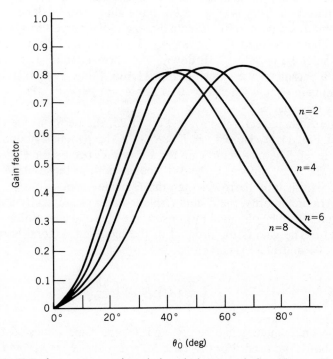

Figure 11.8. Gain factor versus subtended angle for a parabolic reflector. (From Radar Handbook by M. I. Skolnik (ed). Copyright 1970, McGraw-Hill Book Co.)

where $\bar{\delta}^2$ is the mean-square phase error in radians. For low microwave frequencies this is usually not significant, but it ultimately limits the gain of large millimeter wave reflectors.

The desired gain of the antenna is related to the radius by the relation

$$2a = \left(\frac{G_{\text{desired}}}{(\text{Gain factor}) \ (1 - \bar{\delta}^2)\pi} \right) \lambda \qquad (11.22)$$

where a is defined in Fig. 11.6. This relation does not include the effects of nonuniform phase and cross-polarized fields in the aperture plane, blockage due to the feed and feed support mechanism, or attenuation in the feed and its transmission line.

11.4.2   Horn

Horn antennas are realized as a gradual flare from waveguides. Typical configurations are shown in Fig. 11.9. Pyramidal horns have flares in both the E- and H-planes and are frequently used as standard gain (calibrated) antennas. Horns behave like traveling wave structures (having broadband characteristics) and are used to illuminate reflectors and lenses. They are also excellent moderate gain antennas with low sidelobes and backlobes (14).

H-plane sectoral horns (Fig. 11.9a) have almost no sidelobe structure in the H-plane, as shown in the principal plane radiation patterns in Fig. 11.10 with the phase variation $T = A^2/8R_1\lambda_0$ as a parameter (15). In the E-plane, the first sidelobe is only 13 dB down. The corresponding gain curves with R_1/λ_0 as a parameter are shown in Fig. 11.10b. For a given axial length R_1, maximum gain occurs when $T = 0.375$ or $A = (3R_1\lambda_0)^{1/2}$. The 3-dB beamwidth for this case is $78 \ \lambda_0/A$ degrees.

For the E-plane sectoral horn in Fig. 11.9b, the performance curves are shown in Fig. 11.11. Here $U = B^2/8R_2\lambda_0$ is the parameter controlling the sidelobe level. The maximum gain for a given R_2 occurs for $U = 0.25$ or $B = (2R_2\lambda_0)^{1/2}$. For this case, the 3-dB beamwidth in the E-plane is $54\lambda_0/B$ degrees, significantly narrower than the H-plane horn.

For pyramidal horns, the E- and H-plane patterns are nearly independent and therefore can be obtained by considering the solid curves in Figs. 11.10a and 11.11a. The gain is the product of the modified sectoral horn gains in Figs. 11.10b and 11.11b, namely,

$$G = \left(\frac{\pi}{32} \right) (G_H G_E) \left(\frac{\lambda_0^2}{AB} \right) \qquad (11.23)$$

Again for a maximum gain given R_1 and R_2, the same values of A and B as sectoral horns are used. For this case $R_1 \approx L_H$ and $R_2 \approx L_E$, and so $R_H = R_E$. Using these relations in Eq. 11.14, the effective aperture area $A_{\text{eff}} = (1/2)AB$, yielding $G_{\text{pyra}} = 2\pi AB/\lambda_0^2$. Substituting into the geometric relations

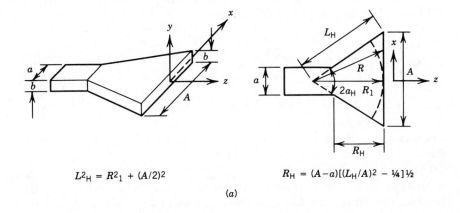

$$L^2_H = R^2_1 + (A/2)^2$$

$$R_H = (A-a)[(L_H/A)^2 - \tfrac{1}{4}]^{\tfrac{1}{2}}$$

(a)

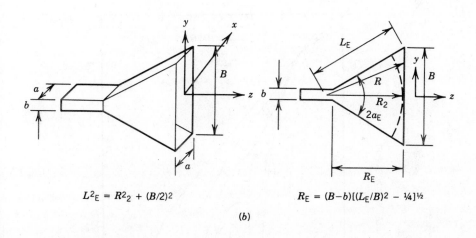

$$L^2_E = R^2_2 + (B/2)^2$$

$$R_E = (B-b)[(L_E/B)^2 - \tfrac{1}{4}]^{\tfrac{1}{2}}$$

(b)

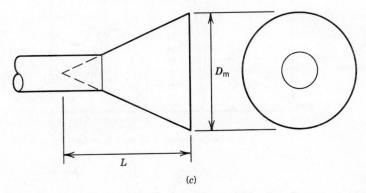

(c)

Figure 11.9. Diagrams of common microwave horn antenna configurations. (a) H-plane sectoral horn; (b) E-plane sectoral horn; (c) conical horn.

in Figs. 11.9a and b for $R_E^2 = R_H^2$ yields

$$\left[\left(\frac{2L_E}{\lambda_0}\right)^{1/2} - \frac{b}{\lambda_0}\right]^2 \left(\frac{2L_E}{\lambda_0} - 1\right)$$

$$= \left[\frac{G_{pyra}}{2\sqrt{2}\,\pi}\left(\frac{\lambda_0}{L_E}\right)^{1/2} - \frac{a}{\lambda_0}\right]^2 \left(\frac{G_{pyra}^2\,\lambda_0}{18\pi^2 L_E} - 1\right)$$

(11.24)

which can be solved for L_E by iterative techniques. Thus given a, b, λ_0, and G_{pyra}, L_E is obtained from Eq. 11.24 and then $R_E = R_H$, A, and B are determined.

Conical horns are used with circular waveguide. The gain of a conical horn for the parameters in Fig. 11.9c is shown in Fig. 11.12 (16). The flare length for a given diameter that yields maximum gain is $0.3D_m^2/\lambda_0$ (17). For

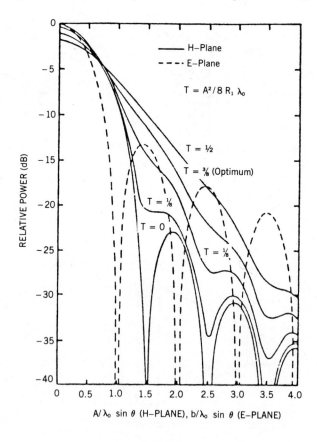

(a)

Figure 11.10. *Performance curves for a H-plane sectoral horn antenna.*

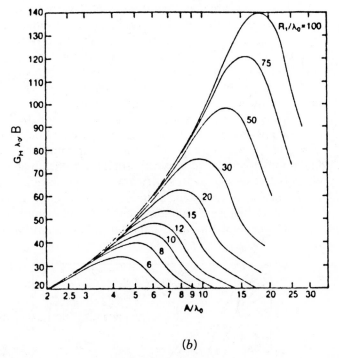

(b)

Figure 11.10. (continued)

this case, the E- and H-plane 3-dB beamwidths are $60\lambda_0/D_m$ and $70\lambda_0/D_m$ degrees, respectively. The conical horn gain is

$$G_{con} \approx 0.52 \left(\frac{\pi D_m^2}{4}\right) \left(\frac{4\pi}{\lambda_0^2}\right) = 0.52 \left(\frac{\pi D_m}{\lambda_0}\right)^2 \qquad (11.25)$$

More complex horns, such as corrugated horns (Fig. 11.13), have been described in Refs. 1, 14, and 18. They have axially symmetric patterns, a fixed phase center, low cross polarization, wide bandwidth and low side- and backlobes resulting in high efficiency. Corrugated horns are the "ideal" feed horn for many applications.

11.4.3   Spiral

No practical antenna's operation is truly independent of frequency, but the spiral antenna belongs to a class termed frequency independent because of its larger than normal bandwidth. Spiral antennas with polarization ellipticity ratios of less than 3 dB over azimuth angles to 60° and an input VSWR of less than 2:1 over three octaves bandwidth have been realized.

Figure 11.14 shows the three configurations that have received the most attention. The logarithmic (equiangular), Archimedian (arithmetic), and rectangular version of the Archimedian spiral are all fabricated from photolithographically etched copper-clad dielectric material (e.g., material used in microstrip circuitry). Either one or two conductors may be used to form a single spiral. The most uniform operation of the spiral occurs when the spiral is excited from a balanced two-wire transmission line.

The analysis of these antennas is available in Refs. 19 and 20. A heuristic presentation of the operation of these antennas is presented below based on Ref. 21. The "band" model agrees well with the experimental results and is easily understood based on intuitive reasoning for the dual-arm spiral antenna. This spiral is envisioned as a two-wire transmission line that is gradually transformed into a radiating structure as shown in Fig. 11.15. Radiation occurs for bands whose circumference is an integral number of wavelengths. As a result, several "modes" exist on the antenna; but only two

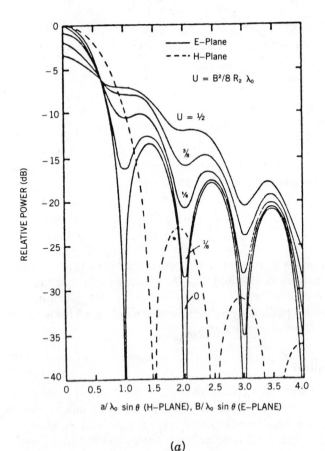

$$U = B^2/8 R_2 \lambda_0$$

(a)

Figure 11.11. *Performance curves for an E-plane sectoral horn antenna.*

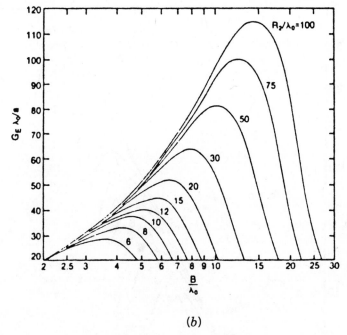

(b)

Figure 11.11. *(continued)*

are described here. The normal mode produces a single lobe of radiation whose maximum is along the axis of the spiral, and the split-beam mode provides a split-beam pattern with a null on the spiral's axis. These two modes are present when the spiral elements are excited 180° out of phase and in phase, respectively.

Consider the normal mode. When the distance from A to A' in Fig. 11.15a is one-half wavelength, the two differential currents A and A' are in phase (180° occurs because of the excitation phase shift and the other 180° due to the phase delay around the spiral). There exists a "band" of finite breadth (e.g., B and B') where nearly the same phase relation exists. The mean radius of this first radiation band is $\lambda/2\pi$. There is also a radiating group (e.g., C and C'), which is in quadrature so that the radiation is circularly polarized.

For the case where the excitation is in phase (see Fig. 11.15b), efficient radiation occurs from a "band" whose mean circumference is 2λ. For this mode, the diametrically opposite differential current elements are out of phase so the radiation pattern has a null along the axis normal to the plane of the spiral. The radiation is circularly polarized near the spiral axis and elliptically polarized for off-axis angles, which correspond to a phase delay of one-half wavelength (see Fig. 11.15c). Nearly all spiral antennas are designed to operate in the normal or fundamental mode rather than the split-beam mode.

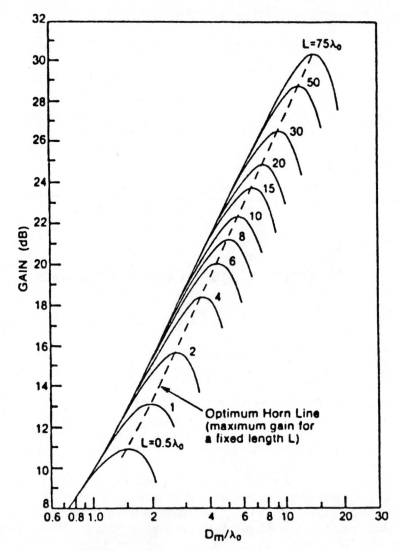

Figure 11.12. *Gain of a conical horn antenna.*

The bidirectional radiation pattern of the metal spiral on a dielectric substrate is not desired for most applications. Therefore, a closed cavity is usually mounted behind the spiral. This provides other design variables, including cavity and spiral diameter, spiral growth rate, conductor spacing and width, substrate dielectric, feed design, and cavity depth.

The cavity and spiral diameters are usually made equal (21). The spirals may either be terminated at their outer circumference or left open so that the energy reflects back into the spiral to be reradiated with the opposite sense of circular polarization. This effect is shown in Fig. 11.16a where the on-axis ellipticity is significantly affected by the termination, while the gain is relatively independent of the spiral outer termination since it is normalized to the average axial ratio. The rate of growth (pitch) of the spiral decreases the axial ratio slightly and increases the radiation efficiency slightly as shown

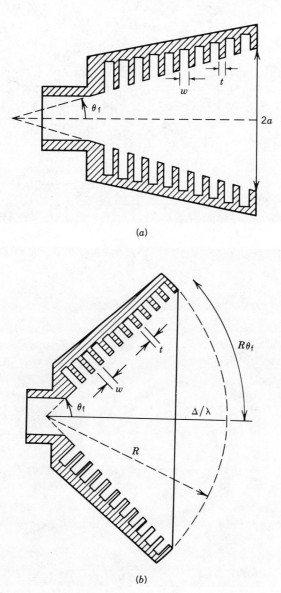

(a)

(b)

Figure 11.13. *Cross sections of corrugated conical horn antenna (notation from ref 18). (a) Small flare angle; (b) large flare angle (From Ref. 18, © 1978 IEEE).*

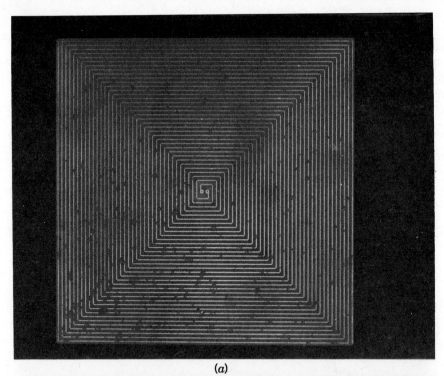

(a)

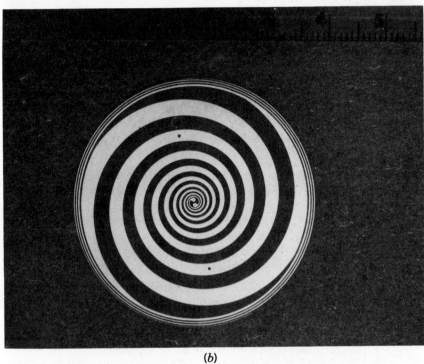

(b)

Figure 11.14. Dual arm spiral antenna configurations. (a) Square Archimedian spiral; (b) logarithmic spiral; (c) dual arm spiral antenna configurations.

Figure 11.14. (continued)

in Ref. 21 for spirals with 0.07λ per turn and 0.035λ per turn operating at 1.445 GHz. The conductor spacing-to-width ratio is generally made one, but there exists no reason for this other than to maximize the number of turns in a given diameter. Higher dielectric substrate material tends to result in a small additional gain because of an apparent increase in the size of the antenna. If the feed point currents are not of equal amplitude (assuming that the feed does not radiate directly), a combination of the normal and split-beam modes of radiation exist resulting in a squinted beam (combine the patterns in Figs. 11.15a and b to show this effect). Squinting can be experimentally verified by reducing the operating frequency so the split-beam mode is not radiated and noting the reduction of the squint. If the squint remains, the feed is probably radiating directly. The cavity depth limits the gain bandwidth of the spiral antenna to 3.3:1. The optimum cavity depth is λ/4 as is readily seen using image theory (1). The performance of square spirals is similar to round units (21).

11.4.4 Lens

Lens antennas are used in applications where the collimating features of a reflector are not adequate and the flexibility of a phased array is not required.

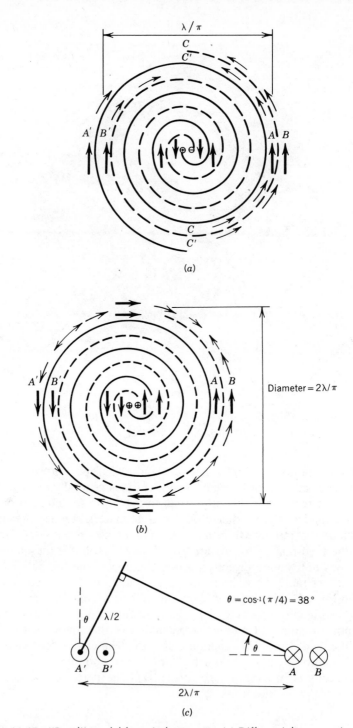

Figure 11.15. "Band" model for spiral antennas. (a) Differential currents in the normal mode; (b) differential currents in the split-beam mode; (c) geometry of first lobe in split-beam mode.

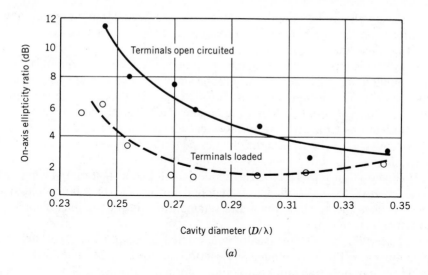

(a)

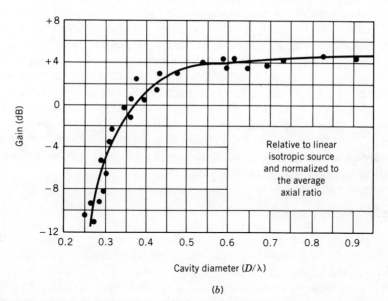

(b)

Figure 11.16. *Axial ratio and gain for terminated and unterminated spiral antennas. (a) Axial ratio; (b) gain.*

Lenses collimate one wavefront into another using the three degrees of freedom: inner surface contour, outer surface contour, and index of refraction n. In most applications, a lens focuses a point (spherical wavefront) or line (cylindrical wavefront) source to infinity. Thus the lens forms the focusing medium between the feed and a plane wave.

At microwave frequencies, n can be greater or less than 1 corresponding to a phase velocity less than or greater than the speed of light, respectively. For high percentage bandwidth (broadband) operation, $n > 1$ designs are used because they are less dispersive (have small variation of n with frequency). This type of lens using material such as Lucite® is commonly used at millimeter wave frequencies where the small size allows edge support and light weight. The design equations for a stepped convex lens are shown in Fig. 11.17a. The stepped lens reduces the thickness of the center region. The reflections from the flat surface that appear as a feedpoint mismatch VSWR $\approx (1 + n)/(1 - n)$ are the chief disadvantage of this design. A method for reducing the reflections involves quarter-wave matching transformers (22) as shown in Fig. 11.17b. In this design sufficient material is removed so $n_{eff} = (n)^{1/2}$ assuming the lens is matched into air. This matching technique also reduces the sidelobe level and retains the gain for the lens compared to a flat surface (23).

Lenses with $n < 1$ are usually constructed of metal plates (parallel and equally spaced) or waveguides as described in Refs. 1, 24, and 15. Lenses of this design only transmit polarizations normal to the metal plate or normal to the broad wall of the rectangular waveguide, i.e., the propagation is TE mode with $n = 1 - (\lambda_l/\lambda_c)$, where λ_l is the wavelength in the lens and λ_c is the cutoff wavelength. The shape design for a stepped concave lens is shown in Fig. 11.18. The shadowing effect shown causes discontinuities in the aperture illumination. Techniques for avoiding this shadow effect are available (1) but are difficult to manufacture. Waveguide lenses have provided wider bandwidth (40–20% for 20–100 wavelength diameter) if all rays through the lens have equal time delay (26).

In general, the mechanical tolerances of lens antennas are not as stringent as reflector antennas (16). This reference also develops bandwidth formulas useful in the design of stepped lenses.

11.4.5 Patch and Slot

Patch (also called microstrip) antennas are compatible with microstrip transmission media, cost little to replicate, and can be made conformal to the surface of the supporting vehicle (e.g., missile cone, aircraft ogive). Their use has been limited due to their limited bandwidth and a tendency to develop surface modes. These antennas are usually configured in a rectangular (see Fig. 11.19) or circular format on a low dielectric substrate. Using a "band-like" model, the electric field lines at the edge of the patch can be resolved into normal and tangential components. Only the net fields radiate and form the far-field pattern for the antenna. For example, referring to Fig. 11.19a, the normal components of the electric field are out of phase because the patch is about $\lambda_g/2$ long, which leads to zero field at a distance. The tangential components are in phase and give a maximum radiated field normal to the plane of the antenna (this is equivalent to the radiation of two slots fed out of phase and radiating into the half-space above the ground plane). In fact,

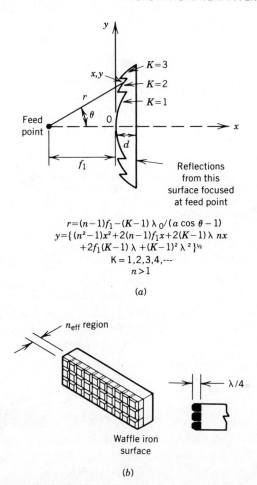

$$r = (n-1)f_1 - (K-1)\lambda_0/(a\cos\theta - 1)$$
$$y = \{(n^2-1)x^2 + 2(n-1)f_1x + 2(K-1)\lambda\, nx$$
$$+ 2f_1(K-1)\lambda + (K-1)^2\lambda^2\}^{1/2}$$
$$K = 1,2,3,4,\cdots$$
$$n > 1$$

(a)

Waffle iron surface

(b)

Figure 11.17. *Stepped convex lens cross section and equations for n > 1. (a) Lens; (b) matching surface. (From Ref. 22, © 1956 IEEE)*

the patch antenna can be accurately represented by equivalent slots representing the fringing fields at the edges of the patch. Units with dual polarization can be obtained by feeding two edges of the patch as shown in Fig. 11.19b, and circular polarization is obtained when the dual excitation includes a 45° phase delay. Two techniques for detailed analysis of these antennas are presented in Ref. 16, whereas a more comprehensive description is found in Chapter 7 of Ref. 1.

The fringe fields from the patch are equivalent to the radiation expected from a slot. For example, Fig. 11.20 shows four slots in a rectangular waveguide, which, when operating in its dominant TE$_{10}$ modes, are excited by the waves within the waveguide and radiate equivalent to the edges of a patch antenna. Like patch antennas, slot antennas can be mounted close to

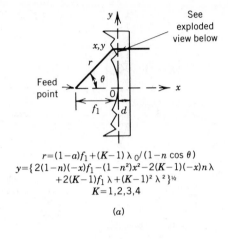

$$r = (1-a)f_1 + (K-1)\lambda_0 / (1-n\cos\theta)$$
$$y = \{\,2(1-n)(-x)f_1 - (1-n^2)x^2 - 2(K-1)(-x)n\lambda$$
$$+ 2(K-1)f_1\lambda + (K-1)^2\lambda^2\,\}^{1/2}$$
$$K = 1,2,3,4$$

(a)

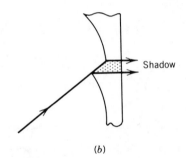

(b)

Figure 11.18. *Stepped concave lens cross section and equations for $n < 1$. (a) Lens; (b) shadow region.*

the surface of missiles, aircraft, and spacecraft with a minimum feedline design. Since the slots radiate, power is removed from the waveguide. Closed-form expressions have been developed (27) for the equivalent conductance (resistance) across the waveguide for the common slot configurations shown in Fig. 11.20a under typical conditions of TE_{10} waveguide mode, high conductivity thin walls, a slot width small compared to its length, and the field in the slot transverse to its length and varying sinusoidally along its length. Reference 28 and Chapter 8 of Ref. 1 are also recommended references.

11.4.6 Phased Array

Spiral, patch, and slot antennas exhibit broad beamwidth patterns (low gain). The radiated energy from an array of these antennas can be combined in space to increase their gain. The antennas become elements in a phased array whose combined performances meets the design goals while being fabricated from simple elements. Arrays can also be designed using reflectors, horn, and lens antenna elements, but usually simpler, planar elements are chosen to keep the cost of fabrication low. Arrays of thousands of elements have been built. Phased array techniques are most often used in radar applications where multiple, rapidly scanned beams are required.

Electronically controlled phase shifters and switches are the heart of phase scanned arrays (time-delay and frequency scanned arrays are also possible). The geometric position and phase of the radiating elements determine the performance of the array. In the array, the combination of the elements is important, whereas the performance of an array is determined by the number of elements rather than their detailed performance. Reference

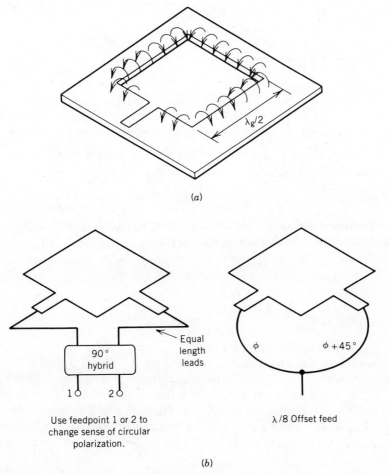

(a)

Use feedpoint 1 or 2 to
change sense of circular
polarization.

$\lambda/8$ Offset feed

(b)

Figure 11.19. *Fields and feeds for rectangular patch antennas. (a) Fringing fields; λ_g, wavelength in media; (b) feed techniques for circular polarization.*

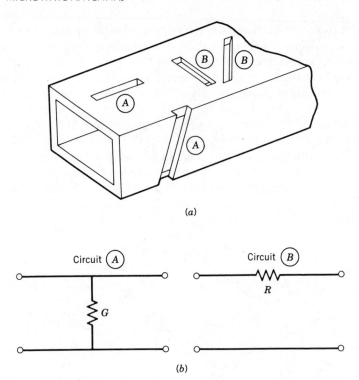

Figure 11.20. *Radiating slots in a rectangular waveguide. (a) Orientation of slots; (b) equivalent loss elements.*

5 gives the following values for an N element phase array with half-wavelength interelement spacing configured to generate a pencil beam:

$$\text{Gain } (\phi) = \pi N \cos \phi \qquad (11.26)$$

$$3 \text{ dB Beamwidth} = \frac{100}{(\pi N \cos \phi)} \qquad (11.27)$$

where ϕ is scan angle off broadside.

The normal to the phase front generated by the phase shifted radiating elements shown in Fig. 11.21a is the direction of propagation. For a scan angle ϕ, the interelement phase shift is ψ (radians) $= (2\pi/\lambda) \, d \sin \phi$. Thus if the phase is independent of frequency, the scan angle ϕ varies with frequency. The phase shifters vary 0 to 2π radians. For the uniformly spaced linear array in Fig. 11.21b, the far-field electric field from the elements is

$$E_{\text{array}}(\phi) = \frac{1}{N^{1/2}} \sum_{n=0}^{N-1} \exp j\left(\frac{2\pi}{\lambda}\right) nd(\sin \phi) \qquad (11.28)$$

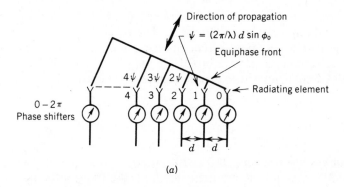

(a)

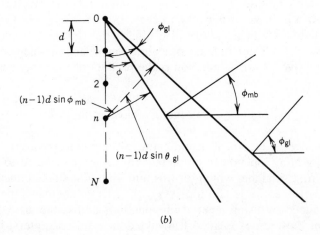

(b)

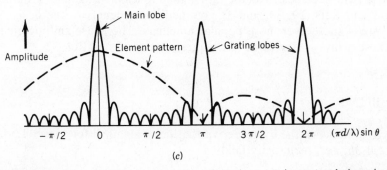

(c)

Figure 11.21. *Linear phased array analysis and results. (a) Schematic of phased array; (b) N-element phased array with uniform spacing; (c) array factor for 10 element. ϕ_{mb} and ϕ_{gl} are directions of main beam and grating lobe respectively.*

where the $N^{-1/2}$ is the electric field strength from each element fed with $1/N$ of the total input power to the array. The receive E_{array} is the same via reciprocity. If the array field is recomputed using the center element of the array as the reference and the broadside gain is set to unity ($E_{array} = 1$), the complex exponential sum can be reduced to the trigonometric relation

$$E_{array} = \frac{\sin[N\pi(d/\lambda) \sin \phi]}{N \sin[\pi(d/\lambda) \sin \phi]} = \text{array factor} \qquad (11.29)$$

This factor is shown in Fig. 11.21c for an array with $N = 10$. The pattern is repeated because of the grating lobes, i.e., the transmitted power would be distributed equally to each grating lobe if the element pattern (see below) or element spacing did not prevent it. These grating lobes occur whenever the phase delay modulo 2π yields a maximum in the array factor, i.e. (d/λ) sin ϕ equals a positive or negative integer.

The radiating element pattern also affects the radiation from the array since the array will not radiate in a direction where the elements do not radiate. The overall array radiation is the product of the array and element patterns

$$E(\phi) = (\text{array factor}) (\text{element factor}) \qquad (11.30)$$

As shown in Fig. 11.21c, if a minimum in the element pattern is located where the grating lobe(s) are predicted, they can be reduced to levels far below the first sidelobe, which for this uniformly illuminated linear array is about -13 dB (see Table 11.1 top). As with reflector antennas, the sidelobes can be reduced by using tapered illuminations across the array elements. Analysis of these cases is almost always done with computers, but follow the same principles outlined above. For two-dimensional arrays, the above formulas are applied in orthogonal coordinates and the direction of maximum electric field becomes a vector rather than a scalar. Chapters 3 and 20 of Ref. 1 and Refs. 29, 30, and 31 give additional details of the design and analysis of these ever more popular antennas. The phase shifting element for these arrays is described in Chapters 10 and 12 and Ref. 32.

PROBLEMS

1. Use network theory to derive that i/v is identical for Figs. 11.3a and 11.3b, i.e., reciprocity holds.

2. Use Eq. 11.16 to show that the magnitudes of the radiated field and the reactive field from a current element are equal at $r = k^{-1}$. If the operating frequency of the antenna is doubled, describe the change in the length of the reactive near-field region.

3. Derive the distance to the transition between near-field and far-field regions assuming phase error = 22.5° and 11.25°.

4. Derive the length of the parallel beam region. What is the edge-to-center path length difference at R?

5. Compute the distance to the far field for 10-, 3-, and 1-m diameter parabolic antennas at 4, 12, 43, and 94 GHz. Compare to size of a 300-m antenna range.

6. Compute the magnitude of the reflection coefficient for a 20-dB gain circular horn mounted on a 3-m diameter circular parabolic antenna with a focal length to diameter ratio of 0.5 at 4 GHz.

7. What is the rms deviation of the surface in centimeters for a 3-m diameter parabola operating at 4 GHz for a 1-dB gain reduction?

8. Derive Eq. 11.29.

REFERENCES

1. R. C. Johnson and H. Jasik, eds., *Antenna Engineering Handbook,* McGraw-Hill, New York, 1984.

2. J. D. Kraus, *Antennas,* McGraw-Hill, New York, 1953.

3. A. W. Rudge, K. Milne, A. D. Olver, and P. Knight, eds., *The Handbook of Antenna Design,* Vol. II., IEE Electromagnetic Waves Series #16, 1983.

4. S. Silver, ed., *Microwave Antenna Theory and Design,* McGraw-Hill, New York, 1949, Vol. 12 of the Radiation Laboratory Series.

5. M. I. Sloknik, ed., *Radar Handbook,* McGraw-Hill, New York, 1970.

6. R. Plonsey and R. E. Collin, *Principles and Applications of Electromagnetic Fields,* McGraw-Hill, New York, 1961.

7. R. E. Collin and F. J. Zucker, *Antenna Theory,* McGraw-Hill, New York, 1969.

8. H. G. Booker and P. C. Clemmow, "The Concept of an Angular Spectrum of Plane Waves, and Its Relation to that of Polar Diagram and Aperture Distribution," *Jrnl. Inst. Elect. Eng.,* Vol. 97, 1950, p. 11.

9. T. T. Taylor, "Design of Line Source Antennas for Narrow Beamwidth and Low Side Lobes," *IRE Trans. Ant. Prop.,* Vol. AP-3, January 1955, p. 16.

10. R. C. Hansen, "Gain Limitations of Large Antenna," *IRE Trans. Ant. Prop.,* Vol. AP-8, September 1960, p. 490.

11. B. Berkowitz, "Antennas Fed by Horns," *Proc. IRE,* Vol. 41, December 1953, p. 1761.

12. A. B. Pippard and N. Elson, "Elimination of Standing Waves in Aerials Employing Parboloidal Reflectors," *Jrnl. Inst. Elect. Eng. (London),* Vol. 93A, January 1946, p. 1531.

13. J. Ruze, "Antenna Tolerance Theory—A Review," *Proc. IEEE,* Vol. 54, April 1966, p. 633.

14. A. W. Love, ed., *Electromagnetic Horn Antennas*, IEEE Press, New York, 1976.

15. W. L. Stutzman and G. A. Thiele, *Antenna Theory and Design*, Wiley, New York, 1981.

16. P. Bhartia and I. J. Bahl, *Millimeter Wave Engineering and Applications*, Wiley, New York, 1984.

17. E. A. Wolff, *Antenna Analysis*, Wiley, New York, 1966.

18. B. M. Thomas, "Design of Corrugated Conical Horns," *IEEE Trans. Ant. Prop.*, Vol. AP-26, March 1978, p. 367.

19. V. H. Rumsey, "Frequency Independent Antennas," *IRE Nat. Conv. Rec., Part I*, 1957, pp. 114–118.

20. W. L. Curtis, "Spiral Antennas," *IRE Trans. Ant. Prop.*, Vol. AP-8, May 1960, pp. 298–306.

21. R. Bawer and J. J. Wolfe, "The Spiral Antenna," *IRE Intl, Conv. Rec.*, 1960.

22. T. Morita and S. B. Cohn, "Microwave Lens Matching by Simulated Quarter-Wave Transformers," *IRE Trans. Ant. Prop.*, Vol. AP-4, January 1956, p. 33.

23. E. M. T. Jones, T. Morita, and S. B. Cohn, "Measured Performance of Matched Dielectric Lenses," *IRE Trans. Ant. Prop.*, Vol. AP-4, January 1956, p. 31.

24. W. E. Kock, "Metal Lens Antenna," *Proc IRE*, Vol. 34, 1946, p. 828 and J. Ruze, "Wide-Angle Metal-Plate Optics," *Proc. IRE*, Vol. 38, January 1950, p. 53.

25. A. R. Dion, "An Investigation of a 110-Wavelength EHF Waveguide Lens," *IEEE Trans. Ant. Prop.*, Vol. AP-20, July 1972, p. 493.

26. A. R. Dion, "A Broadband Compound Waveguide Lens," *IEEE Trans. Ant. Prop.*, Vol. AP-26, September 1978, p. 751.

27. A. A. Oliner, "The Impedance Properties of Narrow Radiating Slot in the Borad Face of Rectangular Waveguide," *IRE Trans. Ant. Prop.*, Vol. AP-5, No. 1, January 1957, p. 4.

28. A. F. Stevenson, "Theory of Slots in Rectangular Waveguide," *Jrnl. Appl. Phys.*, Vol. 19, January 1948, p. 24.

29. R. C. Hansen, ed., *Microwave Scanning Antennas*, Academic Press, New York, 1966.

30. N. Amitay, V. Galindo, and C. P. Wu, *Theory and Analysis of Phased Array Antennas*, Interscience Publishers, Wiley, New York, 1972.

31. L. Stark, Microwave Theory of Phased Array Antenna—A Review," *Proc. IEEE*, Vol. 62, No. 12, December 1974.

32. L. R. Whicker, ed., "Special Issue on Microwave Control Devices for Array Antenna Systems," *IEEE Trans. Micro. Th. Tech.*, Vol. MTT-22, No. 6, June 1974.

12
Linear Passive
Applications of
Microwave Diodes
(Switches, Limiters, and
Digital Phase Shifters)

Most microwave systems use diode control devices. Typical applications are shown in Table 12.1. Microwave semiconductor diode switching was first realized (in waveguide at X-band) in 1956, just before varactors and PIN diodes. TEM diode switching became well understood in 1961. Diode phase shifters began to emerge in 1964.

This chapter describes the qualitative operation of diode switches, attenuators, limiters, and phase shifters; derives equations for switching; and describes diode biasing and attenuation measurement (1). Equations for phase shifters are given in Chapter 6 as an application of matrices.

12.1 THE DIODE

The DC voltage–current (V–I) characteristic of a typical diode is shown in Fig. 12.1. As the bias voltage increases in the forward direction, the diode begins to draw current at $+0.7$ V and reaches 100 mA at 1 V. At reverse bias, the diode draws very little current until the breakdown voltage V_B is reached (defined at 10 μA) and above that provides practically constant voltage for a wide range of current. The values given are typical of a 1N34 general-purpose rectifier.

When the diode is biased to point A sufficiently far into the linear region, the slope of the curve is steep, giving a high differential conductance ($\Delta I/\Delta V$), which corresponds to a low forward resistance, R_s. The subscript s is used because this resistance is dominated by spreading resistance for point

Table 12.1   Microwave Diode Control Applications

System	Application
Radar	Protect receiver from transmitter transients (shared antenna)—switch/limiter
	Generate short pulse for close-in range-high-speed switch
	Automatic gain control—variable attenuator or limiter
	Sensitivity time control for ECCM—variable attenuator
	Prevent transmitter coherence ECCM—high-power switch
	Tester to simulate target—switch
	Phased array antenna—diode phase shifter
Communications	AM modulator—variable attenuator
	Digital phase modulator—phase shifter
	Time domain multiplex—high-speed switch
Electronic warfare	Frequency band selection—high-power switch
	Modulator—switch or phase shifter
Measurements	Comb generator—high-speed switch
	Sampling gate—high-speed switch
	Subcarrier modulator—variable attenuator
	Reference switch—low loss switch
	Reference path control—attenuator, switch, or phase shifter
	Frequency band selector—wideband switch
	Sweeper leveler—variable attenuator
Radar detector	Modulator—waveguide switch

contact diodes on a bulk semiconductor die. At reverse bias point B the slope is practically zero corresponding to a high junction resistance, R_p. The subscript p is used because this resistor is in parallel with the diode junction capacitance. R_p is typically 10k to 1M ohm. By changing the bias point, the diode will alternately appear as a very low resistance R_s and a very high resistance R_p to small signal AC. Therefore, at forward bias the diode approximates a closed switch and at reverse bias an open switch.

A typical diode, Fig. 12.2 has parasitic impedances that are important at microwave frequencies. The junction has a small capacitance C_D in parallel with the back-biased resistance as shown in Fig. 12.3. The subscript D stands

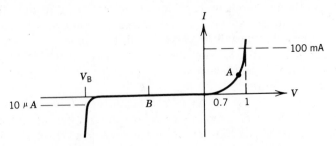

Figure 12.1. Diode voltage–current characteristic.

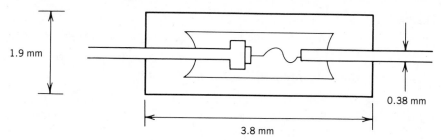

Figure 12.2. *Typical glass diode.*

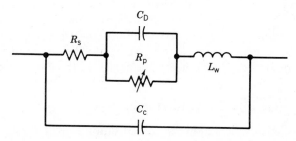

Figure 12.3. *Equivalent circuit of diode with parasitics.*

for depletion region where the carriers have been swept out by the reverse bias. The thin wire shown in Fig. 12.2 going to the junction has an inductance of 2–5 nH. This is called L_w, where w stands for whisker. In old crystal radios a 0.05 mm diameter pointed tungsten (Wolfram) wire was used to make contact at a good detector spot on a Galina crystal, and it was called a cat's whisker. Point contact diodes (1N23) are still made with a pointed tungsten wire. The remaining parasitic is the capacitance between the leads through the glass. For a 1N23 this capacitance is 0.15–0.20 pF, and this cartridge capacitance is labeled C_c. For the glass package shown in Fig. 12.2 it is very low, 0.02–0.05 pF. The resulting equivalent circuit of the diode is shown in Fig. 12.3. As bias is varied, R_p varies from 0 ohms to its high value.

Diodes are also made in pill packages, as beam lead devices, and as chip devices. The same parameters continue to be used but with smaller values to represent bonding wires and mounting capacitance where appropriate.

12.2 p–i–n DIODES AND CIRCUITS

The *p–i–n* diode has heavily doped *p* and *n* regions separated by an intrinsic region of high resistivity. Typical *i* layer thickness is 10–100 μm. If the *i* region of the diode has a light *n*-type dopant, the *i* region is labeled *ν* (nu), and if lightly p doped the i region is labeled π (pi). Note the use of the same first letters as a nmemonic. The diodes are called $p-\eta-n$ or $p-\pi-n$ or PIN.

12.2.1 Doping Profile and Equivalent Circuit

The doping profile for a typical PIN diode is shown in Fig. 12.4. The i region is in series with the $p-i$ and $i-n$ junction regions. As a result, the equivalent circuit includes an additional capacitance and resistance compared to a regular $p-n$ diode as shown in Fig. 12.5, and i-region ($R_i + C_i$) and the $p-i$ and $i-n$ junction combined ($C_d + R_p$). Typical values of the equivalent circuit elements for a microwave switch application are given in Table 12.2.

The diode is operated as a switch with low impedance for small values of forward current and high impedance for zero or reverse bias voltages. It can be operated as a variable loss element, but the power dissipated in the diode must be limited by the heat sink design of the diode and its mount.

12.2.2 Typical PIN Diode Characteristics

The PIN diode is fabricated in the configuration shown in Fig. 12.6. The die is mounted in the ''pill'' package shown in Fig. 12.7 to minimize the package parasitic reactances that tend to negate the performance of the bare diode. Actual diode forward and reverse $I-V$ characteristics are shown in Fig. 12.8. The PIN diode capacitance is nearly indpendent of the reverse bias because of the capacitance of the i layer.

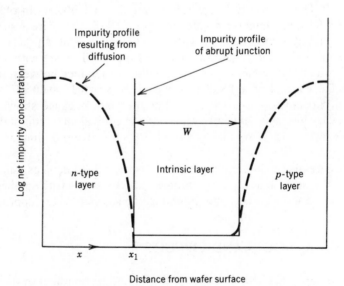

Figure 12.4. *Typical doping profile for a PIN diode.*

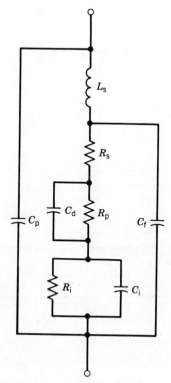

Figure 12.5. Equivalent circuit of a PIN diode.

Table 12.2 Typical PIN Diode Equivalent Circuit Element Values

| Diode Element | Bias | | |
	0 V	50-V Reverse	25-mA Forward
C_d	≈2.5 pF	≈0.5 pF	>10 pF
C_i	0.25 pF	0.05 pF	>0.25 pF
C_f	0.02 pF	0.02 pF	0.02 pF
C_p	0.3 pF	0.3 pF	0.3 pF
L_s	0.3 nH	0.3 nH	0.3 nH
R_s	0.3 ohms	0.3 ohms	0.3 ohms
R_p	≈250 ohms	≈1K ohms	<0.1 ohms
R_i	2500 ohms	10K ohms	0.5 ohms

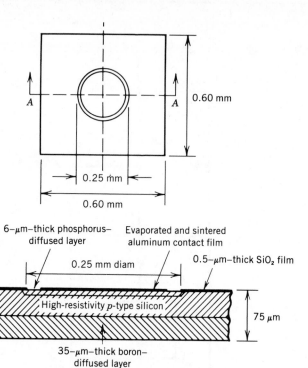

0.60 mm

0.25 mm

0.60 mm

6–μm–thick phosphorus–diffused layer

Evaporated and sintered aluminum contact film

0.25 mm diam

0.5–μm–thick SiO₂ film

High-resistivity p-type silicon

75 μm

35–μm–thick boron–diffused layer

Section A–A (enlarged)

Figure 12.6. Planar PIN diode construction.

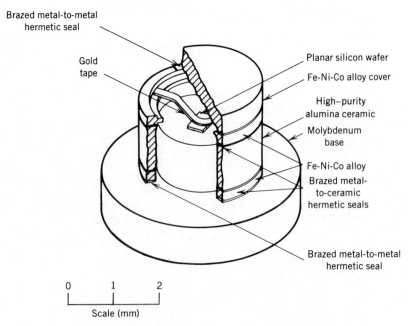

Brazed metal-to-metal hermetic seal

Gold tape

Planar silicon wafer

Fe-Ni-Co alloy cover

High–purity alumina ceramic

Molybdenum base

Fe-Ni-Co alloy

Brazed metal-to-ceramic hermetic seals

Brazed metal-to-metal hermetic seal

0 1 2

Scale (mm)

Figure 12.7. PIN diode in a microwave package.

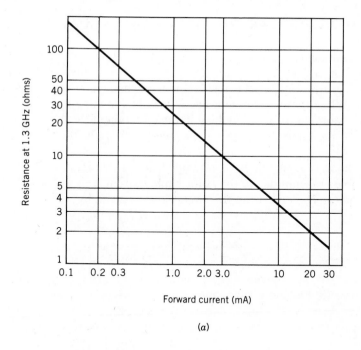

(a)

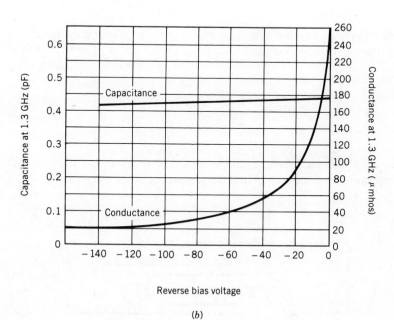

Reverse bias voltage

(b)

Figure 12.8. Resistance and capacitance of a PIN diode. (a) Resistance under forward bias; (b) capacitance and conductance under reverse bias.

12.3   DERIVATION OF THE ATTENUATION EQUATION

When the diode is put in series with the center conductor of a coaxial transmission line, it passes incident power when it is in the conducting (low-impedance) state and reflects incident power when it is reverse biased (open circuit). When the transmission line is matched looking in the direction of the generator and matched looking in the direction of the load as seen from the diode, the circuits can be represented by discrete 50-ohm resistors, as in Fig. 12.9.

The diode is represented as an impedance $Z_D(= R + jX)$. E and I are the complex magnitudes of the voltage and current ($v = E \cos \omega t$), and the load L is Z_0. The current in the circuit loop is given by

$$I = \frac{E}{2Z_0 + Z_D} = \frac{E}{(R + 2Z_0) + jX} \tag{12.1}$$

The power in the load is given by

$$P_L = \tfrac{1}{2}I(I^*)Z_0 = \frac{\tfrac{1}{2}E^2 Z_0}{(R + 2Z_0)^2 + X^2} \tag{12.2}$$

where I^* is the complex conjugate of I. When the diode impedance is zero, all of the microwave power incident on the diode (P_i) goes past it to the load. Therefore,

$$P_i = \frac{(\tfrac{1}{2})E^2 Z_0}{(2Z_0)^2} \tag{12.3}$$

When the diode impedance is not 0, power getting past the diode is reduced in amplitude (attenuated). Attenuation α is measured in decibels and given

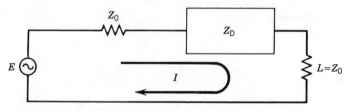

Figure 12.9. *Simple equivalent circuit of series diode in a transmission line.*

by the ratio of incident power to transmitted power:

$$\alpha = 10 \log \frac{P_i}{P_L} = 10 \log \frac{\frac{1}{2}E^2 Z_0/(2Z_0)^2}{\frac{1}{2}E^2 Z_0/[(R + 2Z_0)^2 + X^2]}$$

$$= 10 \log \left[\left(\frac{R}{2Z_0} + 1 \right)^2 + \left(\frac{X}{2Z_0} \right)^2 \right] \tag{12.4}$$

A similar expression can be derived for a shunt diode using Fig. 12.10, namely

$$\alpha = 10 \log \left[\left(\frac{G}{2Y_0} + 1 \right)^2 + \left(\frac{B}{2Y_0} \right)^2 \right] \tag{12.5}$$

The attenuation equation can also be derived from *ABCD* matrices. Recall that the voltage transfer coefficient of an *S*-parameter matrix is

$$S_{21} = \frac{2}{A + BY_0 + CZ_0 + D} \tag{12.6}$$

Attenuation is given by

$$\alpha = -20 \log | S_{21} | = 20 \log | (\tfrac{1}{2}) (A + BY_0 + CZ_0 + D) | \tag{12.7}$$

Recall that a diode impedance in series Z_D is represented by the *ABCD* matrix

$$\begin{bmatrix} 1 & Z_D \\ 0 & 1 \end{bmatrix}$$

$$\alpha = 20 \log \left| 1 + \frac{Z_D}{2Z_0} \right| = 20 \log \left| \left[\left(\frac{R}{2Z_0} + 1 \right) + j \left(\frac{X}{2Z_0} \right) \right] \right| \tag{12.8}$$

$$= 10 \log \left[\left(\frac{R}{2Z_0} + 1 \right)^2 + \left(\frac{X}{2Z_0} \right)^2 \right]$$

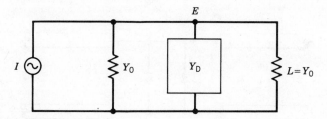

Figure 12.10. Equivalent circuit of shunt diode in a transmission line.

For switching, the lowest attenuation point is taken as the "on" state giving the insertion loss, $\delta(=\alpha_{min})$. The highest attenuation point is taken as the "off" state giving isolation, $\eta(=\alpha_{max})$.

12.4 LIMITERS

Consider two diodes mounted across a 50-ohm TEM transmission line in the same electrical plane and in opposite polarities as shown in Fig. 12.11. The DC characteristics of the two diodes are shown in Fig. 12.12. For incident voltage swings below $\pm$ 0.7 V, the diodes draw very little current and thus appear as an open circuit across the line. The output is equal to the input. For higher input voltage swings, the output clips to $\pm$ 0.7 V giving a square wave at a power level given by Eq. 12.9:

$$P_{LIM} = \frac{E^2}{Z_0} = \frac{(0.7)^2}{50} \approx 10 \text{ mW} \tag{12.9}$$

When the diodes are in the same electrical plane, Eq. 12.5 is used to determine insertion loss and maximum isolation of the limiter. When the diodes are spaced apart, they can be made part of a low-pass filter and provide low insertion loss over a wider bandwidth. The locations of the diodes in the filter structures are shown in Fig. 12.13. In Fig. 12.13a, the diodes are in the same electrical plane and are the only element of the filter. In Fig. 12.13b, a three-element filter is shown (n = 3) and the diodes are elements 1 and 3. These filters are built in TEM transmission lines such as stripline or microstrip. The series inductor is realized by a fine wire, which is a small section of high-impedance transmission line. For the diodes in the same electrical plane (n = 1), Eq. 12.5 is used. For the other structures, the elements are defined by their Chebyshev filter coefficients as given in Table 9.1. The bandwidths that can be achieved with a pair of 1-pF diodes for the different filter structures are shown in Fig. 12.14.

As an example, consider the highest frequency that can be achieved with

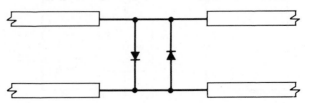

Figure 12.11. Limiter diodes in same electrical plane shunting a transmission line.

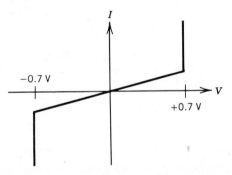

Figure 12.12. *V–I characteristics of a pair of limiter diodes.*

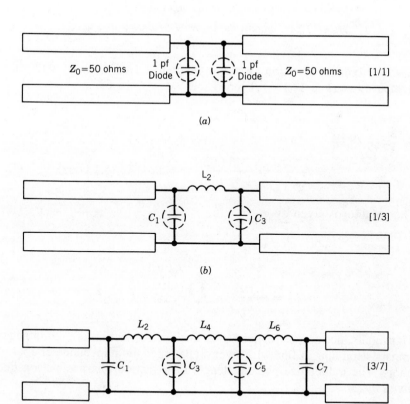

(a)

(b)

(c)

Figure 12.13. *Limiter diodes imbedded in low-pass filter structures.*

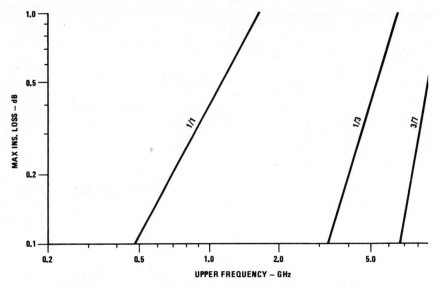

Figure 12.14. Bandwidth of two 1 pF diodes in a limiter.

0.5-dB maximum insertion loss (and the pair of 1-pF diodes), with a three element filter (Fig. 12.13*b*). From Table 9.1, $g_1 = 1.60$.

$$g_1 = \frac{\omega C_1}{Y_0} \tag{12.}$$

$$\omega = \frac{1.6 \, Y_0}{C_1} = \frac{1.6 \, (.02)}{1 \times 10^{-12}} = 32 \times 10^9 \text{ (5.1 GHz)} \tag{12.}$$

The inductor is given by $g_2 = 1.10$. $\tag{12.}$

$$g_2 = \frac{\omega L_2}{Z_0} \tag{12.1}$$

$$L_2 = \frac{1.1 \, Z_0}{\omega} = \frac{1.1 \, (50)}{32 \times 10^9} = 1.7 \text{ nH} \tag{12.1}$$

This inductance can be realized by a small series section of high-impedan transmission line. A fine wire gives a characteristic impedance of about 1 ohms. The inductance per unit length of a section of transmission line given by

$$L = \frac{Z_0}{c} = \frac{150}{3 \times 10^{10}} = 5 \text{ nH/cm} \tag{12.1}$$

where c is the speed of light (3×10^{10} cm/s in air). The desired 1.7 nH can be obtained with 0.34 cm of 150-ohm line. In order to avoid higher-order corrections to this simple model of a transmission line, it must be shorter than $\lambda/10$ at the highest frequency. The length 0.34 cm is $\lambda/10$ at 8.8 GHz and therefore is acceptable at 5 GHz. This length of line also has a capacitance associated with it, which can be represented as end capacitors of a π equivalent circuit:

$$C = \frac{1}{Z_0 c} = \frac{1}{150(3 \times 10^{10})} = \frac{2}{9} \text{ pF/cm} \qquad (12.16)$$

A total of 0.075 pF is associated with this length of line, and for accurate calculation half of it should be part of the 1 pF at each diode location. In other words, each diode can be about $1.0 - 0.04 = 0.96$ pF to provide the indicated 5.1-GHz bandwidth. The speed of light c was taken as 3×10^{10} cm/s indicating that the wire is over air and the ground plane. If there is solid dielectric between the diodes, it tends to be about 2×10^{10} varying inversely proportional to $\sqrt{\epsilon_r}$ of the dielectric.

The bandwidths shown in Fig. 12.14 are inversely proportional to capacitance. A pair of 0.5-pF diodes could therefore provide 0.5-dB maximum insertion loss up to 10.2 GHz. At an increase in the size of the filter (to $n = 7$) more bandwidth is obtainable (16.8 GHz for 0.5 dB maximum for the pair of 0.5-pF diodes).

The PIN diode junction can be bigger because of the increased thickness of the I region. The thicker I region permits a larger area diode to be used for the same value of junction capacitance. The limit level of PIN diodes is determined by diode thickness (h') and frequency (f in GHz) as (1)

$$P_{\text{LIM}} = 0.13 h'^2 f \qquad (12.17)$$

This limit level is shown in Fig. 12.15. Note that when the limit level is above 10 mW as indicated in Eq. 12.17 the limiter tends to have spike leakage (delay in turn-on) as well as a higher limiting threshold level. For the best performance of limiters it is recommended that PIN diodes be specified such that h' satisfies

$$h' = 0.28 f^{-0.5} \qquad (12.18)$$

One GHz could therefore have a 7 μm high I region, whereas 10 GHz could have a 2.2 μm high I region.

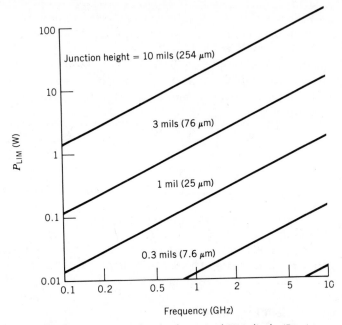

Figure 12.15. *Limit level (plateau) of PIN diode (P_{LIM}).*

12.5   BIASING THE DIODE

The circuit shown in Fig. 12.16 is required to provide bias to the center conductor and not disturb other circuits connected to the center conductor of TEM transmission lines. The reactance of the capacitors should be less than 5 ohms at the operating frequency, and the reactance of the inductors should be greater than 500 ohms. The components are easily realized in

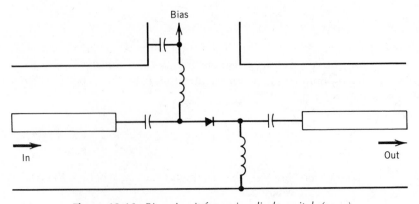

Figure 12.16. *Bias circuit for series diode switch (coax).*

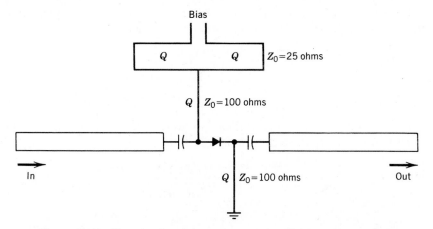

Figure 12.17. *Bias circuit using quarter wavelength elements (stripline).*

stripline using quarter-wavelength resonance principles as shown in Fig. 12.17. All sections labeled Q are a quarter-wavelength long. The capacitors are small solderable chip capacitors. This bias circuit has low VSWR over an octave bandwidth.

At the center frequency (resonance), the 25-ohm stubs convert the RF open circuits at their outside ends to an RF short circuit to ground at the common bias point making the upper 100-ohm stub look like the lower ground return stub. Both grounded 100-ohm stubs transform these grounds to open circuits across the main transmission line producing no loading to (and reflection on) the center conductor. This type of bias circuit was originally developed for diode switching but has been widely adapted to removing the IF from mixers and for biasing microwave transistors.

12.6 POWER RATINGS OF DIODE SWITCHES

The power rating levels for diode switches are dictated by their breakdown voltage, V_B, forward resistance, R_s, the power dissipating rating, P_D. A series diode in the isolation state has to hold off the entire generator voltage since the diode is the highest impedance in the loop (Fig. 12.9). If reverse bias current is to be prevented from flowing for all parts of the RF cycle, then the entire RF voltage swing must be positioned between V_B and 0 V (Fig. 12.1) and the diode must be biased to $V_B/2$ (point B). The maximum instantaneous peak RF voltage is then $V_B/2$. When this is placed in Eq. 12.3, the maximum incident peak power for a series diode is

$$\hat{P}_i = \frac{\frac{1}{2}E^2 Z_0}{(2Z_0)^2} = \frac{V_B^2}{32 Z_0} \qquad (12.19)$$

The peak rating of a shunt diode is for the low loss state at which half of the generator voltage is dropped across it. The power rating is higher by a factor of 4:

$$\hat{P}_i' = \frac{V_B^2}{8Z_0} \tag{12.20}$$

The average power a series diode can dissipate, P_D, is taken from the current of Eq. 12.1 where $X = 0$, and $R = R_s$:

$$P_D = \tfrac{1}{2}I^2 R_s = \frac{E^2 R_s}{2(R_s + 2Z_0)^2} \tag{12.21}$$

Solving for E^2 yields

$$E^2 = \frac{2(R_s + 2Z_0)^2 P_D}{R_s} \tag{12.22}$$

Recalling Eq. 12.3, the average incident power a series diode can control is given by

$$\overline{P}_i = \frac{E^2}{8Z_0} = \frac{P_D Z_0}{R_s} \left(1 + \frac{R_s}{2Z_0} \right)^2 \tag{12.23}$$

The average incident power rating for a shunt diode can be derived using G_s for R_s and Y_0 for Z_0 in Eq. 12.23 to give

$$\overline{P}_i' = \frac{P_D Z_0}{4R_s} \left(1 + \frac{2R_s}{Z_0} \right)^2 \tag{12.24}$$

Since R_s is usually much smaller than Z_0, the term in parenthesis is normally very close to 1.

Notice that the expressions for series and shunt diodes differ only by a factor of 4. The same plot can show the performance of a diode in series or shunt by using two scales for the characteristic impedance variable as shown in Fig. 12.18. Under both curves average power and peak power are both satisfied. To the left of the intersection of the curves peak power is still satisfied but average power is exceeded. Pulse power can be controlled, but care must be exercised that the duty cycle be low enough to prevent exceeding the average power. The areas above the peak power limit are useful only for shunt diodes used as limiters. They are always in conductance when high power is present. Above the average power limit the duty cycle must be considered. A dashed line is also shown in Fig. 12.18 indicating expanded peak power limit. PIN diodes with thick I regions do not respond instantly to incident RF when they are above the rectification guidelines indicated in

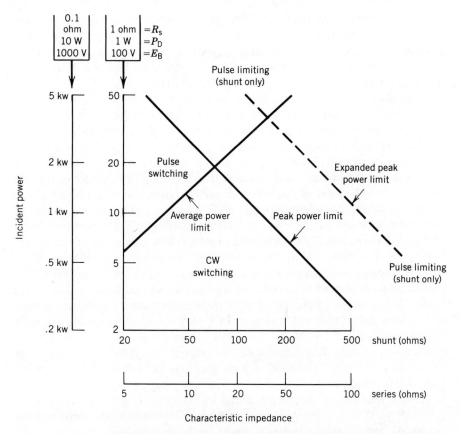

Figure 12.18. *Power ratings of typical diodes.*

Eq. 12.18. Isolation can be sustained by biasing the diode to less the $V_B/2$, usually just a few volts. The peak voltage can then approach V_B giving a peak power limit four times higher as indicated by the dashed line in Fig. 12.18. When biasing to less than $V_B/2$ care must be exercised in the system application because the diode switch tends to generate harmonics and intermodulation products. Very high peak power applications usually result in multiple diodes mounted in a low-impedance configuration. When mounted in parallel (in the same electrical plane), the R_s is divided and the P_D is multiplied by the number of diodes in parallel.

12.7   MEASUREMENT OF SWITCHING

The assumption made for the switch discussion was that the diode sees a matched load looking in both directions. Therefore, if the generator is not well matched it should have a 20-dB attenuator on it, and if the detector is not well matched, it too should have a 20-dB attenuator.

To make the measurement, the two 20-dB attenuators are first connected together to establish a reference power level. The output level of the signal generator or the gain of the indicator is adjusted until 0 dB is indicated. Then the DUT (device under test—Fig. 12.19) is inserted, and the drop in the indicator indicates the insertion loss of the switch.

A number of precautions must be observed to obtain accurate measurements. Drift in power in the generator or detector can cause errors for low insertion loss readings. Warming the equipment up for 1 hr with the 20-dB attenuators connected together (no DUT) and the indicator showing 0 dB reduces turn-on drift. Observing the 0-dB fluctuation gives an indication of the error. Disconnecting and reconnecting the 20-dB attenuators gives an even better indication of the error. Finally, setting the zero (20-dB attenuators together), then inserting the DUT and measuring the low-insertion loss, then rechecking the zero as quickly as possible gives the best indication of low insertion loss. The measured insertion loss should be corrected by half of the drift indicated by the second zero. To facilitate the measurement of all values of attenuation for a diode, the insertion loss is measured by averaging a number of insertion loss measurements obtained by alternating rapidly between the reference (straight through) and the diode in the on state (DUT inserted). Once the insertion loss of the switch has been carefully established, measurements of the other attenuation states can be accomplished by resetting the reference to the insertion loss reading with the diode in the on state. All of the work of reestablishing the reference by removing the DUT is eliminated.

If the capacitor on the output of the switch (Fig. 12.17) is omitted and a large forward bias current is used to turn on the switch, the voltage drop across the ground return stub is imposed on the detector and can cause errors in low-insertion loss measurements. If the generator or detector is well matched, the appropriate attenuators can be omitted. When PIN diodes are biased to their extremes, there are no problems of harmonic generation. However, when point contact or Schottky diodes are biased around the R_p = 50-ohm point, harmonics are generated by the diode and introduce errors in attenuation measurements. To avoid the harmonics, the detector attenuator should be used and a low-pass filter should be placed in front of the detector. The harmonics can be a problem in the system application of the variable diode attenuator, and therefore PIN diodes are favored in variable diode attenuator applications. When the detector is a diode rectifier rather than a power meter, the linearity over the range of attenuation being measured should be checked using calibrated attenuators in place of the DUT.

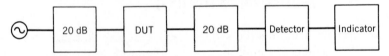

Figure 12.19. Measurement circuit for diode switching.

12.8   PHASE SHIFTERS

Most diode phase shifters are digital giving bits of 180°, 90°, 45°, and $22\frac{1}{2}°$. Seldom are five or more bits used because each bit increases loss and in most systems other effects contribute to errors larger than the fifth $11\frac{1}{4}°$ bit.

The three commonly used ways to make digital diode phase shifters are the switched line, reflection, and loaded line.

For the switched-line phase shifter, two double throw switches are used, one at each end of a pair of lines of unequal length as shown in Fig. 12.20. When the upper line is switched in, the microwave signal travels an extra phase distance $\Delta\phi$, compared with the lower path.

Care must be exercised in the design of switched-line phase shifters so that the effective length of the off line is not $\lambda/2$ in either state. The troublesome regions are shown by the intersections of the lines in Fig. 12.21 (the 10- and 20-dB lines are for the isolation of the diodes if they are used for series switches). When the off line is resonant, it couples to the on line and gives phase error and increased insertion loss.

For the reflection phase shifter, the reflection properties of a 3-dB coupler are used. Power in the in port of the 3-dB coupler shown in Fig. 12.22 divides and emerges equally from the two right-hand ports. When both right-hand ports have identical reflectors, the reflected waves add in the 3-dB coupler so that it all emerges at the out port. When the path length to the shorts is reduced as indicated by the switchable short circuits, the plane of reflection is moved closer to the 3-dB coupler and the path length of the reflected microwave power is shorter by twice the distance between the short circuits (the microwaves have to make a round trip over the path). This type of phase shifter is used almost universally for the 180° bit in digital diode phase shifters and for about half of the applications for the other bits. When it is used for the 180° bit, the diode simply serves as the termination—no other line lengths are required. The difference between an open circuit (reverse biased diode) and short circuit (forward biased diode) is 180° on the Smith chart, which corresponds to the desired 180° phase shift and is relatively independent of frequency.

Moderately wide bandwidths can be achieved with the reflection phase shifter. A single section quarter-wavelength 3-dB coupler normally provides

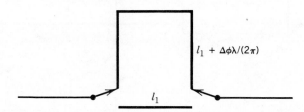

$$l_1 + \Delta\phi\lambda/(2\pi)$$

$$l_1$$

Figure 12.20. Simple schematic of switched-line phase shifter.

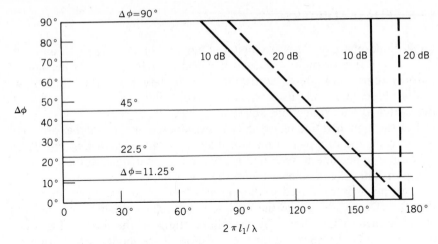

Figure 12.21. *Switched-line phase shifter error regions.*

an octave bandwidth. By controlling the parasitic inductance and capacitance of the switching diodes, the terminations of the 3-dB coupler can be made to provide ± 2° or less phase error over an octave or greater bandwidth as indicated in Fig. 12.23. For example, to make a 45° ± 2° phase shift bit working from 1 to 2 GHz terminating a Z_0 = 50-ohm transmission line, the following calculations would be made:

$$\omega_0 = 2\pi \times 10^9 \tag{12.25}$$

$$Z_0' = kZ_0 = (0.3)\,(50) = 15 \text{ ohm} \tag{12.26}$$

$$L = \frac{Z_0'}{\omega_0} = \frac{15}{2\pi \times 10^9} = 2.4 \text{ nH} \tag{12.27}$$

$$C = \frac{1}{Z_0'\omega_0} = \frac{1}{(15)\,(2\pi \times 10^9)} = 10.6 \text{ pF} \tag{12.28}$$

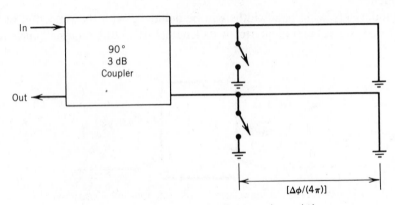

Figure 12.22. *Schematic of reflection phase shifter.*

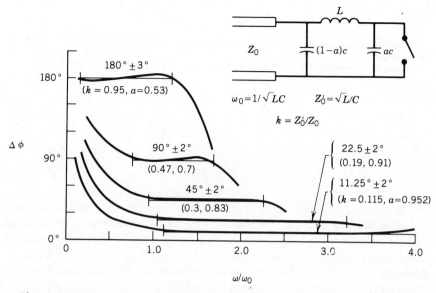

Figure 12.23. Design parameters for broadband reflection diode phase shifters.

$$C_D = aC = (0.83)(10.6 \text{ pF}) = 8.8 \text{ pF} \qquad (12.29)$$

where a is that portion of the total diode capacitance associated with the junction. For this example the cartridge capacitance is

$$C_c = (1 - a)C = (0.17)(10.6 \text{ pF}) = 1.8 \text{ pF} \qquad (12.30)$$

The other parameters of the reflection diode phase shifter can be calculated from the *ABCD* matrix parameters of a single diode:

$$
\begin{vmatrix} A & B \\ \\ C & D \end{vmatrix} =
\begin{vmatrix} 1 - a\left(\dfrac{\omega}{\omega_0}\right)^2 & j\left(\dfrac{\omega}{\omega_0}\right)Z_0' \\ \\ j\left[\dfrac{1}{Z_0'} - (1-a)\left(\dfrac{\omega}{\omega_0}\right)\right]\left(\dfrac{\omega}{\omega_0}\right) & 1 - (1-a)\left(\dfrac{\omega}{\omega_0}\right)^2 \end{vmatrix}
$$

$$(12.31)$$

$$\overline{P}_i = \frac{P_D(\,|\,B\,|^2 Y_0^2 + D^2)}{4R_s Y_0} \qquad (12.32)$$

$$\hat{P}_i = \frac{V_B^2}{32 Y_0}(A^2 Y_0^2 + |\,C\,|^2) \qquad (12.33)$$

$$\delta \approx 10 \log\left[1 + \frac{4R_s Y_0}{|\,B\,|^2 Y_0^2 + D^2}\right] \qquad (12.34)$$

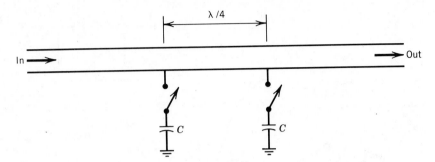

Figure 12.24. *Schematic of loaded-line phase shifter (stripline).*

The loaded line phase shifter consists of two switchable capacitors spaced a quarter wavelength apart as shown in Fig. 12.24. When the capacitors are switched in they load the line, slowing down propagation and increasing phase delay. When they are a quarter wavelength apart, the left-hand one tends to tune out the reactive mismatch introduced by the right-hand one as seen moving away from the load toward the generator. This type of phase shifter is used in the remaining half of the non-180° applications. This structure is especially useful for small phase shifts, giving good bandwidth, low insertion loss, and fair power-handling capacity.

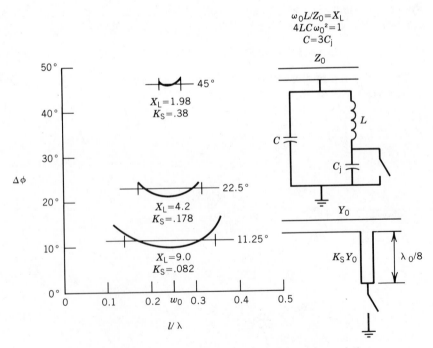

Figure 12.25. *Design parameters for loaded-line phase shifters.*

The parameters for designing these loaded phase shifters can be derived from Fig. 12.25. The curves stop at a VSWR of 1.46, and the ticks indicate the ends of the $\pm 2°$ phase error range. The elements are placed $\lambda/4$ apart at ω_0, and the design parameters are given for lumped element diodes or diodes terminating a $\lambda/8$ stub. The bandwidth is less than an octave for $45°$ and $22.5°$. However, the phase shifter does not require a 3-dB coupler, which makes it desirable for microstrip circuits not requiring wide bandwidth.

The other parameters of the loaded line phase shifter are given by

$$\overline{P}_i = \frac{\overline{P_\mathrm{D}}Z_0}{2R_\mathrm{s}} \left(\frac{4}{K_\mathrm{s}^2} + 1 \right) \tag{12.35}$$

$$\hat{P}_i = \frac{V_\mathrm{B}^2}{16Z_0} (4 + K_\mathrm{s}^2) \tag{12.36}$$

$$\delta = 20 \log \left[1 + 2 \frac{R_\mathrm{s}}{Z_0} \cdot \frac{K_\mathrm{s}^2}{(K_\mathrm{s}^2 + 2)} \right] \tag{12.37}$$

PROBLEMS

1. Given a diode in series in a 50-ohm transmission line with $R_\mathrm{s} = 5$ ohms, $C_\mathrm{D} = 0.2$ pF, $C_\mathrm{c} = 0.05$ pF, $L_\mathrm{w} = 5$ nH, $R_\mathrm{p} = 10$ k ohms, find δ and η at 1 GHz. (*Note:* Short out R_p for δ.)

2. Design a 180° phase shifter by making a sketch equivalent to Fig. 12.17 showing all bias lines and lengths necessary to obtain the desired phase shift.

3. Given a PIN diode with a resistance at forward bias of 2 ohms, a breakdown voltage of 100 V, and a power dissipating capacity of 0.5 W, how much peak and average power can it switch in series in a 50-ohm line? In shunt?

REFERENCES

1. R. V. Garver, *Microwave Semiconductor Diode Control Devices,* Artech House, Waltham, MA, 1976.

13 Nonlinear and Active Applications of Microwave Diodes (Detectors, Mixers, Harmonic Generators, and Oscillators)

13.1 INTRODUCTION

Microwave diodes operate in the same basic manner as their low-frequency counterparts designed for rectification and logic circuitry. Some microwave diodes, because of the materials and the junction region carrier distribution, are able to amplify and oscillate when placed in the proper circuits. The device parameters and the circuitry required to use these devices efficiently are described in this chapter. Descriptions of passive devices (detectors and mixers) and active devices (amplifiers and oscillators) are included.

The majority of active solid-state devices (shown in Fig. 13.1) have two terminals (diodes). However, three-terminal microwave transistors and field-effect transistors (FETs) are also important (see Chapter 14).

13.2 NONLINEAR DIODES

Nonlinear diodes are used as detectors and mixers. Linear applications as switches, limiters, and phase shifters have been described in Chapter 12.

13.2.1 Parameters of Semiconductor Junctions

Most microwave diode $p–n$ junctions are fabricated using diffusion of dopants into a semiconductor. Techniques such as ion implantation are used for very shallow junctions. The other common junction is the metal–semiconductor (Schottky-barrier) junction.

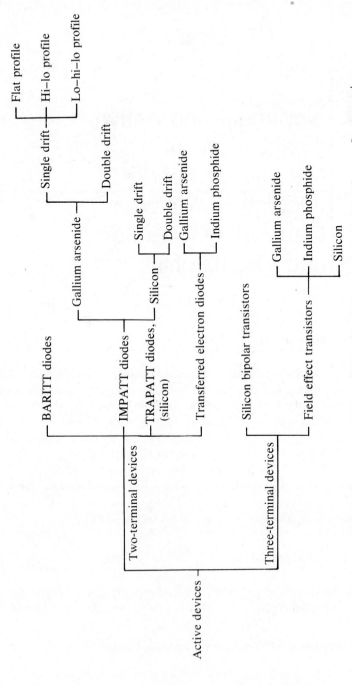

Figure 13.1. *Active solid-state microwave devices. (Courtesy Bert Berson, Berson and Associates, Mountain View, CA)*

Each technique yields different carrier profiles within the junction. This distribution of the ionized impurities is the most important parameter of a diode. Other factors include the hole and electron mobility, the minority carrier lifetime, and the geometry or structure of the junction. The factors combine to determine the terminal characteristics of the diode, which are represented by its equivalent circuit (Fig. 12.3). The junction capacitance is a function of the junction voltage. The reverse voltage across the diode is limited by the breakdown voltage V_b, whereas the maximum forward current is determined by heat dissipation. These parameters are shown for a typical diode in the familiar diode current–voltage (I–V) characteristic shown in Fig. 12.1.

Both a DC resistance and a differential resistance are defined by the I–V curve. The differential resistance is used for small-signal applications. In some cases the variation of the small-signal differential resistance is important. The average operating current–voltage values set the diode's DC operating resistance.

Most diodes used below 10 GHz are mounted in packages. Their equivalent circuit is included as L_w and C_c in Fig. 12.3. For the purpose of designing nonlinear and active devices, the equivalent circuit must be made more detailed as shown in Fig. 13.2. Compared to Fig. 12.3, the junction capacitance is now variable and represented by C_j.

In most applications it is necessary to match the diode impedance to the microwave circuit impedance to optimize the circuit performance. The diode cross section can be reduced to increase resistance and decrease capacitance

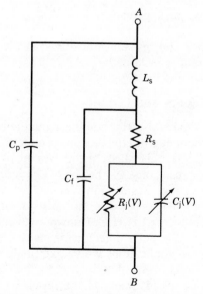

Figure 13.2. *Equivalent circuit of a packaged diode.*

for a given bias voltage. Alternatively, the impedance of the circuit can be modified to match to the diode.

13.2.2   Varistors and Varactors

The nonlinear current–voltage curve (Fig. 12.1) and variation of the junction capacitance with bias voltage lead to the concepts of the varistor (variable resistor) and the varactor (variable reactance). For small signal applications, the resistance varies from R_s for large positive values of forward bias to R_j (in the megohm range) for large values of reverse bias (but less than the reverse breakdown voltage V_b). Near zero bias the value of the resistance varies rapidly at the terminals of the diode for varying bias and signal voltages. This change of resistance with voltage (nonlinear resistance effect) allows diodes to operate as detectors and mixers.

The change of the junction capacitance with voltage $C_j(V)$ corresponds to a change of reactance with voltage $X_j(V)$ through the familiar relation

$$X_j(V) = \frac{1}{2\pi f C_j(V)} \tag{13.1}$$

Because of the other elements shown in Fig. 13.2, namely,

$$R_s = \text{series resistance}$$
$$C_f = \text{fringing capacitance}$$
$$L_s = \text{package bonding strap inductance}$$
$$C_p = \text{package capacitance}$$

the values of reactance and resistance observed at the terminals A–B will be significantly different than R_j and X_j.

13.2.3   Detector Diodes and Circuits

Microwave detector diodes use the nonlinear junction resistance for their operation. Metal-semiconductor contacts perform better than p–n junctions because the diffusion capacitance (associated with the p–n junction) is lower. In addition the series resistance of the metal semiconductor is generally lower than the p–n junction. Besides introducing loss, this series resistance tends to mask the nonlinear resistance. An ancestor of the Schottky diode is the point-contact diode [e.g., H.H.C. Dunwoody's cat's whisker detector (1)], which was the most popular detector for many years and is still used in many millimeter wave applications where extremely small junction areas are desirable.

The backward diode avoids many of the problems associated with the p–n junction diode. Backward diodes are fabricated with extremely heavy

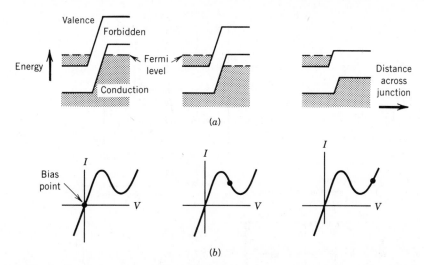

Figure 13.3. *Band structure and backward diode current as a function of bias voltage. (a) Conduction and valence energy band diagrams; (b) corresponding bias points on diode current–voltage characteristic.*

doping on both sides of the junction. The junction is abrupt, and at very low values of forward bias electrons are able to quantum mechanically "tunnel" across the junction region. The backward diode current–voltage characteristics that arise from the conduction and valence band diagrams are shown in Fig. 13.3. For reverse bias conditions, the diode breaks down (V_B almost zero) and current flows. For voltages between the peak and valley of the forward current, the diode exhibits a negative differential resistivity. This effect allows tunnel diodes to operate as oscillators and amplifiers. For the backward diode detector application, the peak current at low voltage is suppressed by proper doping profiles and fabrication techniques to provide a very nonlinear current–voltage characteristic near zero. A comparison of the current–voltage characteristics for the Schottky, point-contact, and backward diodes is shown in Fig. 13.4. Schottky diodes for zero external bias detectors have been developed. These units have reverse breakdown voltages of several volts. Regular Schottky diodes are forward biased when operating as a low-level detector.

The detector diode is used to convert the modulation on a microwave signal to the information-bearing, low-frequency (video) signal and as a square-law detector to measure the power in a signal. The detector's sensitivity is limited by the low-frequency noise (proportional to $1/f$) generated by the diode. This noise masks the information near the carrier of the received signal. For the backward diode, the large nonlinearity near zero bias allows the diode to operate with lower shot noise (due to the small current flow) and lower $1/f$ noise compared to a point-contact diode. Schottky diodes also have low $1/f$ noise if their surfaces are properly conditioned.

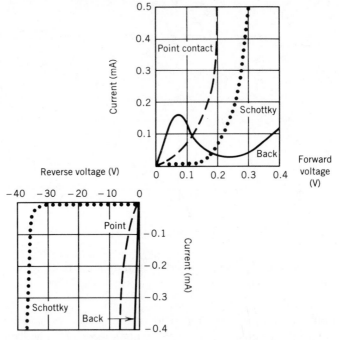

Figure 13.4. *Typical current–voltage characteristics for various diode types. Exact values depend on material, doping profiles, and temperature. (Courtesy of the Narda Microwave Corp., Hauppauge, NY.)*

13.2.3.1 Nonlinear Resistance.

The current–voltage relation of a diode is (using the formulation of 2)

$$I(V) = I_s\{[\exp(eV/nkT)] - 1\} \tag{13.2}$$

For a range of voltage about a bias voltage V_0, this exponential relation can be expressed as a Taylor series

$$I(V) = I(V_0) + \frac{dI}{dV}\Delta V + \frac{1}{2}\frac{d^2I}{dV^2}(\Delta V)^2 + \cdots \tag{13.3}$$

An amplitude-modulated voltage placed on this detector is

$$v = V_p(1 + m\sin\omega_s t)\sin\omega_{rf}t \tag{13.4}$$

where V_p is the peak voltage for the unmodulated RF carrier, m is the modulation index, ω_s is the signal angular frequency, and ω_{rf} is the RF carrier angular frequency. Using the relation for the product of sinusoids

$$v = V_p \sin \omega_{rf}t + \frac{1}{2}V_p m\{\cos(\omega_{rf} - \omega_s)t - \cos(\omega_{rf} + \omega_s)t\} \quad (13.5)$$

which contains the signal voltages associated with the sidebands. Setting v = ΔV and substituting into Eq. 13.3, the term $\frac{1}{2}(d^2I/dV^2)(\Delta V)^2$ yields the relative values of current flowing through the diode at the indicated frequencies in Table 13.1. By suitable filtering, the ω_s term will be read out of the detector circuit yielding the modulation information. The filter bandwidth should be less than $2\omega_s$ to suppress the output from this component of the detector diode current. The DC term is the square law detector output proportional to V_p^2.

The above analysis assumed that the diode operates without any parasitic elements surrounding it. In practice the circuit tuning elements are used to match the RF signal to the diode resistance, R_j.

13.2.3.2 Detector Circuits. Figure 13.5 shows the equivalent circuit of the diode detector between the matching elements to the RF input and the low-pass filter for the video and/or DC output. C_j should be small to avoid shunting the RF current around the diode resistance R_j. A measure of the diode quality involving the factors affecting efficiency is the diode cutoff frequency defined as

$$f_c = \frac{1}{2\pi R_s C_j} \quad (13.6)$$

Table 13.1 Amplitude of Various Frequency Components in a Detector

Frequency	Relative Amplitude
DC	$1 + \dfrac{m^2}{2}$
ω_s	$2m$
$2\omega_s$	$\dfrac{1}{2}m^2$
$2(\omega_{rf} - \omega_s)$	$\dfrac{1}{4}m^2$
$2\omega_{rf} - \omega_s$	m
$2\omega_{rf}$	$\dfrac{1}{2} + \dfrac{1}{2}m^2$
$2\omega_{rf} + \omega_s$	m
$2(\omega_{rf} + \omega_s)$	$\dfrac{1}{4}m^2$

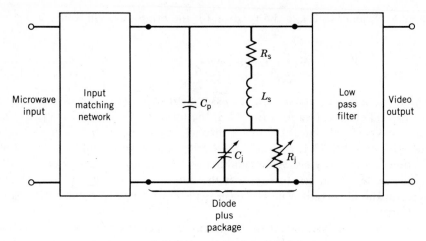

Figure 13.5. Equivalent circuit of a detector.

where R_s represents the loss expected in the diode due to the series resistance of the diode. The cutoff frequency is that frequency at which the reactance of the junction is equal to the spreading resistance.

The sensitivity of a detector to low-level signals is related to the $1/f$ noise, the impedance match of the circuits around the diode, and the diode parasitics. The impedance matching of the diode to the microwave circuit is adjusted by varying the bias current through the diode. Currents of 10–100 μA are common.

13.2.3.3 Selection of Microwave Detectors. Three types of diodes used in detector applications have been described: point–contact, Schottky, and backward diodes. Some guidelines for selection of the best diode type for a particular application follow:

a. Choose a point-contact detector for
   Best frequency flatness (also consider a Schottky diode)
   Extremely high frequencies (above 40 GHz)
   High video output resistance
   Low cost
b. Choose a Schottky detector for
   Best tangential sensitivity
   Maximum power handling
   Highest resistance to burnout
   Integration into microwave integrated circuits (beam lead diodes)
c. Choose a back (backward) detector for
   Smallest variation with temperature
   Wide-video bandwidth (low-video resistance)
   Applications where external biasing is not possible

Typical characteristics for detector diodes are shown in Table 13.2 and Fig. 13.6.

Table 13.2 Typical Characteristics of Microwave Detectors

Category	Point Contact	Schottky	Schottky	Backward
Frequency range commercially available (GHz)	1–100 Broadband	1–18 Octave and Broadband	18–40 Broadband	1–18 Octave and Broadband
Open circuit voltage sensitivity (mV/mW)	400	1000	1000+	700
Tangential sensitivity (dBm)	−35	−55	−50	−50
Power handling capacity (mW)	100	100–150	100 CW 1000 pulsed	30–50
VSWR (max.)	>2	2.2	Fig. 13.6	3
Flatness (dB)	±1–18 GHz	±0.6	Fig. 13.6	±0.8
Temperature sensitivity −54–100°C (dB)	±1.5	±1	±1	±0.5
Video resistance (ohms)	3000–15,000	200–400	350 at 100 μA	50–150
Shock and vibration	Affects sensitivity	No effect	No effect	Fragile

(Courtesy of the Narda Microwave Corp., Hauppauge, NY.)

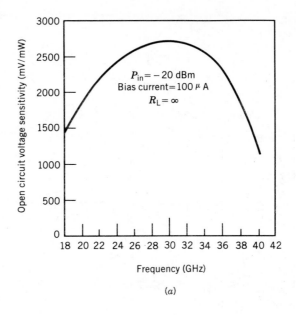

(a)

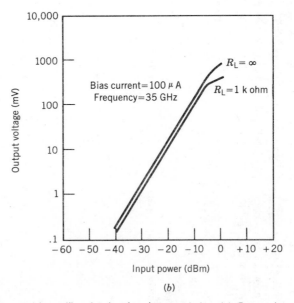

(b)

Figure 13.6. Typical broadband Schottky characteristics. (a) Open circuit voltage sensitivity versus frequency; (b) typical transfer characteristics; (c) typical sensitivity versus bias current; (d) VSWR versus frequency. (Semiconductor Bulletin No. 4221, MA/COM Inc., Dec. 1980.)

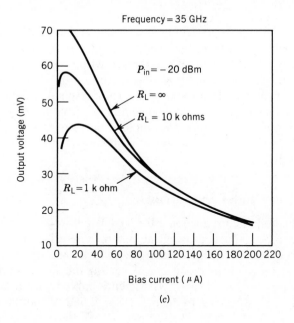

Frequency = 35 GHz

$P_{in} = -20$ dBm

$R_L = \infty$

$R_L = 10$ k ohms

$R_L = 1$ k ohm

Bias current (μA)

(c)

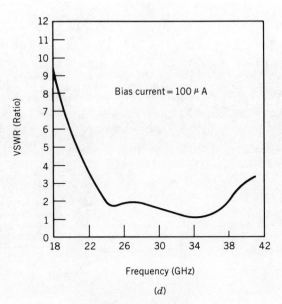

Bias current = 100 μA

Frequency (GHz)

(d)

Figure 13.6. (continued)

The tangential sensitivity technique is used to measure the sensitivity of a detector. Observers obtain the results using a display similar to Fig. 13.7. The noise at the detector output is observed with no signal, as shown at the ends of Fig. 13.7. The signal power is increased until the bottoms of the noise spikes are level with the spike tops with no signal. The power required to obtain this display is termed the tangential sensitivity. The video amplifier bandwidth is important because the amplifier noise contributes to the output. The tangential sensitivity measurement corresponds to a signal-to-noise ratio of about 2.6 dB.

13.2.4 Mixer Diodes and Circuits

The detector diode is not adequate for signals below −50 or −60 dBm. Its sensitivity can be improved using the heterodyne principle used in the superheterodyne receiver. The heterodyne involves mixing two signals in the nonlinear resistance of the diode and separating the signals in the associated circuitry. Normally, the desired RF signal is mixed with a local oscillator voltage, and the difference frequency is amplified in the intermediate fre-

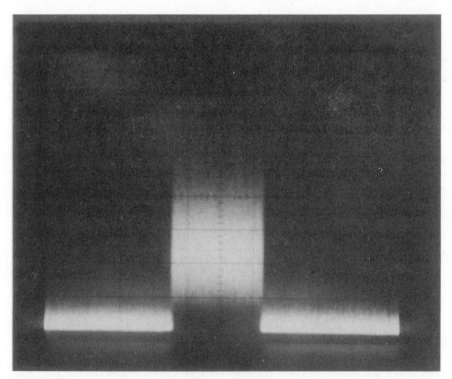

Figure 13.7. Visual display of a tangential sensitivity measurement. (Hirsch, Ronald B., "Know the Meaning of Signal-To-Noise Ratio," Microwaves and RF, February 1984)

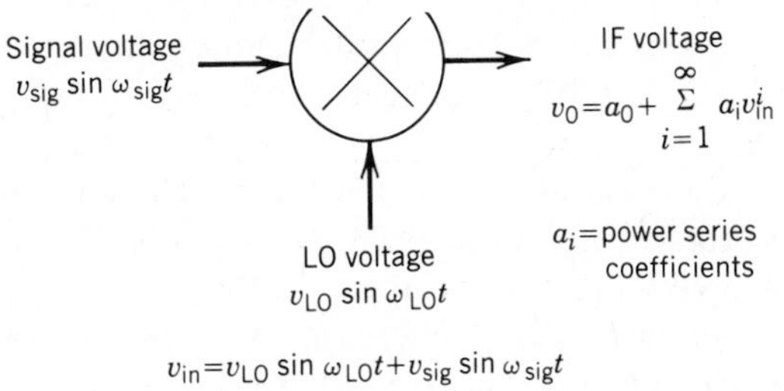

Signal voltage
$v_{sig} \sin \omega_{sig}t$

IF voltage
$v_0 = a_0 + \sum\limits_{i=1}^{\infty} a_i v_{in}^i$

LO voltage
$v_{LO} \sin \omega_{LO}t$

a_i = power series
coefficients

$v_{in} = v_{LO} \sin \omega_{LO}t + v_{sig} \sin \omega_{sig}t$

Figure 13.8. Schematic of a mixer and voltages for one input signal (terms defined in text).

quency (IF) amplifier for processing. The mixer is schematically depicted in Fig. 13.8 along with the voltages for one RF signal. The IF frequency is chosen high enough so the diode $1/f$ noise is negligible and to reduce the effects of signals at the image frequency. The image frequency is the RF on the other side of the local oscillator frequency from the desired signal frequency (image frequency equals RF plus or minus the IF).

13.2.4.1 Nonlinear Resistance. A formulation similar to the detector expansion in Section 13.2.3 can be developed for the mixer. The analysis is cumbersome, so the simplified method using the power series expansion for the output voltage is used. The resulting IF voltage is

$$V_{IF} = a_0 + a_1 v_{in} + a_2 v_{in}^2 + a_3 v_{in}^3 + \cdots \qquad (13.7)$$

where

$$v_{in} = v_{LO} \sin \omega_{LO}t + v_{sig} \sin \omega_{sig}t \qquad (13.8)$$

The first two terms are a DC offset voltage and a replica of the input voltage. The second-order term yields the interesting result

$$a_2 v_{in}^2 = a_2 \{ \tfrac{1}{2} v_{LO}^2 [1 - \cos(2\omega_{LO}t)] + \tfrac{1}{2} v_{sig}^2 [1 - \cos(2\omega_{sig}t)]$$
$$+ v_{LO} v_{sig} [\cos(\omega_{LO} - \omega_{sig})t - \cos(\omega_{LO} + \omega_{sig})t] \} \qquad (13.9)$$

In the above relation, the DC terms and second harmonic terms are proportional to the square of their input amplitudes (power). The heterodyne products are proportional to the input signal level (the local oscillator voltage is assumed constant) and occur at the sum and difference frequencies. The bandwidth of the IF amplifier selects one of these frequencies and rejects the other frequency.

13.2.4.2 *Mixer Performance.* Mixer diodes can be arranged in several circuit configurations (Fig. 13.9) to cancel unwanted output terms (2). The performance of mixers is specified by the following parameters over the frequency of operation and at a specified local oscillator drive level: conversion loss, noise figure, isolation, 1-dB compression level, third-order intercept level, and VSWR. A key parameter is the conversion loss (a_2^2 in the earlier analysis). Typically, the conversion loss (ratio of power in the IF signal to the input RF signal power) is in the range 5–10 dB (yielding values of $a_2^2 = 0.3–0.1$). The value of a_2 depends on the port impedance, whereas the conversion loss is defined in terms of power. The noise figure usually includes the noise contributed by the input circuit of the IF amplifier.

Isolation is a measure of the amount of RF, LO, and IF that leak through to the other ports of the mixer. The 1-dB compression point and the third-order intercept are power levels that specify the amplitude nonlinearity between the RF input amplitude and the IF amplitude, and the RF power level where $a_2 v_1 = \frac{3}{4} a_3 v_1 v_2^2$ (the third-order intermodulation products) equals the regular IF signal power, respectively. The graphical representation of these parameters is shown in Fig. 13.10.

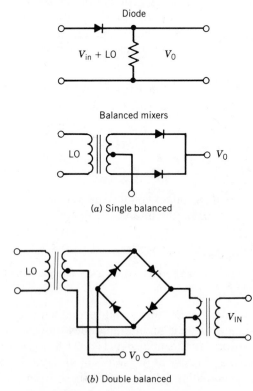

Figure 13.9. *Mixer circuit configurations.*

13.2.5 Varactor Diodes and Circuits

In Section 13.2.2 it was noted that the junction capacitance varies with the bias, resulting in a variable reactance—hence the name varactor. Varactors use junctions with specific doping profiles to provide a specified variation of capacitance with voltage.

13.2.5.1 Varactor Diode Parameters. Varactor diodes are fabricated using silicon (Si) or gallium arsenide (GaAs). Either *p–n* or Schottky-barrier junctions are fabricated. For most applications there are four varactor parameters that should be optimized. These are junction capacitance at zero bias, variation of capacitance with voltage, series resistance, and breakdown voltage. The zero-bias junction capacitance is set to provide a reasonable level of capacitive reactance to the circuit. The breakdown voltage is sufficiently high to allow a safety margin for circuit voltage swings at the specified power level. It is desirable to have as high a capacitive variation and the minimum series resistance the diode's material parameters will allow. The capacitance

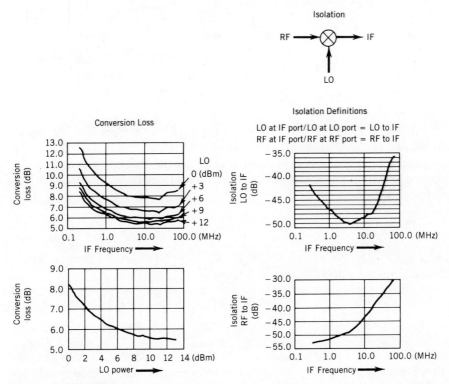

Figure 13.10. *Mixer parameters. (Kotzian, Brian and Al Schmidt, "Understanding, Specifying, and Characterizing Mixers," Hewlett-Packard Co., RF and Microwave Symposium Presentation, June 1982.)*

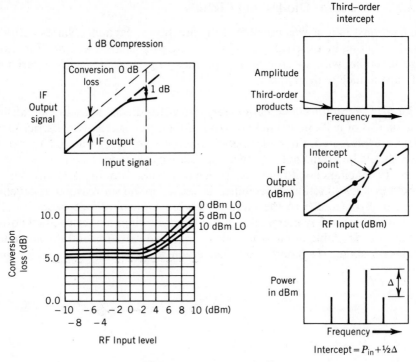

Figure 13.10. (continued)

variations measured at 1 MHz for two types of junction profiles are shown in Fig. 13.11.

Several figures of merit are used to compare the characteristics of diodes for use as varactors. Among these is the cutoff frequency at a specified bias defined as

$$f_{CV} = (2\pi R_s C_{jV})^{-1} \tag{13.10}$$

which at zero bias is written $f_{CO} = (2\pi R_s C_{jO})^{-1}$. The second figure of merit is the dynamic cutoff frequency defined as

$$f_{CV} = (2\pi R_s)^{-1} \left[\left(\frac{1}{C_{j\,min}} \right) - \left(\frac{1}{C_{j\,max}} \right) \right] \tag{13.11}$$

where the value of the capacitance $C_{j\,min}$ is measured at the reverse breakdown voltage and $C_{j\,max}$ is measured at moderate values of forward bias (e.g., 10 or 100 μA). The final figure of merit included here is the quality factor $Q_v = f_{CV}/f$, where f is the measurement frequency and the bias voltage V is specified. This factor indicates how close to the cutoff frequency the varactor is operating. It is desirable to operate with Q_v of order 10.

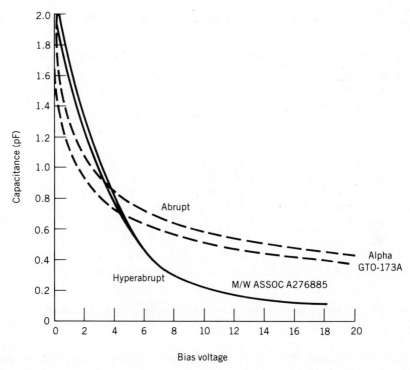

Figure 13.11. Capacitance variation for abrupt and hyperabrupt varactor diodes (two from each batch).

Typical diode parameters for epitaxial and nonepitaxial silicon varactors are given in Table 13.3. An epitaxial diode is one fabricated with either liquid- or vapor-phase layer(s) of controlled material on top of the basic material substrate. Epitaxial films have few impurities and are near-perfect crystals with few unintentional impurities and imperfections. Several resistance terms are used in Table 13.3. R_p is the resistance of the p region, R_n is the resistance of the n region, R_b is the resistance of the bulk, and R_c is the resistance of the metal-semiconductor contacts. Schematically, these are shown in Fig. 13.12 for a mesa varactor. The mesa configuration (named for the flat-topped hills in the West) is used to define the junction area by a chemical etching process. In general, epitaxial diodes perform better than nonepitaxial devices in both Si and GaAs. Schottky (metal–semiconductor) contacts are the most popular type today.

13.2.5.2 Application of Varactor Diodes. Varactors are used for low-noise parametric amplifiers, harmonic frequency generators, and frequency con- trollers because of their nonlinear capacitance with total applied voltage (see Fig. 13.11) and low loss. Parametric amplifiers will not be discussed because they are complex, and GaAs field effect transistors are used in this appli-

Table 13.3   Resistance Contributions Abrupt-Junction Silicon Varactor[a]

Parameter	Nonepitaxial			Epitaxial		
V_B (volts)	10	50	50	10	50	50
C_{jo} (pF)	1	1	2.4	1	1	2.4
C_{jB} (pF)	0.27	0.11	0.27	0.27	0.11	0.27
R_p (ohms)	0.04	0.01	0.005	0.04	0.01	0.005
R_n (ohms)	0.25	5.0	2.1	0.25	5.0	2.1
R_b (ohms)	10.6	42.7	28.1	0.15	0.09	0.056
R_c (ohms)	0.34	0.18	0.13	0.34	0.18	0.13
R_{total} (ohms)	11.2	47.9	30.3	0.78	5.28	2.29
f_{co} (GHz)	14	3.3	2.2	204	31	29
f_{cB} (GHz)	53	30	19	756	274	257

Source: Reference 2.

[a] In this table C_{jo} and C_{jB} are the junction capacitances at zero bias and at breakdown, and f_{co} and f_{cB} are the corresponding cutoff frequencies.

cation. Extensive information on parametric amplifiers is available in the literature (2 and 3).

Harmonic Frequency Generators. Harmonic frequency generators are used to generate high-frequency signals when direct generation of the signal is difficult. An example of a frequency doubler's performance in converting 9–13-GHz signals to the 18–26-GHz (*K*-band) frequency range is shown in Fig. 13.13.

Nonlinear resistance harmonic generators (operating like full-wave rectifiers) typically convert signals with an efficiency less than 100/*n* percent,

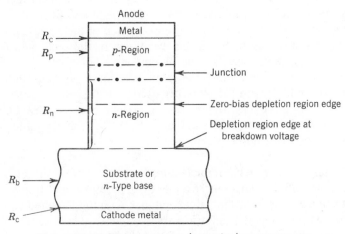

Figure 13.12. *Cross section of a typical mesa varactor.*

where *n* is the harmonic number. Nonlinear capacitance harmonic generators can have efficiencies up to 100%. This is intuitively shown in Fig. 13.14, where low-loss bandpass filters are connected to the varactor. Since the power introduced at ω_0 must be dissipated or reflected, it is converted to $n\omega_0$ and passes into the load. The efficiency is high because the losses are low. The Manley–Rowe relations (2) are used for a detailed analysis.

Frequency Controllers. The varactor diode is used as the frequency con-

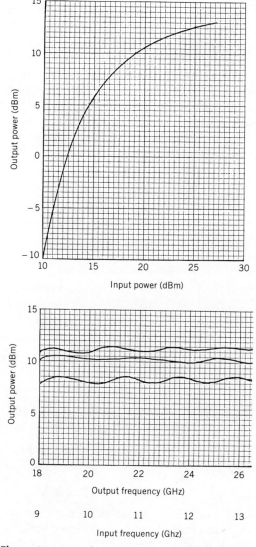

Figure 13.13. *Performance of a frequency doubler.*

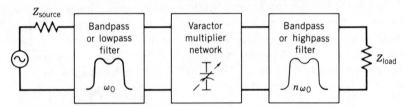

Figure 13.14. *Harmonic generation using a nonlinear capacitance.*

trolling element in varactor-tuned oscillator circuits (called varactor or voltage controlled oscillators—VCOs). Because of the rapid capacitance change, they have found application as frequency-agile local oscillators in electronic countermeasures (ECM) receiving systems and fast modulation "noise" sources in active jamming systems.

Tuning linearity (frequency proportional to tuning voltage) relative to the best-fit straight line of 1/2% has been achieved by selecting the appropriate diode doping density profile. A hyperabrupt varactor with a doping density profile similar to Fig. 13.15a is used. Figure 13.15b shows the profile for an abrupt junction varactor for comparison. The variation of the capacitance with voltage follows the relation

$$C_j(V) = \frac{C_j(0)}{(1 + V/\phi)^\gamma} \tag{13.12}$$

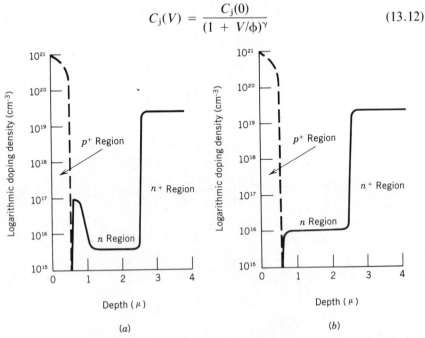

Figure 13.15. *Doping density profiles for varactor diodes. +, Heavily doped (low-resistance) region. (a) Hyperabrupt; (b) abrupt. (Reprinted from Watkins-Johnson Co. Tech Notes, Vol. 5, No. 4, July-August 1978, "Hyperabrupt Varactor Tuned Oscillators.")*

where V is the reverse bias voltage, ϕ is the junction contact potential, and γ is the slope of the varactor's C–V characteristics.

For abrupt and hyperabrupt varactors, the values of γ are approximately 1/2 and 2, respectively. A comparison of the capacitance of the abrupt and hyperabrupt junctions is shown in Fig. 13.11. Generally, the series resistance of the abrupt junction is one-half that of the hyperabrupt junction varactor for an equal capacitance near-zero bias.

The equivalent circuit (Fig. 13.16) for a typical varactor tuned oscillator (VTO) (Fig. 13.17) is used to analyze the voltage dependence of the circuit resonant frequency. The resonant frequency is

$$f_{\text{res}} = (2\pi\sqrt{L_p C_T})^{-1} \tag{13.13}$$

where

$$C_T = \frac{C_t C_j(V)}{C_t + C_j(V)} \tag{13.14}$$

C_t is the transistor input capacitance and L_p is the effective series inductance. By substituting the equation for $C_j(V)$ into the frequency equation and using a device and circuit design where $V \gg O$ and

$$(V + \phi)^\gamma \gg \frac{C_j(0)\,\phi^\gamma}{C_t} \tag{13.15}$$

the frequency relation becomes

$$f_{\text{res}} = \frac{(V^\gamma)^{1/2}}{2\sqrt{L_p C_j(0)\,\phi^\gamma}} = A(V^\gamma)^{1/2} \tag{13.16}$$

For the abrupt junction ($\gamma = 1/2$)

$$f_{\text{res}} = AV^{1/4} \tag{13.17}$$

and for the hyperabrupt junction ($\gamma = 2$)

$$f_{\text{res}} = AV \tag{13.18}$$

The last relation indicates that the resonant frequency is linearly related to the varactor reverse bias voltage.

A comparison of the performance of the hyperabrupt varactor tuned oscillator (HVTO) and the abrupt varactor tuned oscillator (VTO) for tuning, modulation sensitivity, and power output is shown in Fig. 13.18a–c.

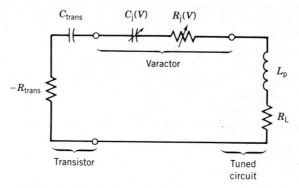

Figure 13.16. Simplified schematic of the varactor tuned oscillator (VTO).

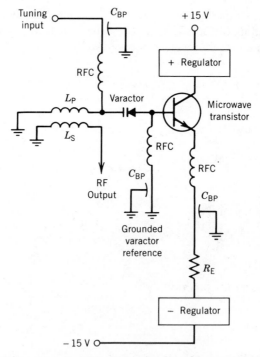

Figure 13.17. VTO schematic. RFC, RF choke; C_{BP}, bypass capacitor; R_E, emitter resistor; L_P, primary inductance; L_S secondary inductance. (Reprinted from Watkins-Johnson Co. Tech Notes, Vol. 5, No. 4, July–August 1978, "Hyperabrupt Varactor Tuned Oscillators.")

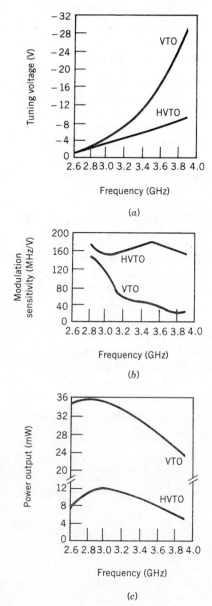

Figure 13.18. Comparison of hyperabrupt and abrupt VTO characteristics. (a) Tuning characteristics; (b) modulation sensitivity characteristics; (c) power output characteristics. (Reprinted from Watkins-Johnson Co. Tech Notes, Vol. 5, No. 4, July-August 1978, "Hyperabrupt Varactor Tuned Oscillators.")

13.3 ACTIVE TWO-TERMINAL DEVICES

13.3.1 Introduction to Active Devices

Microwave devices fabricated to generate microwave power directly from the DC bias power may be operated as either oscillators or amplifiers. As shown in Fig. 13.1 many two-terminal devices have been designed to perform this function, but the two commercially available devices are the transferred-electron (Gunn) and the IMPATT (IMPact Avalanche Transit Time) devices. These devices operate as amplifiers and oscillators over limited bandwidths in the 2 to 100 GHz range.

13.3.2 Transferred-Electron Devices

A transferred-electron device is a bulk semiconductor (contains no *p–n* junction) whose frequency of operation depends linearly on active region length. Gallium arsenide (GaAs) and indium phosphide (InP) are the predominant materials used for fabrication. Both materials exhibit a transferred-electron or negative differential mobility effect.

13.3.2.1 The Transferred-Electron Effect. The transferred-electron effect arises from the movement of the electrons in the conduction band. At room temperature, electrons reside in the valence and central conduction bands, as shown in Fig. 13.19. When the electrons are represented as waves, the wave number k is related to the kinetic energy of the electron. High kinetic energies (velocity) correspond to high wave numbers. The energy gap E_g is 1.43 electron volts at room temperature for GaAs.

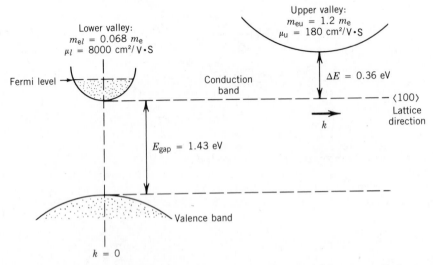

Figure 13.19. *Electron energy versus wave number model for n-type GaAs.*

If an electric field E is applied to the material, a current density J will flow according to the relation

$$J = \sigma E \quad \text{A/cm}^2 \qquad (13.19)$$

The conductivity σ is a function of the electron charge e, the mobility μ of the electrons in the lower conduction band, and the electron density n in that band. Mathematically

$$\sigma = en\mu \qquad (13.20)$$

with mixed MKS and CGS units being commonly used. e has the value 1.6×10^{-19} C, μ has units square centimeter per Volt-second, and n has units per cubic centimeter.

As a larger electric field is applied, electrons gain energies exceeding $\Delta E = 0.36$ eV and *transfer* to the higher-energy conduction band as shown in Fig. 13.20. In this band their mobility $\mu_u = 180$ cm²/V-s compared to $\mu_e = 8000$ cm²/V-s in the lower conduction band for *n*-GaAs. The resulting conductivity is

$$\sigma = e(u_l n_l + \mu_u n_u) \qquad (13.21)$$

since the number of electrons in both bands is

$$n = n_l + n_u \qquad (13.22)$$

This condition applies to both *n*-type GaAs and InP material.

This transferred-electron phenomenon gives rise to the characteristic current density versus electric field curve shown in Fig. 13.21. Other related curves can be derived using the relations

$$v = \text{velocity} = \mu E = \text{(mobility) (electric field)}$$

$$I = \text{current} = JA = \text{(current density)}$$
$$\times \text{(cross-sectional area)} \qquad (13.23)$$

$$V = \text{voltage} = EL = \text{(electric field)}$$
$$\times \text{(device active region length)}$$

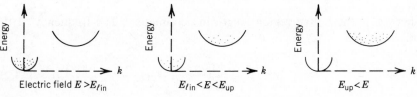

Figure 13.20. *Electron distribution versus current density (electric field) for n-type GaAs.*

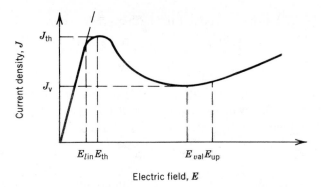

Figure 13.21. *Current density J versus electric field E relation for n-type GaAs.*

yielding the conductivity, mobility, and conductance curves in Fig. 13.22. These alternative representations facilitate relating experimental measurements to the parameters of the active region.

Transferred-electron devices are usually fabricated in a structure similar to Fig. 13.23, with highly doped n^+ regions sandwiching the active n region. Both the active and thin n^+ layers are grown epitaxially (yielding near-perfect crystalline structures) on the n^+ substrate material. The doping density of the n-region is not perfectly uniform at the transition regions or the interior, thus assisting with the development of Gunn domains.

13.3.2.2   Gunn Domains. When the electric field exceeds E_{th} (refer to Fig. 13.21), the electron distribution becomes more nonuniform. Because of the negative differential mobility, this process continues until a mature Gunn domain exists. The domain transits the active region at the saturated electron velocity. For GaAs at room temperature, the saturated electron drift velocity is near 10^7 cm/s.

The current at the terminals (n^+-layers) of the Gunn device is not constant. Each time the domain forms at the cathode the current increases and decreases as the domain transits, as shown in Fig. 13.24. The time t between current pulses is

$$t = \frac{L}{v_s} = \frac{L}{(10^7)} \tag{13.24}$$

where L is the active region length in centimeters. The frequency of the pulses is

$$f = \frac{1}{t} = \frac{10^7}{L} \quad \text{in Hertz} \tag{13.25}$$

For short samples, the current-time waveform is more sinusoidal. For the

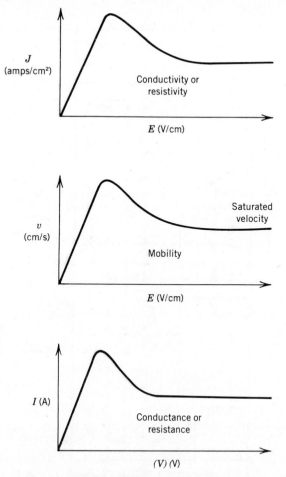

Figure 13.22. *Several presentations of the Gunn device curves.*

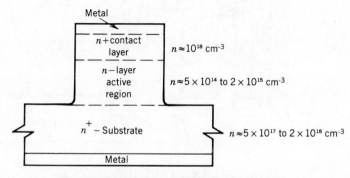

Figure 13.23. *Section of a Gunn device wafer.*

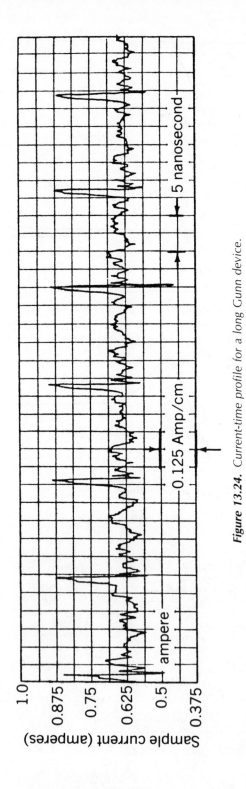

Figure 13.24. Current-time profile for a long Gunn device.

sample shown in Fig. 13.24, the average current is approximately equal to the current associated with the saturated velocity. Normally the valley (saturated) current is not observable at the terminals, so the dropback of the average operating current (Fig. 13.25) is a measure of the efficiency. The larger the dropback, the more efficient the device (typical values range from 1 to 6%).

13.3.2.3 Typical Gunn Device Operation.

Gunn devices designed to operate in *X*-band (10 GHz) use a transit time equal to the period of oscillation. The

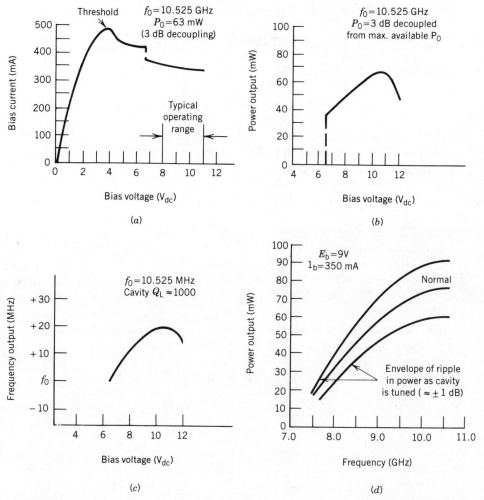

Figure 13.25. *Typical Gunn device bias, power, frequency curves. (a) Typical DC characteristics; (b) typical power output versus bias; (c) typical frequency pushing versus bias voltage; (d) power output versus frequency. (Application Note #ETD-6058, "Bulk Diode Oscillators," General Electric Co., Owenshore, KY, June 1971.)*

active region length is

$$L = \frac{10^7}{f} = \frac{10^7}{10^{10}} = 10^{-3} \text{ cm} \tag{13.26}$$

Typically the operating (DC bias) voltage is three times the threshold voltage, V_{th},

$$V_{th} = E_{th}L \tag{13.27}$$

The threshold electric field, E_{th}, for GaAs is about 3000 V/cm, so

$$V_{th} = (3000)(10^{-3}) = 3 \text{ V} \tag{13.28}$$

and the DC bias voltage is

$$V_{bias} = 3V_{th} = 9 \text{ V} \tag{13.29}$$

In X-band the CW efficiency is typically 3%.

In Gunn devices most of the input power is converted into heat. For a 100 mW, 3% efficiency CW device

$$P_{in} = P_{out}/\text{efficiency} = 3.3 \text{ W} \tag{13.30}$$

Of this P_{in}, 3.2 W are dissipated as heat.

The input bias current is

$$I_{bias} = \frac{P_{in}}{V_{bias}} = 0.37 \text{ A} \tag{13.31}$$

The device's current density J can be estimated assuming a typical electron doping density of 2×10^{15} cm^{-3} using the relation

$$J = nev_s = (2 \times 10^{15})(1.6 \times 10^{-19})(10^7) \tag{13.32}$$
$$= 3.2 \times 10^3 \text{ A/cm}^2$$

Finally, the cross-sectional area of the device is

$$A = \frac{I_{bias}}{J} = 1.16 \times 10^{-4} \text{ cm}^2 \tag{13.33}$$

This area corresponds to a square 10^{-2} cm (0.004 in.) on a side.

The device chip is mounted in a microwave package similar to those shown in Fig. 13.26. The package series inductance and shunt capacitance

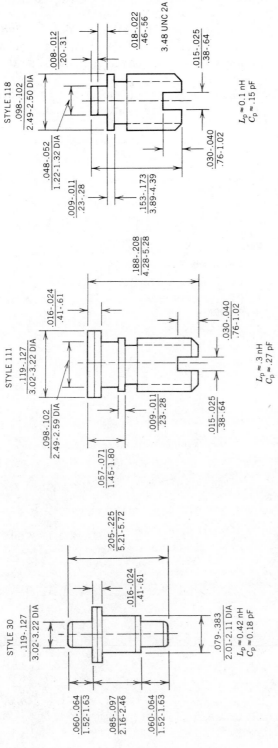

Figure 13.26. *Typical microwave packages used for Gunn devices.* (Dana W. Aitchley Jr. (Ed.), *Gun Diode Circuit Handbook*, MA/COM, 1971.)

are listed below each. For stable oscillations the effective negative resistance of the device is equal to the resistance of the circuit and package combination. The reactive components of the impedances cancel at the operating frequency.

The fabrication of Gunn devices has become a mature art. Therefore, the performance (Fig. 13.27) is not expected to improve significantly in the fore-

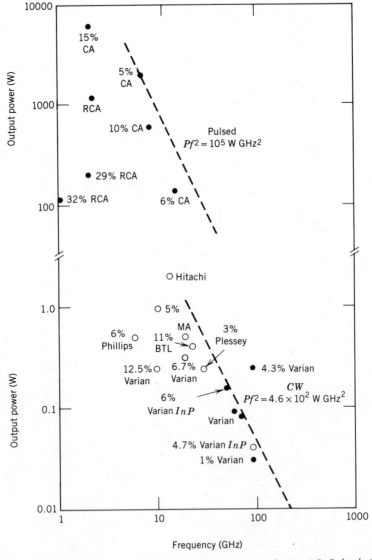

Figure 13.27. *State-of-the-art performance of GaAs Gunn devices.* ● *Pulsed;* ○ *continuous wave; CA, Cayuga; MA, Microwave Associates; BTL, Bell Telephone Labs. (Courtesy Bert Berson, Berson and Associates, Mountain View, CA)*

seeable future. The devices purchased from vendors are typically 3 dB less powerful than the state-of-the-art.

The previous discussion has assumed operation in the oscillator mode. By modifying the impedance presented to the device by the circuit and package, oscillations are suppressed and the device can be used as a reflection-type amplifier. Typical state-of-the-art performance is given in Table 13.4.

13.3.3 IMPATT Devices

13.3.3.1 Introduction to IMPATT Devices.
The generation of microwave power in a reverse-biased $p-n$ junction was suggested in 1958 by Read (4). This suggestion was correct for IMPATT (IMPact ionization Avalanche Transit Time) diodes, which today are commercially available and used in a variety of oscillator and amplifier applications.

13.3.3.2 Theory of Operation

Device Structure and DC Characteristics. The structure of a silicon $p-n$ or metal–GaAs (Schottky) junction IMPATT diode is similar to that of tuning

Table 13.4 State-of-the-Art Performance of Gunn Amplifiers

Center Frequency (GHz)	Bandwidth (%)	Saturated P_{out} (watts)	Small Signal Gain (dB)	Comments	Company
5.7	30	1.9	20.0		RCA
6.0	67	0.1	7.5		RCA
9.3	15		9.0	<14.5 dB NF, cathode notch device	Varian
9.8	40	0.4	25.0	3 stages	RCA
10.2	40		10.5	2 cathode notch devices	Varian
11.8	1	1.0	30.0	4 stages	MA
12.0	50	0.1	6.0		RCA
13.3	22	0.25	10.0		MA
14.1	50		9.5		Varian
16.0	35	0.25	30.0	4 stages	Varian
21.9	16	0.01[a]	15.0	2 stages	Varian
23.5			9.0	10.1 dB NF, InP	Varian
33.0	36	0.01[a]	10.0	Stagger tuned	Varian
35.0	29	0.15	8	14.1 dB noise figure	Varian
36.5	4	0.25	30.0	4 stages	Varian
37.7	5	0.13	13.5		Varian
35.0	10	0.2	25	3 stages, InP	Varian
56.0	8	0.1	25	3 stages, InP	Varian

[a] At 1 dB compression point (Courtesy Bert Berson, Berson and Associates, Mountain View, CA)

and varactor diodes. Figure 13.28 shows a typical IMPATT DC current versus voltage characteristic. IMPATT diodes operate in the avalanche breakdown region with a DC reverse voltage greater than V_b and significant avalanche reverse current.

Figure 13.29 shows a sketch of an IMPATT reverse biased into avalanche breakdown. A depletion region, whose width depends on the magnitude of the voltage, forms in the n-region of the junction. When V_{bias} is greater than V_b, the saturation current creates additional electrons and holes in the region (the process of avalanche multiplication). These additional electrons flow into the drift region. Now a large current flows with only a small increase in voltage (the avalanche breakdown current in Fig. 13.28). The operating voltage for a 10 GHz silicon IMPATT diode will be 70–100 V, depending on the junction profile, temperature, and bias current. Typical bias currents (I_{DC}) vary from 20 to 200 mA yielding a bias terminal impedance of 1000 ohms.

Microwave Properties. A brief description of microwave negative resistance follows (4–6). Assume that RF and DC breakdown voltages exist across the depletion region

$$V_T(t) = V_{DC} + V_D \sin \omega t \qquad (13.34)$$

This voltage (Fig. 13.30) will exist in the case where the diode is operated in a resonant circuit ($Q > 10$). Under the appropriate conditions this RF voltage induces an RF current that is more than 90° out of phase (this means the diode exhibits negative resistance). During the positive half-cycle of the

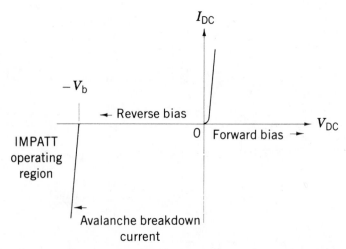

Figure 13.28. Current–voltage characteristics of a p–n junction.

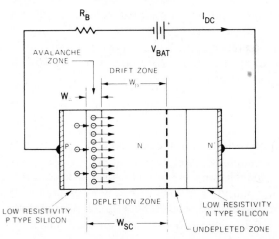

Figure 13.29. Construction and regions of p–n IMPATT diode under reverse bias conditions. (Courtesy Hewlett-Packard Co.)

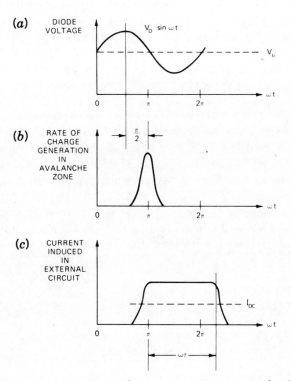

Figure 13.30. Key features of IMPATT operation. (Courtesy Hewlett-Packard Co.)

RF voltage, charge builds up in the avalanche region and peaks at $\omega t = \pi$ (Fig. 13.30). The charge waveform is peaked and lags the RF voltage by 90° because of the nonlinear nature of the avalanche generation process. Then the electrons drift at saturated velocity v_{sat} across the drift region. The traverse time t across the drift region is

$$t = \frac{W_D}{v_{sat}} \tag{13.35}$$

While the electrons drift they induce a current in the external circuit as shown in Fig. 13.30. Examining Figs. 13.30a and c it is seen how the combined delays of the avalanche and the transit cause positive current to flow in the external circuit when the RF voltage is in its negative half-cycle. The diode is exhibiting negative resistance (delivering RF energy to the circuit). Maximum negative resistance occurs when the transit angle

$$\omega t = \frac{3\pi}{4} \tag{13.36}$$

is at the operating frequency.

Diode Chip Equivalent Circuit. A simple equivalent circuit for the IMPATT chip is shown in Fig. 13.31. The active part of the diode is a negative resistance $-R_D$ and a reactance X_D. $-R_D$ includes the parasitic series resistance in the contacts and the undepleted region (Fig. 13.29). Some important properties of this equivalent circuit are that the chip impedance is about X_D at breakdown, X_D is approximately the reactance of $C_j(V_b)$, R_D is small compared to typical transmission line impedances, and $-R_D$ varies with signal level (Fig. 13.32) and with bias current. The decrease of R_D with signal level leads to stable oscillation amplitudes. Oscillation theory predicts that oscillation occurs if the load impedance Z_L connected to the chip is the negative of the chip impedance (Fig. 13.33). Since the chip is capacitive, the load must be inductive. The oscillations will build up if $R_D > R_L$ and remain stable in amplitude if $R_D = R_L$, the stable operating point in Fig. 13.32.

The power delivered to the load is calculated from the operating points $I_D = 0.8$ A and $R_L = 2$ ohms, yielding $P_0 = (\frac{1}{2} I_D^2) R_L = 0.64$ W. For $R_L = 6$ ohms, which does not intersect the R_D versus I_D curve in Fig. 13.32,

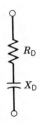

Figure 13.31. RF equivalent circuit for the IMPATT chip.

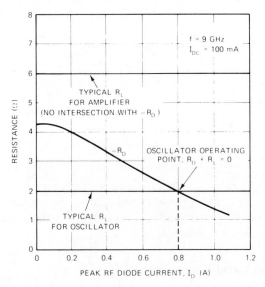

Figure 13.32. *RF current amplitude dependence of R_D. (Courtesy Hewlett-Packard Co.)*

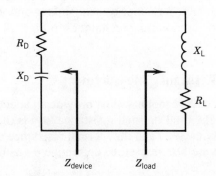

Figure 13.33. *Oscillator circuit for stable operation.*

no oscillation occurs and the circuit behaves as a reflection-type amplifier provided a means of separating the input and output signals (a circulator, see Chapter 10) is provided.

Packaged Diode Equivalent Circuit. The chip equivalent circuit is modified by the package parasitic reactances. At microwave frequencies the entire packaged diode must be evaluated. For simplicity, the package is described by a series inductance and a shunt capacitance as shown in Fig. 13.34. The values of L_P and C_P vary with package style. L_P is also affected by the microwave circuit surrounding the diode (sometimes used to tune the os-

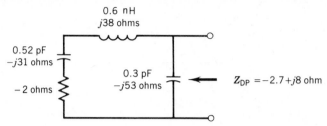

Figure 13.34. Equivalent circuit and typical values for packaged IMPATT at 10 GHz.

cillator). Typical values for the chip and package elements for a diode are shown in Fig. 13.34. The overall impedance of the packaged diode is Z_{DP} = $-2.7 + j\,8$ ohms. Other diodes yield different values, and the manufacturer's data should be consulted.

13.3.4 IMPATT and Gunn Circuits

Several mounting techniques for IMPATT and Gunn devices are shown in Fig. 13.35. For end-mounting devices in coaxial circuits, the collet–clamp–sleeve arrangement in Fig. 13.35*a* is recommended. For permanent, secure mounting while allowing diode replacement, the arrangement in Fig. 13.35*b* is recommended. Finally, for superior heat sinking and vibration and shock protection the microstip mounting techniques in Figs. 13.35*c* and 13.35*d* are recommended.

13.3.5 Thermal Resistance Measurement

IMPATT and Gunn device mounts must provide an adequate heat flow path from the package. The total thermal resistance (θ_T) is the sum of the active region to case resistance (θ_{jc}) and the thermal resistance of the mount (θ_{cav}). The Haitz et al. (7) method is used to determine θ_T using readily available instrumentation.

13.3.5.1 θ_T for IMPATT Diodes. The measurement method uses the temperature variation of the avalanche breakdown voltage as a sensor of the junction temperature while DC power is being dissipated. The thermal resistance is the ratio of the temperature rise to the dissipated power. Referring to Fig. 13.36, at a given value of bias current I_{DC} the voltage drop has three components:

1. The breakdown voltage $V_b(T_0)$ at ambient temperature T_0
2. The rise in breakdown voltage ΔV_b due to junction heating
3. A space-charge resistive voltage drop $I_{DC}R_{SC}$ measured by a short-pulse or high-frequency sine wave

For temperatures between 25 and 100°C, ΔV_b is proportional to ΔT for silicon *p–n* junction IMPATTs. The constant of proportionality is itself proportional to V_b (25°C) by the relation

$$V_b = [\beta V_b(25°C)]\,\Delta T \qquad (13.37)$$

where $\beta = 1.17 \times 10^{-3}\ °C^{-1}$ for silicon diodes.

By definition, ΔT is

$$\Delta T = \theta_T V_{DC} I_{DC} \qquad (13.38)$$

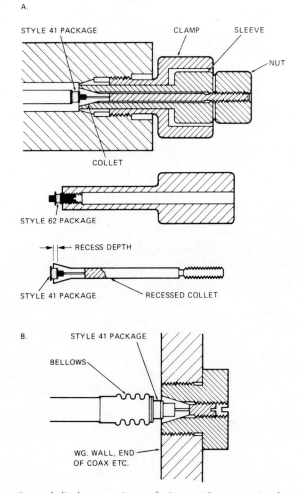

Figure 13.35. *Several diode mounting techniques. (Courtesy Hewlett-Packard Co.)*

C. MICROSTRIP MOUNTING

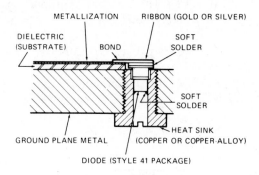

D.

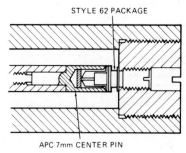

E. MICROSTRIP MOUNTING

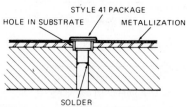

Figure 13.35. (continued)

Therefore, the total voltage V_{DC} across the diodes for $T_0 = 25°C$ is

$$V_{DC} = V_b(25°C) + \theta_T V_{DC} I_{DC} \beta V_b(25°C) + I_{DC} R_{SC} \quad (13.39)$$

Solving for θ_T yields

$$\theta_T = \frac{V_{DC} - V_b(25°C) - I_{DC} R_{SC}}{V_{DC} I_{DC} \beta V_b(25°C)} \quad (13.40)$$

The junction temperature of the diode should not exceed 200°C for long

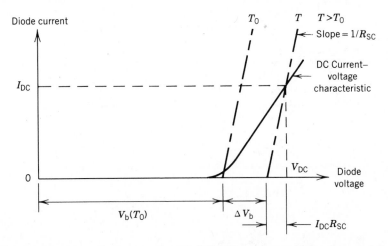

Figure 13.36. *Contributions to the DC IMPATT current–voltage characteristic.*

periods of time. The junction temperature can be calculated from

$$T_j = T_{amb} + (V_{DC}I_{DC} - P_o)\theta_T \tag{13.41}$$

where T_{amb} is the ambient temperature and P_o is the output power.

13.3.5.2 θ_T for Gunn Devices.
A modified Haitz method is used for estimation of Gunn device thermal resistance. First, the below-threshold electrical resistance of the device is measured at several known elevated temperatures in an oven with short, nonheating pulses. Second, the device is DC biased until the resistance stabilizes (indicating a steady-state heat flow) and is correlated to the appropriate isothermal resistance value measured above yielding the active region temperature. θ_T is the input power divided by the active region temperature rise.

13.3.6 Bias Circuits

13.3.6.1 IMPATT Bias Circuits.
The IMPATT behaves as a constant-voltage device (requires a constant-current DC bias supply). An interaction also exists between the microwave and bias port. Under some conditions, oscillations in the bias circuit are possible. A transistor current regulator as shown in Fig. 13.37 is the most practical bias supply. This supply is not resonant and does not present a large capacitance to ground. Some examples of the microwave spectra of IMPATT oscillators with bias circuit instabilities are presented in Fig. 13.38. The bias circuit should be physically close to the bias port and not connected via a coaxial cable. In lower power operation

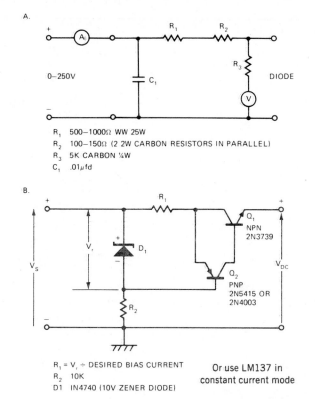

R$_1$ 500–1000Ω WW 25W
R$_2$ 100–150Ω (2 2W CARBON RESISTORS IN PARALLEL)
R$_3$ 5K CARBON ¼W
C$_1$ .01μfd

R$_1$ = V$_r$ ÷ DESIRED BIAS CURRENT
R$_2$ 10K
D1 IN4740 (10V ZENER DIODE)

Or use LM137 in
constant current mode

Figure 13.37. *Current regulated bias supply for an IMPATT.*

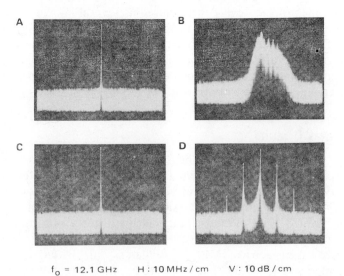

f$_o$ = 12.1 GHz   H : 10 MHz / cm   V : 10 dB / cm

Figure 13.38. *IMPATT RF spectrum for various bias circuits. (Courtesy Hewlett-Packard Co.)*

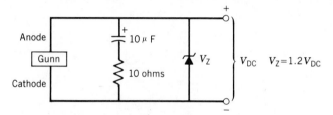

Figure 13.39. *Gunn device bias circuit.*

($P_o < 1/4$ W) the bias circuit is less critical, and a resistive supply may be used without excessive power loss.

13.3.6.2 Gunn Bias Circuits. An operating Gunn device behaves as a constant current element (requires a constant-voltage bias supply; see Fig. 13.39). Inductance in the bias circuit can result in bias circuit oscillations

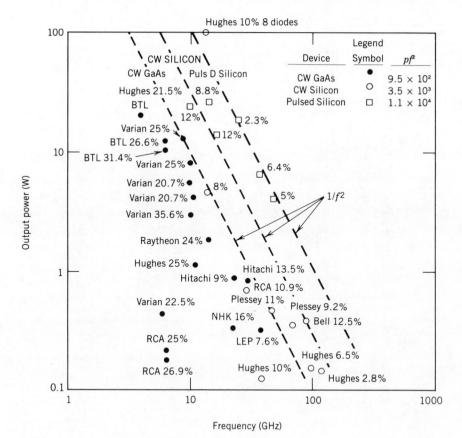

Figure 13.40. *State-of-the-art performance of IMPATT devices. (Courtesy Bert Berson, Berson and Associates, Mountain View, CA)*

Table 13.5 State-of-the-Art Performance of IMPATT Amplifiers

Center Frequency (GHz)	Bandwidth (%)	Saturated Power Out (Watts)	Efficiency n (%)	Gain (dB)	Diode Type	Comments	Company
6.20	9.6	10.0	7.0	31.5	GaAs flat profile	34 dB NF, 3 stages	BTL
6.20	9.6	5.0	10.0– 15.0[a]	37.0	GaAs Read output	3 TEO stages 5 IMPATT stages 25 dB NF	Raytheon
6.70	1.0	4.2	6.0	23.0	Si	3 stages	MA/COM
7.75	1.3	13.0	5.0	36.0	Si	1st stage TEO	Bradley
8.15	6.0	1.5		36.0	Si	7 stages	Hughes
8.25	6.0	4.5	22.0	4.5	GaAs Read	1 stage	Raytheon
9.20	6.5	4.0	13.0	5.0	GaAs	1 stage	Hughes
9.20	1.3	20.0		3.0	Si	16 diodes	Hughes
10.30	1.5	100.0	7.2	11.0	8 Si double drift diodes	Locked pulsed oscillator	Hughes
11.20	2.6	2.0		10.0	Si double drift diodes		HP
36.00	1.4	.2		15.0	Si	2 stages	Hughes
94.0	2.0	.1		18.0	Si	2 stages	Hughes

[a] For output read diode stages. (Courtesy Bert Berson, Berson and Associates, Mountain View, CA)

because of the negative differential resistance (see Fig. 13.23*a*). An RC network and zener diode usually prevents buildup of the oscillations. The Zener also prevents burnout by application of bias with reversed polarity.

13.3.7 Performance of IMPATT Oscillators and Amplifiers

The state-of-the-art performance for IMPATT oscillators and amplifiers is shown in Fig. 13.40 and Table 13.5, respectively. The devices are available today from several vendors, but because of their noise they are not as widely used as Gunn devices. Table 13.6 lists some of the relative advantages and disadvantages between Gunn and IMPATT devices.

Table 13.6 Relative Advantages/Disadvantages of Gunn and IMPATT Devices

Advantages	Neutral Factors	Disadvantages
Transferred Electron Devices (Gunn)		
Oscillator noise far from carrier	CW power and efficiency	Temperature stability of power
Amplifier noise		Oscillator noise near carrier
Bias voltage		
Reliability		
Pulsed power and efficiency		
Range of frequencies of operation		
Amplitude and phase linearity		
Dynamic range		
Tunability		
IMPATTS		
CW power and efficiency	Circuit complexity	Bias voltage
Range of frequencies of operation	Oscillator noise near carrier	Oscillator noise far from carrier
Reliability	Amplifier noise	Reliability (Read GaAs)
Pulsed power	Gain—bandwidth	
Temperature stability of power	Linearity—phase and amplitude	
	Dynamic range	
	Tunability	

PROBLEMS

1. What is the resonant frequency of the packaged device assuming $R_s = R_j(V) = 0$ in Fig. 13.2?

2. Neglecting C_f and R_s and assuming a back-biased condition $[R_j(V < V_B)$ large], what is the new resonant frequency of the packaged diode?

3. Derive Eq. 13.9 using the identities $\sin^2 x = \frac{1}{2}(1 - \cos 2x)$ and $\sin x \sin y = \frac{1}{2}[\cos(x - y) - \cos(x + y)]$.

4. When two signals ω_1 and ω_2 are present at the mixer input and the $a_2 v_{in}^2$ term is expanded, the results shown in Fig. P13.4 are obtained. Derive all these terms. Note the relative amplitudes in Fig. P13.4. Your coefficients should confirm these relative values.

5. Using the identity $\sin^3 x = \frac{1}{4}(3 \sin x - \sin 3x)$, derive the $a_3 v_{in}^3$ terms (third-order intermodulation products) for a mixer with two signals f_1 and f_2 as shown in Fig. P13.5. Note the amplitudes are proportional to $v_1 v_2^2$ or $v_1^2 v_2$. Because of their frequency proximity they are difficult to remove by filtering.

6. Compute the γ of Eq. 13.12 for the abrupt varactor in Fig. 13.11 assuming that $\phi = 1.2$ eV. For simplicity use only the two voltages, $V = 0$ and $V = 10$ V.

7. Derive Eq. 13.16.

8. Estimate the terminal values (voltage–current) for a 250-mW Gunn oscillator at 15 GHz.

9. Estimate the drift region thickness for an IMPATT diode designed for 15 GHz. Compare to the active region length of a Gunn device. Explain the difference. Assume the electron saturated drift velocity is 10^7 cm/s.

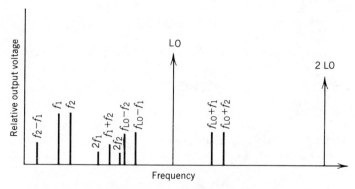

Figure P13.4. *Mixer output spectrum with two input signals.*

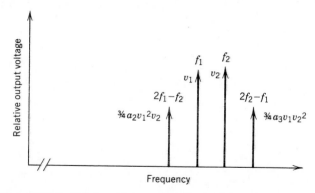

Figure P13.5. *Third-order intermodulation products for a mixer.*

10. Can the following Si 0.7-W IMPATT oscillator operate in a 85°C ambient temperature (military environment) and keep the junction temperature below 200°C? Compute the junction temperature when the parameters are $I_{DC} = 50$ mA, $V_{DC} = 84$ V, $R_{SC} = 31$ ohms, V_b (25°C) $= 76$ V, $R_D = -2$ ohms, $C_j(V_b) = 0.52$ pF.

11. Estimate the intrinsic 3-dB bandwidth of an IMPATT diode in the simple circuit in Fig. 13.33 (neglect the package effects).

REFERENCES

1. D. P. C. Thackeray, "When Tubes Beat Crystals: Early Radio Detectors," *IEEE Spectrum,* March 1983, p. 64.
2. H. A. Watson, ed., *Microwave Semiconductor Devices and Their Circuit Applications,* McGraw-Hill, New York, 1969.
3. S. Y. Laio, *Microwave Devices and Circuits,* Prentice-Hall, Englewood Cliffs, NJ, 1980.
4. W. T. Read, Jr., "A Proposed High-Frequency Negative Resistance Diode," *Bell System Tech. Jrnl.,* Vol. 37, 1958, p. 401.
5. G. I. Haddad, et al., "Basic Principles and Properties of Avalanche Transit Time Devices," *IEEE Trans. Micro. Th. Tech.,* Vol. MTT-18, November 1970, p. 752.
6. D. L. Scharfetter and H. K. Gummel, "Large-Signal Analysis of Silicon Read-Diode Oscillator," *IEEE Trans. Elect. Dev.,* Vol. ED-16, p. 64, January 1969.
7. R. H. Haitz, et al., "A Method for Heat Flow Resistance Measurements in Avalanche Diodes," *IEEE Trans. Elect. Dev.,* Vol. ED-16, May 1969, p. 438.

14 | Three-Terminal Active Devices

14.1 INTRODUCTION

Three-terminal devices have been developed to replace bulky, life-limited vacuum tubes and two-terminal reflection-type amplifiers (see Chapter 13) that lack a signal control port. Three-terminal solid-state devices provide advanced performance, high reliability with reduced maintenance, and cost-effective components for systems. The signal port allows separate, nearly isolated adjustment of the input and output ports for optimum performance and a port to control operation over the environmental conditions.

Two generic devices dominate modern applications. The *bi*polar *t*ransistor (BPT) fabricated in silicon is an extension to 6 GHz of the well-developed low-frequency technology. The *f*ield-*e*ffect *t*ransistor (FET) fabricated in gallium arsenide (GaAs) with a metal-semiconductor junction (MESFET) is the dominant performer for frequencies approaching 100 GHz. Its performance is <1 dB noise figure at 10 GHz for low-noise applications and 10 W CW at 10 GHz with a single chip for a power amplifier. The Si BPT will be replaced by the MESFET except below 4 GHz, where low cost and the extreme reliability of Si fabricated devices is required. For this reason only a brief description of the BPT is presented here to allow room for the presentation of the FET information.

An outgrowth of the development of room-temperature high-speed logic devices has been applied to microwave applications. The device, called the high electron mobility transistor (HEMT), is fabricated using heterojunctions involving tertiary compunds such as $Al_xGa_{1-x}As$ ($x = 0.3$) and GaAs grown by molecular beam expitaxy (1). The source–gate–drain arrangement is similar to the MESFET, but the high mobility yields noise figures as low as 1.4

dB at 12 GHz and an associated gain of 11 dB. The HEMT equivalent circuit is similar to that of the MESFET.

14.2   BIPOLAR TRANSISTOR

Operation of the bipolar transistor involves injection of minority carriers into the base of the transistor. These carriers move under a diffusion process to the collector where they recombine. These processes limit the high frequency and noise performance of BPTs.

14.2.1   Equivalent Circuit

The microwave BPT uses a planar geometry to keep parasitic resistances and reactances small while allowing room for single-surface bonding. For small-signal and power applications, the interdigitated format is used (2), as shown in Fig. 14.1. The equivalent circuit for the transistor operated in the common emitter configuration without package parasitics is developed similar to its low-frequency counterpart and is shown in Fig. 14.2. The problem for the circuit designer is to obtain the values for the equivalent circuit elements. A combination of inspection of the manufacturer's data sheet and inversion of the transistor time constants and corner frequencies may be required. Given these lumped values, the transistor is readily modeled using computer-aided design techniques (see Chapter 7). Alternatively, the manufacturer may provide the packaged or bare chip S-parameters as a function of bias conditions and frequency. Finally, direct measurement of the S-parameters is done with network analyzers (see Chapter 20) that can de-embed the effects of the bias and measurement fixture circuitry. Large signal measurements are more difficult because of the danger of burnout of the analyzer, but they can be performed using attenuators.

14.2.2   Low-Noise Performance

The noise generated within the BPT has three principal sources:

1. Shot noise due to generation and recombination in the emitter-base junction and injection of carriers across the potential barriers
2. Shot noise in the collector-base junction
3. Thermal noise in the base resistance (2)

A series of analyses was developed for the noise factor in BPTs culminating

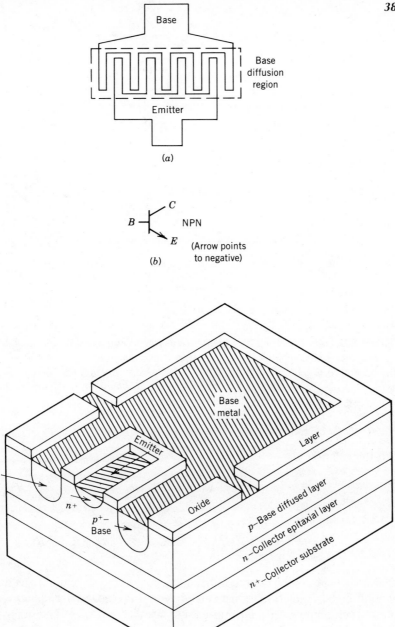

Figure 14.1. Microwave bipolar transistor. (a) Top view of interdigitated bipolar NPN transitor; (b) NPN transistor circuit symbol; (c) cross section of microwave NPN transistor.

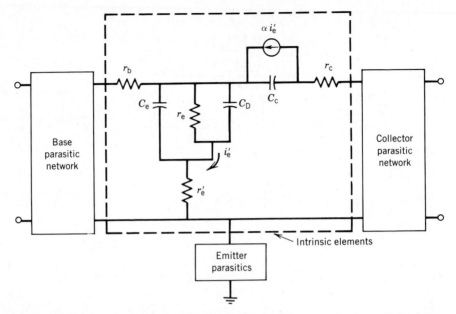

Figure 14.2. Equivalent circuit of bipolar transistor in common-emitter configuration. r_b, Resistance under emitter plus emitter to p^+ resistance plus resistance in p^+ region plus contact resistance of metal to p^+; r_e, emitter-base junction resistance; r'_e, emitter series resistance; r_c, collector series resistance; C_e, C_c, emitter, collector junction capacitances; C_D, forward-biased emitter-base junction diffusion capacitance.

with the relation (3)

$$F = 1 + \frac{R_b}{R_s} + \frac{r_e}{2R_s} + \left(\frac{\alpha_0}{|\alpha|^2} - 1\right)\left(\frac{(R_s + R_b + r_e)^2 + X_s^2}{2r_e R_s}\right)$$

$$+ \frac{\alpha_0 r_e}{2|\alpha|^2 R_s}[\omega^2 C_e^2 X_s^2 - 2\omega C_e X_s + \omega^2 C_e^2 (R_s + R_b)^2] \quad (14.1)$$

where $R_s + jX_s$ is the source impedance, r_e is the differential resistance of the emitter-base junction for a given current, α_0 is the low-frequency asymptote of the common base current gain $\alpha = \alpha_0/(1 + jf/f_b)$. The other terms are defined in Fig. 14.3.

Since the noise factor is a function of the source impedances, the derivative of Eq. 14.1 with respect to R_s and X_s when set equal to 0 yields the values $R_{s,opt}$ and $X_{s,opt}$ for minimum noise factor. The optimum source values are (4)

$$R_{s,opt} = R_b^2 + X_{s,opt}^2 + \left(1 + \frac{f^2}{f_b^2}\right)\left(\frac{r_e(2R_b + r_e)}{\alpha_0 a}\right) \quad (14.2)$$

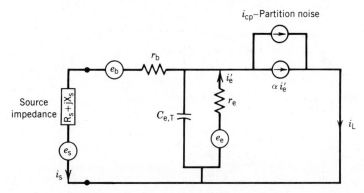

Figure 14.3. Noise equivalent circuit for a bipolar transistor in the common emitter configuration. $\overline{e_s^2} = 4kTR_s$; $\overline{e_b^2} = 4kTr_b$; $\overline{e_e^2} = 4kTr_e$; $\overline{i_{cp}^2} = 2kT(\alpha_0 - |\alpha|^2)/r_e$.

and

$$X_{s,opt} = \left(1 + \frac{f^2}{f_b^2}\right)\left(\frac{2\pi f C_e r_e^2}{\alpha_0 a}\right) \tag{14.3}$$

where

$$a = \frac{1}{\alpha_0}\left[\left(1 + \frac{f^2}{f_b^2}\right)\left(1 + \frac{f^2}{f_e^2}\right) - \alpha_0\right] \tag{14.4}$$

f_b is the base cutoff frequency related to the base delay time τ_b by $f_b = (2\pi\tau_b)^{-1}$, and f_e is the emitter-base cutoff frequency. These values do not include the effects of parasitics which must be included in the circuit design. The minimum noise factor and gain versus collector current for a high-performance n–p–n Si microwave transistor is shown in Fig. 14.4 (4).

14.2.3   High-Power Performance

The Si BPT is a low-cost device suitable for moderate power level (2–10 W) and high reliability up to 4 GHz. The devices described here are packaged so hermetic sealing of the device and its circuit are not required. In the common base mode, the packaged device is soldered directly to the heat sink for minimum electrical and thermal parasitic effects.

For linear power transistors (class A operation), a single-stage 2-GHz amplifier can provide 1-W output power. The S-parameters for a transistor of this type are shown in Fig. 14.5 (5). At 2 GHz the transistor input impedance is $2 + j15$ ohms and the output impedance is $12 - j3$ ohms. The current gain S_{21} is about $1\underline{/5°}$ but this is into a load to source impedance ratio of 6 yielding 8 dB of gain. The S_{12} is only $0.04\underline{/40°}$ at 2 GHz so variations in the

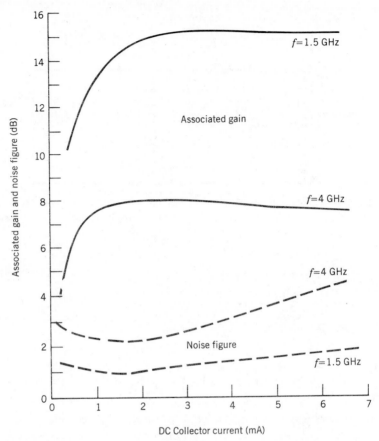

Figure 14.4. *Noise figure and associated gain versus collector current for a low-noise BPT. Note: 1 mA corresponds to a 2000 A/cm² current density. (From Ref. 4, ©1978 IEEE)*

output matching network will have only a small effect at the input of the transistor. For a stage with 8 dB of available gain at 30 dBm output, 22 dBm (160 mW) of drive power is needed. Thus several stages are required for full output from the 0 dBm (typical) output of many signal sources.

When operated class B with nonsinusoidal waveforms, the S-parameters are not defined, so optimum input and load impedance curves are presented as shown in Fig. 14.6 for the circuit in Fig. 14.7 (6). Here the transistor is biased off (both emitter and base are at DC ground) until negative input voltage turns the device on during about one-half the cycle. The efficiency of these devices is quite high (35%) indicating that for an output power of 7 W, the DC input power is 20 W (28 V DC at 0.7 A). The remainder of the DC input power generates junction heat (10.5 W) and raises the junction temperature to (10.5 W) (17 °C/W) = 180°C above the heat sink surface

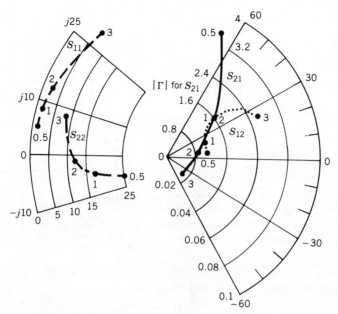

Figure 14.5. *S-Parameters for a bipolar power transistor.* Z_0 = 50 ohms; V_{CE} = 18 VDC; I_c = 0.2 A. *Note: Numbers on curves indicate frequency in GHz; power out =* 1 W at 2 GHz. *[From Data Sheet LP106, Microwave Semiconductor Corp. (MSC), Somerset, NJ, April 1982.]*

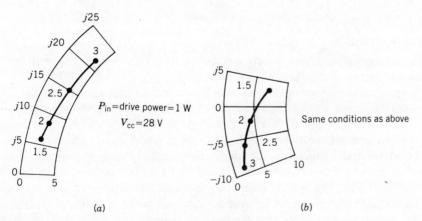

Figure 14.6. *Optimum input and load impedances for a class B BPT. (a) Series input impedance of transistor; (b) optimum load impedance viewed from transistor. (From Data Sheet GP103, Microwave Semiconductor Corp., (MSC), Somerset NJ)*

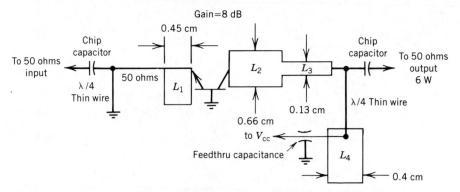

Figure 14.7. *2.2-GHz power amplifier circuit. L_1, open circuit stub, 0.75 cm long; L_2, low impedance line, 0.66 cm long; L_3, low impedance line, 0.66 cm long; L_4, open circuit stub, 1.2 cm long; substrate, Alumina, 0.635 mm thick, $\epsilon_r = 9.9$. (From Data Sheet GP103, Microwave Semiconductor Corp., (MSC), Somerset NJ)*

temperature. When operated in this manner, the gain is 8 dB, thus indicating a drive power of over 1 W is required.

14.3 METAL-SEMICONDUCTOR FIELD-EFFECT TRANSISTOR

14.3.1 Configurations

The MESFET is part of a family of FETs as shown in Fig. 14.8. Similar to the BPT, MESFETs have two subparts—low noise and power. Descriptions of the *p–n* junction FET (JFET) and insulated-gate FET (IGFET) are found in Ref. 7.

14.3.2 Equivalent Circuit

The MESFET is a unipolar device in which nearly all the current flows due to the majority carriers (electrons in *n* GaAs). The electrons flow from the source to the drain through a planar configuration as shown in Fig. 14.9. The epitaxially grown *n* GaAs active layer electronically constricts the electron flow cross section because of the depletion region formed under the metal-semiconductor (Schottky) junction.

The origin of the depletion region can be envisioned as follows. When the metal and semiconductor are in contact and equilibrium, their Fermi levels must be equal. Since in free space the Fermi level of the semiconductor is greater than the metal, when they contact electrons from the semiconductor pour into the metal leaving a bound net positive charge (holes) in the GaAs and an excess negative charge in the metal. This dipole of charge forms a capacitance across the depleted region of the GaAs such that the net voltage around a loop including this junction is zero (no sources are

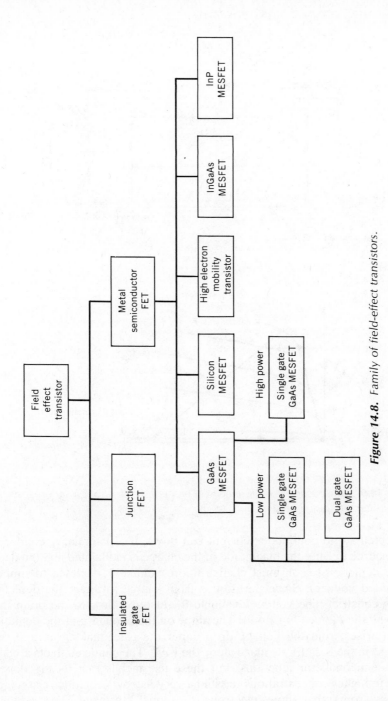

Figure 14.8. *Family of field-effect transistors.*

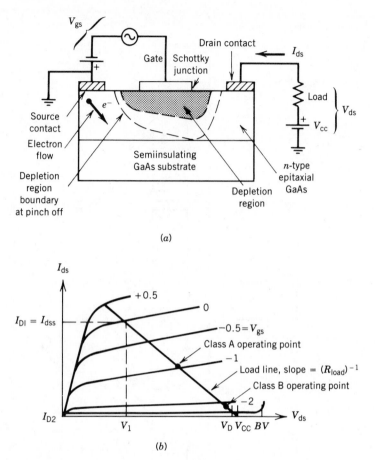

Figure 14.9. *Construction and DC electrical properties of a MESFET. (a) Cross section and bias circuit; (b) DC current–voltage characteristic curve.*

connected so no continuous current can flow). If a potential is connected from source to gate, the magnitude of the charge is modulated and the depth of the depletion region must change (more volume is depleted to support the added voltage). Since the flow is from source to drain, the depletion region constricts the volume available for electron flow to the drain thus reducing the I_{ds} (see Fig. 14.9). The slope on the boundary of the depletion region arises from the voltage drop under the gate, making an apparent increase in the Schottky voltage along the FET. The ohmic contacts are also metal-semiconductor junctions, but these use metals that locally dope n GaAs to higher concentrations to yield a very narrow depletion region (several angstroms) that allows electrons to "tunnel" through. These contacts appear to be ohmic (show a linear current–voltage relationship in both polarities) and have low loss. Low loss is essential here since the microwave current passes through the resistance R_d and R_s of these ohmic contacts.

When the device is biased for small-signal class A operation, the depletion layer is only perturbed by V_{gs}. The lumped-element equivalent circuit for operation up to about 12 GHz can be directly related to the physical parameters of the device as shown in Fig. 14.10a and redrawn in Fig. 14.10b. The

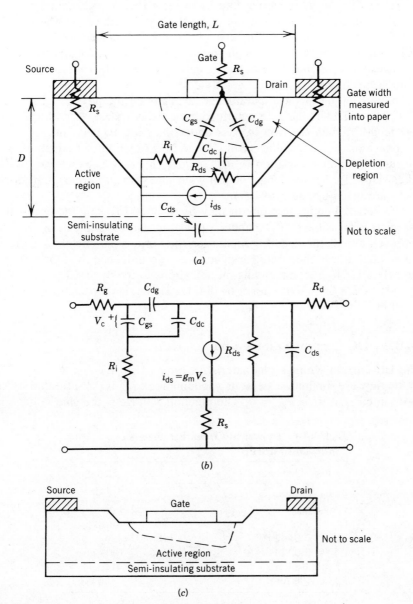

Figure 14.10. *Lumped-element two-part equivalent circuit of a MESFET. (a) Location of lumped-element components; (b) lumped-element equivalent two port; (c) recessed gate configuration for lower noise.*

elements C_{dg} and C_{gs} represent the total gate to channel capacitance, C_{dc} the dipole layer capacitance arising from the lateral stationary depletion layer since the electric fields exceed the threshold for the onset of negative differential mobility (a stationary Gunn domain, see Chapter 13), R_i and R_{ds} are resistances in the channel, and i_{ds} is the RF current generator related to V_{gs} (voltage-controlled current generator) across the C_{gs}. The R_s, R_g, and R_d are the respective parasitic (extrinsic) resistances, and C_{ds} is the drain-source capacitance through the semiinsulating GaAs substrate. The package parasitics must be added to have an equivalent circuit from the terminals available for circuit mounting.

The magnitude of these parameters depends on the physical size of the FET. In particular, the gate length sets the delay time τ_0 (5 ps for a 1 μm gate length) for an electron to pass under the gate. Because of this delay, the phase of the gate voltage will change as a function of frequency. This appears as a phase variation in g_{mo}, namely $g_m = g_{mo}[\exp(-j\omega\tau_0)]$. The other values for a 1 μm long, 500 μm wide gate with 10^{17} donors/cm^3 doped GaAs are listed in Table 14.1 (7).

The detailed equivalent circuit may be avoided by using the terminal S-parameters provided by the manufacturer or measured by the designer. This approach avoids the need for separating the various parameters and is recommended unless the circuit designer seeks to influence the fabrication of the MESFET by recommending modifications to the device designer. Typical MESFET parameters for a 6–18-GHz small signal packaged FET are given in Table 14.2.

14.3.3   DC Current-Voltage Characteristics

The DC current–voltage characteristics for a MESFET are shown in Fig. 14.9b (they are similar to a pentode vacuum tube). I_{dss} is the saturated source to drain current with $V_{gs} = 0$. The pinchoff voltage is the gate voltage re-

Table 14.1   Equivalent Circuit Parameters of a Low-Noise GaAs MESFET[a]

Intrinsic Elements	Extrinsic Elements
$g_m = 53$ mS	$C_{ds} = 0.12$ pF
$\tau_0 = 5$ ps	$R_g = 2.9$ ohm
$C_{gs} = 0.62$ pF	$R_d = 3$ ohm
$C_{dg} = 0.014$ pF	$R_s = 2$ ohm
$C_{dc} = 0.02$ pF	$L_{gate}{}^b = 0.05$ nH
$R_i = 2.6$ ohm	$L_{drain}{}^b = 0.05$ nH
$R_{ds} = 400$ ohm	$L_{source}{}^b = 0.04$ nH

[a] DC operating conditions: $V_{DS} = 5$ V, $V_{gs} = 0$ V, $I_{DS} = 70$ mA.

[b] These inductances are associated with the test fixture.

(From Ref. 7, © 1976, IEEE)

Table 14.2 S-Parameters of a Small-Signal Packaged MESFET[a]

(Avantek, Type AT-10650-1, −3)
COMMON SOURCE
AT-10650-1, −3 V_{CE} = 8V, I_c = 10 mA

Frequency (GHz)	S_{11}		S_{21}			S_{12}			S_{22}	
	Mag	Ang	dB	Mag	Ang	dB	Mag	Ang	Mag	Ang
6.0	0.78	−147	8.1	2.55	49	−17.2	0.138	−17	0.47	−104
7.0	0.74	−174	7.3	2.31	26	−16.8	0.144	−36	0.42	−127
8.0	0.71	165	6.4	2.09	9	−17.1	0.140	−52	0.41	−149
9.0	0.69	147	5.7	1.92	−11	−17.3	0.137	−65	0.41	−172
10.0	0.68	132	4.7	1.71	−29	−17.6	0.132	−76	0.44	172
11.0	0.68	117	4.0	1.58	−47	−17.8	0.129	−88	0.47	157
12.0	0.67	103	3.2	1.45	−64	−17.7	0.130	−100	0.50	142
13.0	0.67	92	2.2	1.29	−78	−18.5	0.119	−111	0.53	130
14.0	0.66	84	1.7	1.21	−88	−18.6	0.118	−119	0.55	119
15.0	0.65	74	1.4	1.18	−104	−18.7	0.116	−129	0.59	109
16.0	0.62	63	1.3	1.16	−119	−18.5	0.119	−136	0.62	99
17.0	0.59	51	0.7	1.09	−135	−18.3	0.122	−147	0.64	89
18.0	0.51	31	0.6	1.07	−153	−18.3	0.121	−162	0.63	77

[a] Packaged performance: 1.5-dB noise figure and 12-dB gain at 6 GHz; 1.8-dB noise figure and 9-dB gain at 12 GHz. Unpackaged (chip) performance: same performance at 6 and 12 GHz; 2.8-dB noise figure and 6 dB gain at 18 GHz. (From Avantek 1985 Semiconductor Device Catalog)

quired to make I_{ds} almost 0. The input resistance of the gate at low frequencies is very high. The low-frequency transconductance g_{mo} is the drain current variation for a given gate voltage variation:

$$g_{mo} = \left.\frac{\Delta I_{ds}}{\Delta V_{gs}}\right|_{V_{ds} \,=\, constant} \quad \text{Siemens (formerly referred to} \quad (14.5)$$
$$\text{as mho} = \text{ohm}^{-1})$$

For linear operation the device is biased as a class A amplifier (middle region of the I–V curves) where there is little variation of g_{mo} over the RF gate voltage range. For low-noise operation, I_{ds} is reduced to 10–20% of I_{dss}. Power amplifiers can be operated class AB or B for higher efficiency.

14.3.4   Corner Frequency

MESFETs are parameterized like low-frequency transistors with corner frequencies. The unity gain frequency f_T occurs when the current through C_{gs} equals the current in the voltage-controlled generator

$$2\pi f_T C_{gs} V_c = V_c g_m \tag{14.6}$$

or

$$f_T = \frac{g_m}{2\pi C_{gs}} \tag{14.7}$$

Because of the impedance transformation between the input and output, power gain can occur for frequencies greater than f_T.

14.3.5   Intrinsic Stability

When the MESFET input is matched with its complex-conjugate network, the device becomes unstable with decreasing frequency. This occurs because the drain voltage is coupled back through the drain-gate capacitance C_{dg} across the input resistance R_{in} (a combination of C_{gs}, R_g, R_i, and R_s, see 7) in Fig. 14.11. The impedance of C_{dg} is $(j\omega C_{dg})^{-1}$ and R_{in} is proportional to ω^{-2} so at some critical frequency f_k the MESFET's C_{dg}–R_{in} voltage divider is less than the voltage gain and oscillation can begin. For the parameters in Table 14.1, $f_k = 6.1$ GHz. This problem will be avoided by proper input circuit design (See Chapter 15) along with other stability considerations.

14.3.6   Noise Sources

FETs exhibit shot noise in the leakage current of the gate-semiconductor junction, thermal noise in the source and drain resistances of the FET, and

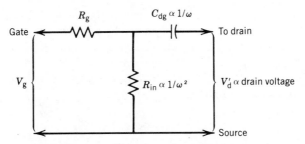

Figure 14.11. Simplified MESFET equivalent circuit.

recombination noise in the depletion region. Modern low-noise MESFETs use recessed gate structures as shown in Fig. 14.10c because they provide lower noise factors. The minimum noise factor for the MESFET increases linearly with frequency (8)

$$F_{\min} = 1 + \left(\frac{2.5f}{f_T}\right) \sqrt{g_m(R_g + R_s)} \qquad (14.8)$$

whereas the corresponding relation for a BPT increases as the square of the operating frequency (see Eq. 14.1).

14.3.7   Upper-Frequency Limits

GaAs is preferred over Si for MESFET fabrication because the mobility (for a typical carrier concentration) and saturated electron drift velocity are greater in GaAs. These relate to lower resistance (less loss) and longer gate lengths (at high velocity the electron spends less time under the gate), which are easier to fabricate. Decreasing the gate length decreases C_{gs} and increases g_m until the gate length is equal to the thickness D of the epitaxial layer. Thus as L is decreased, D is decreased and N_D is increased so that pinchoff does not occur at zero bias. This procedure continues until N_D is near 4×10^{17} donors/cm^3, because above this value junction electric field breakdown (avalanching) occurs before pinchoff.

14.3.8   Power MESFETs

The key parameters for power MESFETs are output power, efficiency, and gain associated with the power output. The estimates of the power output and efficiency for a resistive load line can be made from the current–voltage characteristic and the maximum bias voltage and current ratings. Referring to Fig. 14.9b for class A operation, a bias point midway between V_{CC} (operating drain bias) and the voltage associated with the onset of the saturated current for a given gate bias, V_1 is selected. Note that V_{gs} can be operated slightly positive to counteract the Fermi potential and reduce the depletion region to nearly zero before excessive gate currents are drawn. V_{CC} and V_D must be less than the breakdown voltage V_{BR} minus the pinchoff voltage V_P for all temperatures. The pinchoff current for power FETs is not zero because of leakage. In some cases one-tenth I_{dss} is selected as the pinchoff current. The corresponding drain current swings extend from I_{D1} to I_{D2}. The RMS output power is then

$$P_{\text{out,class A}} = [\tfrac{1}{2}]\left(\frac{(I_{D1} - I_{D2})}{2}\right)\left(\frac{(V_D - V_1)}{2}\right) \tag{14.9}$$

The class A efficiency is

$$\eta \text{ class A} = \frac{P_{\text{out,class A}}}{[V_1 + (V_D - V_1)/2][I_{D2} + (I_{D1} - I_{D2})/2]} \tag{14.10}$$

This never exceeds 50% because $V_1 > 0$ and $I_{D2} > 0$. Typical values range from 40 to 45%.

For class B operation, the load line is twice as steep since a resonant circuit will swing the drain voltage to V_D. The output power is the same as Eq. 14.9, but now the maximum $V_D = V_{BR} - 2V_P$ since the gate voltage also swings to twice pinchoff. The output current swings appear as a half-wave rectifier with an efficiency that cannot exceed 78% because $V_1 > 0$. Typical values are near 60%.

The efficiency terms above assume none of the input signal appears in the output circuit. However, for low gain stages a significant amount of the device power reaches the output. For this reason the power-added efficiency parameter has been applied. The power-added efficiency is defined as

$$\eta_{\text{pa}} = \frac{(P_{\text{out}} - P_{\text{in}})}{P_{DC}} \tag{14.11}$$

where P_{out} and P_{in} are microwave powers and P_{DC} is the average DC input power at the operating point. For higher gain stages, $P_{\text{in}} \ll P_{\text{out}}$ so the usual efficiency used to describe the device is

$$\eta = \frac{P_{\text{out}}}{P_{DC}} \tag{14.12}$$

For power MESFETs this interplay of output power, efficiency, and gain it rather complex, as shown in Fig. 14.12 (9). Clearly no single operating point provides the maximum of all parameters simultaneously. The output impedances are different for maxium output power, gain, efficiency, and linearity. All parameters but gain are strongly dependent on the drain–source

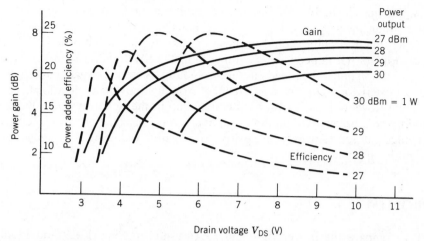

Figure 14.12. *Power MESFET performance curves at a single frequency. Curves are maximum possible for optimally tuned circuit at each bias value. (From Application Note TE-214, Microwave Semiconductor Corp., (MSC), Somerset NJ, 1980)*

voltage. Reduced voltage is desirable because it lowers the channel operating temperature and raises the device lifetime.

The third-order intercept (see Chapters 13 and 21) is usually 10 dB greater that the 1-dB compression power level when operating in the linear region. As the output approaches the 1-dB compression point (see Chapter 21), the distortion departs from the "classical" behavior, so each type of FET should be measured at the desired load and bias point (10).

Pulse operation does not yield significantly higher output powers since operation is limited as described above by V_{BR} and I_{dss}. Depending on the relation between the pulse length and the thermal time constant of the MESFET, the heat sink resistance may be raised during pulsed operation without exceeding the 150°C channel temperature.

Depending on the bias point and supply voltages, the FET may be operated temporily into a mismatched load (increases voltage swings beyond V_{BR}). Generally a load-fault circuit is desirable to reduce last stage bias should the load be disconnected.

14.3.9 Other Considerations

MESFETs with aluminum gates are susceptible to burnout due to static discharges, but gold gates require minimal precautions such as grounding wrist straps and work benches. Curve tracers with bypass capacitors and ferrite beads are desirable for DC testing to minimize low-frequency oscillations. A volt-ohmmeter should not be used to test the junctions.

The gate bias should be applied before the drain bias to prevent operation at I_{dss}. Zener diodes are desirable to prevent transient bias peaks exceeding maximum ratings. The channel temperature should be 150°C or less. The manufacturer usually specifies the thermal resistance to the case surface. The circuit thermal resistance is the responsibility of the circuit designer.

The tuned circuits should be near the device since high-power devices have low-output impedances resulting in high VSWR and loss on a 50-ohm transmission line if the tuned circuit is remotely located. The gate bias should be adjusted for optimum performance. For high-power applications, gate current flows because of forward conduction in the Schottky gate junction and reverse breakdown on high negative voltage peaks. For this reason the average gate current may be positive or negative requiring the bias supply to both supply and sink current.

PROBLEMS

1. Compare the power BPT and FET results to Figs. 13.27 and 13.40. Why are they comparable/not comparable?

2. Compute the low-frequency current gain for the BPT neglecting the capacitors and parasitic networks in the equivalent circuit of Fig. 14.2.

3. Use the following values to compute F, $R_{s,opt}$, and $X_{s,opt}$ for a BPT at 1 GHz. R_b = 15 ohms, r_e = 13 ohms, α_0 = 0.99, C_e = 0.8 pF, $f_b = (2\pi\tau_b)^{-1}$, base cutoff frequency = 20 GHz, and $f_e = (2\pi\tau_e)^{-1}$ = 16 GHz.

4. If the transistor in Problem 3 were matched with only a 50-ohm source impedance, compute the noise factor and compare with the result for $R_{s,opt}$.

5. Compute the reverse gain of the linear BPT in Section 14.2.3.

6. Compute the power gain of the BPT in Fig. 14.5 when conjugately matched at 1 GHz.

7. How many stages are needed to drive the linear BPT in Section 14.2.3 from a 0-dBm source? Assume the gain of the driver stage (160-mW output) is 10 dB, all preceding stages have 14 dB gain, and the interstage matching networks are lossless.

8. Confirm that the series output impedance of the circuit in Fig. 14.7 compares with the optimum values in Fig. 14.6 at 2.2 GHz.

9. Estimate the current density for the low-noise FET in Table 14.1. The gate is 1 μm long and 500 μm wide. The active layer is 10 μm thick. Are the electrons moving at their saturated drift velocity?

10. Estimate g_{mo} in Fig. 14.9b if I_{dss} = 100 mA.

11. Compute F_{min} for the MESFET in Table 14.1 at 4 GHz.

12. Estimate the maximum sinusoidal class A power output for a MESFET with $V_D = 4$ V, $R_{load} = 30$ ohms, and $I_{dss} = 100$ mA. Assume the current peaks are limited to I_{dss}.

13. Compute the class A efficiency for the MESFET in Problem 12.

14. Compute the maximum class B efficiency assuming the drain current is half a sine wave, $I_{D2} = 0 = V_1$, and $V_D = V_{BR}$.

15. Compute the power-added efficiency for the BPT stage in Section 14.2.3.

16. Compute the maximum thermal resistance allowed to keep the channel temperature at 150°C for the power MESFET in Fig. 14.12 operating at 30-dBm output and optimally tuned. What happens to the channel temperature if the tuning is not optimal?

REFERENCES

1. K. Joshin, T. Mimura, M. Niori, Y. Yamishita, K. Kosemura, and J. Saito, "Noise Performance of Microwave HEMT," *1983 IEEE MTT-S Int'l. Micro. Symp. Digest,* May 31–June 3, Boston, MA p. 563.

2. H. F. Cooke, "Microwave Transistors: Theory and Design," *Proc. IEEE,* Vol. 59, No. 8, August 1971, p. 1163.

3. R. J. Hawkins, "Limitations of Nielsen's and Related Noise Equations Applied to Microwave Bipolar Transistors, and a New Expression for the Frequency and Current Dependent Noise Figure," *Solid-State Electronics,* Vol. 20, No. 3, March 1977, p. 191.

4. T. Hsu and C. P. Snapp, "Low-Noise Microwave Bipolar Transistor with Sub-Half-Micrometer Emitter Width," *IEEE Trans. Elec. Dev.,* Vol. ED-25, No. 6, June 1978, p. 723.

5. MSC 80196/80186 Data Sheet LP-106, Microwave Semiconductor Corp., Somerset, NJ, April 1982.

6. Microwave Power Transistor Data Sheet GP-103, Microwave Semiconductor Corp., Somerset, NJ.

7. C. A. Liechti, "Microwave Field-Effect Transistors—1976," *IEEE Trans. Micro. Th. Tech.,* Vol. MTT-24, No. 6, June 1976, p. 279.

8. H. Fukui, "Optimal Noise Figure of Microwave GaAs MESFET's," *IEEE Trans. Elec. Dev.,* Vol. ED-26, No. 7, July 1979, p. 1032.

9. "Care and Feeding of Power GaAs FET's," Appl. Note TE-214, Microwave Semiconductor Corp., Somerset, NJ., 1980.

10. E. W. Strid and T. C. Duder, "Intermodulation Distortion Behavior of GaAs Power FET's," *1978 IEEE MTT-S Int'l. Micro. Symp. Digest,* p. 135, and J. A. Higgins, "Intermodulation Distortion in GaAs FET's," ibid., p. 138.

15 | Microwave Amplifier and Oscillator Design

15.1 INTRODUCTION

Amplifiers and oscillators are the most frequent circuits involving active microwave components that the microwave engineer will be asked to build. Commercially available components frequently are not suitable for subsystems in which small size, light weight, or specialized responses are required.

This chapter only treats FET and bipolar amplifiers and oscillators. Reflection amplifiers and oscillators involving Gunn and IMPATT devices are covered in Refs. 1 and 2.

Modern amplifier design techniques make liberal use of computer-aided design and analysis programs. These are used here but are preceded by a discussion of narrowband amplifiers. In this discussion, the basic ideas implemented via computer are presented. The topics include power gain definitions, amplifier stability considerations, constant gain circles, and constant noise factor circles. These can be hand calculated for certain limiting cases. The other cases are best done via computer-aided design (CAD). Reference 3 is excellent for more in-depth study of amplifier and oscillator design.

15.2 MICROWAVE AMPLIFIERS

Microwave amplifiers involving discrete FETs in microstrip circuits are a common requirement for subsystem development. The basic types of FET amplifiers are narrowband (designed to operate over a bandwidth less than 10% of the center operating frequency), low noise, broadband, and high power. Each type has its own design techniques, but they are interrelated via terminal parameters (*S*-parameters, gain) and stability.

15.2.1 Amplifier Block Diagram

The *S*-parameters for FET and bipolar three-terminal devices have been presented in Chapter 14 and are available from the manufacturer and/or stored in the CAD data bank. On either port of the amplifier a network "matches" or couples the amplifier from its source to its load (possibly another stage of the amplifier chain). The transistor is usually imbedded in matching and bias networks as shown in Fig. 15.1. The bias networks are not included in the analytic studies in this chapter but should be considered for CAD because they frequently affect the overall amplifier performance. Bias networks are included in Fig. 15.1 to remind the designer of their importance.

The input and output reflection coefficients to the transistor are

$$\Gamma_{in} = S_{11} + \frac{S_{12}S_{21}\Gamma_L}{1 - S_{22}\Gamma_L} \quad \text{and} \quad \Gamma_{out} = S_{22} + \frac{S_{12}S_{21}\Gamma_S}{1 - S_{11}\Gamma_S} \quad (15.1)$$

which states that the input reflection coefficient is a function of Γ_L unless $S_{12}S_{21}$ is small or zero. S_{21} could also be made zero, but the transistor would have no gain—a disinteresting case. Similarly Γ_{out} depends on S_{12} and Γ_S. The derivation of these relations is given in Chapter 20 and Ref. 3. To simplify this presentation, the *unilateral* ($S_{12} = 0$) transistor approximation is made. This approximation keeps the hand mathematics tractable, but CAD programs use the full relations in Eq. 15.1. For the unilateral case,

$$\Gamma_{in} = S_{11} \quad \text{and} \quad \Gamma_{out} = S_{22} \quad (15.2)$$

The stability discussion (Section 15.2.3) considers the case in which $S_{12} \neq 0$ for the transistor in order to present more general results.

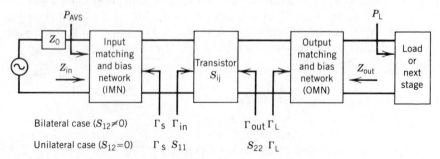

Figure 15.1. *Block diagram of a single-stage transistor amplifier. All reflection coefficients normalized to Z_0.*

15.2.2 Power Gain Relations

Two power gain relations are useful for the gain and noise design procedures. For the gain design, the transducer power gain G_T is the ratio of the power delivered to the load P_L to the power available from the source P_{AVS}. For the unilateral ($S_{12} = 0$) case

$$G_T = \frac{1 - |\Gamma_S|^2}{|1 - S_{11}\Gamma_S|^2} |S_{21}|^2 \frac{1 - |\Gamma_L|^2}{|1 - S_{22}\Gamma_L|^2} = G_S |S_{21}|^2 G_L$$

$$= \begin{pmatrix} \text{Input} \\ \text{parameters} \end{pmatrix} \begin{pmatrix} \text{Unilateral} \\ \text{transistor} \\ \text{gain} \end{pmatrix} \begin{pmatrix} \text{Output} \\ \text{parameters} \end{pmatrix}$$

(15.3)

The maximum transducer gain occurs when $\Gamma_S = S_{11}^*$ and $\Gamma_L = S_{22}^*$. For the noise design, the available power gain G_A is the ratio of power available from the network P_{AVN} to P_{AVS}. For the unilateral case

$$G_A = \frac{1 - |\Gamma_S|^2}{|1 - S_{11}\Gamma_S|^2} |S_{21}|^2 \frac{1}{1 - |\Gamma_{out}|^2}$$

(15.4)

where the available power in the output occurs when $\Gamma_{out} = \Gamma_L^*$.

15.2.3 Amplifier Stability Relations

Amplifier designers may not have control of the source and load impedances so the amplifier must be designed to operate with any Γ_S or Γ_L whose magnitude is less than 1 (passive network). If this is not considered the amplifier may become unstable (oscillate), thus generating a spurious signal. To assure that the amplifier is stable, the input and output port impedance (Z_{in} and Z_{out}, see Fig. 15.1) must not have negative real parts at any frequency. This case is called unconditionally stable. For the other case in which some values of source and load impedances cause a negative real part to Z_{in} and/or Z_{out}, the amplifier is potentially unstable.

It is convenient to represent the impedances as reflection coefficients referenced to the characteristic impedance Z_0 because constant $|\Gamma|$'s appear as circles on the Smith chart. The goal is to show the regions of stability/instability on the Smith chart and then find the requirements on the network impedances. This leads to synthesis of the matching network.

Heuristically, the amplifier begins to oscillate if one port has a negative real part because noise generated in the adjoining network enters the port, the negative resistance generates more noise rather than dissipating the incident noise, and some of this generated noise combines with the incoming noise to input more noise. This positive feedback effect continues until a large voltage is developed at the frequency selected by the reactances in the circuit. Remember that the port is not matched because a passive network does not match to a network whose $\text{Re}(Z_{in}) < 0$. This subject is discussed in Section 15.3.

The conditions for unconditional stability in terms of reflection coefficients are

$$|\Gamma_S| < 1, |\Gamma_L| < 1, |\Gamma_{in}| = |S_{11}| < 1$$

$$\text{and} \quad |\Gamma_{out}| = |S_{22}| < 1, \quad (15.5)$$

for the unilateral case. Since the $|\Gamma|$ is frequency dependent, these conditions must be met at all frequencies. If the transistor $S_{12} \neq 0$, the bilateral case, the required conditions for unconditional stability are

$$|\Gamma_S| < 1 \qquad\qquad\qquad (15.6a)$$

$$|\Gamma_L| < 1 \qquad\qquad\qquad (15.6b)$$

$$|\Gamma_{in}| = \left| S_{11} + \frac{S_{12}S_{21}\Gamma_L}{1 - S_{22}\Gamma_L} \right| < 1 \qquad (15.6c)$$

and

$$|\Gamma_{out}| = \left| S_{22} + \frac{S_{12}S_{21}\Gamma_S}{1 - S_{11}\Gamma_S} \right| < 1 \qquad (15.6d)$$

Note that the Γ's are defined with respect to the full amplifier (transistor plus matching networks) as shown in Fig. 15.2 rather than Fig. 15.1.

The relations in Eq. 15.6 can be solved for the range of values of Γ_L and Γ_S given the values of S_{ij} where $|\Gamma_{in}| = |\Gamma_{out}| = 1$; i.e., the boundary between the stable and unstable regions. This is begun as follows for Γ_L. Rewriting Eq. 15.6c yields

$$|S_{11} - \Gamma_L(S_{11}S_{22} - S_{12}S_{21})| = |1 - S_{22}\Gamma_L| \qquad (15.7)$$

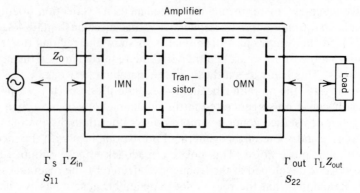

Figure 15.2. *Definition of reflection coefficients for stability analyses.*

Squaring both sides and using the complex arithmetic relations

$$| a + jb |^2 = a^2 + b^2 = (a + jb)(a + jb)^* \qquad (15.8)$$

$$[(a + jb) + (c + jd)(e + jf)]^* = (a + jb)^* + (c + jd)^*(e + jf)^* \quad (15.9)$$

and

$$\Delta = S_{11}S_{22} - S_{12}S_{21} \qquad (15.10)$$

yields

$$\left| \Gamma_L - \frac{(S_{22} - \Delta S_{11}^*)^*}{|S_{22}|^2 - |\Delta|^2} \right| = \left| \frac{S_{12}S_{21}}{|S_{22}|^2 - |\Delta|^2} \right| \qquad (15.11)$$

and similarly for the input

$$\left| \Gamma_S - \frac{(S_{11} - \Delta S_{22}^*)^*}{|S_{11}|^2 - |\Delta|^2} \right| = \left| \frac{S_{12}S_{21}}{|S_{11}|^2 - |\Delta|^2} \right| \qquad (15.12)$$

Equations 15.11 and 15.12 can be interpreted as circles in the output and input reflection coefficient planes. In the output plane

$$| \Gamma_L - (\text{center of circle, } C_L) | = | \text{radius of circle, } r_L | \qquad (15.13)$$

and similarly for the input plane. The center and radius are given in Eq. 15.11 (output plane) and Eq. 15.12 (input plane). Note that for the unilateral amplifier, $S_{12} = 0$ and the radius is zero; thus the bilateral case was considered in this section.

Two typical results are shown in Fig. 15.3. In the first case, $\| c_L | - r_L | > 1$, so all output loads on the Smith chart are stable. These designs are unconditionally stable. A more difficult case occurs when $\| c_L | - r_L | < 1$. Now there are passive load reflection coefficients that will cause oscillation (see Fig. 15.3b), and two cases exist as shown. Similar constructions occur for the input reflection coefficient plane. To determine which of the two cases in Fig. 15.3b applies, consider the following limiting case. If $Z_L = Z_0$, then $\Gamma_L = 0$ and $| \Gamma_{in} | = | S_{11} |$; so if $| S_{11} | < 1$, then $| \Gamma_{in} | < 1$ and the center of the Smith chart is a stable point (left side of Fig. 15.3b). If $| S_{11} | > 1$, the center of the Smith chart is unstable and only load reflection coefficients in the stable region in the right side of Fig. 15.3b yield a stable amplifier. The S-parameters are defined as shown in Fig. 15.2.

Equations 15.6 are the necessary and sufficient conditions for stability. For ease of application, these four equations have been rearranged into several forms (4–6). The simplest form is

$$K = \frac{1 - | S_{11} |^2 - | S_{22} |^2 + | \Delta |^2}{2 | S_{12}S_{21} |} > 1 \qquad (15.14)$$

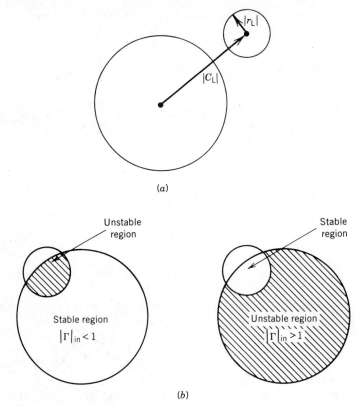

Figure 15.3. *Stable and unstable regions of the L-plane on a Smith chart. (a) Uncon-ditionally stable case; (b) stable and unstable regions.*

and

$$| \Delta | < 1 \qquad (15.15)$$

One of the CAD programs (SUPER-COMPACT®) uses

$$K = \frac{1 - | S_{11} |^2 - | S_{22} |^2 + | \Delta |^2}{2 | S_{12}S_{21} |} > 1 \qquad (15.14a)$$

$$B_1 = 1 + | S_{11} |^2 - | S_{22} |^2 + | \Delta |^2 > 0 \qquad (15.15a)$$

for stable conditions. For the unilateral case ($S_{12} = 0$), K is infinite and there is unconditional stability if $| S_{11} | < 1$ and $| S_{22} | < 1$ for all passive source and load terminations.

If the transistor is potentially unstable (the above conditions in Eqs. 15.14 and 15.15 apply to two ports such as transistors only or transistor and associated matching networks), it can be stabilized using several techniques. For narrowband applications the best technique is to select Γ_S and Γ_L to be in the stable region. The selection of Γ_S and Γ_L is made using the constant gain circles, to be described next.

15.2.4 Unilateral Constant Gain Circles

The unilateral ($S_{12} = 0$) transducer power gain G_T is a function of both the input and output matching networks and the transistor (see Eq. 15.3). The maximum unilateral transducer gain occurs when $\Gamma_S = S_{11}^*$ and $\Gamma_L = S_{22}^*$, yielding for the input transducer

$$
G_{T,max} = \frac{1 - |S_{11}^*|^2}{|1 - S_{11}S_{11}^*|^2}
$$

$$
= \frac{1 - S_{11}S_{11}^*}{(1 - S_{11}S_{11}^*)(1 - S_{11}^*S_{11})} = \frac{1}{1 - |S_{11}|^2}
$$

(15.16)

and similarly for the output. If the input or output reflection coefficient is unity, then $G_T = 0$. Thus G_T varies between 0 and $G_{T,max}$ and allows the definition of normalized transducer power gain factor g_T at the input (source) and output (load) of

$$
g_{T,S} = \frac{G_{T,S}}{G_{T,max,S}} = G_{T,S}(1 - |S_{11}|^2)
$$

(15.17a)

and

$$
g_{T,L} = \frac{G_{T,L}}{G_{T,max,L}} = G_{T,L}(1 - |S_{22}|^2)
$$

(15.17b)

Using the technique in Section 3.4 of Ref. 3, calculate the locus of reflection coefficients Γ_S or Γ_L, which make $g_{T,S}$ or $g_{T,L}$ a constant (constant gain locus). Forming the real and imaginary parts of $\Gamma_S = U_S + jV_S$, $\Gamma_L = U_L + jV_L$, $S_{11} = A_{11} + jB_{11}$, and $S_{22} = A_{22} + jB_{22}$ and substituting these into the appropriate part of Eq. 15.17 yields for the input

$$
\left(U_S - \frac{g_S A_{11}}{D_S}\right)^2 + \left(V_S + \frac{g_S B_{11}}{D_S}\right)^2
$$

$$
= \left(\frac{\sqrt{1 - g_S}\,(1 - |S_{11}|^2)}{D_S}\right)^2
$$

(15.18)

where $D_S = 1 - |S_{11}|^2(1 - g_S)$ and an analogous relation for the output. This equation is a family of circles in $\Gamma_S = U_S + jV_S$ whose centers and radius are a function of g_S. The real and imaginary values of the constant gain contour (circle) centers are

$$U_C = \frac{g_S A_{11}}{D_S} \qquad V_C = \frac{-g_S B_{11}}{D_S} \qquad (15.19)$$

and the radius of the circle is

$$R_S = \frac{\sqrt{1 - g_S}\,(1 - |S_{11}|^2)}{D_S}. \qquad (15.20)$$

At the output, analogous relations hold by replacing the subscripts S by L and 11 by 22.

Both the center of the constant gain circle and the radius of the circle are functions of g_S or g_L. Thus the circles are not concentric. Also when g_S or $g_L = 1$, R_S or $R_L = 0$ and the maximum gain point is located at $A_{11} - jB_{11} = S_{11}^*$ or $A_{22} - jB_{22} = S_{22}^*$. For intermediate values of g_S or g_L the circle centers lie along a line between $\Gamma_S = 0$ and S_{11}^* or $\Gamma_L = 0$ and S_{22}^*. Finally, the 0 dB (G_S or $G_L = 1$) gain circle passes through the center of the reflection coefficient plane (Smith chart).

The constant gain circles and the stability considerations facilitate the design of the narrowband amplifier. Assuming that the constant gain circle lies in the stable region, any point away from the boundary (avoid being near the boundary of the stable region in case the device S-parameters vary somewhat from the manufacturer's values) can be selected as the desired reflection coefficient. On the Smith chart ($Z_0 = 50$ ohms or some other value) this point should be selected to allow movement along constant reactance (susceptance) curves to develop the lumped-element matching circuit with reasonable component values. For a distributed network the constant Γ curves are followed for development of the matching network.

So far in this section it has been tacitly assumed that $|S_{11}| < 1$ and $|S_{22}| < 1$. If this were not true, Eq. 15.16 would be infinite and a passive termination can produce oscillations because the real part of the input (output) impedance is negative, the real part of the source (load) impedance is equal in magnitude and positive, and so the net loop resistance can be zero. For this case a critical value of reflection coefficient exists, namely,

$$\Gamma_{S,C} = \frac{1}{S_{11}} \quad \text{and} \quad \Gamma_{L,C} = \frac{1}{S_{22}} \qquad (15.21)$$

where G_S and G_L in Eq. 15.3 have zero value denominators. Refer to Ref. 3 to design for this case.

15.2.5 Example of Narrowband Amplifier Design

With the concepts of stability and constant gain circles a narrowband (5–10% bandwidth) amplifier can be designed.

For this example the NEC 67383 GaAs FET is selected because it has a maximum stable gain of 16.8 dB needed later. The S-parameters and the noise parameters in a range of frequencies near 4 GHz are given in Table 15.1. Using the S-parameters, the stability of the transistor only is computed at midband, 4 GHz, using Eqs. 15.14 and 15.15. Computing $|\Delta|$ first requires

$$S_{11}S_{22} = (0.88)(0.61) \, \underline{/-79° - 58°} = -0.39 - j0.37 \quad (15.22a)$$

$$S_{12}S_{21} = (0.06)(2.85) \, \underline{/35° + 107°} = -0.13 + j0.11 \quad (15.22b)$$

$$S_{11}S_{22} - S_{12}S_{21} = -0.26 - j0.48 \quad (15.22c)$$

The $|\Delta| = |S_{11}S_{22} - S_{12}S_{21}|$ is

$$[(0.26)^2 + (0.48)^2]^{1/2} = 0.55 < 1 \quad (15.23)$$

so the transistor meets this criteria for stability. Now computing

$$
\begin{aligned}
K &= \frac{1 - |S_{11}|^2 - |S_{22}|^2 + |\Delta|^2}{2|S_{12}S_{21}|} \\
&= \frac{1 - (0.88)^2 - (0.61)^2 + (0.55)^2}{2(0.17)} = 0.46 < 1
\end{aligned}
\quad (15.24)
$$

Table 15.1 S- and Noise-Parameters for NEC 67383 FET

Small Signal S-Parameter

Frequency	S_{11} Mag	/Angle	S_{21} Mag	/Angle	S_{12} Mag	/Angle	S_{22} Mag	/Angle
2 GHz	0.97	/−43°	3.39	/140°	0.04	/161°	0.63	/−32°
4 GHz	0.88	/−79°	2.85	/107°	0.06	/35°	0.61	/−58°
6 GHz	0.84	/−103°	2.57	/81°	0.07	/20°	0.62	/−77°

Noise Parameters

	Frequency	
Parameter	4 GHz	8 GHz
Minimum noise figure, NF_{min}	0.4 dB	0.8 dB
$|\Gamma_s|$ for NF_{min}	0.64	0.55
$\underline{/\Gamma_s}$ degrees for NF_{min}	69	115
Normalized effective noise resistance ($Z_0 = 50$ ohms)	0.38	0.2

Data for the NEC 67383 FET provided courtesy of California Eastern Laboratories Inc.

which does not pass the stability criteria. Thus the transistor is potentially unstable if matched directly to a 50-ohm circuit. For a narrowband design, the Γ_S and Γ_L are selected to present stabilizing source and load reflection coefficients. This procedure is done with the aid of the stability circles.

The source and load stability regions are defined by the reflection coefficient circles in Eqs. 15.12 and 15.11, respectively. Using the notation given earlier for the radii and centers at 4 GHz,

$$r_S = \left| \frac{S_{12}S_{21}}{|S_{11}|^2 - |\Delta|^2} \right| = \frac{0.17}{(0.88)^2 - (0.55)^2} = \frac{0.17}{0.48} = 0.35 \quad (15.25)$$

$$c_S = \frac{[S_{11} - \Delta S_{22}^*]^*}{|S_{11}|^2 - |\Delta|^2}$$

$$= \frac{[(0.17 - j0.86) - (-0.26 - j0.48)(0.32 - j0.52)^*]^*}{0.48} \quad (15.26)$$

$$= \frac{[-j0.86 + j0.29]^*}{0.48} \rightarrow 1.19 \,\underline{/90°}$$

$$r_L = \left| \frac{S_{12}S_{21}}{|S_{22}|^2 - |\Delta|^2} \right| = \frac{0.17}{0.08} = 2.1 \quad (15.27)$$

$$C_L = \frac{[S_{22} - \Delta S_{11}^*]^*}{|S_{22}|^2 - |\Delta|^2} \quad (15.28)$$

$$= \frac{[(0.32 - j0.52) - (0.37 - j0.30)]^*}{0.08} \rightarrow 2.82 \,\underline{/103°}$$

These values include roundoff errors, but since circuits are not designed to operate near the region of instability these round off errors are unimportant. The stability circles are plotted on the Smith chart as shown in Fig. 15.4.

To determine if the stable regions are inside or outside the circles (see Fig. 15.3b), assume $Z_L = Z_0 = 50$ ohms so $\Gamma_L = 0$ and $|\Gamma_{in}| = |S_{11}| < 1$. Therefore, the region outside the source reflection coefficient Γ_S circle is stable. Similarly for the load, the region outside the circle is stable since for Z_S and $Z_0 = 50$ ohms, $\Gamma_S = 0$ and $|\Gamma_{out}| = |S_{22}| < 1$.

In anticipation of the broadband design to be done later in this chapter, it is desirable to stabilize the transistor for all values of source and load reflection coefficients. Resistive elements in series or shunt will stabilize the transistor. Also "negative" feedback to cancel S_{12} will stabilize the transistor, but this circuitry is complex and tends to be frequency sensitive. Therefore, for this design several different resistive stability techniques will be investigated.

Using the constant impedance and conductance curves in Fig. 15.4, the first four configurations as shown in Table 15.2 can be derived using a series

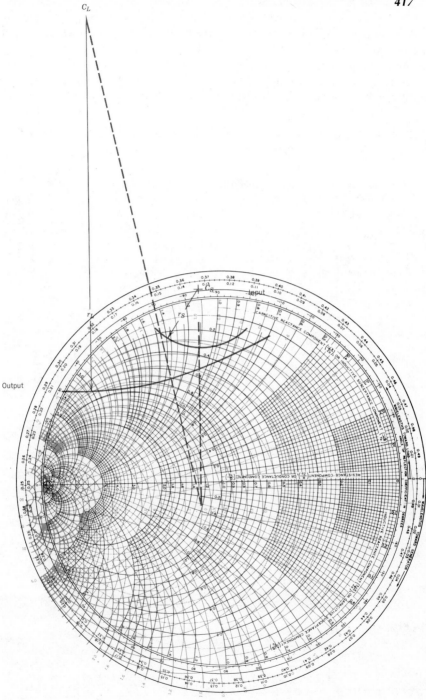

Figure 15.4. Stability circles for NEC 67383 GaAs FET.

or shunt element in the gate and drain circuits. Each of these assures stability for passive source and load circuits but yields different small signal gains and, although not shown here, an increased amplifier noise figure. Comparing techniques (a) through (d), a 3-dB maximum small-signal gain (MSG) difference is observed. Also note that as K increases, the MSG decreases.

Table 15.2 Comparison of Stabilization Techniques for Narrowband Amplifier (FET is NEC 67383 resistors in ohms)

Configuration	r_s	C_s	r_1	C_1	K	Max Small Signal Gain (dB)
(a) 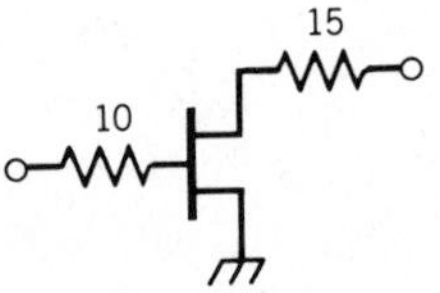	0.35	1.5 /84°	1.3	2.6 /70°	1.6	12.4
(b) 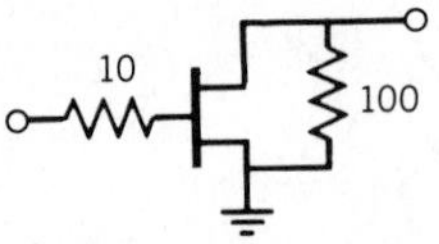	0.14	1.3 /77°	1.7	3.4 /106°	2.2	10.6
(c) 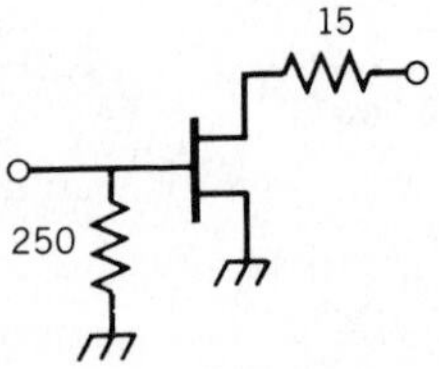	0.36	1.5 /91°	0.67	1.9 /64°	1.6	12.1
(d) 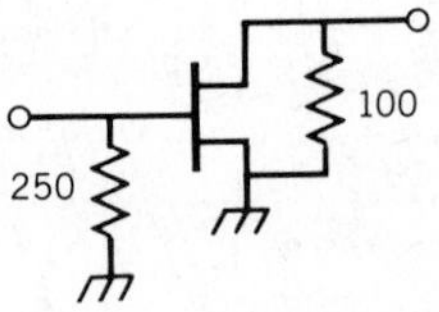	0.16	1.4 /81°	1.2	3.0 /93°	2.9	9.3
(e) 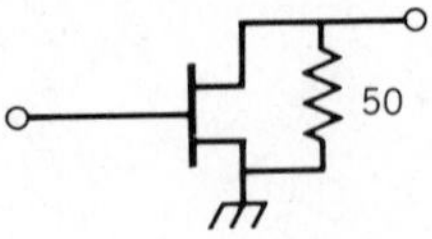	0.08	1.1 /76°	1.8	2.8 /128°	1.1	15.0
(f) 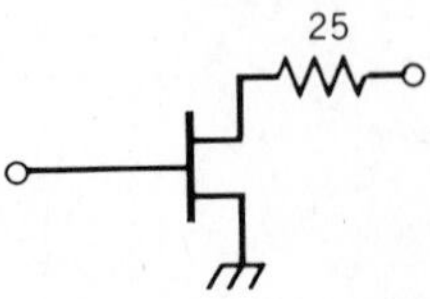	0.26	1.26 /87°	3.7	4.7 /85°	1.0	16.8

Table 15.2 Comparison of Stabilization Techniques for Narrowband Amplifier (FET is NEC 67383 resistors in ohms) (continued)

Configuration	r_S	C_s	r_l	C_l	K	Max Small Signal Gain (dB)
(g) [circuit: 10 Ω series resistor, FET]	0.43	1.4 $\underline{/87°}$	1.1	2.2 $\underline{/82°}$	1.1	15.3
(h) [circuit: FET with 250 Ω shunt resistor]	0.42	1.4 $\underline{/92°}$	0.65	1.7 $\underline{/73°}$	1.0	15.8

However, since S_{12} and S_{21} are nonzero, it is possible that a single loading resistor will stabilize the narrowband amplifier. Entries (e) through (h) show these cases for K values near unity. Note the MSG rapidly approaches the transistor's MSG of 16.8 dB as K approaches unity. Because of variations between transistors and variations with temperature of the S-parameters, operation too close to $K = 1$ is not recommended. The technique with the shunt 50-ohm resistor (e) will be selected for further design because it remains stable over a wider frequency range than some of the others and it $|S_{12}|$ is slightly smaller than the others. The S-parameters for the 50-ohm shunt loaded transistor are

$$S_{11} = 0.91 \,\underline{/-77}, \, |S_{11}| < 1$$
$$= 0.20 - j0.89 \tag{15.29a}$$

$$S_{21} = 1.70 \,\underline{/116} \tag{15.29b}$$

$$S_{12} = 0.04 \,\underline{/44} \tag{15.29c}$$

$$S_{22} = 0.25 \,\underline{/-134}, \, |S_{22}| < 1$$
$$= -0.17 - j0.18 \tag{15.29d}$$

In a real design (using CAD), the bias circuit effects are also considered since they affect the stage stability. Also if stages are cascaded, the overall amplifier S-parameters and stability must be checked. These calculations are tractable with CAD.

The narrowband amplifier design continues with development of the constant gain circles described in Section 15.2.4. In order to allow hand calculation, S_{12} is made zero—only the unilateral case is considered. The amplifier is unconditionally stable since both $|S_{11}|$ and $|S_{22}|$ are less than one.

The MSG (unilateral) available from this circuit is 15 dB and occurs when $\Gamma_S = S_{11}^*$ and $\Gamma_L = S_{22}^*$. The resulting maximum transducer gains (Eq. 15.16) are

$$\text{Input circuit: } G_{T,\text{max}} = \frac{1}{1 - |S_{11}|^2} = \frac{1}{1 - (0.91)^2} = 5.8 \rightarrow 7.6 \text{ dB}$$

$$(15.30)$$

$$\text{Output circuit: } G_{T,\text{max}} = \frac{1}{1 - |S_{22}|^2} = \frac{1}{1 - (0.25)^2} = 1.07 \rightarrow 0.3 \text{ dB}$$

$$(15.31)$$

Note that matching the output has only a minor effect on the overall gain.

Working with the input circuit and using the Smith chart in Fig. 15.5, the S_{11}^* (optimum match) is plotted. The $Z_{11} = 50 (0.11 + j1.26) = 5.5 + j63$ ohms. The constant gain circles will lie along the line from the origin to S_{11}^* with $G_{T,\text{max}} = 7.6$ dB at S_{11}^*.

Assume that only 5 dB (gain ratio of 3.16) of transducer gain is desired, so

$$g_{T,S} = \frac{3.16}{5.75} = 0.55 \tag{15.32}$$

Evaluating Eq. 15.18 requires

$$D_S = 1 - |S_{11}|^2 (1 - g_S) = 0.63 \tag{15.33}$$

so

$$U_C = \frac{0.55(0.20)}{0.63} = 0.17 \tag{15.34}$$

and

$$V_C = \frac{-0.55(-0.89)}{0.63} = 0.78 \tag{15.35}$$

yielding the center of the unilateral constant gain circle in the Γ_S plane. In polar coordinates Γ_S, center = $0.8 \, \underline{/78°}$. Note that the angle corresponds with the line to S_{11}^*. The radius of the circle is computed using Eq. 15.20 as

$$R_S = \frac{\sqrt{1 - 0.55}\,[1 - (0.91)^2]}{0.63} = 0.18 \qquad (15.36)$$

in the Γ_S plane. The 5-dB constant gain circle is shown in Fig. 15.5. To summarize, this circle represents the locus of input reflection coefficient Γ_S, which will yield 5-dB gain for the input matching network compared to no IMN.

To design a lumped element network to match from a point on the 5-dB circle to the origin ($Z_0 = 50$ ohms), paths along constant resistance and conductance lines will be followed from the origin (Fig. 15.5) to point A. First, moving from 0 to B along a constant resistance line would entail adding a series inductor whose reactance is $X_{L1} = 1.3$ (50) ohms at 4 GHz. Second,

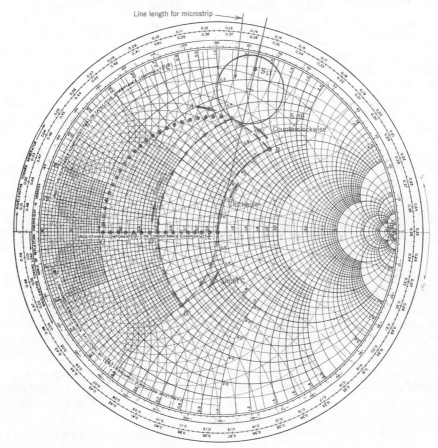

Figure 15.5. *S-Plane for narrowband amplifier.* $Z_0 = 50$ ohms; $f = 4$ GHz.

move in a counterclockwise direction along a constant conductance curve from B to A. This corresponds to adding a shunt inductor so the normalized input admittance changes from $0.37 - j0.48$ at point A to $0.37 - j0.8$ at point B. The difference in susceptance is $0.32/50$ Siemens $= 1/(X_{L2})$. The resulting inductor values are

$$L_1 = \frac{X_L}{2\pi(4 \times 10^9)} = \frac{65}{[2\pi(4 \times 10^9)]} = 2.6 \text{ nH}$$

$$L_2 = \frac{1}{[2\pi(4 \times 10^9)(0.32/50)]} = 6.2 \text{ nH}$$

The network is shown in the upper left corner of Fig. 15.5.

If the gate is to be operated with DC bias, the inductor to ground must terminate in a large capacitor. An alternative circuit involving a capacitor can be derived as follows. Moving from 0 (the impedance looking into the generator) to B' along a constant conductance line is accomplished by adding a shunt capacitor. The resulting $Z_B' = 25 - j25$ ohms or $Y_B' = (1/50)(1 + j1)$. The value of capacitance required at 4 GHz is $C = (1/50)/[2(3.14)(4 \times 10^9)] = 0.8$ pF. To travel on the Smith chart from point B' to A, a series inductor is added. Of course, point B' was originally selected to lie on the constant resistance line that passes through point A. The inductor must add sufficient reactance to go from $-j25$ ohms (point B') to $+j50$ ohms (point A); that is, the inductor must have 75 ohms of reactance at 4 GHz. The resulting value of L is $L = (75)/[2(3.14)(4 \times 10^9)] = 3$ nH. This circuit is shown in the upper right corner of Fig. 15.5. These inductor values are not easily realized at 4 GHz without significant losses. As a result, the transistor would probably be mounted in a microstrip circuit configuration. The microstrip matching circuit design follows.

Noting that the desired impedance (point A) lies on a reflection coefficient circle of radius 0.63 suggests a simple microstrip matching circuit. Namely, use a quarter-wavelength transformer to transform 50 to 11 ohms (point C) and then a length of 50-ohm transmission line to arrive at point A. The characteristic impedance of the $\lambda/4$ transformer should be $Z_{\text{trans}} = \sqrt{(50)(11)} = 23.5$ ohms. Use of this transformer moves the impedance looking into the 50-ohm generator to point C. Now we need only compute the length of 50-ohm line needed to match out the reactance from point A to C. The length of 50-ohm line is read from the chart to be 0.135 wavelengths. Reference to Chapter 8 will indicate the physical dimensions required depending on substrate thickness, material, etc. In actual practice some trimming would be needed because of the junction between the two lines of differing characteristic impedances.

A hybrid (uses distributed and lumped circuit elements) realization would be to add a lumped series inductance ($\omega L = 50$ ohms at 4 GHz) at the end of the 25-ohm transformer ($Z_0 = 35$ for a quarter-wave transformer, point

D). The inductor (2 nH) moves along the constant resistance line from the real axis of the Smith chart to point A in Fig. 15.5.

The input circuit does not match to S_{11} of the transistor, so the input VSWR is not low even at the design frequency. The mismatch ensures less than maximum gain. The 3-dB bandwidth for these designs (analyzed via computer) extends from 3.5 to 4.7 GHz. The amplifier is unstable below 3.6 GHz because the S-parameters change with frequency.

15.2.6 Constant Noise Circles

The noise factor for the amplifier is related to the internal components and source impedance of the amplifier. A digression to describe the noise in linear two ports is needed (7). The noise factor measured at the output terminals of the two port in Fig. 15.6 is

$$F = \frac{\text{Noise power of total circuit}}{\text{Noise power due to source}} \qquad (15.37)$$

$$= \frac{\overline{i_S^2} + |\overline{i + Y_S e}|^2}{\overline{i_S^2}} = 1 + \frac{|\overline{i + Y_S e}|^2}{\overline{i_S^2}}$$

where $\overline{i_S^2}$ is the mean-square source noise current, Y_S is the source admittance, $\overline{i^2}$ is mean-square network noise current, and e^2 is the mean-square network open-circuit voltage. Some of the voltage e is uncorrelated with i, so the remaining correlation current is $i - i_u$, where i_u is the uncorrelated current. A correlation admittance $Y_c = G_c + jB_c$ may be defined so that $i - i_u = Y_c e$, which allows calculation of the ei^* cross-product fluctuation; namely,

$$\overline{ei^*} = \overline{e(i - i_u)^*} = Y_c^* \overline{e^2} \qquad (15.38)$$

in terms of the noise voltage fluctuation $\overline{e^2}$, since $\overline{ei_u^*} = 0$. But this noise is related to an equivalent noise resistor R_n by the relation

$$\overline{e^2} = 4\,kT_0 R_n \,(\text{BW}) \qquad (15.39)$$

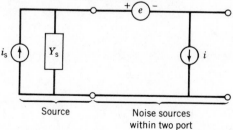

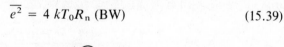

Source   Noise sources
             within two port

Figure 15.6. Noise sources in a linear two port.

where (BW) is the noise bandwidth. Similarly the uncorrelated noise current fluctuation is related to the noise conductance G_u by the Nyquist formula

$$\overline{i_u^2} = 4kT_0G_u \text{ (BW)} \qquad (15.40)$$

and similarly $\overline{i_S^2} = 4kT_0G_S$ (BW) for the source fluctuation current. Writing the total current fluctuations as

$$\overline{i^2} = | \overline{\text{correlated current}} |^2 + | \overline{\text{uncorrelated current}} |^2$$

$$= \overline{(i - i_u)(i - i_u)^*} + \overline{i_u i_u^*}$$

$$= \overline{(Y_c e)(Y_c e)^*} + 4kT_0G_u \text{ (BW)} \qquad (15.41)$$

$$= 4kT_0R_n \text{ (BW)} \overline{| Y_c |^2} + 4kT_0G_u \text{ (BW)}$$

$$= 4kT_0 (| Y_c |^2 R_n + G_u) \text{ (BW)}$$

Substituting these values for the noise fluctuations into the noise factor F in Eq. 15.37 yields

$$F = 1 + \frac{(\overline{(i - i_u)^2} + \overline{i_u^2} + | Y_s |^2 \overline{e^2})}{\overline{i_s^2}} = \frac{(\overline{i_u^2} + | Y_s + Y_c |^2 \overline{e^2})}{4kT_0G_S \text{ (BW)}} \qquad (15.42)$$

$$= 1 + \frac{G_u}{G_S} + \frac{R_n}{G_S} [(G_S + G_c)^2 + (B_S + G_c)^2] \qquad (15.43)$$

which is a function of the four network admittance parameters and the source conductance G_S and susceptance B_S.

Since the noise factor is a function of G_S and B_S, there are optimum values $G_S = G_0$ and $B_S = B_0$ that yield a minimum value for F. These values are

$$G_0^2 = \frac{(G_u + R_nG_c^2)}{R_n}, \qquad (15.44)$$

and

$$B_0 = -B_c \qquad (15.45)$$

yielding the minimum noise factor $F_{\min}$ of

$$F_{\min} = 1 + 2R_n(G_c + G_0) \qquad (15.46)$$

Substituting F_{min} back into Eq. 15.43 for an arbitrary source impedance gives

$$F = F_{min} + \frac{R_n}{G_S} \mid Y_s - Y_0 \mid^2 \tag{15.47}$$

$$= F_{min} + \frac{R_n}{Z_0 G_S} \left| \frac{1 - \Gamma_S}{1 + \Gamma_S} - \frac{1 - \Gamma_0}{1 + \Gamma_0} \right|^2 \tag{15.48}$$

$$= F_{min} + \frac{4R_n \mid \Gamma_S - \Gamma_0 \mid^2}{Z_0 (1 - \mid \Gamma_S \mid^2) \mid 1 + \Gamma_0 \mid^2} \tag{15.49}$$

The three quantities F_{min}, R_n/Z_0, and Γ_0 are the noise parameters provided by the manufacturer, stored in the CAD data bank, or measured experimentally. Experimentally, the Γ_S is varied until F_{min} is found. At this value $\Gamma_S = \Gamma_0$, Γ_S is measured with a network analyzer and the noise figure meter measures F_{min}. The noise resistance R_n/Z_0 is calculated when $\Gamma_S = 0$ (source $Z = Z_0$) and solving Eq. 15.49 for

$$\frac{R_n}{Z_0} = (F_{\Gamma_S = 0} - F_{min}) \frac{\mid 1 + \Gamma_0 \mid^2}{4 \mid \Gamma_0 \mid^2} \tag{15.50}$$

The constant noise circles are derived from Eq. 15.49. Consider a noise factor F' ($F' \geq F_{min}$) whose center position and radius we wish to find. Using Eq. 15.49,

$$\frac{(F' - F_{min})Z_0}{4R_n} \mid 1 + \Gamma_0 \mid^2 = \frac{\mid \Gamma_S - \Gamma_0 \mid^2}{1 - \mid \Gamma_S \mid^2} = N_i \tag{15.51}$$

where N_i is a noise factor parameter. Also using Eq. 15.51,

$$(\Gamma_S - \Gamma_0)(\Gamma_S^* - \Gamma_0^*) = N_i - N_i \mid \Gamma_S \mid^2 \tag{15.52}$$

Rearranging yields

$$N_i = \mid \Gamma_S \mid^2 (1 + N_i) + \mid \Gamma_0 \mid^2 - 2 \operatorname{Re}(\Gamma_S \Gamma_0^*) \tag{15.53}$$

Multiplying by $(1 + N_i)$ finally yields the circle format in terms of Γ_S of

$$\left| \Gamma_S - \frac{\Gamma_0}{1 + N_i} \right|^2 = \frac{N_i^2 + N_i(1 - \mid \Gamma_0 \mid^2)}{(1 + N_i)^2} \tag{15.54}$$

where the center of the circles is located at

$$\frac{\Gamma_0}{(1 + N_i)} \tag{15.55}$$

with radii of

$$\frac{[N_i^2 + N_i(1 - |\Gamma_0|^2)]^{1/2}}{(1 + N_i)} \tag{15.56}$$

As expected, the center of the F_{min} circle is located at Γ_0 and the circle has zero radius.

15.2.7 Example of Low-Noise Narrowband Amplifier Design

The three noise parameters required for the low-noise design have been presented in Table 15.1. The Γ_0 is plotted on the Smith chart in Fig. 15.7. For this value of reflection coefficient, the minimum noise figure of 0.4 dB ($F_{min} = 1.10$) is obtained. To compute the noise circle for a noise figure of

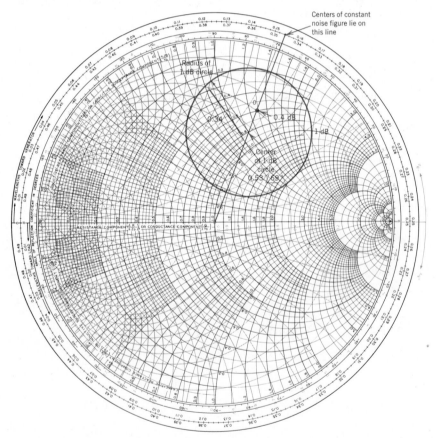

Figure 15.7. *Constant noise figure circle for NEC 67383 narrowband amplifier.*

1 dB ($F' = 1.26$), use Eq. 15.51 to compute the noise factor parameter:

$$N_i = \frac{(1.26 - 1.10)\,(50)}{4\,(19)}\,|\,1 + 0.64\,\underline{/69°}\,|^2$$
$$= 0.11\,|\,1.23 + j0.60\,|^2 = 0.11\,(1.37\,\underline{/26°})^2 = 0.21 \qquad (15.57)$$

The center of the 1-dB circle is (Eq. 15.55)

$$\frac{\Gamma_0}{1 + N_i} = \frac{0.64}{1 + 0.21}\,\underline{/69°} = 0.53\,\underline{/69°} \qquad (15.58)$$

The radius of the 1-dB noise figure circle is computed from Eq. 15.56:

$$\frac{\{(0.21)^2 + (0.21)\,[1 - (0.64)^2]\}^{1/2}}{1 + 0.21} = \frac{0.41}{1.21} = 0.34 \qquad (15.59)$$

If the constant gain curves are also drawn on the Smith chart, the tradeoff between the design of the input matching circuitry for maximum gain and minimum noise figure is clearly evident. Both sets of curves are shown in Fig. 15.8 as transferred from Figs. 15.5 and 15.7. Since S_{11}^* and Γ_0 do not coincide, the maximum gain and minimum noise figure cannot be realized simultaneously. If the minimum noise figure is desired for the front end of a narrowband receiver, for example, the input reflection coefficient Γ_0 is selected. For this case $7.6 - 5 = 2.6$ dB of gain has been sacrificed, but the amplifier noise figure will be near its minimum value. In an actual amplifier the input circuit losses add a few tenths of a decibel to the F_{min}. For the narrowband, low-noise amplifier, a quarter-wavelength transformer to $(0.21)\,(50)$ ohms and a 0.154-wavelength long 50-ohm transmission line matches the 50-ohm input source impedance to Γ_0. The impedance of the quarter-wave transformer is $\sqrt{(0.21)\,(50)\,(50)} = 23$ ohms.

If a maximum gain design is desired, the input reflection coefficient corresponding to S_{11}^* is selected, but the noise figure exceeds 1 dB.

The overall performance of the low-noise, narrowband amplifier has been computed using a CAD program with no output transducer matching. The performance between 2 and 6 GHz is shown in Fig. 15.9. The minimum input VSWR for this amplifier is 5:1.

15.2.8 Example of Broadband Amplifier Design

Broadband design is best done using one of the computer-aided design tools. A full design entails the procedures shown in Fig. 15.10, but here the analytic design process is emphasized before engineering prototypes are evaluated.

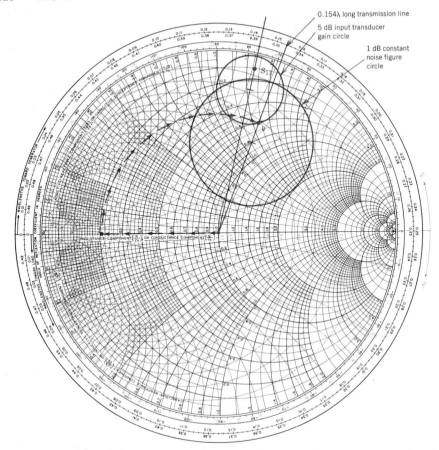

Figure 15.8. Constant gain and noise figure curves for NEC 67383 narrowband amplifier.

The example selected here for illustration is a 3.5–4.5-GHz amplifier with less than a 1-dB noise figure and more than 10-dB gain.

The NEC 67383 has been selected as the active device because it has a low noise figure of 0.45 dB at 4.5 GHz and a high maximum stable gain (MSG) of 16 dB. This MSG is not available because the K factor is too close to 1 and the input circuit gain must be reduced to obtain the desired noise figure. Thus the margin of 6 dB over the requirement will be used for stabilization and input matching network (IMN) tuning.

In the narrowband design, resistive loading was used to stabilize the amplifier. For this broadband design a transmission line is added in the output matching network, as shown in Fig. 15.11, to help flatten the gain response of the amplifier over the band. The SUPER-COMPACT® circuit is shown with the stabilization network STAB after optimization. The circuit was optimized for a stability factor greater than 1 while maintaining the maximum

available gain (GMAX). This optimization is performed by the line K = 1
GT GMAX W = −1, meaning make K > 1 and 1/GMAX a minimum.
Optimizing for a maximum value in SUPER-COMPACT® requires specifying a negative weighting value. The entire program is shown in Fig. 15.11.
The optimization technique and the number of steps are selected by the user
via interaction with the program. The optimum values for the transmission
line TRL are Z_0 = 56.5 ohms with an electrical length of 40.4° at 4.5 GHz.
The resistor value is 19.1 ohms. The entire printout of the S-parameters and
noise parameters is also given in Fig. 15.11. Thus, the OMN has been designed to provide stability and gain flatness. The IMN is used to obtain the
desired gain and noise figure.

Port modeling and synthesis (PMS) is used to model the active device
and determine the real and parasitic components. These values are passed
to the synthesis routines for IMN extraction.

The first step in PMS is to determine the frequency dependence of the
important parameter (minimum noise figure in this case) at the input of the
transistor. As shown in Fig. 15.12, the optimum noise figure (NF_{opt}) nearly
follows the 30-ohm constant resistance circle. Since it follows the constant
resistance circle, this reactive component appears as a series parasitic that
moves counterclockwise around the Smith chart. This corresponds to a series positive reactance that decreases with frequency. A negative capacitor

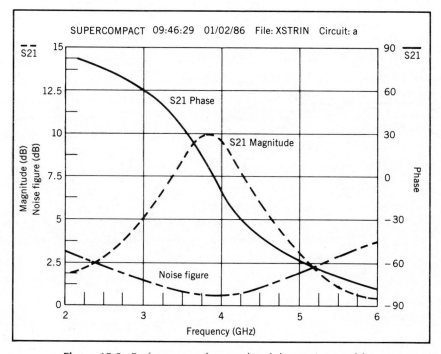

Figure 15.9. *Performance of narrowband, low-noise amplifier*

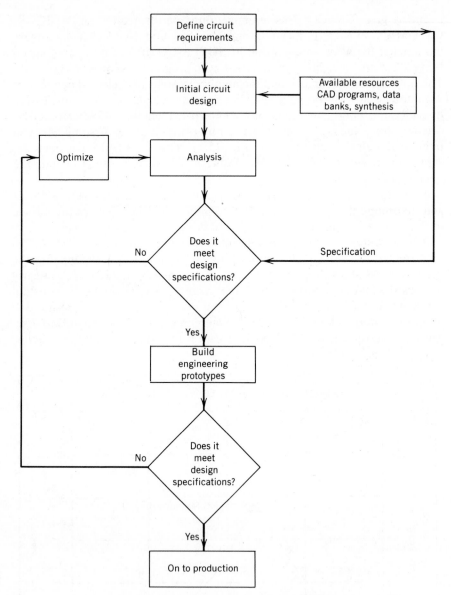

Figure 15.10. *Flowchart of CAD procedure.*

(a hypothetical lumped circuit element) could be used to model the parasitic of the device, since the magnitude of its reactance decreases with increasing frequency. But since a microstrip circuit is desired, the parasitic is modeled as a series open-circuited stub (OS) with negative characteristic impedance. The electrical length of the stub can be approximated from the angle at the high frequency of the optimum noise match divided by two, and the imped-

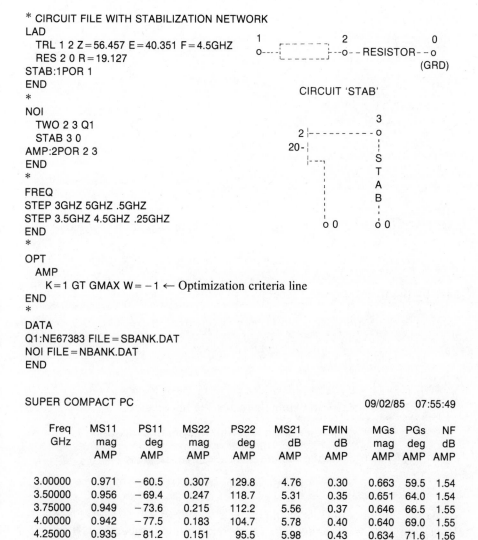

```
* CIRCUIT FILE WITH STABILIZATION NETWORK
LAD
   TRL 1 2 Z = 56.457 E = 40.351 F = 4.5GHZ
   RES 2 0 R = 19.127
STAB:1POR 1
END
*
NOI
   TWO 2 3 Q1
   STAB 3 0
AMP:2POR 2 3
END
*
FREQ
STEP 3GHZ 5GHZ .5GHZ
STEP 3.5GHZ 4.5GHZ .25GHZ
END
*
OPT
   AMP
   K = 1 GT GMAX W = − 1 ← Optimization criteria line
END
*
DATA
Q1:NE67383 FILE = SBANK.DAT
NOI FILE = NBANK.DAT
END
```

CIRCUIT 'STAB'

SUPER COMPACT PC   09/02/85   07:55:49

Freq GHz	MS11 mag AMP	PS11 deg AMP	MS22 mag AMP	PS22 deg AMP	MS21 dB AMP	FMIN dB AMP	MGs mag AMP	PGs deg AMP	NF dB AMP
3.00000	0.971	− 60.5	0.307	129.8	4.76	0.30	0.663	59.5	1.54
3.50000	0.956	− 69.4	0.247	118.7	5.31	0.35	0.651	64.0	1.54
3.75000	0.949	− 73.6	0.215	112.2	5.56	0.37	0.646	66.5	1.55
4.00000	0.942	− 77.5	0.183	104.7	5.78	0.40	0.640	69.0	1.55
4.25000	0.935	− 81.2	0.151	95.5	5.98	0.43	0.634	71.6	1.56
4.50000	0.928	− 84.8	0.120	83.1	6.15	0.45	0.629	74.3	1.57
5.00000	0.915	− 91.4	0.077	39.1	6.45	0.50	0.618	80.0	1.59

Figure 15.11. *NEC 67383 transistor with stabilizing output matching.*

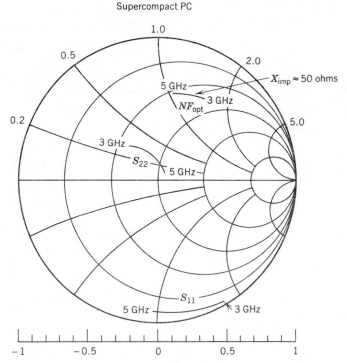

Figure 15.12. S_{11}, S_{22}, and optimum NF match versus frequency.

ance by the relation for the reactance of a lossless OS

$$Z_0 = -j \frac{X_{imp}}{\tan \theta_{elec}} \tag{15.60}$$

where

$$Z_0 = \text{the stub characteristic impedance}$$

$$X_{imp} = \text{the reactance from the Smith chart}$$

$$\theta_{elec} = \text{the half-angle at the high frequency}$$

$$\text{end of the passband (5 GHz)}$$

Using the numbers for this example, $\theta_{elec} = 80/2 = 40°$ at 5 GHz (see Fig. 15.11) and $X_{imp} = 50$ ohms. The characteristic impedance at 5 GHz is calculated as -60 ohms. Extensive calculations are not required since this value will be optimized.

The principle of using negative elements to model the device ports in PMS is termed negative image modeling (8). Negative image modeling uses the fact that a unit transfer coefficient ($S_{21} = 1$) two-port network can be broken into two parts—the positive image and the negative image. For the simple case of a lumped element two port, see Fig. 15.13a in which all the negative values are not necessarily in the same image.

With this modeling technique in the design of the input matching network

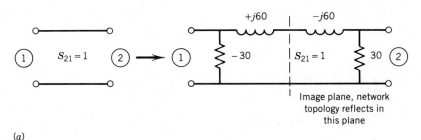

(a)

(b)

(c)

Figure 15.13. Use of negative-image models for amplifier design.

to give a good noise figure, the network shown in Fig. 15.13*b* has been derived using the open-circuited stub. This estimate of the input parasitic network when cascaded with the input of the 67383 transistor yields a good noise figure amplifier. When optimized via CAD, the final values for Z_0 and stub length are determined. The negative image for the negative of the STAB circuit in Fig. 15.11 is shown in Fig. 15.13*b* and is used in the drain circuit. Thus with the addition of these two negative-image circuits, the amplifier's input and output parasitics are matched to about 50 ohms. The total circuit in Fig. 15.13*b* is now optimized to meet the criteria for the overall amplifier whether it be flat gain, good port match, or as in our case, a low noise figure. The result of this optimization is a new set of values for the port image components. The negative images are then transferred back to real images so that input and output matching networks can be derived as shown schematically in Fig. 15.13*c*. In this case, for example, the desired impedance is not the impedance of the device (that would give us a good match but a poor noise figure), but the optimum noise match of the device. The resultant network gives a resistive and parasitic part to be transferred to the Synthesis portion of SUPER-COMPACT®. Because of the fact that the transistor is a bilateral device, applying this input matching network changed S_{22}, thus the CAD simultaneously optimized for a minimum noise figure for the input, flat gain, and a good match with the output.

The next step synthesizes the input matching network. SUPER-COM-PACT® automatically transfers the real and reactive parts from PMS to Synthesis. The Synthesis routines can not only match between two real impedances, but they can automatically extract the parasitic part of the source. Normally when an impedance matching network is designed by hand on a Smith chart the reactive portion of the source is compensated for in the matching network. Synthesis in SUPER-COMPACT® absorbs the parasitic making it the first element of the matching network. The next part of Synthesis calculates the Chebyshev response function (insertion loss) for the network based on the calculations in PMS. The input reflection coefficient R is computed using the relation (based on conservation of energy)

$$| R(\omega) |^2 + \left| \frac{1}{IL(\omega)} \right|^2 = 1 \qquad (15.61)$$

The complex variable p is introduced using the relation $p = -j\omega$ for lumped element circuits and $p = j \tan(\beta l)$ for distributed circuits. βl is the electric length of the circuit. $| R(p) |^2$ is then factored to yield $R(p)$ and $R(p)^*$. Finally $R(p)$ is used to compute the input impedance

$$Z_{in}(p) = R_S \left(\frac{1 + R(p)}{1 - R(p)} \right) \qquad (15.62)$$

where R_S is the source resistance specified by the user (50 ohms is selected otherwise). The realization of $Z_{in}(p)$ is done by a partial fraction expansion technique that "EXTracts" the elements of the network (9).

```
***** SUPER-COMPACT SYNTHESIS *****
INPUT MATCHING NETWORK *****
F1(GHz), F2(GHz)? (X,Y/⟨ 3.500 4.500⟩):
RT? (X/⟨ 50.0⟩):
```

The first two lines of data are automatically transferred from the PMS routine. The option of choosing manual synthesis allows a greater flexibility, but in this example the computer decides:

```
Automatic Synthesis? (⟨Y⟩/N):
You have the following synthesis options:
  1) Structure with electrical length = 48.1270 DEGS
  2) Structure with specified length less than 90 DEGS
  3) Bandpass structures with quarter-wave length lines
Option? (x/⟨1⟩):1
```

At this point the operator can choose the length of the distributed elements to use in the network. All elements that are extracted will be of the same length. The 48.127° comes from the length of the parasitic, and the operator normally tries that length first if the value is between 30° and 60°. If the length is very short or close to 90°, it is difficult to extract realizable elements. For instance, to extract 50 ohms of shunt reactance with a 45° long stub requires a 50-ohm stub, but to achieve the same impedance with a 10° long stub requires a 283-ohm stub, which is normally not realizable in microstrip. In this example the default in brackets (⟨ ⟩) is tried first. A review of the inputs follows, then the computer tries to optimize the structure to reduce the total loss, which is the sum of the minimum insertion loss (MIL) and the ripple (RIP) while at the same time absorbing the parasitic as the first element of the network:

```
RL = 50.00000 Ohms    FREQ =    3.50000 to 4.50000 GHz
RS = 29.36200 Ohms    Parasitic(s):
                OS = 57.23000 Ohms at 48.12700 Degs
Gain-Bandwidth Calculation
N = 2    Rip = 0.35142 dB    Mil = 0.08567 dB
N = 4    Rip = 0.05264 dB    Mil = 0.01254 dB
The minimum structure has 2 HPE    1 LPE & 1 TRL
```

In this case the program decided that four elements would be necessary. In manual synthesis the operator can override the defaults and choose values. The computer has decided on a topology of 2 High Pass Elements (HPE), 1 Low Pass Element (LPE), and a 1 TRansmission Line (TRL). In the next step the parasitic is extracted. Since it is an Open-circuited Stub (OS) it is a high pass element, one of the HPEs is used up, and the network consists of three elements. The CAD indicates

```
Parasitic(s) have been extracted:
   OST    SE    ZO = 57.22143 Ohms    ELEN = 48.12700    F = F2
   Elements to be extracted are: 1 HPE,    1 LPE, & 1 TRL
```

The next section of synthesis is the extraction process in which the individual elements of the matching network are extracted. The elements used in this example are Short-circuited Parallel stubs (SP), Open-circuited Parallel stubs (OP), and Unit Elements (UE), or transmission lines as they are commonly called. The Short-circuited Series stub (SS) is normally not chosen since it is not realizable in microstrip. The CAD displays follow as the synthesis progresses:

```
EXT⟩        (EXT = extract)
      Select one of the following elements:
         SP    ZO = 86.0324 Ohms
         SS    ZO = 62.0483 Ohms
         UE    ZO = 72.5321 Ohms
      Element?
 SP
 EXT⟩
      Select one of the following elements:
         SS    ZO = 222.5709 Ohms
         UE    ZO = 462.2197 Ohms
      Element?
 UE
 EXT⟩
      Select one of the following elements:
         OP    ZO = 497.6858 Ohms
         UE    ZO = 312.7628 Ohms (non-minimum)
      Element?
 OP
      Termination = 841.74289 Ohms
 EXT ⟩ SAV
 Network saved.
```

The termination impedance is not 50 ohms. Fifty ohms is obtained using a Kuroda transform, which SUPER-COMPACT® does interactively to achieve the proper termination impedance. Figure 15.14 shows several Kuroda transforms. The circuit below gives a high-pass Kuroda transform on the Short-circuited STub (SST) and the TRansmission Line (TRL):

```
SYN ⟩ TFM
1            R1 = 29.36200 Ohms
2  SST PA  ZO = 86.03235 Ohms    ELEN = 48.12700  F = F2
3  TRL SE  ZO = 462.21970 Ohms   ELEN = 48.12700  F = F2
4  OST PA  ZO = 497.68579 Ohms   ELEN = 48.12700  F = F2
5            R2 = 841.74292 Ohms
```

Add, Split, Combine, Exchange, Transform, Print, Restart, SYN, Help
Command? (A/S/C/E/T/P/R/SYN/H/⟨H⟩): T
Element numbers (x,y)? 2 3
Transformation between 841.74292 and 20.72729
Desired termination (x/⟨ 50.00000⟩)?

1			R1 =	29.36200 Ohms		
2	SST	PA	ZO =	203.65721 Ohms	ELEN = 48.12700	F = F2
3	TRL	SE	ZO =	112.65325 Ohms	ELEN = 48.12700	F = F2
4	SST	PA	ZO =	36.30427 Ohms	ELEN = 48.12700	F = F2
5	OST	PA	ZO =	29.56281 Ohms	ELEN = 48.12700	F = F2
6			R2 =	50.00000 Ohms		

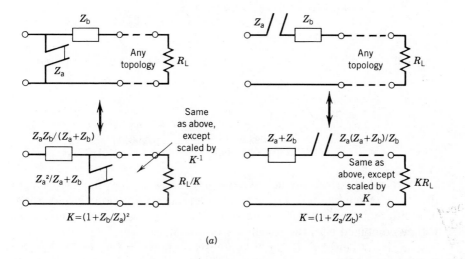

(a)

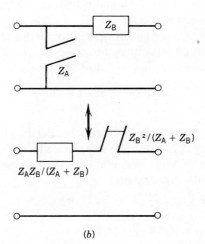

(b)

Figure 15.14. *Kuroda identities. (a) High-pass Kuroda identities; (b) low-pass Kuroda identity (does not change impedance ratio but may yield realizable element values).*

The proper termination impedance has been achieved, but the values of the first and second components are too high to be realized. The next step is to go back through the synthesis and extraction process trying another electrical length to see if better results can be achieved. In the next example 60° was tried with better results, but the first element was still too high in impedance. Practical microstrip impedances range from 20 to 100 ohms (see Chapter 8). The results follow:

```
1              R1 =   29.36200 Ohms
2 SST PA   ZO = 114.63267 Ohms   ELEN = 60.00000   F = F2
3 TRL SE   ZO =  94.87827 Ohms   ELEN = 60.00000   F = F2
4 SST PA   ZO =  31.26567 Ohms   ELEN = 60.00000   F = F2
5 OST PA   ZO =  54.07270 Ohms   ELEN = 60.00000   F = F2
6              R2 =   50.00000 Ohms
```

Finally a length of 65° was tried and acceptable results were found:

```
1              R1 =  29.36200 Ohms
2 SST PA   ZO = 90.15117 Ohms   ELEN = 65.00000   F = F2
3 TRL SE   ZO = 89.09707 Ohms   ELEN = 65.00000   F = F2
4 SST PA   ZO = 29.51438 Ohms   ELEN = 65.00000   F = F2
5 OST PA   ZO = 71.60835 Ohms   ELEN = 65.00000   F = F2
6              R2 =  50.00000 Ohms
```

There are other ways to achieve the proper results; they include changing the ripple or minimum insertion loss, different transmission zero locations, or a different topology for the network. The network is now complete and will be combined with the remainder of the circuit.

The circuit performance after synthesis is shown in Fig. 15.15. The synthesis has yielded adequate gain and a low noise figure so no optimization is necessary at this point.

The next step in the design process is to synthesize the physical dimensions of the distributed structures from the electrical parameters. SUPER-COMPACT® has routines that will do this automatically:

```
TRLCF > TRL MS H = 25 ER = 9.9 MET1 = RC 1 TAND = .0001 F = 4.5 GHZ
MIL
```

Substrate information that the program needs to synthesize the transmission line structures is entered using closed form (TRLCF) formulas. The dimensions default to mils. The next step is to enter the characteristic line impedance and synthesize the line:

```
TRLCF > ZO = 50
TRLCF > SYN
Microstrip Single Line Synthesis
ZO = 50.0
H = 25.000MIL ER = 9.90 TAND = 0.00010 T/H = 0.0540
Freq = 4.5 GHz
W = 22.799 MIL Keff = 6.688
```

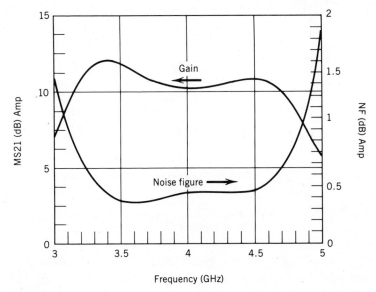

Figure 15.15. *Gain and noise figure with synthesized matching network.*

A 50-ohm line on this substrate is 22.799 mils wide and has an effective dielectric constant of 6.688. The remainder of the lines are now synthesized. It is not necessary to enter all the substrate information again since the program remembers the previous data:

```
TRLCF > ZO = 71.6
TRLCF > SYN
Microstrip Single Line Synthesis
ZO = 71.6
H = 25.000MIL ER = 9.90 TAND = 0.00010 T/H = 0.0540
Freq = 4.5 GHz
W = 8.898 MIL Keff = 6.315
```

The remaining examples follow a similar format. The results are given in Table 15.3.

The physical values of Table 15.3 are used to update the SUPER-COM-PACT® file with physical dimensions for the distributed structures.

The effects of the junctions where the lines and stubs come together must be taken into account. In this example a CROSs junction (where four lines meet) and two TEE junctions (where three lines meet) are required. A plot of gain and noise figure is given in Fig. 15.16. Some interesting effects can be seen here. The minimum noise figure (FMIN) and the K factor have both increased due to losses in the distributed structures. However, all the specifications have been met since gain is greater than 10 dB and the noise figure is less than 1 dB. This completes the CAD of this amplifier.

Table 15.3   Microstrip Single Line Synthesis[a]

Characteristic Impedance, Z_0 (ohms)	Linewidth (W-mils)	Effective Dielectric Constant (K_{eff})
71.6	8.90	6.32
29.5	59.51	7.36
89.1	3.97	6.14
90.2	3.76	6.13
56.5	17.23	6.55
50.0	22.8	7.36

[a] Substrate parameters: H = 25 mil; ER = 9.9; TAND = 0.0001; T/H = 0.054. Frequency: 4.5 GHz.

CAD programs other than SUPER-COMPACT® are available to perform these analyses.

15.2.9   Additional CAD Analyses

CAD allows additional analyses that are useful for manufacture and pricing of microwave circuits. Given that the basic performance criteria have been satisfied, a statistical analysis of component sensitivity and the microstrip layout can be done with CAD.

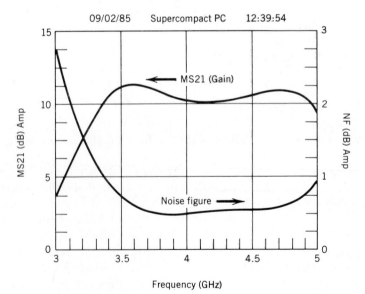

Figure 15.16. *Gain and noise figure of amplifier in a microstrip realization.*

15.2.9.1 Statistical Analysis. To perform a statistical analysis, apply tolerances to the components and see if the circuit yield is greater than some specified number. For example, specify that the linewidths can vary by up to $\pm 10\%$ due to over- and/or underetching. The substrate dielectric constant and height can vary by $+/-5\%$ due to manufacturing tolerances. Also the S-parameters of the transistor can vary $\pm 5\%$. Specify that the gain of the amplifier must not go below 10 dB nor may the noise figure exceed 1 dB. The SUPER-COMPACT® file is shown in Fig. 15.17.

The statistical analysis function first does an analysis to determine the worst-case results. This is followed by a statistical analysis in which each value is perturbed in a random fashion and the performance is analyzed to see if the desired specifications are met. A review of the distribution of the specified responses determines which ones and at which frequency problems are occurring.

A worst-case analysis followed by a yield analysis of 100 random trials was done. As shown in Fig. 15.18, although the original circuit met all the design criteria, less than one out of five would have met the standards on a production basis. This is far short of the (usually 90%) yield desired. Reviewing Fig. 15.18, observe that the noise figure meets all the design criteria, but the gain, especially at 4.1 and 4.3 GHz, has caused most of the failures. Almost half the analyses are below 10 dB.

Without the aid of a microwave CAD analysis package, the only way to determine the circuit yield is to build circuits and hope they work—in this case building 19 amplifiers and 81 paper weights. The production criteria would not have been met, resulting in an unprofitable product line, cost overruns, and late deliveries.

Since the problem has been identified before the circuits were built, the designer can reoptimize the circuit to increase the gain. The gain can be increased across the band if the stabilization network is optimized while increasing the gain specification from a goal of just being greater than 10 dB to a goal of 11 dB as shown in Fig. 15.19. Noise figure is not affected. Statistically analyzing this new circuit (see Fig. 15.20) the yield has increased to 100%. This is a yield increase of about 400% for 5 min of CAD work.

15.2.9.2 Layout. The last step in the design process is to lay out the circuit. Until recently this consisted of an engineer submitting a hand sketch to a draftsman. After drafting the engineer would review the layout and make corrections. This iterative process can be time consuming in the design process. The basic problem is that someone is laying out an MIC who doesn't understand microwave design and has not been involved in the design process.

With the advent of microwave layout programs for the engineer, the process can be improved considerably because now the person closest to the design can do the layout. Some companies are taking the design process one

SUPER-COMPACT Version 1.7 + 001 04/09/84

WID1:?22.8MIL 10%? .
WID2:?8.9MIL 10%?
WID3:?3.97MIL 10%?
WID4:?59.5MIL 10%?
WID5:?3.76MIL 10%?
BLK
  TRL 1 2 W = WID1 P = 100MIL SUB
  CROS 2 3 4 5 W1 = 2ID1 W2 = WID2 W3 = WID3 W4 = WID4 SUB
  OST 3 0 W = WID2 P = 188.5MIL SUB
  SST 5 0 W = WID4 P = 174.6MIL SUB
  TRL 4 6 W = WID3 P = 191.2MIL SUB
  TEE 6 7 8 W1 = WID3 W2 = 5MIL W3 = WID5 SUB
  SST 8 0 W = WID5 P = 191.3MIL SUB
INPUT:2POR 1 7
END
WID6:?22.8MIL 10%?
WID7:?17.2MIL 10%?
BLK
  TEE 1 2 3 W1 = 5MIL W2 = WID6 W3 = WID7 SUB
  TRL 3 4 W = WID7 P = 113.9MIL SUB
  RES 4 0 R = 19.127
  TRL 2 5 W = WID6 P = 100MIL SUB
OUTPUT:2POR 1 5
END
NOI
  INPUT 1 2
  TWO 2 3 Q1
  OUTPUT 3 4
AMP:2POR 1 4
END
FREQ
STEP 3GHZ 5GHZ .5GHZ
STEP 3.5GHZ 4.5GHZ .2GHZ
END
OUT
  PRI AMP SK
END
STAT
  AMP
  F = 3.5GHZ 4.5GHZ MS21 = 10DB GT NF = 1DB LT
END
DATA
Q1:NE67383 FILE = SBANK M = ?5% 5% 5% 5%? P = ?5% 5% 5% 5%?
NOI FILE = NBANK
SUB:MS H = ?25MIL 5%? ER = ?9.9 5%? MET1 = RC 1 TAND = .0001
END

Figure 15.17. *Input file for statistical analysis.*

# Circuit	Frequency	Parm	Hist_Low	Nominal	Hist_High	Pass	Low	High
1:AMP	3.500GHz	MS21	9.369	9.91	12.618	93.0%	7.0%	0.0%
2:AMP	3.500GHz	Nf21	0.606	0.734	0.934	100.0%	0.0%	0.0%
3:AMP	3.700GHz	MS21	9.824	11.409	12.676	98.0%	2.0%	0.0%
4:AMP	3.700GHz	Nf21	0.491	0.522	0.601	100.0%	0.0%	0.0%
5:AMP	3.900GHz	MS21	9.629	10.395	11.983	79.0%	21.0%	0.0%
6:AMP	3.900GHz	Nf21	0.477	0.485	0.521	100.0%	0.0%	0.0%
7:AMP	4.000GHz	MS21	9.076	10.232	11.492	70.0%	30.0%	0.0%
8:AMP	4.000GHz	Nf21	0.483	0.496	0.538	100.0%	0.0%	0.0%
9:AMP	4.100GHz	MS21	9.074	10.373	10.957	55.0%	45.0%	0.0%
10:AMP	4.100GHz	Nf21	0.492	0.513	0.557	100.0%	0.0%	0.0%
11:AMP	4.300GHz	MS21	9.172	10.641	11.013	57.0%	43.0%	0.0%
12:AMP	4.300GHz	Nf21	0.519	0.54	0.579	100.0%	0.0%	0.0%
13:AMP	4.500GHz	MS21	9.41	10.914	11.438	80.0%	20.0%	0.0%
14:AMP	4.500GHz	Nf21	0.539	0.552	0.583	100.0%	0.0%	0.0%

Yield = 19/100 (19.000000 %)
COMMAND> Yield Histogram Last Next Penplot
Yield: perform yield analysis on current circuit.

Figure 15.18. *Results of statistic analysis of broadband amplifier.*

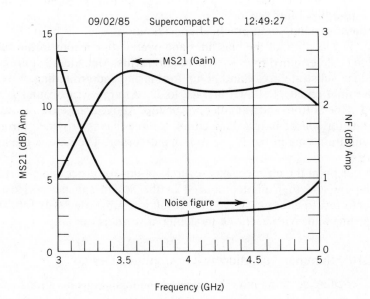

Figure 15.19. *Results of reoptimization (G ≥ 11 dB, NF ≤ 1 dB).*

# Circuit	Frequency	Parm	Hist—Low	Nominal	Hist—High	Pass	Low	High
1:AMP	3.500GHz	MS21	11.699	12.276	12.755	100.0%	0.0%	0.0%
2:AMP	3.500GHz	Nf21	0.396	0.734	0.93	100.0%	0.0%	0.0%
3:AMP	3.700GHz	MS21	11.433	12.198	13.043	100.0%	0.0%	0.0%
4:AMP	3.700GHz	Nf21	0.491	0.522	0.597	100.0%	0.0%	0.0%
5:AMP	3.900GHz	MS21	10.827	11.437	12.066	100.0%	0.0%	0.0%
6:AMP	3.900GHz	Nf21	0.477	0.485	0.519	100.0%	0.0%	0.0%
7:AMP	4.000GHz	MS21	10.651	11.14	11.639	100.0%	0.0%	0.0%
8:AMP	4.000GHz	Nf21	0.482	0.496	0.538	100.0%	0.0%	0.0%
9:AMP	4.100GHz	MS21 ·	10.593	10.952	11.366	100.0%	0.0%	0.0%
10:AMP	4.100GHz	Nf21	0.494	0.513	0.556	100.0%	0.0%	0.0%
11:AMP	4.300GHz	MS21	10.452	10.908	11.468	100.0%	0.0%	0.0%
12:AMP	4.300GHz	Nf21	0.522	0.54	0.574	100.0%	0.0%	0.0%
13:AMP	4.500GHz	MS21	10.553	11.183	11.818	100.0%	0.0%	0.0%
14:AMP	4.500GHz	Nf21	0.54	0.552	0.588	100.0%	0.0%	0.0%

Yield = 100/100 (100.000000 %)
COMMAND> Yield Histogram Last Next Penplot
Yield: perform yield analysis on current circuit.

Figure 15.20. *Statistical results of reoptimized amplifier.*

step further and cutting the rubylithes* in the engineering lab—especially for the prototype designs. This reduces the design time considerably.

The first step is to convert the circuit description to a preliminary layout; the result is shown in Fig. 15.21. The only interaction required is to specify the dimensions of the pad to mount the transistor. In Fig. 15.21 the transistor pad is the square to the right of center. The multilevel square in the lower right is the thin film resistor.

From this point some practical considerations need to be taken into account. The location of the bias lines and overall dimensions of the substrate need to be determined from system requirements. Meandered stubs are used to reduce substrate area and line up for biasing and grounding.

The final result is shown in Fig. 15.22. A large wraparound ground is added at the bottom and a smaller one on top. Above and below the transistor are pads for source bypass capacitors. Source bias is provided through the top and drain bias in the lower right-hand corner. The gate is grounded in this case.

At this point the engineer has several options for processing. A rubylithe can be cut and then photo reduced or the design can be passed on to a photoplotter. In either case the final result is a substrate ready for assembly and testing and, with accurate modeling, a working amplifier.

15.2.10 Summary of Microwave Amplifier Design

The preceding sections have highlighted the main considerations in the design of microwave amplifiers. The economics of the design strongly en-

---

* Rubylith is a photomask used in photoproducing microstrip circuits.

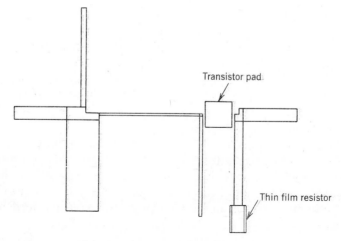

Figure 15.21. *Initial Autoart® layout.*

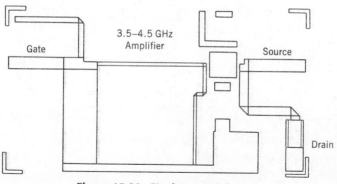

Figure 15.22. *Final Autoart® layout.*

courages the use of CAD. In this chapter SUPER-COMPACT® has been used because of the authors' familiarity with the program. TOUCHSTONE® and other programs can perform the same analyses.

For further study the student is encouraged to read Ref. 3 and the application notes prepared by the transistor manufacturers. These are particularly helpful with high-power amplifier designs in which large-signal S-parameters are needed or experimental tuning techniques are used.

15.3 MICROWAVE OSCILLATOR

Oscillator design for two-port devices requires knowledge of the S-parameters and the use of two matching networks as in the case of amplifiers. For two-port oscillators, one port is used as the frequency-determining network (called the load network), and the second port is used to couple power to the load (called the terminating network). These networks are shown sym- bolically in Fig. 15.23 for the two common FET configurations—common gate and common source. In the low-power configuration, the inductor in the gate (base for a bipolar junction transistor) increases the instability of the transistor by increasing $|\Gamma_{in}|$ and $|\Gamma_{out}|$. For the high-power configu- ration the capacitor (either intentional or parasitic) provides the feedback to increase the instability of the transistor.

The design technique follows procedures similar to those used for am- plifiers. First, select a transistor that is potentially unstable at the operating frequency f_0 and has a sufficient power rating. Second, design the termi- nating network so that $|\Gamma_{in}| > 1$ at f_0 and couples the transistor to the load (usually 50 ohms). Use of an inductor or capacitor in the feedback loop

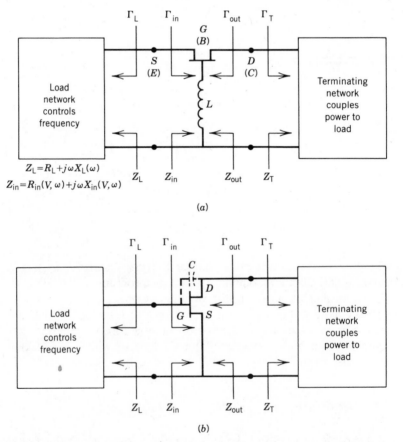

Figure 15.23. *Two common oscillator configurations. (a) Common-gate configuration used for low power (also used for bipolar junction transistors; (b) common-source con- figuration used for high power.*

increases $|\Gamma_{in}|$. Finally, the frequency-determining network (load) is designed so the reactances at the input resonate at f_0

$$X_L(f_0) = -X_{in}(f_0) \tag{15.63}$$

and the real part of the load is a fraction n of the small-signal (ss) value of $|R_{in}|$ at f_0

$$R_L = n \, | R_{in} \, (\text{ss}, f_0) \, | \tag{15.64}$$

Since $|\Gamma_{in}| > 1$, the $R_{in} < 0$. $R_{in}(V, f_0)$ is a function of the amplitude V of the oscillation so the oscillator continues to oscillate with uniform amplitude. Generally, $\partial \, |R_{in}(V, f_0)| \, /\partial V < 0$ so that the oscillator is stable. n values range from one-fourth to one-half with one-third being the usual design starting value. This parameter allows the readily available small-signal values to be used in the design of oscillators that operate with nonconstant $|R_{in}|$ and large signals. R_L is also varied to optimize the oscillator's power output.

Heuristically, since the negative resistance looking into the transistor is a maximum at small-signal levels, noise in the circuit is amplified before oscillations build up. Because the circuit is resonant at f_0, the most amplification occurs at that frequency resulting in a sine wave at f_0. The amplitude continues to grow until $|R_{in}|$ decreases to where the energy lost per cycle to the termination and losses equals the energy added by the transistor. If the oscillation exceeds the stable operating amplitude, the $|R_{in}|$ decreases, less energy is added, and the amplitude decreases, thus increasing the value of $|R_{in}|$. The stable amplitude occurs when $|R_{in}(V, f_0)| = R_L$.

The stable operating frequency f_0 occurs when

$$X_{in}(V, f_0) = -X_L(f_0) \tag{15.65}$$

X_{in} is a function of the amplitude of the oscillation. As a result, a circuit designed using small-signal S-parameters may need to be "tweaked" on frequency when operated. This is another way of stating that the frequency of oscillation is not stable. Kurokawa (10) has derived the condition for stable oscillation of a one port (the two-port oscillator may be considered a one port when the terminating network and transistor are considered a negative-resistance oscillator) to be

$$\left(\frac{\partial R_{in}(V, f)}{\partial V} \bigg|_{V=V_0} \right) \left(\frac{\partial X_L(f)}{\partial f} \bigg|_{f=f_0} \right)$$
$$- \left(\frac{\partial X_{in}(V, f)}{\partial V} \bigg|_{V=V_0} \right) \left(\frac{\partial R_L(f)}{\partial f} \bigg|_{f=f_0} \right) > 0 \tag{15.66}$$

For most resonant circuits the $\partial R_L(f)/\partial f = 0$, so the stability condition becomes

$$\left(\frac{\partial R_{in}(V,\,f)}{\partial V}\,\bigg|_{V=V_0}\right)\left(\frac{\partial X_L(f)}{\partial f}\,\bigg|_{f=f_0}\right) > 0 \qquad (15.67)$$

which says the same as the heuristic discussion given earlier.

An example of the CAD of a low-power 4-GHz oscillator using the NEC 67383 transistor follows. The common-gate configuration in Fig. 15.23*a* is used. The first step in the design is to convert the common-source (Table 15.1) to common-gate *S*-parameters. This is done in SUPER-COMPACT® using the BLK format as shown in Fig. 15.24 with the proper numbering of

```
.LIST XTRSD@TAB
* FILE NAME XTRSTAB
* THIS FILE ANALYZES THE STABILITY OF A NEC67383 TRANSISTOR
+ WITH SEVERAL FEEDBACK NETWORKS.
******
BLK
TWO 1 2 3 Q1
IND 1 0 L=0.001NH
CAP 2 1 C=0.0001PF
A:2POR 3 2
END
*****
FREQ
STEP 2GHZ 6GHZ 0.5GHZ
END
******
OUT
PRI A S
END
*****
DATA
Q1: S
2GHZ .97 -43 3.39 140 .04 61 .63 -32
4GHZ .88 -79 2.85 107 .06 35 .61 -58
6GHZ .84 -103 2.57 81 .07 20 .62 -77
END
* THIS ENDS THE PROGRAM TO ANALYZE THE STABILITY DUE TO FEEDBACK
EDIT >

.ANA

CIRCUIT:  A
S-MATRIX, ZS =   50.0+J   0.0  ZL =   50.0+J   0.0
```

Freq	S11		S21		S12		S22		S21
GHZ	Mag	Ang	Mag	Ang	Mag	Ang	Mag	Ang	dB
2.00000	0.328	176.7	1.352	-11.5	0.159	32.4	0.908	-11.4	2.62
2.50000	0.326	176.5	1.357	-14.6	0.178	36.2	0.921	-14.4	2.65
3.00000	0.326	176.2	1.367	-17.6	0.197	38.9	0.938	-17.4	2.72
3.50000	0.325	175.7	1.381	-20.6	0.216	40.7	0.960	-20.3	2.81
4.00000	0.325	174.8	1.400	-23.6	0.234	42.2	0.985	-23.1	2.92
4.50000	0.324	173.7	1.422	-26.5	0.253	43.3	1.015	-25.8	3.06
5.00000	0.322	172.2	1.451	-29.4	0.271	44.2	1.049	-28.4	3.23
5.50000	0.321	170.4	1.488	-32.3	0.289	45.1	1.091	-31.0	3.45
6.00000	0.319	168.1	1.538	-35.3	0.309	46.0	1.143	-33.4	3.74

```
PLOT, PRINT OR QUIT? (PL/PR/<Q>):
.PR
OUTPUT CODE, 'HELP' OR 'CONT'? (S/SK/Y/Z/VG/S2/S4/S5/S6/NOI/G/<H>/C):
.S2

CIRCUIT:  A
                      STABILITY CIRCLES
             *---- INPUT PLANE -----* *---- OUTPUT PLANE ----*  GA MAX   K  Sgn
```

Freq	LOCATION		RAD STB	LOCATION		RAD STB	OR MSG		B1
GHZ	Mag	Ang	REG	Mag	Ang	REG	dB		
2.00000	1.157	134.9	1.639 IN	1.288	14.5	0.366 OUT	9.3	0.72	+
2.50000	1.213	134.3	1.568 IN	1.300	18.2	0.411 OUT	8.8	0.63	-
3.00000	1.243	134.7	1.484 IN	1.309	21.7	0.455 OUT	8.4	0.56	-
3.50000	1.261	135.8	1.400 IN	1.315	25.2	0.495 OUT	8.1	0.49	-
4.00000	1.276	137.1	1.322 IN	1.316	28.6	0.532 OUT	7.8	0.42	-
4.50000	1.291	138.5	1.250 IN	1.312	31.8	0.564 OUT	7.5	0.36	-
5.00000	1.306	140.1	1.183 IN	1.302	35.0	0.591 OUT	7.3	0.29	-
5.50000	1.321	141.8	1.119 IN	1.285	38.0	0.613 OUT	7.1	0.23	-
6.00000	1.337	143.7	1.054 IN	1.261	41.0	0.630 OUT	7.0	0.15	-

Figure 15.24. *Common-gate CAD input file, S-parameters and stability.*

the TWO and the two-port A:2POR 3 2 lines. The common-gate S-parameters for $Z_0 = 50$ ohms in Fig. 15.24 are for small values of feedback L and C. Note that the $|S_{11}|$ (source) is less than 1, whereas $|S_{22}| \approx 1$. This indicates that the inductor L should be increased to make these values larger. Continuing with 50-ohm source and load impedances, the value of L was increased to 1, 2, and 5 nH. At 5 nH the $|S_{11}|$ and $|S_{22}|$ peaks at 4 GHz. The resulting values for the series-gate inductance S-parameters and the impedance levels are given in Fig. 15.25. Noting that the (source) $Z_{11} = 22 + j71$ ohms and (drain) $Z_{22} = -176 - j147$ ohms at 4 GHz, the source is selected as the terminating port and the drain is the frequency-selective load port. Note that even though $|S_{11}| > 1$, Z_{11} is positive because the computer computes the bilateral values corresponding to Eq. 15.1.

The next step in the design is to synthesize the terminating network to match from Z_{11} to 50 ohms. A suitable network consists of a 0.56 pF capacitor to cancel the 71 ohms of inductive reactance and a quarter-wave transformer at 4 GHz of 33-ohms characteristic impedance to transform the

```
ANA

CIRCUIT:  A
S-MATRIX, ZS =   50.0+J   0.0   ZL =   50.0+J   0.0

  Freq        S11           S21           S12           S22         S21
  GHZ     Mag    Ang    Mag    Ang    Mag    Ang    Mag    Ang      dB
 2.00000  0.569  176.1  1.610  -12.9  0.080   58.6  1.008  -10.8   4.13
 2.50000  0.788  174.7  1.868  -17.7  0.078  118.9  1.127  -13.2   5.43
 3.00000  1.230  173.7  2.414  -23.1  0.216  170.4  1.378  -14.9   7.66
 3.50000  2.430  175.3  3.964  -27.5  0.724 -169.1  2.082  -14.0  11.96
 4.00000 11.794 -147.5 16.304    1.9  5.033 -122.2  7.520   22.0  24.25
 4.50000  3.874  -34.1  4.712  105.6  2.148   -2.2  1.769  133.1  13.46
 5.00000  1.912  -29.8  2.162   99.1  1.300    7.2  0.590  132.6   6.70
 5.50000  1.311  -30.6  1.465   87.6  1.053   10.8  0.249  120.4   3.32
 6.00000  1.018  -32.6  1.184   76.3  0.943   12.7  0.114   88.4   1.46

PLOT, PRINT OR QUIT? (PL/PR/<Q>):
.PR
OUTPUT CODE, 'HELP' OR 'CONT'? (S/SK/Y/Z/VG/S2/S4/S5/S6/NOI/G/<H>/C):
.Z

CIRCUIT:  A
                        Z-Matrix (Ohms)
  Freq          Z11/Z21               Z12/Z22
  GHZ        Re        Im          Re        Im
 2.0000     32.23    -32.28       24.18    -28.73
           -392.1   -642.7       -299.9   -689.5

 2.5000     29.11     -.9558      22.17     2.884
           -340.1   -417.0       -257.9   -470.5

 3.0000     26.21     25.67       20.13     29.65
           -296.2   -269.2       -223.7   -325.6

 3.5000     23.84     49.37       18.42     53.42
           -260.8   -166.6       -196.9   -223.6

 4.0000     21.93     71.15       17.02     75.21
           -232.7    -90.76      -176.0   -147.3

 4.5000     20.33     91.67       15.83     95.69
           -210.4    -31.24      -159.7    -86.63

 5.0000     18.89    111.3        14.74    115.3
           -192.4     18.12      -146.9    -35.86

 5.5000     17.54    130.5        13.68    134.3
           -177.6     60.99      -136.7     8.529

 6.0000     16.20    149.4        12.61    153.0
           -165.0     99.65      -128.2    48.77
```

Figure 15.25. *Series gate inductor (L = 5 nH) S-parameters and impedance values for $Z_0 = 50$ ohms.*

22 ohms to 50 ohms. This circuit's input impedance at 4 GHz is $22 - j71$ ohms, the conjugate match to Z_{11}. In microstrip, an open-circuited stub can replace the lumped 0.56-pF capacitor.

Attaching this circuit to the transistor as shown in the input file in Fig. 15.26 yields the one-port (drain) results for $L = 5$ nH. Note that the drain impedance values have changed significantly (compare to Fig. 15.25) because the transistor's source load is not 50 ohms. Also note that the maximum value of $| \text{Re} Z_{22} |$ occurs near 3.5 GHz. It is desirable to move this closer to the 4 GHz operating frequency, so the value of the gate inductor L is reduced to raise the resonance frequency of Z_{22} (drain). By using the CAD, $L = 4$ nH places the resonant frequency of the drain port at 4 GHz with a port impedance of $R_{in} + jX_{in} = -343 + j44$ ohms. The load termination is approximately $114 - j44$ ohms using Eqs. 15.63 and 15.64. Also from Eq. 15.67 check that $\partial X_L(f)/\partial f \, |_{f \, = \, 4 \, \text{GHz}}$ is negative for the network that yields this value of load impedance.

The implementation of the network is a series C, quarter-wave transmission line of 75-ohm characteristic impedance and a 50-ohm resistor. For this case $(\partial X_L/\partial f)_{f \, = \, 4 \, \text{GHz}} = -26$ ohms/GHz. Using another configuration with longer lines and a DC path to ground consisting of a 0.28 wavelength 50-ohm line terminated by a quarter-wavelength 31-ohm line and a 50-ohm resistor provides more dispersion $(\partial X_L/\partial f)_{f \, = \, 4 \, \text{GHz}} = -85$ ohms/GHz. Both oscillators are stable (see Eq. 15.67), but the higher dispersion tends to make the oscillator's frequency more controlled by the circuit and less influenced

```
***********
*. FILE NAME OSCCKT
* THIS CIRCUIT IS USED TO ANALYZE THE
* NEC67383 AS AN OSCILLATOR
***********
BLK
RES 1 0 R=50
TRL 2 1 Z=33 E=90 F=4GHZ
CAP 3 2 C=0.56PF
TWO 5 4 3 Q1
IND 5 0 L=5NH
A:1POR 4 0
END
**********
FREQ
STEP 2GHZ 6GHZ 0.5GHZ
END
********
OUT
PRI A S
END
**********
```

CIRCUIT: A
ONE-PORT, ZS =   50. +J   0.

Freq	RHO		SWR	RET L/G	Ohms		Millisiemens	
GHZ	Mag	Ang		dB	R	X	G	B
2.00000	1.075	-8.3	27.50:1	0.63	-275.8	-548.8	-0.7	1.5
2.50000	1.129	-9.0	16.56:1	1.05	-307.6	-398.1	-1.2	1.6
3.00000	1.199	-7.4	11.04:1	1.58	-366.4	-259.0	-1.8	1.3
3.50000	1.260	-1.1	8.68:1	2.01	-431.2	-35.2	-2.3	0.2
4.00000	1.184	11.7	11.87:1	1.47	-241.1	288.9	-1.7	-2.0
4.50000	0.869	23.0	14.31:1	-1.22	78.4	218.4	1.5	-4.1
5.00000	0.567	20.4	3.62:1	-4.93	131.0	76.4	5.7	-3.3
5.50000	0.413	7.3	2.41:1	-7.67	118.2	14.9	8.3	-1.0
6.00000	0.352	-5.6	2.09:1	-9.06	103.6	-8.2	9.6	0.8

Figure 15.26. Drain port input parameters for the common-gate oscillator ($L = 5$ nH).

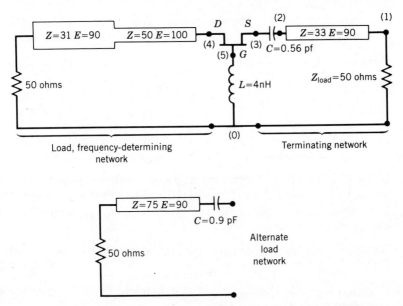

Figure 15.27. *Microwave schematic of a 4-GHz fixed-frequency oscillator. (Numbers in parentheses correspond to node numbers in Fig. 15.26.)*

by the temperature and operating bias variations of the transistor. The overall oscillator circuit diagram is shown in Fig. 15.27.

The above example was for a fixed-frequency oscillator. For a voltage-controlled oscillator a varactor diode (see Chapter 13) is connected into the load network. Here the lumped alternate series C implementation may be desirable. For a current-controlled yttrium–iron–garnet tunable oscillator, the resonator is connected into the load network. Review Ref. 3 for additional design details.

PROBLEMS

1. The common-source S-parameters for the Avantek AT-12570-5 GaAs FET are $S_{11} = 0.78 \,\underline{/-126}$, $S_{21} = 3.19 \,\underline{/66}$, $S_{12} = 0.106 \,\underline{/3}$, and $S_{22} = 0.38 \,\underline{/-97}$ at 4 GHz. For this transistor compute Γ_{in} and Γ_{out}. Compare these results with Γ_{in} and Γ_{out} for the unilateral approximation. Assume source and load are conjugately matched to S_{11} and S_{22}. What happens if $\Gamma_S = \Gamma_L = 0$?

2. Compute the transducer power gain for the AT-12570-5. Compare to $|S_{21}|^2$.

3. Show that if $Re Z_{in} > 0$, $\Gamma_{in} < 1$.

4. Use Eq. 15.19 to show that the distance from the $\Gamma_S = 0$ point to the center of the constant gain circle is $g_S \, | \, S_{11} \, | \, /D_S$ and the angle from the $\Gamma_S = $ positive real axis is

$$\tan^{-1}\left(\frac{-B_{11}}{A_{11}}\right)$$

   Is this angle a function of g_S? Why do all the constant gain circle centers lie along the line from $\Gamma_S = 0$ to S_{11}^*.

5. Show that the $G_S = 1$ constant gain curves always pass through the center of the Smith chart ($\Gamma_S = 0$). Hint: Show that the distance from $\Gamma_S = 0$ to center of the circle (see Problem 4) equals the radius of the circle.

6. Show that the transducer gain is a maximum when $\Gamma_S = S_{11}^*$ and $\Gamma_L = S_{22}^*$.

7. Design a lumped element output matching circuit to 50 ohms for the stabilized NEC 67383 at 4 GHz.

8. Derive a distributed element output matching circuit for the stabilized NEC 67383 at 4 GHz.

9. Compute the stability conditions for AT-12570-5 at 4 GHz. Is it stable?

10. Compute the parameters of the AT-12570-5 stability circles. On a Smith chart draw the parts of the circle where $| \, \Gamma \, | < 1$ cause instability at 4 GHz.

11. Compute the maximum stable gain assuming $S_{12} = 0$ for the AT-12570-5. Compute the constant gain circle for the input for 12 dB overall gain and plot it on a Smith chart at 4 GHz assuming no output matching. Plot the match for maximum stable gain.

12. At 4 GHz the AT-12570-5's optimum noise match is $\Gamma_0 = 0.6 \, \underline{/121°}$ with $R_n = 10$ ohms. Plot this point on the constant gain Smith chart in Problem 11 and derive a distributed input circuit. What is the stage gain at optimum noise match?

13. What is the input VSWR of the amplifier when matched for optimum noise?

14. Rewrite the appropriate lines in Fig. 15.17 to make the magnitude and phase of S_{21} vary by 10% and the linewidths of lines greater than 10 mils vary $\pm 5\%$.

15. Write a CAD program to compute the common-gate S-parameters for the Avantek AT-12570-5 at 4 GHz in 50-ohm circuits.

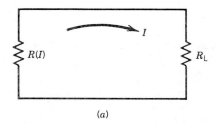

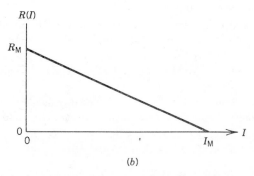

Figure P15.16. Equivalent circuit and current dependence for an oscillator. (a) Equivalent circuit at resonance; (b) peak circuit dependence of device resistance.

16. This problem derives why $n = \frac{1}{3}$ in Eq. 15.64 based on a simple equivalent circuit at resonance shown in Fig. P15.16a. Assume the device resistance is $-R(I) = R_M(1 - I/I_M)$ as shown in Fig. P15.16b. Current dependence is used rather than voltage dependence to keep the algebra simple. Compute the value of $R(I)$ for maximum power to the load R_L. Estimate R_M and find the n value.

17. Write CAD programs to determine the terminating and load circuits for the AT-12570-5 used as an oscillator at 6 GHz.

REFERENCES

1. W. Wagner and F. K. Manasse, "Simplify High-Power Gunn Oscillator Design," *Microwaves,* Vol. 14, No. 5, May 1975, p. 43, and F. N. Secki and D. Zieger, "A New Design Technique for Transferred-Electron Oscillators," *Microwave Journal,* Vol. 16, No. 4, April 1973, p. 47.

2. "Microwave Power Generation and Amplification Using IMPATT Diodes," Hewlett-Packard Application Note 935, Palo Alto, CA 94304, 1971.

3. G. Gonzalez, *Microwave Transistor Amplifiers,* Prentice-Hall, Englewood Cliffs, NJ, 1984.

4. D. Woods, "Reappraisal of the Unconditional Stability Criteria for Active 2-Port Networks in Terms of S-Parameters," *IEEE, Trans. Circ. Systems,* Vol. CAS-23, No. 2, February 1976, p. 73.

5. K. Kurokawa, "Power Waves and the Scattering Matrix," *IEEE Trans. Micro. Th. Tech.,* Vol. MTT-13, No. 2, March 1985, p. 194.

6. T. T. Ha, *Solid-State Microwave Amplifier Design,* Wiley, New York, 1981, Sect. 2.3.

7. H. A. Haus, "Representation of Noise in Linear Twoports," *Proc. IRE,* Vol. 48, No. 1, January 1960, p. 69.

8. M. W. Medley and J. L. Allen, "Broad-Band GaAs FET Amplifier Design Using Negative-Image Device Models," *IEEE Trans. Micro. Th. Tech.,* Vol. MTT-27, No. 9, September 1979, p. 784.

9. M. E. Van Valkenburg, *Network Analysis,* Prentice-Hall, Englewood Cliffs, NJ, 1955.

10. K. Kurokawa, "Some Basic Characteristics of Broadband Negative Resistance Oscillator Circuits," *Bell Syst. Tech. Jrnl.,* Vol. 48, No. 7, July 1969, p. 1937.

16 Monolithic Microwave Integrated Circuits

16.1 INTRODUCTION

Monolithic microwave integrated circuits (MMICs) are an important ingredient in the design and fabrication of future sophisticated radar and electronic countermeasure (ECM) systems and are rapidly replacing existing "hybrid" circuits. Figure 16.1 shows a module composed of hybrid circuits that have been fabricated by the assembly of a number of lumped elements, such as transistors, capacitors, inductors, diodes, and resistors, onto several sapphire substrates on which distributed microstrip lines have been formed. The elements for the hybrid circuit are each placed in turn onto the substrate by bonding or soldering, and the circuit is usually "tweaked" to provide optimum performance. For high-frequency circuits, the performance becomes critically dependent on the position and size of the various circuit elements, and this causes the cost of assembly of the components to increase significantly (1, p. 35).

The relative positions of the elements can be defined very accurately if all circuit components, including the active elements such as transistors, are formed as part of a common semiconductor substrate and interconnected by metal lines that are formed on the substrate surface. This allows the parasitic capacitances and inductances that are such a problem at high frequencies to be controlled and exploited in the design of the circuit (2, p. 108). In addition to accurate placing of the components, the characteristic parameter values of those components, such as transistor transconductances, are more uniform and predictable since they lie on adjacent and very

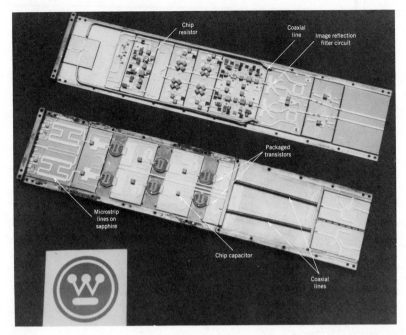

Figure 16.1. *Microwave hybrid circuits with lumped elements and microstrip lines on sapphire substrates.*

similar regions of the semiconductor. Finally, the methods of manufacture of the circuits use batch-processing techniques whereby tens or hundreds of circuits are formed simultaneously, and this, more than any other feature, reduces the cost of each circuit. This method of circuit realization is known as *monolithic integration.*

Monolithic circuits can be digital or analog (linear) in nature, but this discussion is confined to analog circuits because their design and fabrication contain features not found in digital devices. Digital circuits are made using similar techniques; the features that distinguish the two types are the frequency range of operation, the size of the individual transistors, and the interconnection schemes. Digital circuits are limited mainly to operation below 5 GHz, whereas the linear circuits that are of interest all operate at frequencies above 4 GHz and up to 110 GHz. FETs are generally quite small in digital circuits with a single gate stripe of 100 μm or less. Analog circuits have FETs with multiple fingers the individual lengths of which are from 30 to 200 μm depending on the frequency at which they are operating. Finally, the interconnection paths for digital circuits are designed without consideration of their microwave performance as microstrip distributed elements. In monolithic circuits this interconnection is critical to avoid loss and impedance mismatches.

There are disadvantages to the monolithic approach, however (1, p. 13):

1. The device-to-chip area ratio is small, which means the yield and hence the cost of the devices depend on the yield of the active devices, typically the poorest yield of any of the components on the circuit.

2. The circuit cannot be tuned or "tweaked" because it is too small and fragile. This operation would also raise the cost of the circuit by at least an order of magnitude.

3. Trouble shooting or debugging the circuit is difficult because of the difficulty in extracting signals from it. This leads to a philosophy in which the design is not sensitive to the manufacturing tolerances in the active devices, which means a compromise in the ultimate performance of the device. A great deal of help can be gained from the use of CAD in this respect. The use of microwave probes (1, p. 481) has helped to some extent in the analysis of wafers before they are diced.

4. Since the circuits are so small, there is a danger of coupling (crosstalk) between the components. Line spacings of twice the substrate thickness will reduce coupling between adjacent lines to tolerable levels. This coupling is a factor that tends to increase the size of the chip and hence its cost.

5. High-power elements such as IMPATTs cannot be readily integrated into monolithic structures. This is largely because the heat sinking required by these devices cannot be accomplished because of the thickness of the substrate on which the monolithic circuit is formed. There is a tradeoff between the microwave performance and the thermal impedance, since to minimize the thermal impedance it is necessary to make the substrate as thin as possible; however, a thin wafer requires thin microstrip lines and these have high resistive losses. In addition, the thinner wafer introduces greater parasitic capacitance to the grounded metallization on the back of the wafer.

6. Narrow-band high-Q filters are not easily realized in monolithic form due to the previously discussed conductor losses and parasitic capacitances. Such elements are required in low-noise, stable voltage controlled oscillators (VCOs).

7. Tight control of material quality and fabrication tolerances is required if monolithic circuits are to reach their full potential. Device yield depends on the quality of the substrate and the active layer as well as variations in purity, defect density, and traps and in the reproducibility of fabrication processes.

In spite of these disadvantages, a great deal of effort is being devoted to the development of monolithic circuits because of their ability to provide gain and power over wide frequency ranges.

16.2 MATERIALS AND FABRICATION

Gallium arsenide is the present semiconductor of choice for monolithic circuits for two main reasons: The first is that electrons move through the material with high saturation velocities (1.2×10^7 cm/s), which, because of the high electron mobility (5000 cm^2/V·s), can be achieved with modest electric fields, which then leads to small transit times for electrons across devices and high-frequency response for a given dimension. Second, gallium arsenide can be prepared in a semi-insulating form with a resistivity of 10^8 ohm-cm, which allows interconnections in the form of microstrip lines to be made between transistors in a monolithic circuit. This ability to form distributed networks allows a tight packing of the components on the "chips." In some cases, lumped elements are required where the circuit design requires a component value not readily achieved with microstrip. An example of a monolithic circuit with microstrip lines and lumped capacitors on the same chip is shown in Fig. 16.2.

One of the features of the above circuit is the use of "vias," which are connections from the front to the back (ground plane) of the wafer in order to provide low inductance connections to ground for the sources of the FETs and the grounded pads of the capacitors. Wet chemical etching is used to form holes in the arsenide after it has been thinned to 100 μm, and they are then metallized when the gold of the ground plane is deposited. Vias also play a role in improving the thermal resistance of the FETs since gold has a higher thermal conductivity than gallium arsenide and can readily conduct heat from the FETs to the heat sink in contact with the back metal ground plane. The vias are placed under the large via pads shown at each side of the FET in Fig. 16.2.

Gallium arsenide is grown by two main methods. The earlier one is the Bridgman method, which produces D-shaped wafers. The more recent method is the Czochralski technique, which produces round wafers of the type illustrated in Fig. 16.3. The round wafers are preferred for the fabrication of monolithic circuits because they most readily benefit from the extensive experience and equipment from silicon device fabrication.

Recent developments in gallium arsenide growth have concentrated on the reduction of dislocation densities from the 10^5 cm^{-2} for chromium-doped or undoped material to almost zero, which is achieved when a few percent of indium is added to the melt (3). The effect of dislocations is not well understood, and there is some controversy as to whether they affect the threshold voltage of digital FETs. Their effect on analog devices with their higher threshold voltage (2–6 V) is less marked.

The fabrication of monolithic integrated circuits involves a variety of tech-

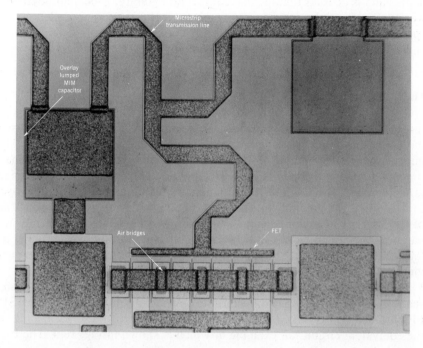

Figure 16.2. Microstrip lines and lumped capacitors on the same monolithic circuit.

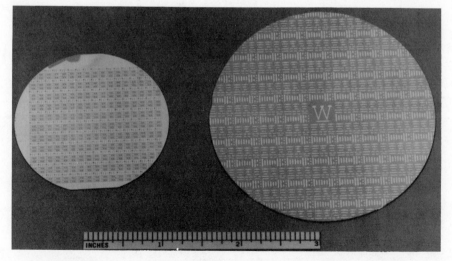

Figure 16.3. Two- and 3-in. wafers grown by the Czochralski technique.

niques and processes that include contact lithography, *E*-beam lithography, wet-chemical and dry (reactive-ion) etching, metal evaporation, plating, and sputtering. It is beyond the scope of this book to describe the applications and limitations of these techniques in detail (4, p. 279).

16.3 DESIGN TECHNIQUES

It is necessary to know how the active elements (the FETs) will behave over the operating frequency range in order to design a monolithic circuit. This information can be obtained by fabricating test FETs in a special preliminary run and then fully evaluating them as a function of frequency using a variety of techniques. The DC and RF parameters thus obtained can then be used to determine the component values in an equivalent circuit representation of the FET. These component values can then be related to the physical parameters of the device. Such a relationship is important in controlling the fabrication of FETs in the monolithic circuit to ensure that they will be identical to those previously fabricated.

One of the methods used to evaluate FETs is to measure their small-signal scattering-parameters (*S*-parameters) (5). Modeling programs such as SU-PER-COMPACT™ and SPICE can then be used to fit these parameters to the equivalent circuit, an example of which is shown in Fig. 16.4. The FET

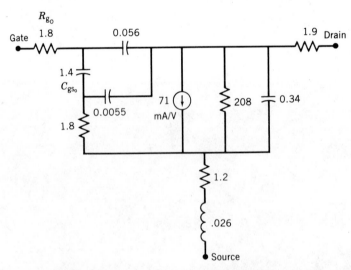

Figure 16.4. *Small signal lumped-element equivalent circuit of 1200 μm direct ion-implanted FET. Nominal gate length = 1 μm = L_{go}; nominal doping = 1.1 × 10^{17}/ cm^3 = N_{Do}; doping change: $C_{gs} = C_{gso}(N_D/N_{Do})^{1/2}$; gate length change: $C_{gs} = C_{gso}(L_g/L_{go})$; $R_g = R_{go}(L_{go}/L_g)$.*

that has been modeled was fabricated by ion implantation into semi-insulating gallium arsenide and has a periphery of 1200 μm composed of eight gate stripes of length 1 μm and width 150 μm. The model is applicable only to small-signal behavior since, under power conditions, some of the parameters change within the period of the microwave signal. An example of the change is that of the gate capacitance, which has a maximum value when the gate is biased to zero or slightly positive and has a minimum value when the gate voltage reaches or exceeds the pinchoff voltage. For large signal modeling, several methods are available:

1. Load–pull using tuners (6)
2. Load–pull using feedback (4, p. 489)
3. High-power S-parameters

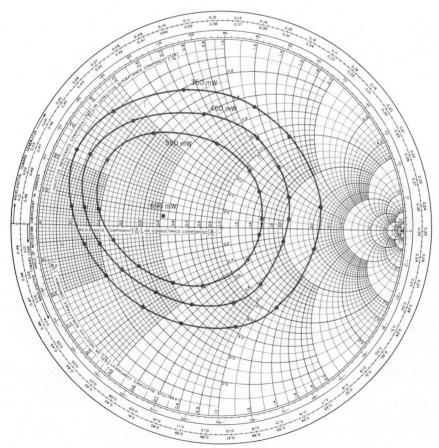

Figure 16.5. Load–pull contours at 5 GHz.

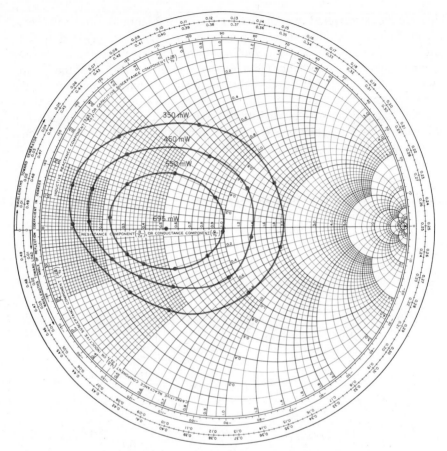

Figure 16.6. Load–pull contours at 7 GHz.

Load–pull consists of feeding a fixed amount of power into the FET and then, using tuners on the output, plotting on a Smith chart the contours of all the values of impedance and reactance that will give a constant amount of gain. Such a plot is shown in Fig. 16.5, which was taken at 5 GHz and shows contours for various powers from 350 to 815 mW. The contour for this highest power is a single point, indicating that there is a single value of impedance that can fulfill the load condition. Similar plots for 7 and 10 GHz are shown in Figs. 16.6 and 16.7, respectively. From these can be plotted the contours for a given amount of output power (560 mW) for various frequencies, as shown in Fig. 16.8. Any circuit that is designed as a match to the output of this FET, to deliver 560 mW, must have an impedance locus as a function of frequency that crosses these loci at the appropriate frequency. This is illustrated in Fig. 16.9.

Westinghouse® has used this technique to design the two stage amplifier that is shown in Fig. 16.10. This amplifier was designed to deliver 1 W from 8 to 12 GHz with 9 dB gain. The success of the design technique is illustrated in Fig. 16.11, which shows the output power as a function of frequency for five such amplifiers.

An alternative to the use of tuners in the output stage is to feed part of the input power to the output of the device (4, p. 279). By varying this signal in phase and amplitude, arbitrary load impedances can be generated. This eliminates the extra time required by manual tuners and allows orthogonal tuning, which may not be possible with some tuners. Reproducibility of the technique is also improved. A problem in determining the extracted power may be encountered if the output reflection coefficient is close to unity, which can happen for large FETs. This technique is particularly well suited

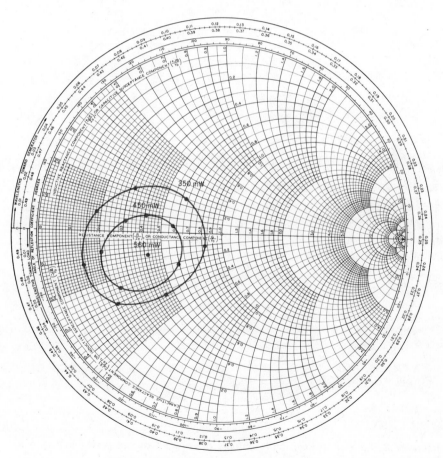

Figure 16.7. *Load–pull contours at 10 GHz.*

to computer control since it does allow orthogonal tuning and can be automated.

The measurement of large-signal *S*-parameters for FETs has not been successful because a 50-ohm load is used to measure the output of the device, and the output power depends nonlinearly on this impedance unlike the small-signal case. Willing et al. (7) have derived a series of small-signal *S*-parameters at various points and have used this to model the large-signal behavior. The method was only applied to a low-power device (200 mW/mm) and was not extended to a wide range of devices. Models that can be implemented on SPICE aid in the design of power amplifiers, oscillators, mixers, and fast-switching digital integrated circuits.

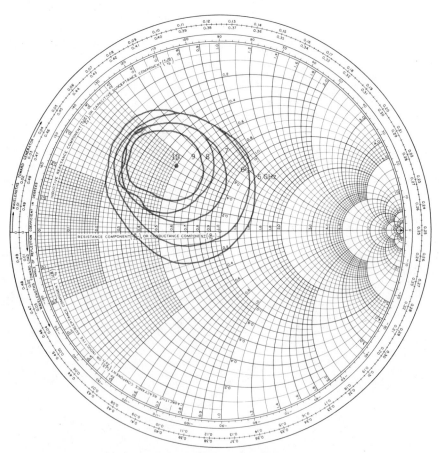

Figure 16.8. *Composite load–pull contours of constant output power and gain for 1200-μm FET.*

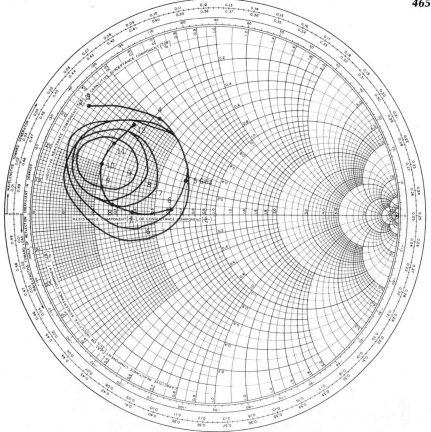

Figure 16.9. *Impedance locus for output circuit overlaid on load-pull contours.*

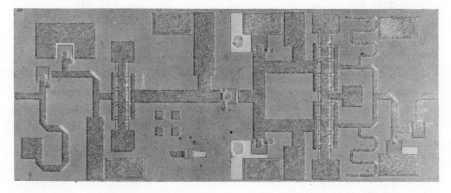

Figure 16.10. *Two-stage, 1-W, 8–12-GHz amplifier.*

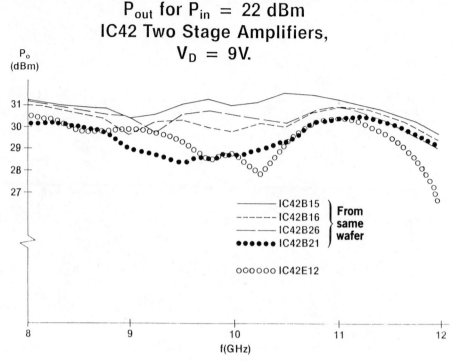

Figure 16.11. Output power of the two-stage amplifier as a function of frequency.

16.4 MMIC TYPES

The two-stage circuit shown in Fig. 16.10 is a power amplifier, which is the most important and most difficult to fabricate of all the forms of monolithic circuits. The difficulties arise in part because of the high thermal impedance (4, p. 571) of the device through the semi-insulating substrate, and in part because of the problem of delivering power from the output of the device. The voltage swing that can be achieved at the output is limited by the breakdown voltages of the FET (4, p. 285), and the drain operating voltage is generally kept below 15 V. In order to achieve powers of several watts, it is necessary to fabricate FETs that have a large amount of periphery. This requirement has led to the development of the "cluster-matched" (1, p. 218) FET, which uses clusters of identical stages to build up to a parallel combination of a large number of FET cells. An example of such an amplifier is shown in Fig. 16.12.

The power output from a FET at *X*-band is between 0.5 and 0.7 W/mm and, hence, to obtain powers of greater than 3–4 W requires peripheries greater than 5–6 mm. Figure 16.12 shows a four-stage amplifier designed to deliver 3 W of output power from 8 to 12 GHz. This amplifier has a final-

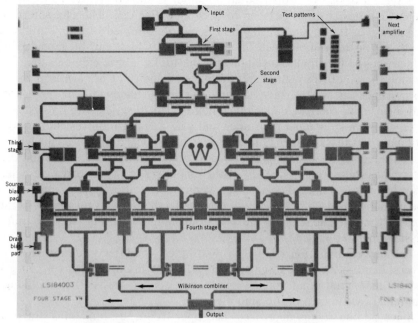

Figure 16.12. *Four-stage, 3-W, 8–12-GHz amplifier.*

stage periphery of 6.4 mm, and the power delivered to the output terminal through the Wilkinson combiner on the output is 2.2 W across the 8–12-GHz band with 25 dB of associated gain. The major reason that the amplifier does not meet its design goals are losses of from 1 to 2 dB that occur in the power-recombining circuit due to losses in the gallium arsenide semi-insulating substrate. These losses are greater than would be predicted from a simple consideration of conductor loss on a dielectric substrate (1, p. 13) and may be due in part to currents injected into the substrate by the large voltage swings that occur on the output lines.

A form of monolithic amplifier that combines FETs in a different manner is the distributed amplifier. This idea is not new but has been revived in several forms (1, p. 252) because of its very large bandwidth potential. The concept of the amplifier is that an artificial transmission line is created using lumped inductors (hybrid form) as series elements and the input capacitance of the FETs for the shunt elements. In this way, transconductances of the active devices can be added without increasing the parasitic capacitance associated with the gates of large devices. A similar transmission line is constructed at the output terminals of the active device. If the phase velocity of the input and output lines are equal, a signal traveling down the input line will be amplified. In addition to the large (decade) bandwidth of this device, the input and output VSWRs are very low, thereby allowing easy cascading of several amplifiers.

A typical design of a distributed amplifier is shown in Fig. 16.13. Four GaAs FET devices can be seen in each amplifier cell. The characteristic impedance of the input transmission line is determined by the width and length of the high-impedance microstrip lines (which can be modeled as inductors) and the input capacitance of the FETs. These elements also determine the cutoff frequency and the phase velocity of the artificial transmission lines. Ideally, as stages are added, gain and bandwidth are increased. However, since the gate capacitance in a FET is in series with a resistance of several ohms (channel resistance and gate metallization resistance), the artificial transmission line is quite lossy in a practical amplifier. This loss at some frequency offsets the gain added by an additional FET stage, and this determines the cutoff frequency of the amplifier. In order to realize equal input and output impedances, as well as phase velocities, some capacitative compensation is used on the drain terminal so that the drain capacity is the same as the gate capacity. In the case of the amplifier in Fig. 16.13, a transmission line stub is used.

The power limitation of such a distributed amplifier is determined by several factors: the total gate periphery, the gate-drain breakdown voltage, and the maximum input voltage that can be applied across the input stage multiplied by the gain of the amplifier.

Since adding stages decreases the upper frequency limit of a practical amplifier, it is not possible to add stages indefinitely to increase the power output at a high frequency. Similarly, increasing the size of each FET stage increases the input capacitance and lowers the upper frequency limit. Although these devices have power limitations, the distributed amplifier demonstrates very wide bandwidth in a monolithic form (8).

Other forms of amplifiers include feedback and those for which the inputs are matched using lossy elements. These are described and compared very well in a paper by Niclas (1, p. 231). Many of the amplifiers described so far operate in class A mode, but the efficiency of these amplifiers (typically

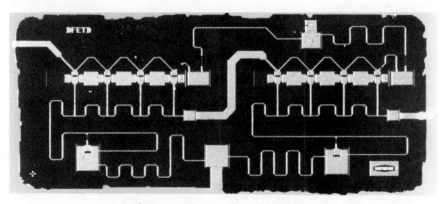

Figure 16.13. Distributed amplifier.

no more than 20%) is much less than the 35% that has been achieved in class B mode. In class B operation, the device only conducts for half of the RF cycle and hence it is best operated in the push–pull configuration (4, p. 505). This form of circuit also has the advantage that second-order harmonics are reduced. In system applications, particularly for airborne radar and ECM systems, class B is an absolute necessity because of the weight saved as a result of the higher efficiency and therefore the lower prime power and heat dissipation requirements. In fact, the second of these is more important because, typically, it requires 5 kg of equipment to generate 1 kW of prime power and it requires 25 kg of equipment to dissipate 1 kW of heat. Improvement in efficiency also leads to lower device temperatures and longer lifetimes for the components.

We have concentrated on power amplifiers because they represent the greatest challenge in the monolithic arena. Low-noise devices and amplifiers have also benefited from monolithic integration, and excellent results been reported (1, p. 160). Low-noise FETs require low pinchoff voltages and low parasitic resistances. The technology known as HEMT (9), which is addressed below, is providing the greatest advances in this aspect of monolithic design.

Amplifiers are not the only circuits that benefit from monolithic integration. Figure 16.14 shows a phase shifter that operates in a digital mode by switching between transmission lines of various lengths. Figure 16.15 shows an analog phase shifter that uses a high-band/low-band lumped circuit to produce the phase shift. These phase shifters can be integrated with amplifiers to produce higher levels of integration. The yield of the combined circuitry determines the degree of integration that is practical.

Broadband (2–18 GHz) attenuators have also been implemented monolithically (1, p. 424). The insertion loss of such circuits is 2–3 dB with an attenuation up to 12 dB that is controlled by a voltage. This kind of performance compares favorably with PIN diode/Lange coupler attenuators.

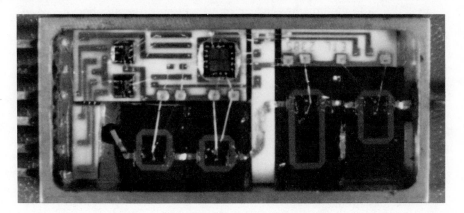

Figure 16.14. Digital phase shifter.

Figure 16.15. Analog phase shifter.

16.5 YIELD AND COST DISCUSSION

Since the cost of processing a semiconductor wafer is approximately constant and independent of the size, shape, and complexity of the circuits that are fabricated on it, the cost of each circuit is inversely proportional to the number of circuits on each wafer. The cost is therefore proportional to the size of each circuit, and hence the goal of the circuit designer is to make the circuit as small as possible consistent with high yield and sufficient electrical isolation between the components. A good example of a very tight design is the distributed amplifier built by Avantek (1, p. 252) and shown in Fig. 16.16, which occupies a chip 0.75 by 0.85 mm yielding a potential of over 2500 on a 2-in. diameter wafer.

In addition to minimizing the size of each circuit, the number of "good" devices depends on the yield, which itself depends on many factors; for example,

Material quality
  Purity
  Dislocation density
  Trap density
Fabrication procedures
  Gate lithography
  Overlay capacitor formation
  Via formation
  Wafer breakage
  Wafer thinning
Mounting and bonding

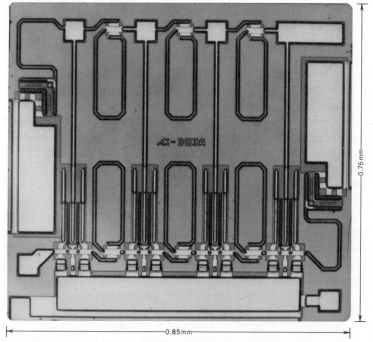

Figure 16.16. *Avantek distributed amplifier.*

Control over many of these factors is increasing. The use of dislocation-free gallium arsenide starting material has improved ion-implanted device electrical characteristics, and the increasing use of cassette-to-cassette fabrication equipment will decrease breakage and help improve the uniformity of processing (1, p. 83).

16.6   ADVANCED CIRCUITS AND TECHNOLOGIES

The main considerations for the future are increasing the frequency, efficiency, and power and lowering the noise of monolithic circuits. To achieve these goals, several new technologies are being developed. One of the most interesting is the device that relies for its carriers between source and drain on a two-dimensional electron gas that is generated at the interface between materials such as GaAs and AlGaAs because of the difference in bandgap. The devices go under the various names of high electron mobility transistor (HEMT), selectively doped heterojunction transistor (SDHT), modulation-doped FET (MODFET) (10), and two-dimensional electron gas FET (TEG-FET), depending on which group of experimenters is describing the device. The electrons in a HEMT have a high mobility because the electrons are physically separated from the scattering fields of their parent donor atoms, and hence the device has high gains at microwave frequencies. The prospect for high powers is less certain since the amount of charge in the two-dimensional electron gas is limited by the band-bending attainable. However, good power results have been published on devices that use two two-dimensional electron gases in parallel (11).

Indium phosphide has been used as an alternative to gallium arsenide to fabricate monolithic circuits, but its material technology is not as well advanced and the device has had limited success. One of the greatest promises for this material appeared when it was discovered that an insulating layer with good electrical properties could be placed under the gate of an FET. This allows a MOS (metal-on-semiconductor) type of structure and permits the gate on the FET to be biased positively with respect to the source, thus allowing an increased number of electrons to flow in the channel and a correspondingly higher current. A FET made in this manner has yielded a power of 4.3 W/mm of periphery, which is four times that obtained for conventional gallium arsenide depletion-mode devices.

The presence of surface states under most insulating layers on gallium arsenide has not allowed this type of MOS structure. Recently, however, the use of aluminum gallium arsenide and aluminum arsenide has been successful in producing power and could be a turning point in producing devices that can tolerate large voltage swings on the gate.

The growth of epitaxial layers has made significant advances in recent years. The fabrication of the HEMT device described above could not have been accomplished without the development of molecular beam epitaxy. This technique is essentially a high-purity high-vacuum evaporation system that is capable of depositing materials a monolayer at a time. The abruptness of the layers approaches 5 Å and allows a variety of sophisticated structures to be formed. A technique with similar abilities is organometallic chemical vapor deposition (OMCVD or MOCVD). This is a form of vapor phase epitaxy (VPE) that uses organometallic compounds (such as trimethyl aluminum) to transport compounds (such as aluminum) that cannot be used in conventional VPE. Both of these techniques are beginning to find widespread use in monolithic circuits.

16.7   CONCLUSIONS

Radar and ECM systems of the 1990–2000 time period will require a multitude of individual modules, each of which must deliver from 2 to 10 W of power across broad spectra of microwave frequencies in the range from 1 to 60 GHz. At present the use of gallium arsenide FETs and monolithic amplifiers is the only way of achieving these goals at a reasonable cost. The efficiency of these amplifiers will be one of the more important properties, and it is likely that class B operation will be the preferred mode.

Advances in materials and technology will push the frequencies up into the millimeter range, and systems are already being planned that will use three-terminal devices operating at 60 GHz. Monolithic microwave circuits are a technology for the future.

PROBLEMS

1. What are the advantages of monolithic integrated circuits over hybrid integrated circuits? What are some of the problems with monolithic integration?

2. Why is gallium arsenide chosen as the material for the majority of microwave monolithic integrated circuit substrates?

3. Why are "vias" necessary in monolithic circuits? What alternatives are there?

4. What are the main features of a monolithic distributed amplifier?

5. How are "load–pull" measurements made? What is their application?

REFERENCES

1. R. A. Pucel, ed., *Monolithic Microwave Integrated Circuits,* IEEE Press, New York, 1985.

2. J. Frey and K. B. Bhasin, eds., *Microwave Integrated Circuits,* Artech House, Norwood, MA, 1985.

3. D. L. Barrett, S. McGuigan, H. M. Hobgood, G. W. Eldridge, and R. N. Thomas, "Low Dislocation, Semi-Insulating In-Doped GaAs Crystals," *Journal of Crystal Growth,* Vol. 70, 1984, p. 179.

4. J. V. DiLorenzo and D. D. Khandelwal, eds., *GaAs FET Principles and Technology,* Artech House, Norwood, MA, 1982.

5. R. S. Pengelly, *Microwave Field-Effect Transistors—Theory, Design and Applications,* Research Studies Press, Wiley, New York 1982, p. 141.

6. M. C. Driver, G. W. Eldridge, and J. E. Degenford, "Broadband Monolithic Integrated Power Amplifiers in Gallium Arsenide," *Microwave Journal,* Vol. 25, November 1982, p. 87.

7. H. A. Willing, C. Raushcer, and P. deSantis, *IEEE Trans. Microwave Theory and Techniques,* Vol. MTT-26, 1978, p. 1013.

8. R. Pauley, P. Asher, J. Schellenburg, and H. Yamasaki, "A 2 to 40 GHz Monolithic Distributed Amplifier," *IEEE Gallium Arsenide Integrated Circuits Symposium Technical Digest,* 1985.

9. A. K. Gupta, E. A. Sovero, R. L. Pierson, R. D. Stein, R. T. Chen, D. L. Miller, and J. A. Higgens, "Low-Noise High Electron Mobility Transistors For Monolithic Microwave Integrated Circuits," *IEEE Electron Device Letters,* Vol. EDL-6, No. 2, February 1985, p. 81.

10. P. M. Solomon and H. Morkoc, "Modulation-Doped GaAs/AlGaAs Heterojunction Field-Effect Transistors (MODFETs), Ultrahigh Speed Device for Su-

percomputers," *IEEE Trans. Electron Devices,* Vol. ED-31, 1984, p. 1015.

11. P. M. Smith, U. K. Mishra, P. C. Chao, S. C. Palmateer, and J. C. M. Huang, "Power Performance of Microwave High-Electron Mobility Transistors," *IEEE Electron Device Letters,* Vol. EDL-6, No. 2, February 1985, p. 86.

17 | Microwave Tubes

17.1 INTRODUCTION

Several types of vacuum tubes are used for microwave power generation. Advances made by solid-state devices will not eliminate the need for microwave tubes. Figure 17.1 indicates the state-of-the-art for both types of devices.

Tubes can be divided into two classes—crossed-field and linear beam devices (sometimes called "m" and "O" types after the abbreviated French terms used to describe them). In a crossed-field tube, electron flow occurs in a space with perpendicular magnetic and electric fields. In the linear beam, the two fields are parallel and the electron flow forms a beam traversing the length of the tube. Table 17.1 indicates the most important types of tubes in both categories.

The practical exploitation of microwaves began with the magnetron, which was the first device to produce microwave power at useful levels. It existed in various rudimentary forms, such as a single cylinder or a split anode magnetron in which the cylinder is in two pieces, in the 1920s and 1930s and with an external resonant circuit was capable of sustaining oscillations at UHF (400–900 MHz). Its power and efficiency were extremely low, and it was generally regarded as a scientific curiosity. The World War II requirement for narrow radar beams (for high angular resolution and low altitude detection) stimulated the magnetron's further development as a source of microwave power. The breakthrough came first in England with the introduction of the resonant circuit into the vacuum envelope in the form of cavities (resonant at the required frequency and taking the place of the

475

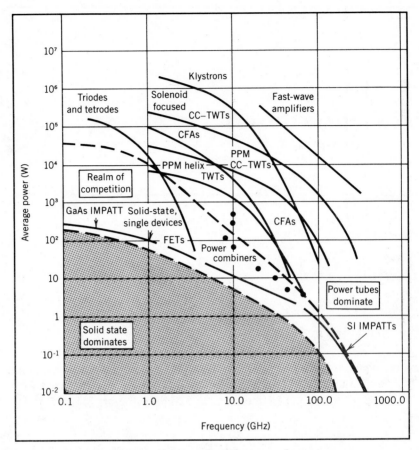

Figure 17.1. Tubes and solid-state performance.

Table 17.1 Microwave tube categories

Crossed Field (m)	Linear Beam (0)
Magnetron	Klystron
Crossed-field amplifiers	Traveling wave tube
i. Backward wave	Extended interaction klystron
ii. Forward wave	Backward wave oscillator
iii. Injected beam	

Table 17.2   Magnetron Capability

Frequency (GHz)	Peak Power (MW)
1	10
3	5
10	1
35	0.1

original cylinder), and second with the replacement of the thoriated tungsten filament with an oxide-coated cathode [capable of producing very high currents for very short (microsecond) periods of time]. This magnetron oscillated in a pulse mode at 3 GHz at 100-kW peak power (1).

Continued development has led to the capability summarized in Table 17.2. Meanwhile, radar systems have become more sophisticated in their extraction of data from radar returns and in their operation in adverse conditions such as heavy clutter and ECM environments. These performance improvements require frequency and phase coherence between transmit and receive signals. This is impossible to achieve with the magnetron since it is purely a power oscillator and has no coherence from pulse to pulse. Although attempts were made to overcome this by "injection locking" and other stabilization techniques, it was ultimately necessary to adopt the master oscillator power amplifier (MOPA) configuration for radar transmitters. The development of amplifying devices such as klystrons, traveling wave tubes, and crossed-field amplifiers has provided sufficient gain and power for the low-power master oscillator radar transmitter (2–5).

Most new radar systems (with the possible exception of millimeter wave radars) use amplifying tubes in their transmitters. Magnetrons are still found in many thousands of existing radars around the world. In recent years they have also found many commercial applications as the RF source used in microwave cookers.

17.2   TUBE ELEMENTS

17.2.1   General

All microwave tubes depend on the interaction between an electron stream and an electric field supported by a microwave circuit that results in a basic amplifying mechanism. Many of the component parts used in this process are common to the various types of tubes.

17.2.2   Electron Guns

The electron gun generates an electron beam for injection into the interactive space. There are two main types—the spherical cathode solid beam gun

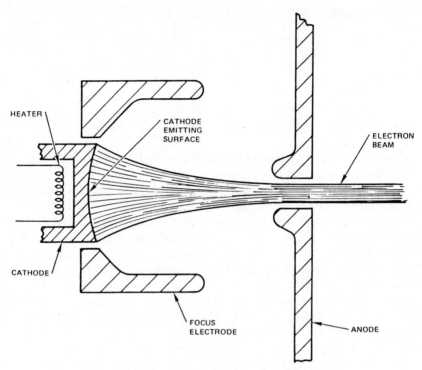

Figure 17.2. Typical electron gun (simplified form).

used in all linear beam tubes and the magnetron injection gun originally used in some types of crossed-field amplifiers and more recently in some of the new high-power fast wave millimeter wave devices such as the peniotron and gyrotron.

The gun is comprised of a cathode (the electron source), a focus electrode, a modulating electrode, and an anode. Since the current density required in the beam is usually very much higher than the allowed emission current density from the cathode, the cathode area is much larger than the beam cross-sectional area. The electron flow from the cathode must be convergent to produce the smaller beam diameter. This is illustrated in Fig. 17.2, which shows a schematic of a solid beam electron gun. The cathode is held at a high negative potential with respect to the anode, which is usually grounded.* Electrons are produced by thermionic emission at the cathode surface and accelerated toward the anode, which is a metal plate with a hole in it. The shape of the anode and the spherical shape of the cathode together with the focus electrode, which is at cathode potential, result in an electron

---

* This arrangement is common to all types of tubes and is used so that the RF output and input ports that are connected to the circuit, which in turn is at the same potential as the anode, may be at ground potential.

lens whose field pattern produces the required covergence of the electrons. These pass through the anode hole as a beam of the required cross section. The ratio of the cathode to beam areas is known as the convergence ratio and is usually in the range of 10 to 60. The current produced by the gun is given by the expression

$$I = KV^{3/2} \tag{17.1}$$

where V is the voltage between anode and cathode and K is a constant known as the perveance. Perveance depends on the gun geometry and has values from 0.1×10^{-6} to 2×10^{-6}. The higher the value, the more difficult it is to achieve good convergence.

If it is desired to use the gun in a pulsed mode, a modulating electrode is introduced in the region between the cathode and anode. By the application of positive and negative voltages with respect to the cathode, it is possible to control the electron flow.

The magnetron injection gun will only be discussed briefly here. Apart from the injected beam crossed-field amplifier (CFA), crossed-field devices do not use electron guns such as described above. They do use cathodes and in many cases modulating electrodes. Anodes are usually not separate electrodes but the microwave circuit itself.

17.2.2.1 Cathodes. Cathodes are the electron source for all microwave tubes. They emit electrons either by thermionic emission, secondary emission, or a combination of both. Thermionic emitters use heat to raise the energy of free electrons in the conduction band to a level high enough to overcome the potential barrier at the surface of a conductor and thus enter the adjoining space. This potential barrier is known as the *work function* and is measured in electron volts. It is a figure of merit for comparing surfaces, with low numbers being the most desirable.

Two types of emitting surfaces are used. The oxide type consists of a layer of barium oxide on a nickel base. It operates in the region of 900–950°C and is capable of very high emission currents (10–12 A/cm^2) for very short periods of time. The emission current drops rapidly, however, as the pulse length is increased. Its DC capability is 0.3 A/cm. It is also subject to damage from arcing and ion bombardment and can easily be "poisoned" by contamination from gas and other metals. It was used in all receiving tubes and many multimegawatt microwave power tubes but is gradually being replaced by the impregnated tungsten cathode also known as the Phillips type B and type M.

The type B emitting surface consists essentially of a tungsten plug or matrix sintered to about 80% density mounted in a molybdenum cylinder, which also contains a heater or filament. The resultant pores left in the tungsten are impregnated with a mixture of barium, calcium, and strontium carbonates, which when heated in a vacuum produce a very thin film of barium on the surface to become the active emitter. As barium is slowly evaporated from the surface during operation, it is continuously replaced by the mixture stored in the tungsten matrix. This results in a long life and high resistance to poisoning and arc damage. A cross section of this type of cathode is shown in Fig. 17.3. It is usually operated in the region of 1020–1100°C for emission current densities in the region of several A/cm². Its DC capability is the same as its pulsed capability. The type M variation is the same as the type B except that it has a thin layer of iridium or osmium on the active surface to lower the work function still further. It therefore produces twice the current density for the same temperature as a type B or the same current density for 25–50°C less temperature, thus conserving life.

Typical emission characteristics showing dependence on temperature and voltage are depicted for sample cathodes in Fig. 17.4. The lower part of the curve where the emission is independent of voltage is termed the temperature-limited region. It is where all the electrons emitted are drawn immediately to the anode. The quasiflat portion is known as the space charge limited region and is independent of temperature. As the temperature is increased, a cloud of electrons or space charge is formed near the cathode, which results in a retarding field so that no more electrons can reach the anode until the voltage is raised. The characteristics are described by the Richardson–Dushman equation for the temperature limited region:

$$J = \left(\frac{4\pi m_e ek^2}{h^3}\right) T^2 e^{-e\phi/kT} \tag{17.2}$$

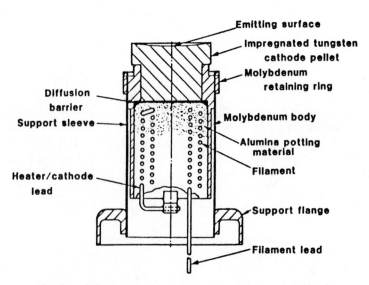

Figure 17.3. Typical type B impregnated dispenser cathode.

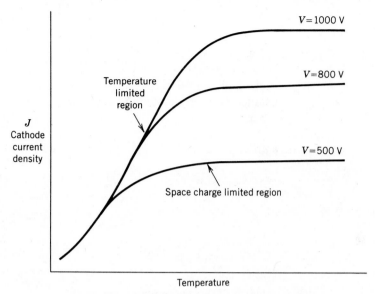

Figure 17.4. Cathode emission characteristics.

where

$$
\begin{aligned}
J &= \text{current density} \\
m_e &= \text{electron mass} \\
e &= \text{electron charge} \\
k &= \text{Boltzmann's constant} \\
h &= \text{Planck's constant} \\
\phi &= \text{work function} \\
T &= \text{temperature (°K)}
\end{aligned}
$$

This can be reduced to

$$ J = A_0 T^2 e^{-e\phi/kT} \text{ A/m}^2 \tag{17.3} $$

where $A_0 = 1.2 \times 10^6$ A/m^2 degrees^2. The space charge limited region where a cathode is normally operated is governed by the Child–Langmuir law, which for the planar diode is

$$ J = \frac{4}{9} \epsilon_0 (2\eta)^{1/2} \frac{V^{3/2}}{x^2} \tag{17.4} $$

where

$$J = \text{current density}$$
$$\epsilon_0 = \text{dielectric constant of free space}$$
$$\eta = \text{electron charge to mass ratio}$$
$$x = \text{distance between anode and cathode}$$
$$V = \text{voltage}$$

If A is the area of the cathode, the total current $I = JA$ can be written as

$$I = K(V)^{3/2} \tag{17.5}$$

where

$$K = \left(\frac{4}{9}\right) \frac{A\epsilon_0}{x^2} (2\eta)^{1/2} \tag{17.6}$$

is the perveance (dependent only on geometry).

The life of a cathode is determined solely by temperature, as shown in Fig. 17.5. Its operating point is therefore a compromise between required current density and life. If the projected current density is too high for good

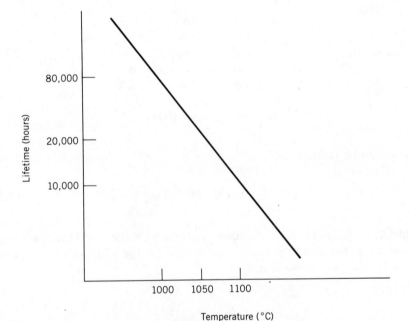

Figure 17.5. Typical cathode life.

life, it is necessary to increase the cathode area and consequently the area convergence ratio of the gun for a given beam diameter and current.

In forward wave crossed-field amplifiers, the impregnated tungsten cathode is used and operated at nonelevated temperatures. Emission is caused by bombardment of the surface by electrons with sufficient energy to release several more, which in turn are driven back into the cathode by the RF and magnetic fields to release more. The buildup to the operating current is extremely rapid, starting with a few field emission electrons when the RF is first applied and reaching full value in approximately one-half of an RF cycle. In higher-power tubes, the back bombardment of the cathode is so intense and local heating is so high that cooling is required to prevent the cathode from overheating.

A variation of this type of operation is usually found in magnetrons and backward wave CFAs. Emission is started by thermionic means. As back bombardment starts, secondary emission takes over and the resulting heating makes it necessary to reduce the application of heating power.

Future developments include the mixed matrix cathode, which is similar to the type M but with the added material embedded in the tungsten instead of on the surface. Another is the field emission cathode, which eliminates the need for heaters with the cathode maintained at ambient temperature.

17.2.2.2 *Beam Modulating Electrodes.*

For pulsed operation it is necessary to turn the electron beam on and off. This is done by the insertion of a control electrode between the anode and cathode. The simplest approach is to use a control grid similar to that used in triodes and tetrodes. Unfortunately, the positive voltage required to turn the beam on results in considerable beam current interception and, except for low-power devices, excessive power dissipation on the grid. The shadow grid was adopted to avoid this problem and is the most widely used control electrode. It consists of two grids—the control grid and a second grid connected to the cathode and placed between it and the control grid. The elements of both grids are exactly aligned so the shadow grid shields the control grid from the beam and deflects electrons away from it, resulting in little or no control grid interception.

A shadow grid's disadvantage is its tendency to degrade the focussing of the beam, producing scalloping at its edges. This results in higher beam interception by the RF circuit wasting power and increasing dissipation. For applications where this cannot be tolerated, other methods of beam control are used. These other methods include the modulating anode, used for high-power applications, and focus electrode modulation.

Figure 17.6 shows the three types of electrodes, and Table 17.3 gives a summary of their characteristics. The terms μ and μ_c refer to the ratio of cathode voltage to the electrode operating and cutoff voltages, respectively. A higher value means lower switching voltages, and in this respect the shadow grid is the most desirable.

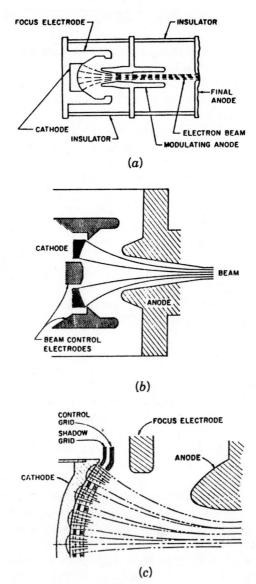

Figure 17.6. *Electron gun control electrodes. (a) Modulating anode gun; (b) control focus electrode gun; (c) shadow gridded gun.*

17.2.3 Focusing

After the electron beam is formed in the gun, it passes through the anode aperture into the interaction space. If left to itself, the beam would not remain a narrow beam for very long but would rapidly diverge due to mutual repulsion of the electrons. A magnetic field is applied in the longitudinal direction parallel to the beam to keep the beam confined. Three methods of doing this are in use. Solenoid focusing is used in high-peak and average power tubes. Periodic permanent magnets (PPM) are used at lower power levels. Permanent magnets are usually used in all crossed-field devices and sometimes in linear beam tubes when the interaction length is short. These beam focusing methods are shown in Figs. 17.7*a*, *b*, and *c*.

17.2.4 Collectors

When a spent electron beam leaves the interaction space, it is necessary to provide an anode electrode to collect it and complete the circuit back to the power supply. In a crossed-beam device, the electrons land on the anode, which is also the interactive circuit. This electrode consists of a grounded copper bucket with high heat dissipation capability. If instead of connecting the collector to the ground it is connected to a negative voltage 30–40% of the cathode voltage, energy can be recovered from the beam as the electrons are slowed by the retarding voltage. Multiple collectors can be used to enhance this effect. The earlier ones are sometimes vanes with the last one the bucket. Voltages applied are in the region of 40–80%. More energy still may be recovered by this method. A diagram of multiple depressed collectors is shown in Fig. 17.8.

17.3 LINEAR BEAM DEVICES

17.3.1 The Interaction Process

There is a theorem in elementary electrostatics on which all microwave amplification depends. It states that "Work is done when two like charges are brought together and may be recovered when they are allowed to separate."

It further states: "When two unlike charges are separated or when two like charges are brought together, the electrical potential energy of the system is increased and the change in electrical potential energy is defined as the work done to effect the separation or approach."

Table 17.3   Typical Characteristics of Control Electrodes

Type	μ	μ_c	Capacity (pF)	Grid Interception (%)	Focusing at Low Voltage
Modulating anode	1–3		50	0	Good
Control focus electrode	2–10	2–10	100	0	Poor
Intercepting grid	50	100	50	15	Fair
Shadow grid	30	300	50	0.1	Fair

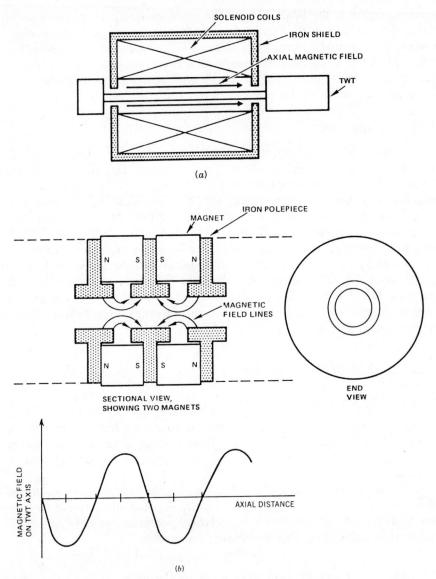

Figure 17.7. Focusing methods (a) Solenoid focusing; (b) periodic permanent magnet (PPM) focusing; (c) permanent magnet focusing.

This is nothing more than the law of conservation of energy. For microwave tubes it can be restated as follows: "When an electron possessing kinetic energy enters an accelerating field it acquires more energy but when it enters a retarding field it gives up energy to the field." *

* Note that this change in energy will generally be in velocity. However, the electron mass will also change if the velocities involved are relativistic.

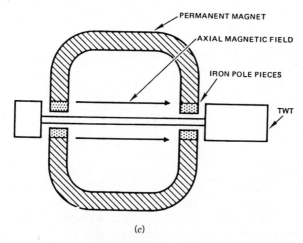

(c)

Figure 17.7. (Continued)

The application of this principle can be explained by reference to Fig. 17.9. An electron beam is made to pass through gaps in two cavities separated by a distance L_0. An RF signal is introduced into cavity 1, which is configured so the alternating RF voltage appears across its gap. This produces a longitude alternating electrical field across the gap. Electrons entering the gap during the positive or negative half-cycle of the RF field will be accelerated or decelerated, respectively, by an amount dependent on the moment during the RF cycle that they entered the gap. Since the gap length is very short, the electrons keep this changed velocity as they leave the gap and travel to the second cavity. This is known as velocity modulation, and because of it the faster electrons will gradually catch up with slower ones that

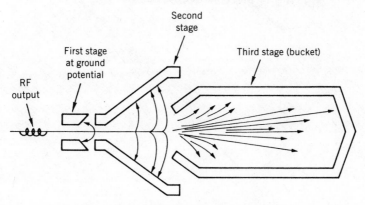

Figure 17.8. *Multiple depressed collector.*

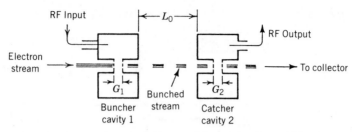

Figure 17.9. Basic interaction of electrons with circuits G_1 and G_2 are gap lengths.

were originally ahead of them and were slowed down during a previous RF cycle. This is illustrated in Fig. 17.10, known as an Applegate diagram, which is a plot of distance versus time for electrons after they leave the first gap. The slope of these lines represents velocity, and they all are different and vary in synchronism with the applied RF voltage. At some distance and time later, the electrons are all bunched together, which is depicted by the heavily shaded areas. If a second cavity is placed at this point, the electron bunches, which represent a strong RF current in the beam, will induce a strong RF field in the cavity. Since the RF energy in the beam is much greater at the second cavity than at the input cavity, the field in the second cavity is also much greater and power is amplified.

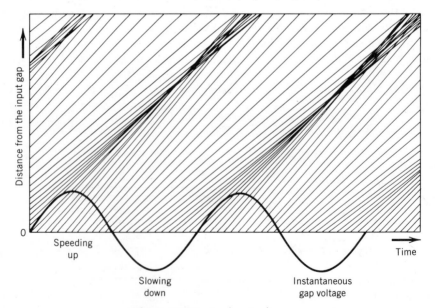

Figure 17.10. Applegate diagram.

17.3.2 Klystrons

The velocity modulation process is used in the klystron. With only two cavities the gain is limited to approximately 20 dB. In the practical klystron, gain is increased by the addition of one or more intermediate resonant cavities between the first two (usually called the buncher and the catcher). The mechanism is the same. The bunched beam produces an amplified RF voltage across the first intermediate cavity, which in turn produces increased bunching of the beam. The process is repeated at the other intermediate cavities until the final or catcher cavity where RF power at a highly amplified level is extracted. Figure 17.11 depicts a four-cavity klystron amplifier. Although more cavities are possible, there is a limit of about 90 dB to the amount of stable gain that can be achieved in one package.

The individual cavities are tunable, usually by mechanical means. For highest gain they are all synchronously tuned, which results in a very narrow bandwidth. To increase bandwidth, stagger tuning is used with some loading on the intermediate cavities to reduce their Q. Gain, however, is lower in this case.

Klystrons have high gain and are capable of extremely high peak (30 MW at S-band) and average (including CW) powers (tens of kilowatts). Their efficiency (RF output power/DC input power) is moderate; it is in the region of 35–45% without depression.

17.3.3 The Traveling Wave Interaction

In the traveling wave interaction process the RF input wave is allowed to travel the length of the electron beam and continuously interact with it in-

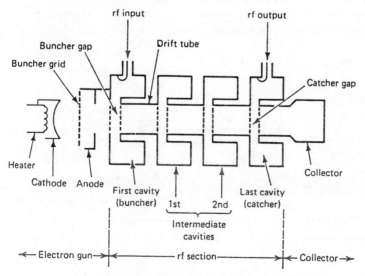

Figure 17.11. Four-cavity klystron.

stead of just at the buncher cavity as in the klystron. This is shown in Fig. 17.12 in which a beam of electrons moves longitudinally through a helical conductor to which an RF signal is applied. Although the RF signal travels around the helix wire at the speed of light, its velocity in the axial direction is slowed down by the helix to

$$v = c \sin \theta \qquad (17.7)$$

where θ is the pitch angle of the helix.

This velocity is approximately equal to that of the electron beam, an essential condition for this type of interaction. The RF field pattern on the helix is shown in Fig. 17.12. Electrons entering the RF field undergo alternate acceleration and deceleration, producing velocity modulation and bunching. The bunches initially tend to be concentrated in the areas marked A in the diagram, but since their velocity is slightly higher than that of the field they soon enter a decelerating region where they give up energy to the field, increasing the voltage on the helix. A little farther on they again enter an accelerating field where the increased field strength increases their bunching so that they overtake the next decelerating region and give up even more energy and increase the field strength still further. This process goes on continuously down the length of the helix resulting in a net transfer of DC energy from the electron beam to the RF field, as illustrated in Fig. 17.13. The process continues until the average velocity of the electron beam is reduced to that of the RF field.

The velocity relationships that must be maintained can be characterized by use of the ω–β diagram shown in Fig. 17.14. This is a plot of angular frequency versus propagation constant for the helix. The slope of the curve at any point is the group velocity of the wave and the velocity of the power

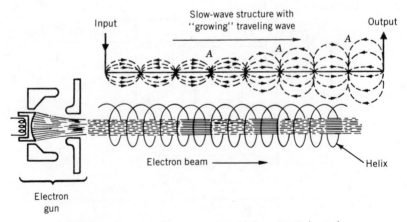

Figure 17.12. *Traveling wave interaction.* θ, *pitch angle.*

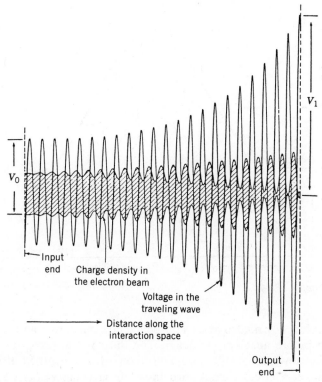

Figure 17.13. *Voltage and charge buildup in a TWT. (From Reich, et al.,* Microwave Theory and Techniques. *D. Van Nostrand, Princeton, N.J., 1953, p. 794. Reprinted by permission of Wadsworth Publishing Co., Belmont, Calif.)*

flow and is given by

$$v_{gp} = \frac{\Delta \omega}{\Delta \beta} \tag{17.8}$$

The quantity ω_0/β_0 at any point is known as the phase velocity. The straight line through the origin represents the electron velocity. Its slope depends on the voltage, and when it intersects the ω–β curve with an equal slope, amplification can occur. For the voltage designated by line V_1 this occurs over a considerable frequency range resulting in very wide bandwidth operation. For voltage V_2, however, the intersection occurs at a point where the group velocity is negative. Amplification can occur at this point, although the power flow is in the reverse direction. This is known as backward wave amplification. The concept can be explained by the following example: Consider a crowded expressway with a closely spaced stream of traffic moving at constant velocity. Suddenly a dog runs in front of one car. The driver brakes slightly to avoid it, then accelerates again. The driver behind him,

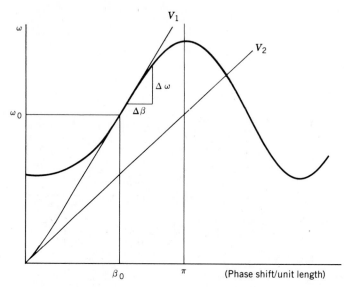

Figure 17.14. ω-β Diagram (forward wave fundamental circuit).

seeing him slow down, also brakes but this time even harder and slows down even more before accelerating again. The third has an even more violent reaction and slows down still more, and so on, back down the traffic stream. We now have in effect a backward traveling growing velocity modulated wave even though all the cars are still traveling forward. In practice, with this system the amplification is so high that the negative modulated velocity becomes equal to the forward velocity of the cars and at about the tenth or twelfth car the traffic stream comes to a screeching halt.

If backward wave operation is desired, it is more usual to use a circuit with a different dispersion curve, as shown in Fig. 17.15 in which the interaction occurs in the fundamental mode rather than a higher one as in the case of the helix.

17.3.4   Traveling Wave Tubes

There are numerous configurations called slow wave structures that can serve the same purpose as the helix to slow down the RF wave to a velocity close to that of the electron beam. The two most commonly used are the helix and the coupled cavity.

17.3.4.1   Helix Tubes. A traveling wave tube (TWT) using a helix is shown in Fig. 17.16. The helix is usually supported by three or four insulating rods holding it away from the body. The helix may be in two or three sections to interrupt the feedback path caused by reflections from the output termination that could cause oscillation. During the interruption in the helix,

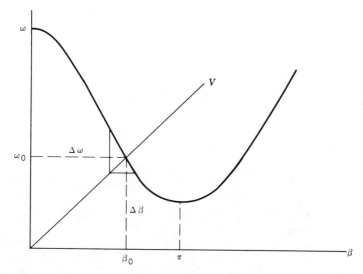

Figure 17.15. ω-β *Diagram (fundamental backward wave circuit).*

the forward power is still carried by the beam and amplification restarts after the sever, as it is known (in Fig. 17.16 this is shown as an attenuator). The total loss in forward gain is about 6 dB per sever.

Helix tubes are capable of octave bandwidths. Peak power is limited to 2–3 kW, at which point a backward wave oscillation occurs. This happens at a frequency where the phase shift between turns is equal to π radians and is voltage dependent. This phenomenon is used in the construction of voltage tuned helix backward wave oscillators.

17.3.4.2 Coupled Cavity TWTs. Coupled cavity TWTs are used when higher power than that available from a helix is required. The circuit consists

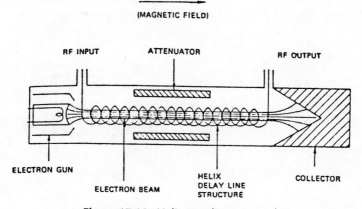

Figure 17.16. *Helix traveling wave tube.*

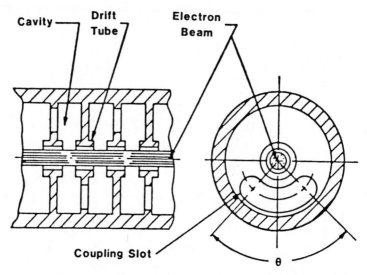

Figure 17.17. *Coupling slot between TWT cavities.*

of a number of cavities similar to those of a klystron but with a coupling slot between them as indicated in Fig. 17.17. The electron beam travels through a hole in the center of each cavity, interaction taking place at each cavity gap. Severs are also used with this circuit (see Fig. 17.18), and velocity relationships are the same as in the helix, with the beam traveling slightly faster than the wave. The structure is usually made of solid copper, has greater thermal capacity than the wire of a helix, and is thus capable of supporting much higher average and CW powers. Its peak power capability is in the region of 2 or 3 MW at *S*-band, much less than that of the klystron. It makes up for this, however, in bandwidth, which may be as high as 10%. Its efficiency is low, as is that of the helix, usually about 18–20%. This can be enhanced to approximately 50% by the use of multiple depressed collectors and a velocity step or taper in the last stages of the circuit.

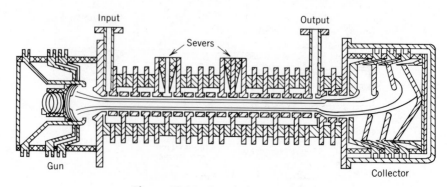

Figure 17.18. *Coupled cavity TWT.*

17.4 CROSSED-FIELD DEVICES

17.4.1 The Crossed-Field Interaction

Figure 17.19 indicates the possible paths an electron might take in a planar diode in the presence of DC magnetic and electric fields. If the magnetic field is zero, the electron will travel straight from cathode to anode in the y-direction. As the magnetic field B is gradually applied, the electron will move in a curved path in the x–y-plane as shown by trajectory a. The stronger the field, the greater the curvature, and ultimately a field strength is reached where the electron just grazes the anode and returns to the cathode as indicated in trajectory b. A further increase in magnetic field results in trajectory c in which the electron never reaches the anode but moves in a cycloidal path with frequency that is proportional to the magnetic field strength B. In a cylindrical configuration, the electrons move in similar trajectories. A particular case in which the electron leaves the cathode at zero velocity is shown in Fig. 17.20. The electron moves closer to or farther from the anode, depending on the relative strengths of the two fields, up to some critical ratio at which the electron just touches the anode and current flows. This point is known as the Hall cutoff condition.

Consider an RF field superimposed on the DC fields. Electrons follow paths similar to that shown in Fig. 17.21. The anode is shown divided into segments, each alternating in polarity at some RF frequency. The electron designated E_1 leaves the cathode at such a time that it enters an accelerating field, receives energy, and is returned to the cathode. E_2, on the other hand, enters a retarding field and gives up energy to the RF field, which it previously obtained from the DC field. If the RF frequency and the electron cycloidal frequency are the same, E_2 will stay in the retarding field for several cycles before reaching the anode. Since there is always a resultant current flow, the total number of electrons following paths similar to E_2 exceeds those following E_1, resulting in a net energy transfer or conversion from DC to RF. This principle is used in magnetrons and crossed-field amplifiers.

17.4.2 Magnetrons

The anode in a magnetron consists of a solid block of metal in which there are at least eight hole and slot structures similar to that shown in Fig. 17.22.

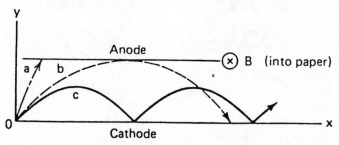

Figure 17.19. *Electron paths in a planar diode.*

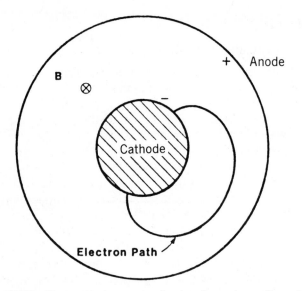

Figure 17.20. *Electron trajectory in circular electrode configuration.*

These are cavities all resonant at the same frequency and capable of supporting an RF field of the type shown in Fig. 17.23. In operation a space–charge cloud of electrons rotates around the cathode. Electrons that follow a path similar to that of E_1 in Fig. 17.21 are returned to the cathode, producing an indentation in the space charge, whereas those following the E_2 path reach the anode and produce an arm or branch projecting from the space–charge. The resulting space–charge pattern, shown in Fig. 17.24, consists of spokelike clouds of electrons rotating around the cathode in synchronism with the RF field and delivering current to the anode. Since the

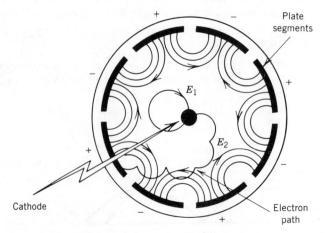

Figure 17.21. *Electron paths in the presence of a microwave field.*

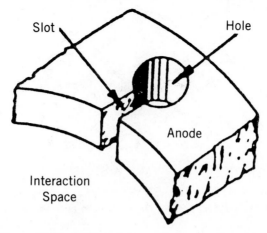

Figure 17.22. Hole and slot structure of a magnetron.

total energy delivered to the RF field is greatly in excess of the losses in the resonant structure, oscillation is sustained and builds up to where RF power can be extracted by a coupling loop in one of the cavities.

Alternating segments of the anode in Fig. 17.23 are 180° or π radians out of phase, giving rise to the term π-mode to describe this mode of operation. It is the preferred mode of operation, but unfortunately there are many others that can easily be excited depending on the number of times the RF field

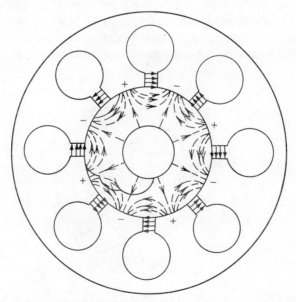

Figure 17.23. RF field in a magnetron.

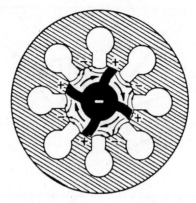

Figure 17.24. *Rotating space charge wheel in an eight-cavity magnetron.*

pattern is repeated in going around the anode once. In the π-mode this is equal to half the number of cavities, and to enhance its excitation straps are usually connected between segments of the anode with the same phase as shown in Fig. 17.25.

17.4.3 Crossed-Field Amplifiers

The crossed-field amplifier (CFA) is a logical extension of the magnetron, which is an oscillator, into an amplifier. It exists in both a linear and circular format. The latter is the most common and is depicted in Fig. 17.26. Instead of the hole and slot structure used in the magnetron, the circuit shown is known as a strapped-vane network. Its function, however, is similar but in this case is interrupted to provide an input and output port. The RF field is initiated by the input and the electron clouds circle around in synchronism

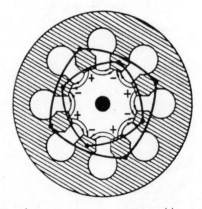

Figure 17.25. *Alternate segments connected by strapping rings.*

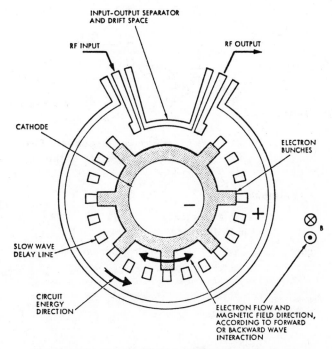

Figure 17.26. Crossed-field amplifier—circular format.

delivering energy to the RF field, which is ultimately extracted at the output port. To minimize feedback and possible oscillation, there is a drift space between the input and output, which allows the electron bunches to smooth out before reentering the active area.

As in the case of the traveling wave tube, operation is possible in both the backward and forward wave modes. However, the voltage to maintain synchronism is frequency dependent.

Cathodes in CFAs are usually "cold" secondary emitters for high-power devices and thermionic or a combination of both for the medium- and low-power devices. Cold cathode tubes have a unique advantage over all other tube types since the tube is keyed on by the RF input pulse, thus eliminating the need for a high-power pulse and allowing the use of a DC power supply. At the end of the pulse it is necessary to turn off the cathode current so the tube will not continue generating high-power noise. This is generally done by applying a low-power positive pulse (which brackets the trailing edge of the RF pulse) to a special cutoff electrode as shown in Figs. 17.27 and 17.28.

CFAs have high-efficiency (greater than 50%), high-power capability (into the multimegawatt region) and wide bandwidth (5–10%). Having high perveance they operate with much lower voltages than equivalent linear beam devices. Because of their limited interaction length, their saturated gain is

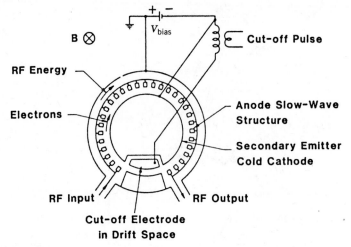

Figure 17.27. *Crossed-field amplifier with cutoff electrode.*

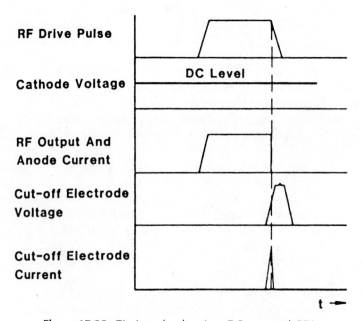

Figure 17.28. *Timing of pulses in a DC-operated CFA.*

low, usually in the region of 13–15 dB. They are generally small, compact, and lightweight, features that make them particularly attractive for use in mobile systems.

PROBLEMS

1. Figure P17.1 shows a simplified schematic of a depressed collector TWT. If the helix intercepts 10% of the beam current and the RF output is 500 W, calculate the overall efficiency (neglecting filament power).

2. It is desired to design a TWT for an RF output of 50 kW peak at a duty cycle of 0.15. The expected undepressed efficiency is 20% and the gun perveance will be no greater than 1×10^{-6} pervs. Calculate
   a. The required cathode voltage
   b. The operating beam current

3. In Problem 2, if the collector is now depressed 35% and the body interception is 8%, calculate
   a. The new efficiency
   b. The total DC input power

4. An electron gun is required to produce a beam of 2 A with a diameter of 0.1 in. If the temperature of the cathode is held at 1000°C, calculate the required cathode area and the convergence ratio. (Assume the cathode work function = 1.9 eV.)

5. At a frequency of 9.4 GHz the phase shift through a TWT is 7 radians/cm. Calculate the voltage that must be applied to the tube to maintain synchronism between the electron beam and the RF wave. *Note*: Electron velocity may be calculated from the kinetic energy relationship

$$Ve = (\tfrac{1}{2})m_e\bar{v}^2$$

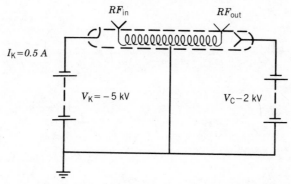

Figure P.17.1. Simplified depressed collector TWT.

where

$$V = \text{applied field in volts}$$

$$e = \text{electron charge}$$

$$m_e = \text{electron mass}$$

$$v = \text{electron velocity.}$$

REFERENCES

1. A. H. Gilmour, *Microwave Tubes,* Artech, Dedham, MA, 1986.
2. G. W. Ewell, *Radar Transmitters,* McGraw-Hill, New York, 1981.
3. J. T. Coleman, *Microwave Devices,* Reston Publishing Co. (Prentice-Hall).
4. Proc. IEEE, Special Issue on Microwave Tubes, March 1973.
5. Reich, Ordung, Krauss, and Skolnik, *Microwave Theory and Techniques,* D. Van Nostrand, Princeton, NJ, 1953.

Part Three
Microwave Measurements

18 Practical Microwave Components and Power Measurements*

18.1 INTRODUCTION

Microwave measurements must be performed using interconnecting hardware with known and repeatable performance. The measuring equipment should be connected to the device under test (DUT) in a way that mimics the system in which the DUT will be used. This requires repeatable connectors that can be removed and reinstalled, and yield the same VSWR and insertion loss. Although some computer-controlled equipment (see Chapter 20) can correct for connector inconsistency, most measurement systems assume that connectors are repeatable. This chapter discusses practical transmission lines, common connector types, typical performance, and practical techniques to minimize connector wear.

The attenuator is a component frequently used for measurements. The performance of several types of attenuators used for coax and waveguide will be presented.

The chapter concludes with a discussion of the measurement of microwave power. Because current and voltage along a transmission line vary with position, power is a more fundamental measurement.

18.2 PRACTICAL TRANSMISSION LINES

Coaxial cables using low loss dielectrics are commercially available for operation to 45 GHz and are being designed for even higher frequencies. Two

---

* If a laboratory accompanies the course, this chapter can follow Chapter 6.

basic types are in use—flexible and semirigid. Flexible coax is desirable in situations in which frequent bending is required such as across the gimbal of a pointing platform. Semirigid coax can be bent a few times and has lower loss and higher isolation (less leakage or radiation) than flexible coax. Semirigid cable should not be bent more than a dozen times with a radius less than 10 times its outer diameter. Semirigid cables should be used wherever possible.

The operating temperature range for cables is from −65°C to +80°C unless constructed for a wider range. Military specifications covering cables include MIL-C-17D, -23806, and -22667.

18.2.1   Flexible Cables

Flexible cables are used to interconnect microwave subsystems temporarily, especially during test periods. The selection of cables is based on many engineering and cost-related factors. The common engineering parameters are characteristic impedance, power-handling capability, attenuation, velocity of propagation, and mechanically related parameters such as diameter, flexibility, weight, operating temperature, and ease of attaching connectors. Other parameters sometimes considered are cutoff frequency, capacitance per unit length, stability of impedance, maximum operating voltage, isolation, self-generated noise when flexed, and resistance to environmental conditions such as sunlight, vapors, underground burial, and flame.

The power-handling capability of cables decreases with increasing frequency because the current flows only in the surface of the conductors. Since the cross-sectional area of the current-carrying portion of the center conductor is less than the outer conductor, the power rating is determined by the power required to maintain the center conductor at 80°C for polyethylene dielectric and 200°C for polytetrafluoroethylene (PTFE) when the ambient temperature is 40°C at sea level. The power rating and attenuation loss for three common 50-ohm cables with solid polyethylene (RG-214, Radio Guide-214), foam polyethylene (FM-8 manufactured by Times Wire and Cable, Wallingford, CT), and solid PTFE (RG-225) are shown in Fig. 18.1. Modern flexible cables use porous Teflon dielectric yielding somewhat lower losses than shown for the RG cables of equal dimensions.

18.2.2   Semirigid Cables

Semirigid cable design uses both solid center (core) and outer conductors. The design yields lower insertion loss and higher isolation than flexible coax. The cable, being uniform because of precision fabrication of the seamless outer tubing and solid core, can be used up to 90% of its lowest higher-order mode frequency (see Chapter 5). If sharp bends (radius less than six times the outer diameter) are used, the maximum operating frequency should be reduced. The common sizes for 50-ohm semirigid cable are 2.16 mm (0.085

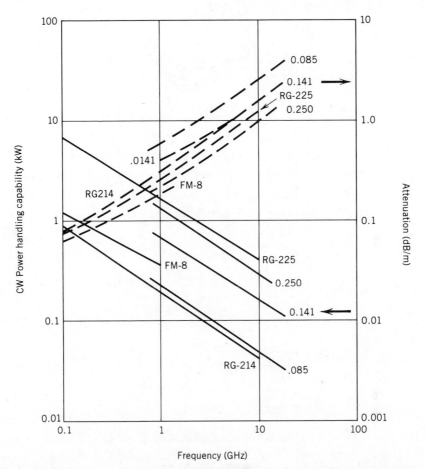

Figure 18.1. *CW power-handling capability and attenuation versus frequency for common coaxial cables. ---, Attenuation; — CW power.*

in.), 3.58 mm (0.141 in.), and 6.35 mm (0.25 in.) outside diameter. The attenuation (decibels per meter) and power handling for these sizes are shown in Fig. 18.1. Note that for a given loss and power-handling capability semirigid cable is usually smaller (and therefore lighter) than standard flexible cables. Connectors are readily available for the three sizes mentioned above.

When working with semirigid cable, a jeweler saw or carborundum wheel (for a Cu-clad steel center conductor) should be used to cut the cable to length. The outer conductor is cut back by scoring the tubing and snapping it off. After the dielectric is trimmed with a razor blade, the cable is ready for connector mounting with minimum use of heat to prevent swelling of the dielectric. When bending the cable, a mandrel with a groove the size of the cable is recommended. If wrinkles appear on the outer conductor, the cable should be remade because the impedance has been affected by changing the spacing of the conductors.

Table 18.1 Rectangular and Double-Ridged Waveguide Specifications

Letter Designation	EIA[a] Waveguide Designation (WRXX)	JAN[b] Waveguide Designation (RG-XX/U)	Inside Dimensions (mm)	Recommended Freq. Range TE$_{10}$ Mode	TE$_{10}$ Cutoff Frequency (GHz)	CW Breakdown Power (kW) at Std. Temp. Pres.	Flange Numbers UG-XXX/U	Rectangular MIL-F-3922/ XX-XXX
K	42	53	10.7 × 4.32	18–26.5	14.0	160–240	594	54-001
Ka	28	96	7.1 × 3.56	26.5–40	21.1	95–145	599 or 381	54-003 or 67B-005
D	22	97	5.7 × 2.81	33–50	26.3	60–90	383	67B-006
U	19	—	4.78 × 2.4	40–60	31.4	50–65	383/MOD	67B-007
V	15	98	3.76 × 1.88	50–75	39.9	30–40	385	67B-008
E	12	99	3.1 × 1.55	60–90	48.4	20–30	387	67B-009
W	10	—	2.54 × 1.27	75–110	59.0	15–20	387 MOD	67B-010
F	8	138	2.03 × 1.02	90–140	73.8	9–14	387 MOD	74-001
D	7	136	1.65 × 0.83	110–170	90.8	6–9	387 MOD	74-002
G	5	135	1.30 × 0.65	140–220	115.7	4–6	387 MOD	74-003
Double-Ridged Waveguide								
—	WRD 180	—	7.32 × 3.40 each ridge 1.83 × 0.98	18–40	14.9	?	—	MIL-F-39000

[a] EIA, Electronic Industries Association; WR, Waveguide Radio.
[b] JAN, Joint Army/Navy; RG, Radio Guide.

18.2.3   Waveguide

Waveguide is widely used above 18 GHz. Several designations for rigid rectangular waveguide are used, as shown in the left three columns of Table 18.1. Each size is specified for a ±22% frequency range between cutoff and the frequency at which the first higher-order mode is propagated.

The peak power-handling capability for waveguide is limited by the break-down electric field of the inside air. For dry air at standard temperature and pressure, the breakdown field is 3×10^6 V/m. If more peak power is needed, the waveguide can be sealed with dielectric windows and pressurized with a dry gas.

The average CW power transmission capability of a waveguide is limited by the losses in the walls. Typically, rectangular waveguide is rated for Cu walls in the TE_{10} mode for a 110°C temperature rise above the 40°C ambient temperature. If cooling fins or liquid cooling is used, several times more power can be carried. The calculation of the wall temperature rise for a given power transferred has been computed in Ref. 1. The power that heats the sidewalls per unit length is

$$\Delta P = \alpha(\text{Np/m})P \qquad (18.1)$$

where α is the attenuation ratio. The average power-handling capability (2) and attenuation (decibels per meter) (3) for Cu rectangular waveguides are shown in Fig. 18.2.

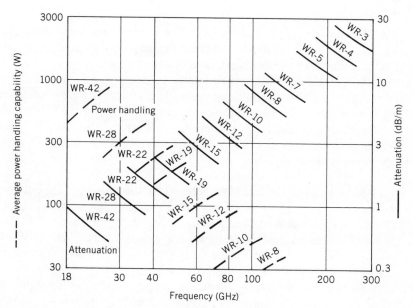

Figure 18.2. *Average power handling capability and attenuation versus frequency for common rectangular waveguides.*

All the rectangular waveguides except WR42 have a $2:1$ inside dimension ratio. As a mnemonic, WR42 is 0.42 in. across the broadwall. Reference 4 is useful for rectangular and circular guides, whereas Ref. 5 discusses over-moded waveguides that exhibit less loss than the fundamental mode. Waveguide bandwidth can be increased to over an octave by placing centrally located ridges in the rectangular guide, as shown in Table 18.1. A source for more detail on ridged waveguides is Ref. 6.

18.3 CONNECTORS AND ADAPTORS

18.3.1 Coaxial Connectors

Coaxial connectors provide the connect/disconnect capability needed for measurement setups and assembly/repair of microwave subsystems. High repeatability during numerous measurement cycles is a requirement that is provided by the 7-mm precision connector. Repeatable cycles are also required for assembly/repair of microwave subsystems. A typical coax connector is the subminiature (SMA) connector. Many other connectors are used such as the older standard type N, the lower-frequency BNC and TNC, and the higher-frequency Omni-Spectra OSSM and Wiltron K. Photographs of some common types are shown in Fig. 18.3. An historical summary of the development and nomenclature of microwave connectors is found in Ref. 7.

18.3.1.1 7 mm Precision Connector. The 7-mm precision connector matches to 50-ohm 7-mm inside diameter airline (air dielectric coax). In addition the connector is sexless (any two connectors can be mated by simply retracting one retaining ring), which makes it ideal for attaching to measurement setups such as network analyzers. The sexless feature is obtained because these connectors use butting center and outer contacts. This avoids bending the center conductor pin when they are joined. Because they are butting, the surfaces should not be rotated with respect to each other while they are being torqued together.

18.3.1.2 Subminiature Coaxial Connector. The subminiature coaxial connector dominates in its use between microwave integrated circuit subassemblies since it is lighter and smaller than most connectors for use to 18 GHz. The male and female mechanical dimensions are controlled by MIL-C-39012. The outer conductor is a butt mating design that can be seated finger tight 0.11–0.34 N-m (1–3 in.-lb) or with a torque wrench 0.68–1.2 N-m (6–10 in.-lb). For field applications having shock and vibration, the nut is torqued to the higher values quoted. The center conductor uses a "wiping" action to assure low-resistance contacts. In addition the female center conductor springs give some relief when the connectors are mated off axis, since the

Figure 18.3. Common microwave connectors. (a) F-mm Precision sexless, (b) type N (large) and SMA (small); (c) BNC.

(c)

Figure 18.3. (Continued)

center pins contact before the outer conductor and locknut align the connector axially. The VSWR of the female connector is generally lower than the male connector, so most manufacturers use this type on their chassis. When subassemblies are connected in series, consideration should be given to alternating chassis connector sexes since this reduces the number of adaptors between chassis.

18.3.2 Waveguide

Waveguides are interconnected with flanges rigidly attached to either rigid or flexible waveguide. If flange face surfaces are protected from scratches (plastic covers should be installed when stored to prevent damage to the flange face), repeatable interconnections are possible with VSWR < 1.1:1 and phase angle standard deviations of order 1°.

Two general types of waveguide flanges are used. The flat contact flange is suitable for most applications, but in situations in which minimum VSWR due to the crack in the waveguide wall is required, a choke flange is used. The choke flange consists of a shorted, one-half wavelength, series branching transmission line as shown in Fig. 18.4. At the design frequency, the crack between the choke and mating flat flange is broken at a low current point, thus minimizing any contact-related problems. The choke is made circular for ease of manufacture.

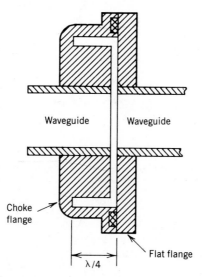

Figure 18.4. *Choke and flat flange designs for rectangular waveguide. Cross-sectional view.*

18.4 ATTENUATORS AND ATTENUATION MEASUREMENTS

18.4.1 Attenuators

Attenuators are linear, passive components inserted in a matched circuit to reduce the incident power. They are specified by their characteristic insertion loss in a matched circuit as shown in Fig. 18.5. For the matched case the characteristic insertion loss equals the attenuation. If the source and/or load are not matched, the attenuator will not perform as expected.

Four classes of attenuators are defined in MIL-A-3933 specification depending on their precision. Class I is a primary standard, whereas class IV attenuators are much less precise and stable. Class II attenuators are used in most laboratories.

Attenuators are used to prevent front-end overloading (generation of nonlinearities) in receivers, spectrum analyzers, and amplifiers. They are also commonly used to mask the effects of mismatches in networks. Attenuators have both average and peak power dissipation ratings that must not be exceeded to retain accuracy.

Both coaxial and waveguide formats are commonly used. The coaxial units are usually fixed in value and fabricated with lumped, compensated resistors in the π or T configuration for frequencies to 3 GHz. Above 3 GHz a distributed resistive card replaces the lumped elements. Reactive (such as PIN diode) attenuators are also used, even though they are not matched to the line. Waveguide attenuators use a resistive card inserted through the

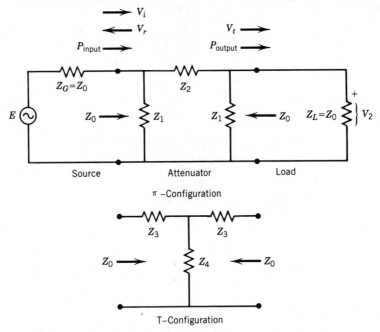

Figure 18.5. *Schematic of an attenuator in a matched circuit.*

broad wall of the waveguide parallel to the electric field of the dominant mode (see Fig. 18.6). Heating in the vane card reduces the energy in the wave with a minimal mismatch due to reflections from the edge of the card. For frequencies at which modes propagate without inducing currents in the card, the attenuator is ineffective, which can cause errors. This condition should be suspected if, for example, a signal source rich in harmonics is being used.

18.4.2 Atttenuation Measurements

Attenuation measurements are usually performed by substituting the network (a two port) for a lossless or nearly lossless network and noting the difference in output power. The power may be measured in several ways, namely, at RF, IF, or audio, depending on the type of detector circuitry used. Mismatch uncertainties can be important if high accuracy is desired.

A common precision attenuation measurement method uses an amplitude-modulated (typically at 1 kHz) signal generator with the barretter detector shown in Fig. 18.7. The barretter microwave detector is operated in its square law region (30-dB dynamic range), and the 1-kHz modulation is attenuated, amplified (constant gain during the measurement period), and detected at the same level with and without the attenuator. The characteristic insertion loss in decibels is one-half the change in the precision audio attenuator. Accuracies of the order of 0.05 dB per 10-dB change are obtainable over a 30-dB range. The characteristic insertion loss is defined when the generator internal impedance and the barretter input impedance are Z_0 (usually 50 ohms). For the general case (see Fig. 18.5), the insertion loss $(\text{IL}_{Z_G \neq Z_L \neq Z_0})$ is

$$
\begin{aligned}
\text{IL}_{Z_G \neq Z_L \neq Z_0} &= 10 \log_{10} \frac{P_{\text{input}}}{P_{\text{output}}} \, 10 \log \left| \frac{V_i^2}{Z_G} \right| \left| \frac{Z_L}{V_t^2} \right| \\
&= 10 \log \left| \frac{1}{S_{12}} \right|
\end{aligned}
\tag{18.2}
$$

where a generalized scattering parameter is used (8). Multiplying numerator and denominator by $1 - |S_{11}|^2$ yields

$$
\text{IL}_{Z_G \neq Z_L \neq Z_0} = -10 \log(1 - |S_{11}|^2) - 10 \log\left(\frac{|S_{12}|^2}{1 - |S_{11}|^2}\right)
\tag{18.3}
$$

$$
= \text{loss due to reflections} + \text{loss due to transmission}
$$

Figure 18.6. Vane attenuator.

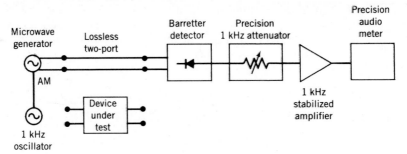

Figure 18.7. Audio substitution attenuation measurement setup.

The characteristic insertion loss ($IL_{Z_G = Z_L = Z_0}$) is the case in which $|S_{11}|^2$ = 0. The techniques for measurement of $|S_{11}|$ are given in Chapter 20.

18.5 MEASUREMENT OF MICROWAVE POWER

Crystal detectors, mixers, thermally sensitive materials, and calorimeters are used to measure microwave power. Crystal detectors (video detectors) and mixers were discussed in Chapter 13. Crystal detector power meters are discussed again briefly at the end of this chapter. Calorimetry is used in standards laboratories for absolute measurements since it is the most fundamental method. In the calorimetric technique the temperature rise of a known liquid volume is measured as CW microwaves are absorbed for a given period of time. The common portable microwave power meters use thermally sensitive materials such as thermocouples, barretters (thin wires), and semiconductor thermistors. The power meter usually consists of the head that transforms microwave energy to the temperature-sensitive element and an indicator device with associated circuitry that introduces a DC or low-frequency AC heating current to rebalance the bridge. It is this "substituted" current that is monitored and indicated on the meter.

18.5.1 Thermocouple

Thermocouple power heads use the Seebeck effect between two metals (typically antimony and bismuth) as their temperature-sensitive element to generate a voltage. The microwave power heats the Sb–Bi junction and varies the Seebeck voltage with respect to the host metal (usually Au). Au–Sb and Au–Bi junctions cancel the effects of the ambient temperature variations. The power indicator is driven by a chopper-stabilized amplifier with proper gain. Typical coaxial power heads work over the range 0.5–100 mW (50-dB dynamic range) and the frequency range from 10 MHz to 18 GHz. Because the head uses a thermal effect, response times of several seconds at low power levels are typical. At higher levels the response time decreases to less than 100 ms. Their maximum CW power level is 150 mW, with more power allowed during pulsed operation, but these meters must be used with caution when pulsed.

Thermocouple mounts are calibrated using an effective efficiency. This efficiency is the ratio of the reference power in the Wheatstone bridge to the microwave power absorbed by the thermocouple element. Errors except the mismatch error between the power head and the transmission line (P_{refl} = $1 - |S_{11}|^2$) are accounted for in this calibration technique.

18.5.2   Bolometer

Barretters and thermistors are temperature-sensitive resistor elements used in bolometer power meters. The resistive sensing element is electrically connected in one of the four legs of a Wheatstone bridge.

18.5.2.1   Barretter. Barretters are thin wires of platinum drawn down to as small as 1.25 μm in diameter or thin metal films mounted on glass or mica substrates. In the power head the microwaves are absorbed by this small volume of positive-temperature coefficient wire (film), and the Wheatstone bridge circuitry drives the indicator. Power ranges from 10 nW to 100 mW are possible with higher linearity than thermistor elements.

18.5.2.2   Thermistor. Thermistor heads use semiconductor films instead of the barretter wire for the sensing element in the Wheatstone bridge. The low frequency or DC resistance of the thermistor is maintained at 100 or 200 ohms by bias from the detector circuitry, but it is matched to 50 ohms for the microwave input power. The thermistor dynamic range is 30 dB with 10-mW maximum average power. Commercial thermistor power heads include a calibration factor that varies with frequency and includes the effects of the different power dissipation distributions between the microwaves and the DC or low-frequency AC power substituted by the bridge circuit. The mismatch loss between the transmission line and the detector is included in the calibration factor.

18.5.3   Crystal

The crystal detector power meter uses the video output from a nonlinear semiconductor diode to drive the indicating meter. As described in Chapter 13, at high power levels the diode rectifier bridge circuit (usually full wave) responds to the peak voltage of each cycle (rectifies) and therefore follows the instantaneous microwave voltage. The rectified voltage appears across a precision load resistor. This crystal detector element covers the 1 nW to 10 mW range (70 dB). CW burnout is about 300 mW, thus making it less susceptible to operator error. Its response time to pulse signals is limited by its video filtering on the rectified output, and the average power of pulsed signals is measurable up to 20-mW peak (square-law region of the diodes). When used with two or more signals, caution is needed above 20-mW power levels because erroneous readings are possible since the detector measures peak-to-peak voltages. For example, if two signals $V_1 \sin \omega_1 t$ and $V_2 \sin \omega_2 t$ are present, a crystal detector will measure $(V_1 + V_2)^2/2Z_0$ W, whereas a total power absorbing (RMS) meter will measure $(V_1^2 + V_2^2)/2Z_0$ W.

18.5.4   Power Measurement Accuracy

Nonideal circuits, sources, and power detectors can yield significant power measurement errors. The general techniques for analyzing complex circuits (flow diagrams) are presented in Chapter 6. In this chapter the basic equations are presented and derived in a heuristic fashion. Three sources of error are considered: mismatch error due to nonideal impedances at microwave frequencies, substitution error due to the difference of the heating effects between the microwave power and the AC or DC substituted power, and instrumentation error.

18.5.4.1 Mismatch Error. The maximum power transfer between a source and a load occurs across a specified interface when the impedances looking into the two are complex conjugates. The frequency-dependent impedances can be represented by $R + jX$ or reflection coefficients Γ referred to as a characteristic impedance Z_0. The proportion of power available from a source with internal reflection coefficient Γ_S into an impedance Z_0 is $1 - \Gamma_S\Gamma_S^*$ at a given frequency. The proportion of power absorbed from a source Z_0 into a load of reflection coefficient Γ_L is $1 - \Gamma_L\Gamma_L^*$. The uncertainty in the power transfer is $(1 - \Gamma_S\Gamma_L)(1 - \Gamma_S^*\Gamma_L^*)$, which accounts for the multiple reflections when mismatched circuits are connected together. The ratio of the power transferred between source and load is

$$\frac{(1 - |\Gamma_S|^2)(1 - |\Gamma_L|^2)}{|1 - \Gamma_S\Gamma_L|^2} \tag{18.4}$$

where $|\Gamma|^2 = \Gamma\Gamma^*$. The phase of the reflection coefficients is frequently not known (only VSWR is known), so the uncertainty term limits are used for calculating the limits

$$\max|1 - \Gamma_S\Gamma_L|^2 = (1 \pm |\Gamma_S||\Gamma_L|)^2 \tag{18.5}$$

of the mismatch error. Figure 18.8 presents a chart of the maximum and minimum mismatch loss versus source and load VSWR. Remember $|\Gamma| = (\text{VSWR} - 1)/(\text{VSWR} + 1)$.

As a simple example, suppose a signal generator output power is to be measured with a power meter. At 18 GHz the source VSWR of the generator

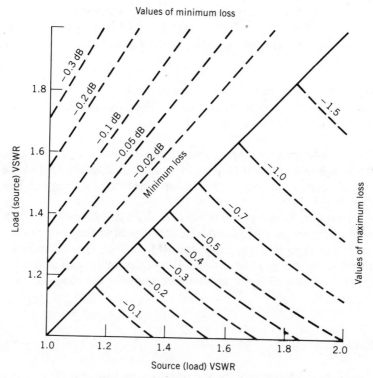

Figure 18.8. Conjugate mismatch loss chart.

is 1.8 and power meter input VSWR is 1.6. Using Fig. 18.8, the minimum mismatch loss is −0.015 dB and the maximum is −1.15 dB, an uncertainty range of 1.135 dB. The effective source VSWR of the signal generator can be reduced significantly with an external power leveling circuit or an isolator.

18.5.4.2 AC–DC Substitution Error.
Bolometers usually substitute DC or low-frequency AC (in the kilohertz range) for compensating the heating due to microwave heating. The spatial distribution of power and resistance in the thermal element is different for microwaves and low frequencies so the thermal paths are different and a slight error is introduced. Two calibration techniques are used leading to the effective efficiency and the calibration factors of the mount.

Effective Efficiency Factor. The effective efficiency factor is defined as the ratio of the power substituted to the dissipated microwave power. Note that mismatch errors are not included. Typical values range from 94 to 99%.

Calibration Factor. The calibration factor includes the detector mismatch and is the ratio of power substituted to the incident microwave power. This factor is more useful in correcting power meter readings. Note that when this factor is used, the $1 - |\Gamma_L|^2$ term in Eq. 18.4 is removed in computing the actual power delivered to the load. $|\Gamma_L|$ is still needed for the uncertainty calculation in Fig. 18.8.

18.5.4.3   Instrumentation Errors. Instrumentation error arises in the indicating circuitry and is a combination of nonlinearity and noise. Typical values are $\pm 1\%$ of full-scale reading.

18.5.4.4   Combined Power Measurement Errors. The peak power measurement error is a combination of the mismatch, substitution, and instrumentation error sources in cases in which the detector reflection coefficient is only considered once.

The uncertainty in the power level is defined as either peak or RMS. It is common practice to consider the peak value of the mismatch error and instrument errors to be three times the RMS value. The calibration factor is assumed to be 1% RMS at a single frequency or several percent over a broadband. The overall mean power uncertainty is the root-sum-square of these values.

PROBLEMS

1. Calculate the frequency for the onset of the lowest higher-order mode for 6.35 mm (0.25 in.) semirigid coax whose dielectric outside diameter is 5.33 mm (0.21 in.) and inside diameter is 1.63 mm (0.0641 in.). The relative dielectric constant is 2.07. The higher-order modes form when the mean circumference of the dielectric equals a wavelength.

2. Prepare a rough plot of semirigid coax attenuation versus diameter at 1 and 10 GHz. Use the data in Fig. 18.1.

3. Prepare a rough plot of semirigid coax CW power-handling capability versus diameter at 1 and 18 GHz using data in Fig. 18.1. Plot the power-handling capability of WR42 waveguide at 18 GHz. Why are they different?

4. From Chapter 5 compute the maximum electric field across WR42 at 22 GHz with 100 W of power to a matched load. Compare to the practical breakdown voltage in air (1.5×10^6 V/m).

5. The power-handling capability of WR10 waveguide at 94 GHz is 40 W. Compute the amount of power dissipated per unit length to heat the walls to 150°C given a 40°C ambient temperature.

6. Compute Z_1, Z_2, Z_3, and Z_4 for the attenuators in Fig. 18.5 when $Z_L = Z_G = Z_0 = 50$ ohms and $P_{\text{output}} = P_{\text{input}}/10$.

7. Assume Z_L has a VSWR = 2:1. Compute the characteristic insertion loss assuming the 10-dB T-configuration attenuator was designed for a 50-ohm system.

8. How much power must Z_1, Z_2, Z_3, and Z_4 dissipate if the input power is 1 W to a 10-dB attenuator in a matched circuit? Which configuration distributes the power more evenly between the resistors?

9. Compute the minimum and maximum power measurement errors due only to a source VSWR = 1.2:1 and a load VSWR = 1.5:1.

10. Compute the range of power errors for the conditions in Problem 9 if the effective efficiency factor is 95%.

11. Repeat Problem 9 if the calibration factor is 95%.

12. If the instrumentation error is 1%, compute the total measurement error for the detector in Problems 9 and 11.

REFERENCES

1. H. E. King, "Rectangular Waveguide Theoretical CW Average Power Rating," *IRE Trans. MTT,* Vol. MTT-9, No. 4, July 1961, p. 349.
2. Hughes Electron Dynamics Division Brochure, 3100 W. Lomita Blvd., Torrance, CA 90509.
3. P. Bhartia and I. J. Bahl, *Millimeter Wave Engineering and Applications,* Wiley, New York, 1984, Chap. 5.
4. N. Marcuvitz, *Waveguide Handbook,* Vol. 10 of MIT Radiation Laboratory Series, McGraw-Hill, New York, 1951.
5. T. N. Anderson, "State of the Waveguide Art," *Microwave J.,* Vol. 25, December 1985, p. 22.
6. T. Chen, "Calculation of the Parameters of Ridge Waveguides," *IRE Trans. MTT,* Vol. MTT-5, January 1957, p. 12, and S. Hopfer, "The Design of Ridged Waveguides," *ibid.,* Vol. MTT-3, October 1955, p. 20.
7. J. H. Bryant, "Coaxial Transmission Lines, Related Two-Conductor Transmission Lines, Connectors and Components: A U.S. Historical Perspective," *IEEE Trans. MTT,* Vol. MTT-32, No. 9, September 1984, p. 970.
8. E. L. Ginzton, *Microwave Measurements,* McGraw-Hill, New York, 1957, Chap. 11.

19 | Time Domain Measurements

19.1 INTRODUCTION

Direct observation of the microwave voltages in a network is not possible with the gain-bandwidth limitations of modern analog oscilloscopes. Waveforms through 12 GHz have been observed using sampling techniques, but distortion on these waveforms is not observable because the higher harmonics are not recorded by the sampler. Thus, if the third harmonic contains meaningful information, operation is limited to signals below 4 GHz. Since sampling scopes are seldom used in microwave development, they are not discussed here, and the reader is referred to Application Notes prepared by Hewlett-Packard Co., Palo Alto, CA and Tektronix, Beaverton, OR.

A frequent application of time domain measurements in networks is the time domain reflectometer and its modern equivalent—the vector network analyzer that produces a Fourier transformed frequency response (described in Chapter 20).

19.2 TIME DOMAIN REFLECTOMETRY

19.2.1 Applications

Time domain reflectometry is used to display the reflected response of systems over a frequency range from DC to several gigahertz and to identify discontinuities in a network so they can be tuned for broadband operation. For example, the time domain reflectometer (TDR) is ideal for measuring network impedance versus distance, electrical length, and location of loss. Typical networks include transmission lines, adaptors from one transmission line mode to another, and loads. These same concepts are used with optical fiber TDRs.

19.2.2   TDR System

The TDR uses a pulsed step generator and a monitor (oscilloscope) in a radar system configuration as shown in Fig. 19.1. The voltage step propagates into the network at the group velocity, and echoes are generated wherever the network impedance varies. These echoes are monitored with a probe located near the output of the generator, which measures the total voltage $E_t(t) = E_i(t) + E_r(t) =$ incident plus reflected voltages.

19.2.3   Typical TDR Measurements

19.2.3.1   Resistive Load Measurements. The shape of the reflected wave from an impedance discontinuity indicates the impedance (resistance and reactance) magnitude along the network. Recall from Chapter 5 that the reflection coefficient is

$$\Gamma(t) = \frac{E_r(t)}{E_i(t)} = \frac{Z_L(x = vt/2) - Z_0}{Z_L(x = vt/2) + Z_0} \tag{19.1}$$

where $\Gamma(t)$ is the reflection coefficient and $Z_L(t)$ and Z_0 are the impedances along the transmission line. The time dependent impedance $Z_L(x = vt/2)$

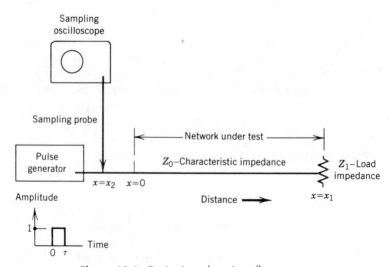

Figure 19.1. *Basic time domain reflectometer.*

indicates the spatial dependence of the line impedance. For the cases in which $Z_L(x_1) = Z_0, 0$, or infinity, $\Gamma(t) = 0, -1$, or $+1$ for all t greater than twice the transit time to the load $Z_L(x_1)$. Consider the transmission line in Fig. 19.2 whose impedance at distance $x = x'$ reduces to $(\frac{1}{3})Z_0$. The time dependent waveform shows a voltage decrease by one-half the magnitude of the beginning at the time t' corresponding to the roundtrip transit time to x'. For a resistive load, the waveform consists of horizontal lines.

19.2.3.2 Reactive Load Measurements. If the load impedance is a combination of resistance and reactance, the TDR waveform will correspond to that expected from a differentiating or integrating network. The waveforms are best derived using the Laplace transform (1) technique. In this case the

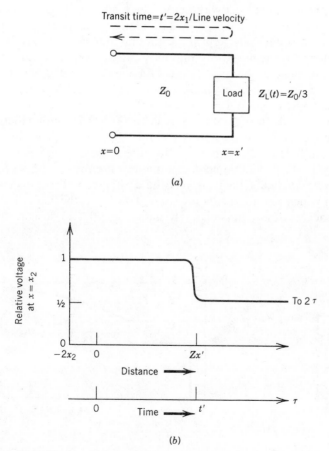

(a)

(b)

Figure 19.2. TDR Display of a mismatched transmission line. (a) Transmission line terminated with $Z_L = \frac{1}{3}Z_0$; (b) amplitude-time response.

$\Gamma(s)$ is found from the $Z_L(s)$ of the load and using Eq. 19.1 above. For example, in Fig. 19.3a the series RC (integrator) presents a load $Z_L(s) = R + (1/sC)$, which is substituted into Eq. 19.1, normalized to Z_0, multiplied by $E_i(s) = E_i/s$, and the inverse transform is taken yielding the time dependent waveform shown in Fig. 19.3b. The Laplace transformations could be avoided by inspection of the problem at $t = 0$, $t = t_1$, and $t =$ infinity, which will give the levels independent of the value of C (assumed to have no charge at $t = 0$) and then assuming that the C charges exponentially through the resistor $R + Z_0$. In performing the measurement, the time constant $(R + Z_0)C$ must be estimated from the monitor display. This is easily accomplished by noting that for a single exponential the voltage reaches one-half its final value in 0.69 times the time constant, i.e., $0.69(R + Z_0)C$ in Fig. 19.3b.

19.2.3.3 Discontinuities at Intermediate Points in a Network.
If a discontinuity exists within the network before the load as shown in Fig. 19.4, the measurement can be divided into two problems since the reflection from the intermediate discontinuity appears at the monitor point before that arising from the load. Typically, a discontinuity arises from each connector in a network. Fortunately the reflection coefficients associated with connectors are small, so the value of the incident voltage exciting the remainder of the network is approximately E_i.

If the intermediate discontinuity is large ($\Gamma > 0.1$), the incident voltage at the second transmission line is $(1 + \Gamma_1)E_i$, where Γ_1 is the reflection coefficient at the first discontinuity. Now the reflected voltage from the load is $\Gamma_L (1 + \Gamma_1)E_i$. If additional discontinuities are present, this process can be iteratively applied. This is usually the case in complex transmission line designs in which several media are used. To repeat, the TDR allows one to observe each discontinuity relatively independently. In design, the discon-

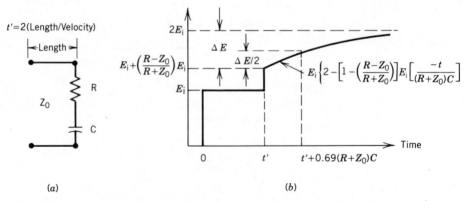

Figure 19.3. TDR Response of a reactively terminated network—Series RC. (a) Z_0 line terminated with series RC; (b) amplitude-time waveform.

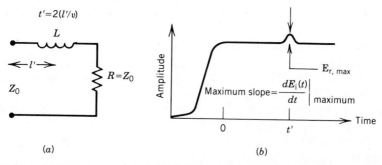

Figure 19.4. *TDR Response of a network with a small series inductance. (a) Z_0 line terminated with series RL; (b) amplitude-time waveform.*

tinuities are removed starting with the one closest to the monitor, the network is remeasured, the second discontinuity is compensated, and the process is repeated.

19.2.3.4 Transmission Loss. The TDR can be used to measure the absolute loss per unit length for a transmission media. Although a more accurate method is presented in Chapter 20, the TDR can compare the loss between "identical" units very rapidly.

If the loss arises due to shunt losses across the transmission media, the E_i has an exponentially decaying dependence (2). Intuitively this is reasonable because the input pulse is seeing a lower impedance as it travels farther into the network. Shunt losses might be associated with poor dielectric in a microstrip or coaxial transmission line. For the series loss case, the TDR display shows an exponentially rising value because the incident voltage travels through more resistance as it moves into the network. The TDR rapidly identifies the type of loss, thus implying the loss mechanism.

19.2.4 Source-Dependent Effects

19.2.4.1 Source Impedance. If the source impedance of the step generator is not equal to Z_0 of the line near the monitor, the reflected voltage wave E_r rereflects and combines with the incident pulse to modify E_i. This condition complicates the TDR display and should be avoided. Analysis of this effect follows the resistive load procedure above. To test for this condition, terminate the transmission line in Z_0 and look for reflections before the network under test is attached. Most commercial TDRs assume $Z_0 = 50$ ohms.

19.2.4.2 Rise Time. The rise time of the source pulse limits the spatial resolution and the ability to measure small reactances with the TDR. If two discontinuities are spaced a distance corresponding to about one-fourth of the pulse rise time, they will not be resolved. This is not an exact relation because the display also depends on the magnitude of the reflection coefficients.

Another ramification of the finite rise time of the source is the inability of the TDR to measure small values of reactance. For the case of small series inductances (see Fig. 19.4a) and parallel capacitances, an estimate of the L and C is possible. Applying Eq. 19.1 for ω corresponding with the highest frequency component in the rise time and the fact that $j\omega L \ll Z_0$, yields $E_r = (L/2Z_0)[j\omega E_i(\omega)]$. Also the E_i may be Fourier analyzed into its sinusoidal components so that

$$E_i(\omega) = E_i(t) \exp(j\omega t) \tag{19.2}$$

Multiplying both sides of Eq. 19.2 by $j\omega$ is equivalent to taking the Fourier components of the derivative of $E_i(t)$, i.e., $j\omega E_i(\omega) = dE_i(t)/dt$. Substituting the derivative relation back into the relation above for E_r in the time domain, yields

$$E_r(t) = \frac{L}{2Z_0}\left(\frac{dE_i(t)}{dt}\right) \tag{19.3}$$

The equation states that E_r is proportional to L and the derivative of E_i. Figure 19.4b shows how the maximum value of $dE_i(t)/dt$ is found on the leading edge of the pulse and maximum values are used to obtain L. Values of $L < 1$ nH are measured with commercial TDRs.

19.3   TIME DOMAIN RESPONSE

19.3.1   Introduction and Applications

This section discusses the time response of a microwave network and/or parts thereof using frequency-dependent measurements to generate the time response. For example, by measuring the input reflection coefficient of a network in the frequency domain and performing the inverse Fourier transform, the time response to a step input (similar to a TDR) can be generated. However, this technique has the added flexibility of moving the reference plane "into" the network, thus allowing observation of the step response independent of the previous and following responses.

Detlefson (3) showed that the inverse Fourier transform of the scattering coefficient $S_{11}(\omega)$ equals $S_{11}(t)$. As a result, the reflected wave

$$b_1(t) = \mathscr{F}^{-1}[S_{11}(\omega) \cdot a_1(\omega)] = S_{11}(t) * a_1(t) \tag{19.4}$$

i.e., the convolution. Note that $a_1(t)$ may be a step response (TDR response) that requires a DC component or an impulse response. If the velocity of propagation within the network is known, the time response is readily converted into distance. If the velocity is not known, the location within the network is usually determined based on the location of known discontinuities in the network. Examples of the frequency and time responses for common circuit elements are shown in Fig. 19.5.

Analog implementations of the Fourier responses were developed in the late 1960s and 1970s (4), but the power of the technique wasn't available until high-speed digital computers interfaced with the microwave network analyzer were developed in the 1970s (5 and 6). Whereas these early implementations were limited because of frequency range, number of data points, and speed of calculation, modern systems can measure from 45 MHz to 26.5 GHz at up to 401 points with modified transforms (Chirp-Z). Since the measurements in the frequency domain do not go to DC, the data are extrapolated

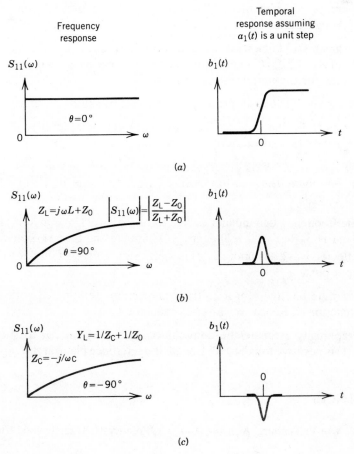

Figure 19.5. *Frequency and temporal responses for lumped circuit elements. (a) Resistive network; (b) series inductive network; (c) shunt capacitive network.*

for $f < 45$ MHz in order to generate the response to a step input (TDR input). When working with transforms having limited frequency inputs, the time domain impulse will show a ringing (see Chapter 20 and Ref. 7). This effect can be reduced by filtering the high-frequency domain data (8) with some increase in the pulse width.

The actual calculations involved here require a computer, and only the simplest problems are tractable via hand calculation. Some of these cases are highlighted in the problems.

PROBLEMS

1. Draw the voltage–time profile (TDR response) for a resistive load ($Z_{load} > Z_0$) with VSWR $= 2$.

2. Compute and plot the TDR response for a parallel RC network terminating a transmission line Z_0. Assume the initial voltage on C is zero.

3. A transmission line with two discontinuities (Γ_1, Γ_2) is terminated with a load $Z_1 = 2Z_0$. Compute the voltage incident on Z_L and the reflected voltage for the initial reflection from Z_L.

4. Derive the TDR response for a shunt loss transmission line.

5. Compute E_i due to a source impedance 10% greater than Z_0 arising from the first reflection of a discontinuity with VSWR $= 2$.

6. If $E_{r,max} = 0.1 \, E_i$ is observable for the network in Fig. 19.4, what is the minimum $dE_i(t)/dt$ to measure $L = 0.1$, 1, and 10 nH in a 50-ohm TDR system.

7. The frequency dependence is $S_{11}(w)$ of a network at a given reference plane is $S_{11}(\omega) = 0.1 \, \underline{/0 \text{ degrees}}$. Compute the time response of the reflected wave assuming $a_1(t)$ is a unit step function (similar to a TDR input pulse).

8. Compute $| S_{11}(\omega) |$ for a Z_0 transmission line terminated with a shunt capacitor C. Repeat for a series inductor L.

9. Graphically estimate the convolution of $S_{11}(\omega) \cdot a_1(\omega)$ to show that $b_i(t)$ is negative for all $t > 0$. Locate the reference plane at the capacitor.

REFERENCES

1. M. E. Van Valkenburg, *Network Analysis,* Prentice-Hall, Englewood Cliffs, NJ, 1955.
2. "Time Domain Reflectometry," Hewlett-Packard Application Note 62, 1964.
3. J. Detlefson, "Frequency Response of Input Impedance Implies the Distribution

of Discontinuities of a Transmission-Line System," *Electronics Letters,* Vol. 6, No. 3, February 5, 1970, p. 67.

4. P. I. Somlo, "The Locating Reflectometer," *IEEE Trans. Micro. Th. Tech.,* Vol. MTT-20, No. 2, February 1972, p. 105.

5. M. Hines and H. E. Stinehelfer, "Time-Domain Oscillographic Microwave Network Analysis Using Frequency-Domain Data," *IEEE Trans. Micro. Th. Tech.,* Vol. MTT-22, No. 3, March 1974, p. 276.

6. B. Ulriksson, "A Time Domain Reflectometer Using a Semiautomatic Network Analyzer and the Fast Fourier Transform," *IEEE Micro. Th. Tech.,* Vol. MTT-29, No. 2, February 1981, p. 172.

7. E. O. Brigham, *The Fast Fourier Transform,* Prentice-Hall, Englewood Cliffs, NJ, 1974.

8. F. J. Harris, "On the Use of Windows for Harmonic Analysis with the Discrete Fourier Transform," *Proc. IEEE,* Vol. 66, No. 1, January 1978, p. 51.

20 | Frequency Domain Measurements

20.1 INTRODUCTION

The three basic frequency-dependent measurement instruments in the design of microwave and millimeter wave circuits and subsystems are the frequency meter, spectrum analyzer, and network analyzer. The operating principles and limitations of modern measurement equipment used in industrial laboratories are presented below. The use of older-style equipment appears in the literature (1–3). Manufacturers provide extensive literature on the measurement techniques and the accuracy of their equipment.

20.2 CW FREQUENCY MEASUREMENT

Frequency is a fundamental parameter needed for characterization of microwave networks. Before the advent of electronic frequency counters, CW frequency measurements were made using absorptive cavities calibrated to better than one part in a thousand (considering all error terms).

A frequency counter operates as a superheterodyne receiver with a multiple of an accurate clock as the LO frequency. With the advent of microprocessor controllers, these units have become very easy to use with power levels in the range −30 to +5 dBm and frequencies to 18 or 26.5 GHz. The basic operation of a popular counter (EIP, Models 575 or 578, San Jose, CA) is described. Referring to Fig. 20.1, the microwave portion of the counter consists of the input tuned YIG bandpass filter that precedes a broadband mixer. The LO for the mixer is derived from a phase-locked comb generator

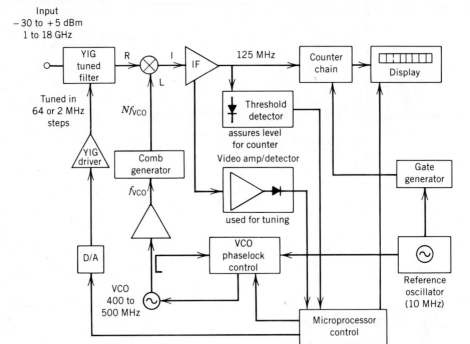

Figure 20.1. *Block diagram of an electronic frequency counter (typical values are shown).*

with comb spacing between 400 and 500 MHz. A 125-MHz IF amplifier increases the signal level to the counter and display. The counter's gate period is derived, as is the phase-locked VCO frequency, from a stable temperature-compensated, crystal-controlled oscillator (10 MHz). The microprocessor controller is the "brain" of the counter since it controls the bandpass center frequency of the YIG filter and other counter circuitry.

When searching, the microprocessor increments the YIG filter in 64-MHz steps through the entire range. All signals are detected and stored via the video amplifier/detector. The largest signal is selected and the YIG filter tuned in 2-MHz steps until the center of the signal f_{YIG} is determined. The value of N^1 is estimated from $N^1 = (f_{YIG} - IF)/(\text{highest VCO frequency}) = (f_{YIG} - 125)/500$ and rounded up to the next highest integer N. The $f_{VCO} = (f_{YIG} - 125)/N$, $f_{VCO} > 400$ MHz. If $f_{VCO} < 400$ MHz, it is recalculated for the other sideband using $f_{VCO} = (f_{YIG} + 125)/N$. The VCO frequency is then shifted in 50-KHz increments to center $f_{IF} = 125$ MHz by using the counter. At this point the counter chain reads and displays (following microprocessor computations) the input frequency. All of the above operations are completed in a fraction of a second. The actual measurement time is 1 ms for 1 kHz resolution [$t_{meas} = (f_{resolution})^{-1}$] and 1 s for 1 Hz resolution. The overall accuracy is ±1 count plus the time base errors that cause the VCO to be off frequency.

The error due to the time base (10-MHz clock) is estimated by assuming that the percentage error of the clock is equal to the negative of the percentage error of the indicated frequency. It is equivalent to speeding up time (clock frequency above 10 MHz) or slowing down time (clock frequency less than 10 MHz). For example, if the clock is 1 Hz too high in frequency, a 10-GHz signal will be read as 10 GHz $(1 - 1/10^7) = 9.999,999,000$ GHz (a 1-KHz error). Most counters allow use of external clock sources, which in turn can be locked to a time reference (such as radio station WWVB at 10 MHz) for precise frequency measurements.

20.3 SPECTRUM ANALYSIS

20.3.1 Overview

Spectrum analyzers were developed during World War II to monitor the development of radar transmitters and receivers. They were used to measure the frequency spreading of the pulsed signals and to show frequency pulling effects (due to rotary joint and other impedance variations) on the pulses and the effects of magnetron (transmitter) misfirings. The spectrum analyzer measures the signal energy (power density in a given bandwidth) as a function of frequency. Today, spectrum analyzers are used to measure modulation, distortion, swept frequency response (also performed by network analyzers), noise, and electromagnetic interference.

20.3.1.1 The Basic Spectrum Analyzer. The spectrum analyzer is a frequency-scanning receiver with a video output proportional to the power (voltage) in the bandwidth of the receiver. The simple spectrum analyzer is implemented with a tunable bandpass filter (YIG filter) followed by a microwave detector, and video and display circuitry as shown in Fig. 20.2. A

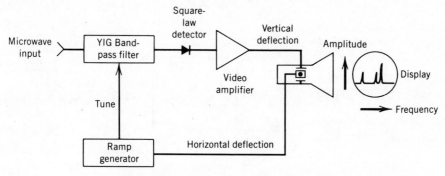

Figure 20.2. *A simple spectrum analyzer.*

commercial unit (Nytek, Model 8011, Los Altos, CA) similar to this design had the advantage of minimal spurious (nonexistent) signals compared to mixer-input units (today this problem is avoided with tracking filters). The sensitivity of the unit was limited because of the use of a diode detector.

The spectrum analyzer displays signals in the amplitude–frequency domain, whereas an oscilloscope views signals in the amplitude–time domain. The three-dimensional diagram in Fig. 20.3 shows the response of these

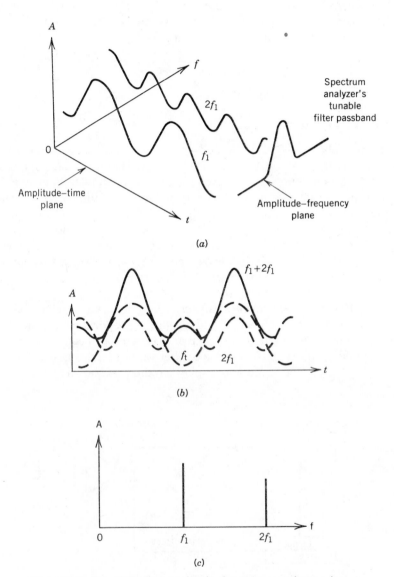

Figure 20.3. *Signals in the amplitude, frequency, and time domains.*

equipments to two signals, $v_1 = A \cos 2\pi f_1 t$ and $v_2 = B \cos(4\pi f_1 t + \pi)$. The oscilloscope observes $v_1 + v_2$ (heavy line) in Fig. 20.3b, while the tunable filter of the spectrum analyzer resolves the two frequencies f_1 and $2f_1$ as shown in Fig. 20.3c. Since the signals are amplitude detected, the phase information of the signals is lost. As a result, the spectrum analyzer's vertical display (proportional to the amplitude of the signal) shows the two signals in Fig. 20.4 as one signal at angular frequency ω (solid curve) while the oscilloscope displays $X + Y$ displaced 45° in phase from each signal X and Y.

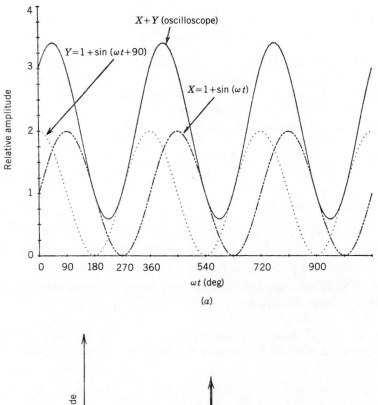

Figure 20.4. Spectrum analyzer display for two signals differing in phase.

20.3.1.2 *Fourier Analysis.* A properly operating spectrum analyzer approximates (less phase information) the Fourier spectrum of the input signal. Therefore, Fourier analysis is required to interpret the output properly. A brief, nonrigorous review of this area of mathematics follows. The application to typical microwave signal structures is given below.

The spectrum analyzer displays the magnitude of the Fourier transform $|F(\omega)|$, where

$$|F(\omega)| = [F_{re}^2(\omega) + F_{im}^2(\omega)]^{1/2} \tag{20.1}$$

and

$$F(\omega) = F_{re}(\omega) + jF_{im}(\omega) \tag{20.2}$$

The phase information $\phi(\omega) = \tan^{-1}[F_{im}(\omega)/F_{re}(\omega)]$ is lost. For example, low modulation AM and FM spectra will look identical because the negative amplitudes (180° out of phase) are folded positive. However, the displays are still very useful in signal characterization.

The Fourier transform is defined as

$$F(\omega) = \int_{-\infty}^{\infty} f(t)e^{-j\omega t} \, dt \tag{20.3}$$

and the Fourier integral is

$$f(t) = \frac{1}{2\pi} \int_{-\infty}^{\infty} F(\omega)e^{j\omega t} \, d\omega \tag{20.4}$$

These two relations couple the time- and frequency-dependent responses for linear microwave circuits.

Carrier Frequency Shift. The spectrum of a function $f(t)$ when amplitude modulating a carrier ω_c is shifted to frequency ω_c. Mathematically,

$$F_c(\omega) = \int_{-\infty}^{\infty} f(t)e^{j\omega_c t}e^{-j\omega t} \, dt = \int_{-\infty}^{\infty} f(t)e^{j(\omega_c - \omega)t} \, dt$$
$$= F_c(\omega - \omega_c) \tag{20.5}$$

Therefore, $F_c(\omega)$ could be found by computing

$$\int_{-\infty}^{\infty} f(t)e^{-j\omega' t} \, dt \tag{20.6}$$

and substituting $\omega - \omega_c$ for each ω'.

Time Product Spectrum. Consider the pulse-modulated sinusoid from a non-chirped radar. This signal is composed of the product of two functions: $h(t) = f(t)g(t)$, where

$$f(t) = V \text{ during the pulse}$$

$$f(t) = 0 \text{ between pulses}$$

$$g(t) = \cos \omega_c t$$

Using the property that the product of two time functions is the convolution of their frequency spectra, yields,

$$H(\omega) \propto F(\omega)*G(\omega) = \frac{1}{2\pi} \int_{-\infty}^{\infty} F(\omega')G(\omega - \omega')d\omega' \qquad (20.7)$$

$F(\omega)$ is the transform of a pulse train as shown in Fig. 20.5a. The resulting

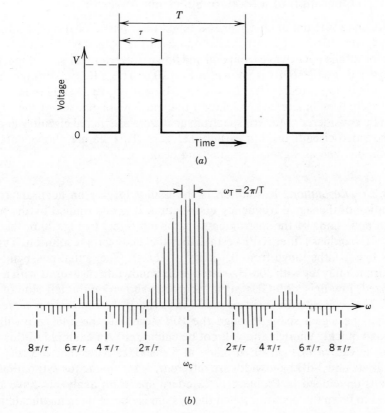

Figure 20.5. *Pulse train envelope and Fourier transform for a modulated carrier.*

$F(\omega)$ for the pulse train is

$$F_{\text{pulse train}}(\omega) = V\left(\frac{\tau}{T}\right)\left(\frac{\sin(\omega\tau/2)}{(\omega\tau/2)}\right) \sum_{n=-\infty}^{\infty} \delta\left[\omega - \left(\frac{2\pi n}{T}\right)\right] \quad (20.8)$$

which is the familiar $\sin X/X$ function for a pulse train. The Fourier transform for $g(t)$ is

$$G(\omega) = \pi[\delta(\omega - \omega_c) + \delta(\omega + \omega_c)] \quad (20.9)$$

The resulting spectrum centered around $+\omega_c$ is shown in Fig. 20.5b. Since the spectrum analyzer measures $|H(\omega)|$ and $G(\omega) = 0$ only where $\omega' = \omega - \omega_c$ and $\omega' = \omega + \omega_c$, the $H(\omega)$ is readily evaluated.

Tables of the Fourier functions are readily available in mathematics books, but a particularly useful subset of these is shown in Table 20.1.

20.3.2 Operation of a Modern Spectrum Analyzer

A modern spectrum analyzer uses a tuned filter ahead of the first mixer to remove the effects of intermodulation distortion (Chapters 13 and 21). An example of the microwave front end and downconverters for such a spectrum analyzer (the HP8569B) is shown in Fig. 20.6. The YIG tuned filter (YTF) operates from 1.7 to 22 GHz. Below 1.8 GHz and above 22 GHz (when using the external millimeter wave mixers) the spurious responses of the mixers must be considered. Modern spectrum analyzers use digital circuitry in place of the video circuits used in earlier analyzers. This feature allows data manipulation such as signal averaging and peak readings.

20.3.2.1 Resolution. Resolution is the ability to give an accurate representation of the signal frequency distribution. It is determined by the bandwidth and shape of the instruments' passband filter. In Fig. 20.6, the 21.4 MHz IF bandpass filter (BPF) establishes the analyzer's resolution. Typical 3-dB bandwidths range from 0.1 kHz to 1 MHz. The actual passband of a spectrum analyzer with 300-Hz stated 3-dB bandwidth measured with a synthesized, low-noise 2.0-GHz signal source is shown on the left side of Fig. 20.7. The scan rate for this trace is 50 ms/kHz. If the scan rate is increased so that the signal scans through the BPF too fast, the analyzer will not respond quickly enough and will not be calibrated. An example of this condition is shown in Fig. 20.7b made at 5 ms/kHz. Note that the response amplitude and 3-dB bandwidth are different. A technique for estimating this error is developed in Problem 17. Modern spectrum analyzers have techniques to warn the operator when the system is operating in an uncalibrated setting.

The high resolution allows the analyzer to resolve two signals such as

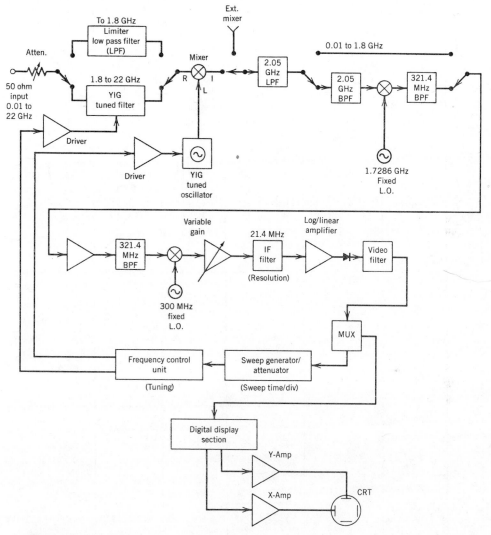

Figure 20.6. *Simplified schematic of a spectrum analyzer (typical values are shown).*

AM sidebands close to the carrier. For example, in Fig. 20.8*a*, a signal 40 dB down (40 dB below the 2-GHz carrier, −40 dBc) will be observable 1 kHz off the carrier. This same condition must be checked for pulsed signals to assure two spectral lines are not significantly exciting the BPF of the analyzer. The operator can check for this error by changing the bandwidth of the BPF and suitably changing the scan speed and resolution to assure that the response remains the same. This effect is shown in Fig. 20.9 where for part A the filter bandwidth is narrow compared to the pulse repetition frequency (PRF), thus allowing only one line to be in the filter. For short

Table 20.1 Useful Fourier Transforms

Time Functions		Frequency Functions (Linear Scales)	(Log Ampl.—Log Freq.)
	$f(t) = \cos \omega_0 t$	$F(\omega) = \pi[\delta(\omega + \omega_0) + \delta(\omega - \omega_0)]$	
	$f(t) = \sum_{-\infty}^{\infty} \delta(t - n\tau)$	$F(\omega) = \dfrac{2\pi}{\tau} \sum_{-\infty}^{\infty} \delta\left(\omega \cdot n \ \dfrac{2\pi}{\tau}\right)$	
	$f(t) = \lim_{\tau \to 0} \begin{cases} \dfrac{1}{\tau}, & |t| < \dfrac{\tau}{2} \\ 0, & |t| > \dfrac{\tau}{2} \end{cases} \begin{pmatrix} \text{delta} \\ \text{function} \end{pmatrix}$ $= \delta(t)$	$F(p) = F(\omega = 1)$	
	$f(t) = \int_{-\infty}^{t} \delta(\lambda)d\lambda = \begin{cases} 0, & t < 0 \\ 1, & t > 0 \end{cases}$ $= u(t) \begin{pmatrix} \text{unit} \\ \text{step} \end{pmatrix}$	$F(p) = \dfrac{1}{p}$	
	$f(t) = \int_{-\infty}^{t} u(\lambda)\,d\lambda = \begin{cases} 0, & t < 0 \\ t, & t > 0 \end{cases}$ $= s(t) \begin{pmatrix} \text{unit} \\ \text{slope} \end{pmatrix}$	$F(p) = \dfrac{1}{p^2}$	
	$f(t) = \begin{cases} 0, & t < 0 \\ 1 - e^{-\alpha t}, & t > 0 \end{cases}$	$F(p) = \dfrac{\alpha}{p(p + \alpha)}$	

Time domain	$f(t)$	$F(\omega)$	Frequency domain
	$f(t) = \begin{cases} \cos \omega_0 t, & \|t\| < \dfrac{\tau}{2} \\ 0, & \|t\| > \dfrac{\tau}{2} \end{cases}$	$F(\omega) = \dfrac{\tau}{2}\left[\dfrac{\sin\left(\dfrac{\omega-\omega_0}{2}\right)\tau}{\left(\dfrac{\omega-\omega_0}{2}\right)\tau} + \dfrac{\sin\left(\dfrac{\omega+\omega_0}{2}\right)\tau}{\left(\dfrac{\omega+\omega_0}{2}\right)\tau}\right]$	
	$f(t) = \dfrac{\sin\left(\pi \dfrac{t}{\tau}\right)}{\pi \dfrac{t}{\tau}}$	$F(\omega) = \begin{cases} \tau, & \|\omega\| < \dfrac{\pi}{\tau} \\ 0, & \|\omega\| > \dfrac{\pi}{\tau} \end{cases}$	
	$f(t) = \begin{cases} 1, & \|t\| < \dfrac{\tau}{2} \\ 0, & \|t\| > \dfrac{\tau}{2} \end{cases}$	$F(\omega) = \tau\,\dfrac{\sin(\omega\tau/2)}{(\omega\tau/2)}$	
	$f(t) = \begin{cases} 1 - \dfrac{\|t\|}{\tau}, & \|t\| < \tau \\ 0, & \|t\| > \tau \end{cases}$	$F(\omega) = \tau\,\dfrac{\sin^2(\omega\tau/2)}{(\omega\tau/2)^2}$	
	$f(t) = \exp\left[-\dfrac{1}{2}\left(\dfrac{t}{\tau}\right)^2\right]$	$F(\omega) = \tau\sqrt{2\pi}\,e^{-(1/2)(\tau\omega)^2}$	

(Courtesy Hewlett-Packard Co.)

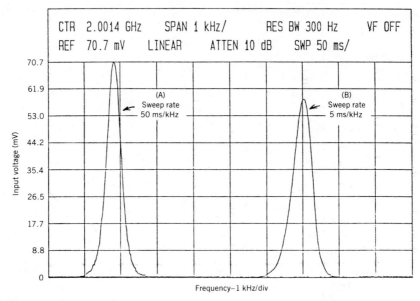

Figure 20.7. *IF response of a spectrum analyzer to two sweep speeds—linear scale.*

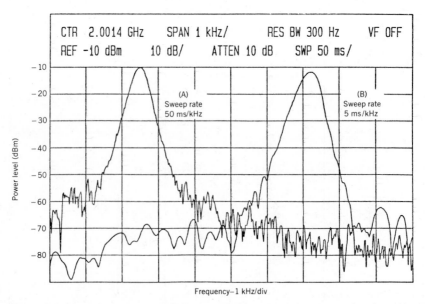

Figure 20.8. *IF response of a spectrum analyzer to two sweep speeds—logarithmic scale.*

RF pulses at a low PRF, it is desirable to have the resolution bandwidth small compared to the inverse pulse width; thus the analyzer responds to several lines simultaneously as shown in (b). For low pulse PRFs, the response will be the IF BPF response excited by varying amplitude pulses, which are averages of the input's spectral lines (c). The number of "pulses" is a function of the PRF and scan time.

If the scanning of the spectrum analyzer is stopped, the unit becomes a superheterodyne receiver, and the time domain of the pulse waveform can be observed from the video output. A trained operator uses the combinations of resolution bandwidth, scanning speed, and scan bandwidth to observe these various patterns (such as Fig. 20.9), thus assuring proper interpretation of the display given knowledge of the signal.

A limitation of a spectrum analyzer to resolve two closely spaced signals arises from incidental FM on the analyzer's internal local oscillators. If one LO has peak-to-peak deviation of Δf, the effective IF passband will be broadened by this amount. In severe cases, this effect can be observed as a flat top to the IF response curve in Fig. 20.7.

20.3.2.2 Sensitivity. The sensitivity of a spectrum analyzer is the minimum input power required to yield a discernible signal out of the video amplifier. This can be defined in terms of the system noise figure (see Chapter 2). The noise factor for the front end of the spectrum analyzer is (refer to Fig. 20.6)

$$F_{SA} = F_{loss} + L_c(N_r + F_{sc} - 1) \tag{20.10}$$

where

F_{loss} = the losses ahead of the mixer (input attenuator plus YIG-tuned filter) expressed as a ratio greater than 1

L_c = the conversion loss for the mixer

N_r = the mixer noise ratio (see Chapter 13)

F_{sc} = the noise factor of the second converter (this value is found by reapplying the above relation up to the 321.4-MHz amplifier in the third converter)

The noise figure is $10 \log_{10} F_{SA}$.

The equivalent input noise power is $P_{in} = F_{SA}kTB$, where B is the effective noise bandwidth of the IF section. Mathematically,

$$B = \int_{-\infty}^{\infty} \frac{[H(f)]^2}{H_0^2} \, df \tag{20.11}$$

where $H(f)$ is the response of the IF section (see Figs. 20.7a and b) and H_0

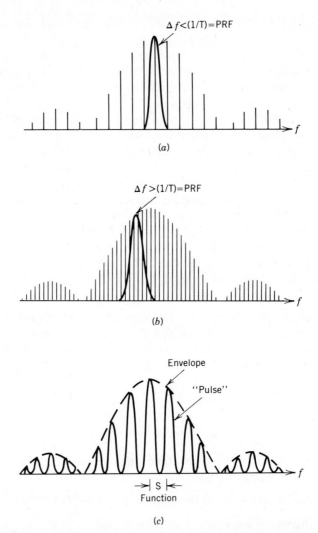

Figure 20.9. *Response of a spectrum analyzer to a pulsed sinusoid depending on the IF bandwidth. (a) one line; (b) several lines; (c) spectral density plot where "S" is a function of scan rate and PRF. (Courtesy Hewlett-Packard Co.)*

is the response amplitude at the center frequency. For a Gaussian IF response, B is approximately equal to the 3-dB bandwidth. At room temperature, $kT = -114$ dBm/MHz bandwidth.

A common technique for measurement of spectrum analyzer sensitivity is to insert a signal X dB above the noise level and then estimate the noise level as X dB below the calibrated input signal power. This technique results in an error because the response of the detector and the logarithmic amplifier to a coherent signal and random noise is not the same. The actual noise level is 2.5 dB greater than is estimated by this technique. The derivation of this correction is found in Problem 8.

20.3.2.3 Dynamic Range.

The dynamic range of a spectrum analyzer can be limited by the range of the logarithmic amplifier that drives the display. Usually, log amplifiers provide over 60 dB of dynamic range, and the mixer limits the dynamic range. When the input signal power approaches the mixer's LO power level, the mixer is no longer linear (see Chapter 13). The second-order and third-order products in the mixer begin to generate signal levels above the average noise level (corrected for impulse noise effects mentioned earlier). The magnitudes of the second(third)-order products are delineated by the second(third)-order intercepts (SOI, TOI), where the input power to the mixer yields equal output for the second(third)-order signals and the up- or down converted signal. These intercepts are described mathematically (see Chapter 13) as

$$\text{Second-order:} \quad (\tfrac{1}{2})v_{sig}^2 = v_{sig}v_{LO} \qquad (20.12)$$

$$\text{Third-order:} \quad (\tfrac{3}{4})a_3v_1v_2^2 = v_1 = v_2 = (\tfrac{3}{4})a_3v_1^2v_2 \qquad (20.13)$$

Typical values are SOI = 30 dBm and TOI = 5 dBm. The maximum dynamic range (DR_{max}) occurs when the higher-order input signal products equal the average noise level yielding

$$DR_{max}(\text{second-order}) = (\tfrac{1}{2}) \mid \text{average noise level (dBm)} - \text{SOI (dBm)} \mid$$
$$(20.14)$$

$$DR_{max}(\text{third-order}) = (\tfrac{2}{3}) \mid \text{average noise level (dBm)} - \text{TOI (dBm)} \mid$$
$$(20.15)$$

To operate the analyzer at the maximum dynamic range, the input attenuator may need to be adjusted to set the second (third)-order signals equal to the average noise level. The values of attenuation required are

Attenuation (second order) dB = input power level (dBm)
$$- \text{mixer conversion loss} - SOI/2 - \text{avg. noise level} \quad (20.16)$$

Attenuation (third order) dB = input power level (dBm)
$$- \text{mixer conversion loss} - 2TOI/3 - \text{avg. noise level}/3 \quad (20.17)$$

20.3.2.4 Other Considerations.

Spectrum analyzer input circuitry can be damaged by the application of DC voltages, AC voltages from low source impedances, and high peak or average powers. It is always advisable to operate analyzers from a 50-ohm signal source. Reference to the operating manual is advised to prevent front end damage.

20.3.3   Typical Measurements with a Spectrum Analyzer

The operator is usually confronted with interpreting the spectrum to determine signal parameters. To assist with this process the AM, FM, and pulse spectra should be readily recognized.

20.3.3.1   AM Modulation. Sinusoidal modulation of a carrier results in two sidebands located in frequency above and below the carrier. The spectrum analyzer is ideally suited for measurement of low levels of AM (4).

20.3.3.2   FM Modulation. The angular frequency of a single tone FM signal is

$$\omega_{FM}(t) = \omega_0 + \Delta\omega \sin \omega_m t \qquad (20.18)$$

where ω_0 is the carrier angular frequency, ω_m is the modulation angular frequency, and $\Delta\omega$ is the peak angular frequency. The phase history of the signal is

$$\phi(t) = \int_{-\infty}^{t} \omega(\tau)d\tau = \omega_0 t + \frac{\Delta\omega}{\omega_m} \cos \omega_m t \qquad (20.19)$$

where $m = \Delta\omega/\omega_m$ is the modulation index. The Fourier transform of this waveform involves Bessel functions and is derived in communication textbooks (5). The easiest measurement of the FM spectrum occurs when $m = 2.4$ because the carrier is suppressed as shown in Fig. 20.10. The frequency separation between peaks equals $\omega_m/2\pi$, and since $m = 2.4$, $\Delta\omega = 2.4\omega_m$. Some caution must be used here because the carrier also nulls for higher values of m ($m = 5.5, 8.6, 11.8$, etc). To confirm the results, the significant sidebands occur over a frequency range of $2(\Delta\omega + \omega_m)/2\pi$.

20.3.3.3   AM + FM Modulation. If one sideband is larger than its opposite sideband, both AM and FM exist at the same frequency and phase. The difference in amplitude (displayed on a linear scale) equals twice the magnitude of the FM sideband since it adds to one AM sideband and subtracts from the other. The AM modulation index is derived from the average height of the two sidebands, as shown in Fig. 20.11*f*.

20.3.3.4   Pulse Microwave Signals. The parameters of the pulsed signal were described earlier (see Fig. 20.9). The resolution bandwidth must be considered, but pulse repetition frequency, pulse width, duty cycle (an analytic factor equal to pulse width times pulse repetition frequency), and peak power are available from the display.

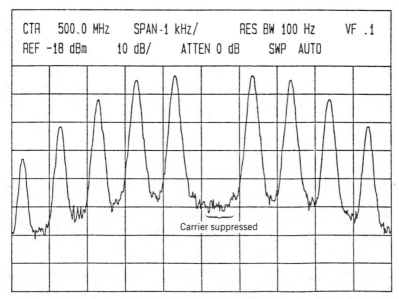

```
CTR    500.0 MHz    SPAN-1 kHz/        RES BW 100 Hz      VF .1
REF -18 dBm        10 dB/      ATTEN 0 dB        SWP   AUTO
```

Carrier suppressed

Figure 20.10. *Spectrum of sinusoidally modulated FM with m = 2.4.*

20.3.3.5 Noise Power Density, Signal-to-Noise Ratio, Noise Figure, AM and FM Noise

Noise Power Density Measurement. The noise power density from a device is measured with the spectrum analyzer taking into consideration both the 2.5 dB correction from the nonlinear detection process and the noise bandwidth (typically noise bandwidth equals 1.2 times the resolution bandwidth, but this should be checked as described above). The video bandwidth should be set to 0.01 or less of the IF bandwidth to provide a good average value. If the noise generated by the device under test does not raise the spectrum analyzer's output by at least 6 dB, it is desirable to add a low-noise amplifier ahead of the spectrum analyzer. This assures that the spectrum analyzer plus amplifier contribute less than 1 dB to the noise density measurement.

Carrier-to-Noise Ratio Measurements. The carrier-to-noise ratio measurement is very similar to the noise power density measurement from random noise. The steps required are

1. Measure the carrier power (C) dBm.
2. Measure the random noise power in bandwidth B_n and apply corrections.

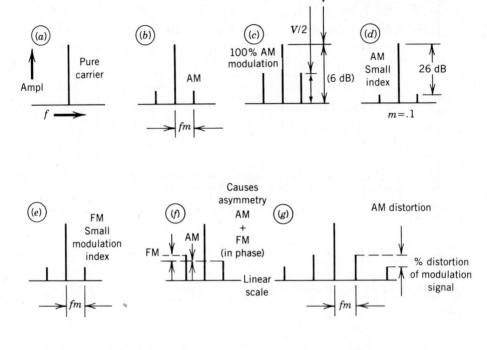

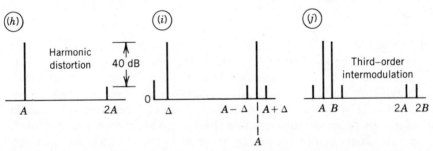

Figure 20.11. Modulation spectra for linear and nonlinear processes. (Courtesy of Hewlett-Packard Co.)

3. Calculate the noise power in the desired bandwidth, B'

$$P_n(B')\text{dBm} = P_n(B_n)\text{dBm} + 10 \log_{10}\left(\frac{B'}{B_n}\right). \qquad (20.20)$$

4. Compute C/P_n or C/N in bandwidth B' in decibels.

Noise Figure Measurements. The noise figure is 10 times the logarithm base

10 of the noise factor (see Chapter 2). Mathematically,

$$NF = 10 \log_{10} F_{DUT} = 10 \log_{10}\left[\left(\frac{S_i}{N_i}\right)\bigg/\left(\frac{S_0}{N_0}\right)\right]$$

$$= 10 \log_{10}\left[\left(\frac{S_i}{kTB_{DUT}}\right)\bigg/\left(\frac{S_i G_{DUT}}{N_0}\right)\right] \quad (20.21)$$

$$= 10 \log_{10}\left[\left(\frac{N_0}{G_{DUT}kTB_{DUT}}\right)\right]$$

If during the measurement the noise bandwidth of the spectrum analyzer (B_n) is less than the DUT's bandwidth (B_{DUT}) the result is

$$NF = \text{noise power output} + \text{detection correction} - \text{gain}$$

$$- \text{noise input in bandwidth } B_n$$

$$= 10 \log_{10}(F_{DUT}G_{DUT}kTB_n) + 2.5 \text{ dB} - 10 \log_{10} G_{DUT}$$

$$- 10 \log_{10} kTB_n$$

(20.22)

In the above relation, the spectrum analyzer measures the noise power output plus correction factor while connected, as shown in Fig. 20.12. The steps in the measurement process are as follows:

1. Measure the total gain of the DUT plus preamplifier by noting the signal level change for the switches in the two positions.
2. Terminate the device's input in its characteristic impedance.
3. Set the spectrum analyzer's input attenuator to 0 dB.
4. Read the display level at a given resolution bandwidth and with sufficient video averaging.
5. Substitute the values into Eq. 20.22.

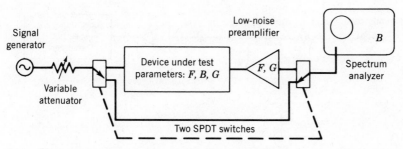

Figure 20.12. *Noise figure test setup using a spectrum analyzer.*

If the gain of the DUT is less than 10 dB, it is necessary to consider the noise contributed by the preamplifier. To do this, make the preamplifier the DUT and compute NF_{preamp} and F_{preamp}. The noise factor of the DUT plus preamplifier is then

$$F_{total} = F_{DUT} + \frac{(F_{preamp} - 1)}{G_{DUT}} \qquad (20.23)$$

If the preamplifier's gain is also low so that the spectrum analyzer's noise is not swamped out, make the preamplifier the DUT and reapply the above procedure.

Residual AM Noise. The noise figure represents the noise contributed by linear devices. For oscillators (output with no input signal), the important noises are the AM and phase modulation (PM) displayed in the sidebands. The noise in the sidebands for high stability oscillators (minimum frequency deviation) is randomly generated assuming no discrete lines appear on the spectrum analyzer (lines represent modulation of the oscillator, not noise).

For the measurement of AM noise only, an envelope detector is used since it ignores PM sidebands. The circuit in Fig. 20.13 is a typical setup in which a crystal detector coupled to DC is used. The calibration source is used to establish the sensitivity of the low-frequency spectrum analyzer. The source should be tunable above the oscillator frequency, i.e., $f_{cal} = f_{osc} + \Delta f$. When P_{cal} (the power introduced by the calibration source) is more than 26 dB below the oscillator power P_{osc}, the signals are equivalent to equal amplitude and coherent phase modulation of a carrier. This case, shown in Fig. 20.11f, yields the same spectrum analyzer display as f_{osc} and $f_{osc} + \Delta f$, since the lower sidebands cancel and the upper sidebands add. As a result, the actual AM noise level is 6 dB below the calibration source output level. The linearity of the detector should be checked using the at-

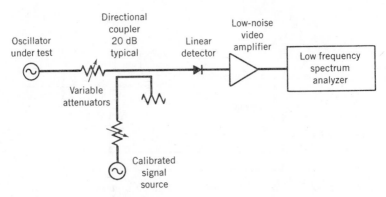

Figure 20.13. *Amplitude modulation noise test setup using a spectrum analyzer.*

tenuator (i.e., a 10-dB increase in attenuation should show a 10-dB change) to assure sufficient oscillator (carrier) power is reaching the diode. The low noise detector amplifier should have a high input impedance and an output impedance to match the spectrum analyzer. Note that this spectrum analyzer is usually not the microwave unit described in Fig. 20.6, but a low-frequency unit since it measures the noise at Δf off f_{osc}(4).

To make the measurement perform the following steps:

1. Introduce a calibration signal -50 to -90 dBc (dB below carrier) at a frequency Δf above f_{osc}.
2. The equivalent AM noise is 6 dB lower than this signal.
3. If measuring a line spectrum (due to AM modulation of the oscillator), measure directly off the analyzer for a given Δf.
4. If measuring random noise, record the noise level ($B_{video} \leq 100\ B_n$) at Δf in bandwidth B_n. Now add the 2.5-dB correction for the nonlinear detection process.

PM Noise Measurement. Microwave oscillator PM spectrum can be measured using the delay line discriminator technique. This circuit typically allows measurements of phase noise -100 dBc/Hz for frequencies exceeding 1 kHz from the carrier. The AM rejection of this discriminator is usually greater than 20 dB (6).

The block diagram of the delay line discriminator phase modulation setup is shown in Fig. 20.14. The phase shifter (variable delay line) is used to adjust the two mixer inputs for 90° phase difference at the carrier frequency, thus nulling out the carrier-generated DC component (see Chapter 13), and yields the maximum output voltage for signals removed from the carrier frequency.

The design of the delay line involves both its length and loss. The length

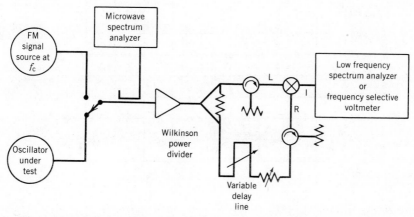

Figure 20.14. *Phase modulation noise test setup using a spectrum analyzer.*

of the line provides 90° phase delay at the carrier frequency. For accurate measurements, the offset frequency Δf range should be

$$(\tfrac{1}{2}\pi) < 2\pi\Delta f t_{\text{delay}} < 0.48 \qquad (20.24)$$

The lower limit assures that the two mixer inputs are decorrelated ($\Delta f > t_{\text{delay}}^{-1}$) while the upper is set by the increased curvature of the discriminator curve away from the carrier (center) frequency. The IF output voltage also depends on the delay line loss (proportional to delay line length) and there-

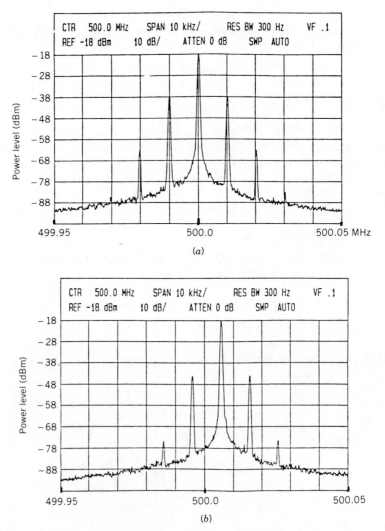

Figure 20.15. Spectrum of FM signals with modulation indices (a) m = 0.2 and (b) m = 0.1.

fore on t_{delay}. The output is maximized when the total delay line loss equals the inverse of the attenuation factor; that is, delay line loss equals 1 Np or 8.69 dB.

The measurement process is very similar to that for AM noise, except for the calibration scheme. The preferred calibration technique involves inserting a FM signal with $\Delta\omega/\omega_m = m$ less than 0.2 (see Section 20.3.3.2). A sinusoidal FM signal with $m = 0.2$ is shown in Fig. 20.15b. Note that the first sidebands are -20 dBc and the second sidebands are -45 dBc. Therefore, by monitoring the power and modulating frequency of this FM source, the spectrum analyzer can be calibrated over the range where Δf is valid. Note again that the FM sideband measurements can be made directly on the spectrum analyzer, but t$^{\backslash}$e random noise measurements require a 2.5-dB correction.

20.4   NETWORK ANALYSIS

20.4.1   Scalar and Vector Network Analyzers

Two types of network analyzers are commonly used in modern laboratories. The scalar network analyzer measures the magnitude of the ratio or the absolute power of the signals in a network. The vector network analyzer measures both the magnitude and phase of the signals in the network. Scalar analyzers are generally cheaper than vector analyzers but are limited in their measurement capability.

Both types of analyzers have two basic parts: the detector receiver and the test unit or signal processor that separates the incident, reflected, and transmitted signals through a device under test. These are the signals associated with the S-parameters and thus analyzers measure S_{ii} and S_{ij}, where i, j indicates the port number.

20.4.2   Scalar Network Analyzer

20.4.2.1 Configuration and Overview. The scalar analyzers measure the magnitude of the incident, reflected, and transmitted waves (voltages or power depending on the detector type) for a DUT versus frequency. Typical parameters derived from these wave magnitudes are gain, attenuation, insertion loss, reflection coefficient, return loss, VSWR, isolation, and directivity.

Depending on the measurement desired, the test unit is configured to provide signals proportional to the magnitude of the desired waves. A typical test unit configuration for return loss and transmission measurements for a two port is shown in Fig. 20.16. The test unit is implemented with a group of directional couplers. The first coupler following the swept frequency generator is used to level the power into the loop to minimize the dynamic range

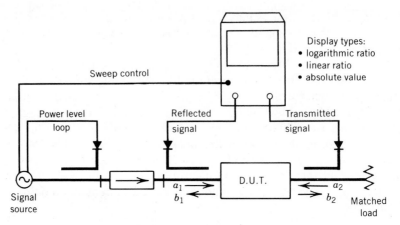

Figure 20.16. *Typical reflection/test unit for a scalar network analyzer.*

errors of the detectors. The isolator provides needed decoupling between the waves reflected from the DUT and the generator (7). The second coupler samples the reflected wave b_1 from the DUT. The final coupler samples the transmitted wave b_2 from the DUT before it is absorbed by the matched load. The final coupler could be replaced by the detector, but generally detectors are not as well matched as loads and so minimum error is obtained with the transmission coupler. If the couplers have equal coupling, the reflected and transmitted wave detectors see nearly the same signal level during the calibration sequence (DUT removed), thus minimizing errors.

The use of these analyzers over a broad swept frequency range is ideal for adjusting circuits while observing the effect on performance. Some of the key relations are (for logarithmic displays)

$$\text{Return loss (dB)} = -20 \log_{10} |\Gamma| \qquad (20.25)$$

$$|\Gamma| = \left| \frac{b_1}{a_1} \right| = \frac{\text{VSWR} - 1}{\text{VSWR} + 1} \qquad (20.26)$$

The return loss (dB) is the difference between the calibration with a short (usually) or open circuit replacing the device and the return loss with the DUT in place. The magnitude of the return loss is limited by the dynamic range of the detectors, the directivity of the couplers, instrumentation errors (nonlinearity and settling time), and noise (influences low-level readings). Typical detector dynamic ranges are delineated by the power ranges below:

Mixers:	Fundamental	-10 to -100 dBm
	Harmonic	-10 to -80 dBm
Diode:	Point contact	$+10$ to -40 dBm
	Schottky	$+10$ to -50 dBm
Thermal:	(Thermocouple, thermistor, barretter)	$+10$ to -30 dBm

The sweep generator power level is adjusted to assure that the detectors operate in their optimum range. The instrumentation errors arising from nonlinearities in the power level calibration of the detectors (frequency errors are eliminated during the calibration procedure) and the settling time of the detector (sweep speeds must allow the detector to measure the signal level at the current frequency) are usually less than ± 0.1 dB (see instrument's operating manual). The thermal noise sets the lower limit for operation of the detector. This noise floor is readily measured by switching off the signal source and observing the signal levels. Operation at least 10 dB above this floor is recommended.

Usually the directivity of the couplers and the mismatches in the calibration procedure limit the scalar analyzer performance. These errors can be estimated using the signal flow graph analysis described in Chapter 6.

20.4.2.2 Analysis of Scalar Network Analyzer Errors. The network in Fig. 6.19 may be considered to be the DUT between the imperfect generator and load in a scalar analyzer with high directivity couplers in its test unit. The measured insertion loss is derived in the problems. If the DUT is removed, to calibrate the system, the insertion loss is simply

$$\frac{b_2}{E_g} = \frac{1}{1 - \Gamma_g \Gamma_1} \qquad (20.27)$$

Taking the ratio of the two insertion loss (IL) measurements yields

$$\frac{IL_{meas}}{IL_{calib}} = \frac{S_{21}(1 - \Gamma_g \Gamma_1)}{1 - \Gamma_g S_{11} - \Gamma_1 S_{22} - \Gamma_g S_{21} \Gamma_1 S_{12} + \Gamma_g S_{11} S_{22} \Gamma_1} \qquad (20.28)$$

$$= S_{21}|_{\Gamma_g = \Gamma_1 = 0}$$

The upper and lower bounds are computed as described earlier. Note that both Γ_g and Γ_1 must be zero to yield the actual insertion loss, S_{21}.

To measure the reflection coefficient b_1/a_1 for the scalar analyzer with high directivity couplers in its test unit,

$$\Gamma_{calib} = \Gamma_1 \qquad (20.29)$$

and Γ_{meas} is given in Eq. 6.49. The resulting

$$\frac{\Gamma_{meas}}{\Gamma_{calib}} = \frac{S_{11}(1 - S_{22}\Gamma_1) + S_{21}S_{12}\Gamma_1}{\Gamma_1(1 - S_{22}\Gamma_1)} \qquad (20.30)$$

which is not defined for $\Gamma_1 = 0$. In an analyzer, this case is masked by the directivity of the couplers designed for insertion loss and reflection coefficient (return loss) measurements.

If logarithmic displays are used for these ratios, the difference between the measurement and calibration displays is the actual measurement. If a display normalizer (a unit that adjusts the calibration signal to zero and adjusts all other values by the same amount) is used, the difference [measured (dB) − calibration (dB)] is displayed directly.

As noted earlier, the worst-case uncertainty in the measurements can be estimated by making the numerator and denominator terms as divergent as possible. However, since these parameters are independent complex parameters, the magnitude of their randomly distributed phase contributions can be estimated using a root sum of squares (RSS) approximation. To perform this analysis, assign a magnitude to each of the error terms; then perform the RSS process. For Eq. 20.28, the worst-case uncertainty in the IL is

$$U_{IL} = \frac{(1 \pm \Gamma_g \Gamma_l)}{1 \pm \Gamma_g S_{11} \pm \Gamma_l S_{22} \pm \Gamma_g S_{21} \Gamma_l S_{12} \mp \Gamma_g S_{11} S_{22} \Gamma_l} \qquad (20.31)$$

The RSS uncertainty is (using the approximation $(1 \pm X)^{-1} = 1 \mp X$, $X \ll 1$)

$$U_{IL} \cong 1 + \sqrt{\Gamma_g^2 \Gamma_l^2 + \Gamma_g^2 S_{11}^2 + \Gamma_l^2 S_{22}^2 + \Gamma_g^2 S_{21}^2 \Gamma_l^2 S_{12}^2 + \Gamma_g^2 S_{11}^2 S_{22}^2 \Gamma_l^2}$$

$$(20.32)$$

And the RSS uncertainty in decibels is

$$U_{IL}(dB) = 20 \log_{10}(U_{IL})$$

The magnitudes of the Γ_g, etc., need not be the worst-case values (quoted by the manufacturer to assure compliance with the specification) if the actual values are known based on previous measurements. Actual values will significantly reduce the magnitude of the uncertainty error. Also, hand selecting the components (couplers, detectors, etc.) for best performance reduces the errors significantly.

The overall measurement uncertainty is now an RSS over all the contributions: test unit errors, instrumentation error, noise; that is,

$$U_{IL} = \sqrt{(\text{test unit errors})^2 + (\text{inst. errors})^2 + (\text{noise})^2} \qquad (20.33)$$

When measuring the reflection coefficient of a well matched DUT (see Fig. 20.16), the leakage in the reflection coupler can introduce a significant error. As shown in the same figure, this error voltage can be significant [since coupler directivities usually range from 25 dB ($\Delta \rho = 0.06$) to 45 dB ($\Delta \rho = 0.006$)] when the DUT S_{11} is comparable. At a given frequency, this coupler directivity effect can be corrected from the measurements by replacing the DUT with an excellent Z_0 load and estimating the directivity (use of good connectors is required here to assure a matched load). For swept measurements, only an estimate of the errors can be made by measuring the directivity and applying the estimated error to the measurements.

20.4.3   Vector Network Analyzer

The vector network analyzer characterizes networks containing both active and passive components in terms of the small signal S-parameters (S-parameters can be related to the impedance and admittance parameters commonly used at lower frequencies). As described earlier, these terminal S-parameters allow the engineer to predict the network's performance when placed in a circuit. The vector analyzer's reflection measurements yield parameters such as S_{11}, S_{22}, VSWR, return loss, and impedance. In the transmission mode the analyzer yields S_{12}, S_{21}, gain, insertion loss, and attenuation. The phase information is required for specialized measurements such as phased-array radar antennas, interferometer direction-finding antennas and receivers, and communication system filters where uniform group delay is required to minimize distortion.

20.4.3.1   Operation of the Vector Network Analyzer. The total vector analyzer consists of the dual- (or more) channel receiver and the reflection/transmission test unit.

The test unit configuration determines the specific measurements to be performed by the dual-channel receiver (ratiometer). The quality of the components (i.e., directivity of the couplers) contributes to the error budget for uncorrected vector analyzer measurements. The schematic of a reflection–transmission test unit for operation in the 2–12.4 GHz region is shown in Fig. 20.17. By suitably switching the microwave relays, the S_{11} and S_{21} parameters are measured. The DUT test ports are reversed to measure S_{22} and S_{12}. The line stretcher allows the reference plane for the measurements to be extended beyond the front connectors of the test unit.

The dual-channel receiver is typically a double-conversion unit with relative amplitude and phase outputs suitable for driving several display options. A simplified block diagram of the popular Hewlett-Packard 8410/11/12/14 unit is shown in Fig. 20.18. The frequency converter downconverts the 110–18,000 MHz signals from the test unit to the high IF (20.278 MHz) by tracking a voltage tuned oscillator (VTO). The VTO drives the harmonic mixers and is offset by the IF from the sweep oscillator's input frequency. A phase lock circuit maintains this offset frequency constant. Therefore, the leads between the frequency converter and the remainder of the dual channel receiver carry at most 20.278 MHz signals.

For the proper input levels to the mixers, the amplitude and phase of the microwave signals in both channels are preserved through the mixers (see Chapter 13). Some care must be taken to ensure that the same harmonic of the local oscillator is used for sequential measurements; otherwise some

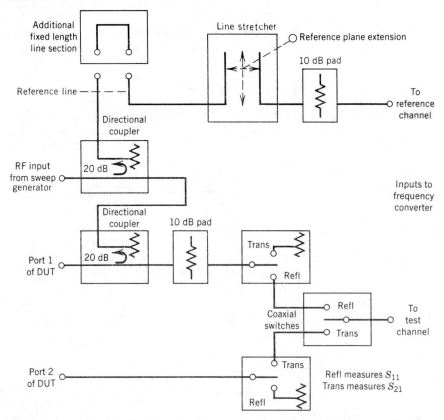

Figure 20.17. *Simplified schematic of vector analyzer test unit. (Courtesy Hewlett-Packard Co.)*

error is introduced due to the variations in the mixer as a function of local oscillator frequency. The 20.278 MHz AGC IF amplifiers keep the reference channel output level constant so the output of the matched test channel amplifier is independent of the input signal level and the ratio of these levels is proportional to the magnitude of the two-channel signal amplitudes. Linear mixers provide 278 kHz outputs for the phase and IF substitution attenuation measurements.

A particularly useful display mode, the polar display, is obtained by suitably detecting the 278 kHz conditioned signals and creating $|\rho|\cos\phi$ and $|\rho|\sin\phi$. This display, when properly adjusted for $|\rho| = 1$ on the circular perimeter, allows use of a Smith chart overlay for ready estimation of the impedance/admittance at the reference plane of the test unit.

The inputs to the frequency converter must be within specified limits to maintain the system accuracy. The power input to the reference channel should be between -16 and -44 dBm. The test channel input may range from -78 to -10 dBm, but it should not be more than 20 dB higher than the reference channel.

20.4.3.2 Typical Measurements

S-Parameter Measurements. The vector analyzer directly measures the ratio and relative phase between two microwave signals. For reflection measurements, the analyzer measures E_r/E_i to yield S_{11} for the port of interest with all other ports terminated with their characteristic impedance. The corresponding input impedance is $Z_{in} = Z_0(1 + S_{11})/(1 - S_{11})$. For transmission measurements, the S_{21} is found from E_t/E_i.

For DUTs with well-matched Z_{in} and high attenuation, the gain of the test channel can be increased. However, eventually the directivity crosstalk between the reference and test channels will begin to introduce errors. To remove these errors manually, a precision sliding load is used to establish the actual location of the analyzer's Z_0. The load actually has a small reflection coefficient, as shown in Fig. 20.18, but the airline in the load is Z_0. For fixed frequency the slide traces a circle, the center of which is the location of the actual $\Gamma = 0$. By moving the vertical and horizontal position controls, this point is located at the center of the polar display. At the test frequency, the analyzer is now calibrated. For other frequencies, the same tedious procedure must be done. By coupling a moderate-sized computer to the analyzer, these effects can be removed analytically (called an automatic network analyzer, ANA).

The power level constraints between the test and reference channels can be exceeded when measuring S_{21} for devices with gain (amplifiers). The input to the mixer can be maintained in a suitable range by adding attenuation in the test channel between the test unit and the frequency converter. The maximum dynamic range of the analyzer is required when measuring the S_{21} for devices with high attenuation (e.g., skirts of a filter). This condition is obtained by adding attenuation in the reference input to the frequency converter.

Q-Measurements. The polar display with a suitable Smith chart overlay can be used to determine the three Q-parameters:

$$Q_0 = \text{unloaded } Q = 2\pi \frac{\text{Energy stored}}{\text{Energy lost per cycle in the resonant circuit}}$$

$$Q_{ext} = \text{external } Q = 2\pi \frac{\text{Energy stored}}{\text{Energy lost per cycle into the input and output circuits}} \qquad (20.34)$$

$$Q_L = \text{loaded } Q = 2\pi \frac{\text{Energy stored}}{\text{Total energy lost per cycle from the resonant circuit}}$$

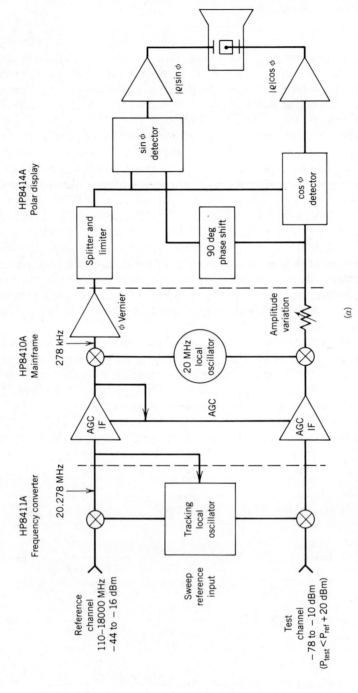

Figure 20.18. Hewlett-Packard microwave vector network analyzer; (a) simplified schematic; (b) polar display.

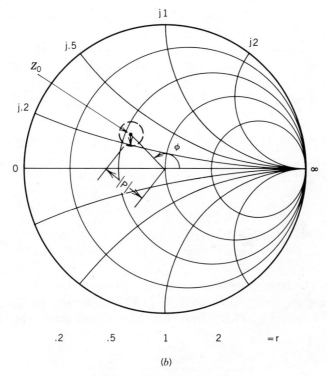

(b)

Figure 20.18. (*Continued*)

These Qs are really a statement of the conservation of energy because they are related to the energy stored and lost in the various internal and external parts of the resonant structure. Q is a "quality factor" for the resonant structure and is related to the bandwidth Δf to resonant frequency f_{res} ratio ($Q = f_{\text{res}}/\Delta f$.) Inverting these Qs and multiplying by the resonant frequencies yields the power relation

$$\frac{f_{\text{res}}}{Q_{\text{L}}} = \frac{\text{Total power lost to the circuit}}{\text{Energy stored}} = \frac{f_{\text{res}}}{Q_{\text{ext}}} + \frac{f_{\text{res}}}{Q_0}$$

$$= \frac{\text{Power lost in input and output circuits}}{\text{Energy stored}}$$

$$+ \frac{\text{Power lost in the resonant circuit}}{\text{Energy stored}} \qquad (20.35)$$

or

$$\frac{1}{Q_{\text{L}}} = \frac{1}{Q_{\text{ext}}} + \frac{1}{Q_0} \qquad (20.36)$$

The relation between the Q-parameters and the impedance parameters on a Smith chart is derivable as follows. Consider the unloaded Q equivalent circuit for a resonant cavity shown in Fig. 20.19a. For this case compute the energy stored, namely, $E_{ind} = (\frac{1}{2})LI^2$, where I is the RMS current through the inductor.

The power lost per cycle in the resistor is

$$P_{disp/cycle} = I^2R \int_0^T \sin^2(\omega_{res}t)dt = \frac{I^2R}{\omega_{res}} \int_0^{2\pi} \sin^2(\omega_{res}t)d(\omega_{res}t)$$

$$= \frac{I^2R}{\omega_{res}} \tag{20.37}$$

so

$$Q_0 = 2\pi \frac{(\frac{1}{2})LI^2\omega_{res}}{\pi I^2R} = \frac{\omega_{res}L}{R} = \frac{X_L}{R}$$

As a result Q_0 for a single resonant cavity is a plot where $R_{cav} = \pm X_{cav}$. The resulting arc for Q_0 is shown in Fig. 20.20. The arcs for Q_{ext} and Q_L are also shown in Fig. 20.20. Thus, for the resonant cavity shown in Fig.

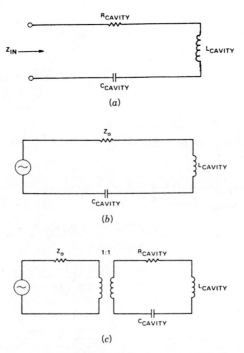

Figure 20.19. Cavity equivalent circuits used to determine Q_0, Q_{ext}, and Q_L. (Courtesy Hewlett-Packard Co.).

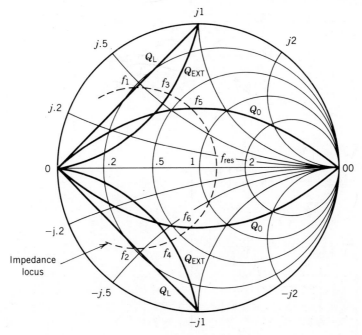

Figure 20.20. Loci of Q_0, Q_{ext}, and Q_L on a Smith chart.

20.20, the f_{res}, f_1, f_2, f_3, f_4, f_5, and f_6 values result in the values

$$Q_0 = \frac{f_{res}}{(f_6 - f_5)} \tag{20.38}$$

$$Q_{ext} = \frac{f_{res}}{(f_4 - f_3)} \tag{20.39}$$

$$Q_L = \frac{f_{res}}{(f_2 - f_1)} \tag{20.40}$$

In 1984, Hewlett-Packard Company announced a new generation of automated vector network analyzer (Series 8510) with capabilities from 45 MHz to 26.5 GHz. A typical test unit and the downconverter block diagram are shown in Fig. 20.21. The unit uses four channels and dual downconversion to 20 MHz and 100 kHz with a 50–300-MHz voltage controlled oscillator. The significant amount of data processing in the 8510 provides an opportunity to compute the time domain response (inverse Fourier transform) of a network (similar to the time domain reflectometer display described in Chapter 19). This option is particularly helpful for locating a discontinuity and, by gating out the effect in the time domain, performing the Fourier transform to estimate the frequency response without the discontinuity.

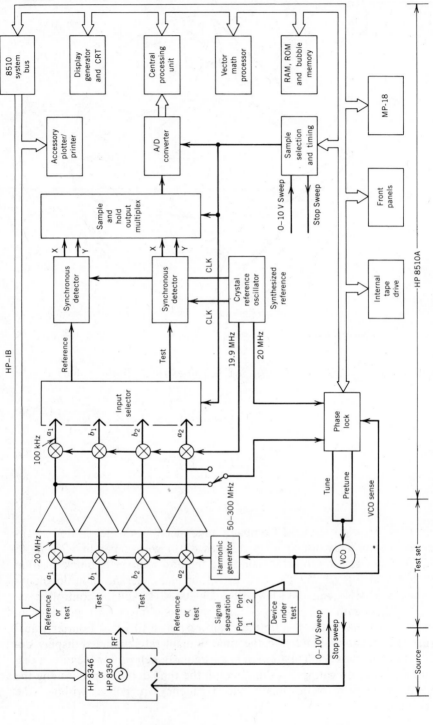

Figure 20.21. Simplified block diagram of the HP8510 four-channel vector network analyzer.

20.4.3.3   *Errors in Vector Analyzer.*

Like scalar analyzers, vector test units include errors that need to be measured and eliminated from measurements to provide high accuracy. The technique used in Hewlett-Packard equipment for doing this is to generate a *fictitious* four-port error network (two ports for a reflection measurement) immediately adjacent to the DUT as shown in Fig. 20.22. The remainder of the test unit is assumed to have perfect performance. The combined networks have 12 error terms for the forward measurements ($S_{11M} = b_0/a_0$, $S_{21M} = b_3/a_0$) and 12 terms for the reverse measurements ($S_{22M} = b_3'/a_3'$, $S_{12M} = b_0'/a_3'$). Six of the 12 error terms for each direction dominate, so a total of 12 terms is considered for both directions leading to the "twelve-term error model." The six dominant error terms for each direction are the port match errors e_{11} and e_{22}, the directivity error e_{00}, the leakage error e_{30}, and the transmission ($e_{10}e_{32}$) and reflection ($e_{10}e_{01}$) frequency response terms. The flow graphs for the forward and reverse direction of measurement are given in Fig. 20.22*b* and *c*. The terms $e_{12} = e_{21} = e_{02} = e_{20} = e_{31} = 0$ and the corresponding reverse terms have been assumed to be zero. The solution to these flow graphs relates the actual parameters to the measured parameters (8) yielding

$$\frac{b_0}{a_0} = S_{11M} = e_{00} + (e_{10}e_{01})\frac{S_{11A} - e_{22}\text{Det}[S_A]}{1 - e_{11}S_{11A} - e_{22}S_{22A} + e_{11}e_{22}\text{Det}[S_A]}$$

(20.41)

where

$$\text{Det}[S_A] = \text{determinant of } S_A = S_{11A}S_{22A} - S_{12A}S_{21A}$$

(20.42)

$$\frac{b_3}{a_0} = S_{21M} = e_{30} + (e_{10}e_{32})\frac{S_{21A}}{1 - e_{11}S_{11A} - e_{22}S_{22A} + e_{11}e_{22}\text{Det}[S_A]}$$

(20.43)

These solutions are obtained by matrix multiplication or use of the signal flow graph techniques.

To obtain the actual network values from the measurement, the effects of the error matrices are removed. This is accomplished by computing the actual network node voltages a_i, b_i in terms of the S_{ij}, e_{ij}, and a_0. The results are

$$a_1 = e_{10}a_0\frac{(S_{11M} - e_{00})}{e_{10}e_{01}} \qquad a_2 = e_{10}a_0\frac{(S_{21M} - e_{30})}{e_{10}e_{32}}$$

(20.44a,b)

$$b_1 = e_{10}a_0\left(1 - e_{11}\frac{S_{11M} - e_{00}}{e_{10}e_{01}}\right) \qquad b_2 = e_{10}a_0e_{22}\frac{S_{21M} - e_{30}}{e_{10}e_{32}}$$

(20.45a,b)

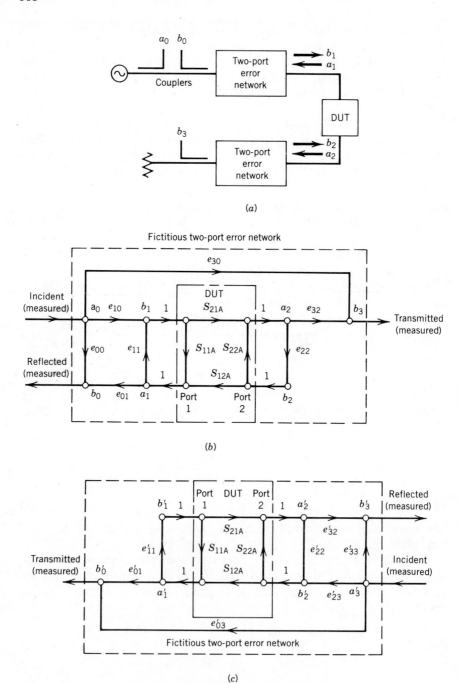

Figure 20.22. *Twelve-term error models for a three-port measurement system. (a) DUT and error networks in a test setup; (b) forward error network; (c) reverse error network.*

where

$$S_{11A} = \frac{a_1}{b_1} \quad S_{21A} = \frac{a_2}{b_1} \quad S_{12A} = \frac{a_1}{b_2} \quad S_{22A} = \frac{a_2}{b_2} \quad (20.46a\text{-}d)$$

By symmetry and analysis, similar results are obtained for the reverse network. These equations involving complex numbers are readily manipulated via computer, except that the values of e_{ij} are needed. These error values are derived from measurements made during calibration.

A standard calibration procedure is described below, but there are others that may be better for specific applications. To calibrate up to port 1, place a short ($S_{11A} = -1$), a calibrated open ($S_{11A} \approx 1$, not equal to one because the fringing capacitance must be taken into account), and a load $S_{11A} = 0$. These three measurements allow evaluation of the terms e_{00}, $(e_{10}e_{01})$ and e_{11}. The leakage e_{30} is measured by placing two calibrated loads on ports 1 and 2 so all $S_{ijA} = 0$. The resulting $S_{21M} = e_{30}$ by direct substitution into Eqs. 20.44b and 20.45b. Now, knowing e_{00}, $(e_{10}e_{01})$, e_{11}, and e_{30}, the two ports are connected together ($S_{12A} = S_{21A} = 1$ and $S_{11A} = S_{22A} = 0$). The reverse transmission term ($e_{10}e_{32}$) is derived from S_{21M}. The procedure is repeated for the six terms related to the reverse error network (Fig. 20.22). The accuracy of the technique is dependent on the values of the load, open, and short, so calibrated standards should be used here whose connectors are clean and yield repeatable results.

PROBLEMS

1. Show via trigonometric identities that when two sine waves, differing only in phase, are added, they yield a sine wave. For what amplitudes do two sine waves differing in phase yield a sine wave of constant amplitude?

2. Compute $F(\omega)$ for a sinusoidal amplitude modulation

$$v(t) = V_0(1 + m \cos \omega_m t) \cos \omega_c t.$$

The result is

$$F_v(\omega) = 2\pi V_0\{\tfrac{1}{2}[\delta(\omega - \omega_c) + \delta(\omega + \omega_c)$$
$$+ (m/4)[\delta(\omega - \omega_c - \omega_m) + \delta(\omega - \omega_c + \omega_m)$$
$$+ \delta(\omega + \omega_c + \omega_m) + \delta(\omega + \omega_c - \omega_m)]\}$$

Now compute $|F_v(\omega)|$ as would be observed on a spectrum analyzer.

3. Derive the Fourier transform for the time function $A \sin \omega_c t$. *Hints:*

$$\sin X = \frac{(e^{jX} - e^{-jX})}{2j}$$

and do integration in polar coordinates over 0 to 2π. Also,

$$\int_{-\infty}^{\infty} \delta(x)dx = 1.$$

4. Evaluate $H(\omega)$ and $|H(\omega)|$ for $f(t)g(t)$ in the section on carrier frequency shift. How does the spectrum analyzer display differ from Fig. 20.5*b*?

5. Estimate the 3-dB bandwidth of the left-hand IF filter response in Fig. 20.7.

6. Estimate how close the left-hand curve in Fig. 20.7 can be approximated by a Gaussian response,

$$H(\omega) = H_0 \exp[-(\omega - \omega_0)^2/2\Delta^2]$$

What is the relation between Δ and the 3-dB bandwidth?

7. Compute the noise bandwidth for a Gaussian filter with

$$H(\omega) = \exp\left(\frac{-\omega^2}{2\Delta^2}\right),$$

where ω is the angular frequency off the filter center frequency and Δ is related to the 3-dB bandwidth as derived in Problem 6.

8. The envelope of narrowband white noise is described by the Rayleigh distribution

$$p(A) = \frac{A}{\langle A^2 \rangle} \exp -\left(\frac{A^2}{2\langle A^2 \rangle}\right)$$

where A is the amplitude of the noise and $\langle A^2 \rangle$ is the mean square amplitude (5, Chapter 7). For 1 W of power into a 1 ohm resistor $\langle A^2 \rangle = 1$, so the normalized Rayleigh distribution is

$$p(A) = A \exp\left(\frac{-A^2}{2}\right)$$

The detection process in the spectrum analyzer is peak detection (provided by the envelope detector), log amplifying, and video averaging (typically the video filter is 100 times narrower than the IF bandwidth). Stated mathematically (integration done numerically)

$$\overline{V}_{noise} = \int_0^\infty p(A)\ (\ln A)dA = 0.058 \text{ Np}$$

To compute the effect of a signal through this detection process, find the A envelope (peak) value of a signal of unit RMS value. Since 1 Np = 8.68 dB, show that the difference is 2.5 dB.

9. Derive Eqs. 20.14 and 20.15. *Hint:* Graph the power variation of the desired signal ($f_{sig} - f_{LO}$) versus input signal power and graph the second(third)-order signal powers versus input signal power on logarithmic plots. The dynamic range for a given input power is the difference between these power levels. Note the slopes and intercepts for each of these three signals. The maximum dynamic range occurs when the second(third)-order signal powers equal the average noise level.

10. Compute the maximum dynamic ranges for a spectrum analyzer with average noise level = -100 dBm, SOI = 30 dBm, and TOI = 5 dBm.

11. Derive the attenuation relations given in Eqs. 20.16 and 20.17.

12. For a -20-dBm signal level and the values given earlier, compute the input attenuation to yield the analyzer's maximum dynamic range assuming a 7-dB mixer conversion loss.

13. Derive the ratio sideband level/carrier level for the AM signal $v(t) = (1 + m \cos \omega_m t) \cos \omega_c t$. If $m = 0.1$ how many decibels below the carrier ω_c is the upper sideband $\omega_c + \omega_m$? Compare to the dynamic range in Problem 12.

14. Compute and plot the allowable range of Δf for the delay line discriminator versus carrier frequency to 20 GHz.

15. Based on the calibration technique described above, the PM noise power can be expressed as a noise-to-carrier ratio or as a modulation index. Derive the relation (assuming narrowband FM)

$$(N/C)^{FM} = (\tfrac{1}{2})\left(\frac{\Delta f_{rms}}{f_m}\right)^2$$

in one sideband where Δf_{rms} is the RMS noise deviation that would be measured in a bandwidth B_n from a discriminator output driven by the oscillator and f_m is the modulation frequency equal to Δf.

16. If N/C = -100 dBc, compute Δf_{rms} for $f_m = \Delta f$ over the allowable range for a 10-GHz oscillator.

17. This problem develops the response to be expected when a signal is scanned through the passband of a spectrum analyzer at both high and low speeds. Assume the frequency scanning through the filter is

$$f(t) = (\text{sweep rate})t = \left(\frac{\text{Sweep width}}{\text{Sweep time}}\right)t = \left[\frac{F_s}{T_s}\right]t$$

using the notation of Ref. 4. Therefore, the input is

$$s(t) = \exp(j\omega t) = \exp\left(j\pi \frac{F_s}{T_s} t^2\right)$$

passing through a filter whose (assumed Gaussian) response is

$$H(\omega) = \exp\left(\frac{-\omega^2}{2\Delta^2}\right)$$

centered at the filter frequency (see Problem 6). The response of the filter is the product of the spectra of the filter with the spectrum of the signals

$$Y(\omega) = H(\omega)S(\omega)$$

(This is the inverse of the product of two time functions described earlier.) The time response of the filter is the convolution

$$y(t) = \int_{-\infty}^{\infty} h(\tau)s(t - \tau)d\tau$$

For this problem, the responses are found in Table 20.1. Compute $S(\omega)$ from Table 20.1, form the product $Y(\omega)$ and use the table to reconvert back to $y(t)$. The answer is

$$y(t) = \frac{1}{\left(1 - j\dfrac{2\pi F_s}{T_s\Delta^2}\right)^{(1/2)}} \cdot \exp\left(-\frac{1 - j(\Delta^2 T_s/2\pi F_s)}{1 + (T_s\Delta^2/2\pi F_s)^2} \frac{\Delta^2 t^2}{2}\right)$$

Again compute the envelope of $y(t)$ to yield $|y(t)|$. The result is

$$|y(t)| = \frac{1}{[1 + (2\pi F_s/T_s\Delta^2)^2]^{(1/4)}} \cdot \exp\left(-\frac{\Delta^2 t^2/2}{1 + (T_s\Delta^2/2\pi F_s)^2}\right)$$

For low sweep rates $| T_s/2\pi F_s | \gg (1/\Delta^2)$,

$$| y(t) | = \exp\left[-(\tfrac{1}{2})\left(\frac{2\pi F_s}{\Delta T_s}\right)^2 \right] t^2$$

which is the response of the IF filter on a time scale, as shown in Fig. 20.7a. The equivalent results for a logarithmic detector are shown in Fig. 20.8a (50 ms/kHz) and Fig. 20.8b (5 ms/kHz). Note the significant broadening of the 3-dB bandwidth for the faster scan speed.

18. Show, using the nontouching loop rule, that

$$\frac{b_2}{E_g} = \frac{S_{21}}{1 - \Gamma_g S_{11} - S_{22}\Gamma_1 - \Gamma_g S_{21}\Gamma_1 S_{12} + \Gamma_g S_{11}S_{22}\Gamma_1}$$

for the network in Fig. 20.19.

19. Redo Problem 18 using the graphical technique.

20. Plot the magnitude of the effective reflection coefficient and the effective VSWR versus directivity over the range 25–45 dB. For VSWR values lower than this curve, the measurement error is over 100% (dominated by the test unit coupler). Assume a 10-dB coupler.

21. For the test unit shown in Fig. 20.17, compute the sweep generator output power (RF input) range to provide the -44 to -16 dBm reference channel levels. Assume lossless transmission lines and ideal couplers.

22. Compute the attenuation to be added to the test channel and to the amplifier output to provide maximum dynamic range for an amplifier with 30-dB gain. The reference channel input level is -30 dBm, and the noise floor of the analyzer is -78 dBm (see Fig. 20.17).

23. Compute the attenuation to be added to the reference channel to provide maximum dynamic range for the measurement of filter skirt attenuation in Fig. 20.17.

24. Derive the $S_{22M} = b_3'/a_3'$ for the reverse network shown in Fig. 20.22. Use either the signal flow graph or matrix manipulation technique. The solution is of the same form as Eqs. 20.41 and 20.42.

25. Use Eqs. 20.44, 20.45, and 20.46 to derive the relation between S_{11M} and S_{11A} given the calibration load conditions for a short, open, and matched load. Assume the capacitance of the open is zero.

26. Compute the error in the appropriate terms (e_{ij}, see Problem 25) if the "calibration" load has a reflection coefficient magnitude of 0.05 instead

of 0. What is the range of impedance values of the load, return loss, and VSWR if $Z_0 = 50$ ohms?

REFERENCES

1. E. L. Ginzton, *Microwave Measurements,* McGraw-Hill, New York, 1957.
2. S. F. Adam, *Microwave Theory and Applications,* Prentice-Hall, Englewood Cliffs, NJ, 1969, Chaps. 3, 5, and 6.
3. T. S. Laverghetta, *Microwave Measurements and Techniques,* Artech House, Dedham MA, 1976, Chaps. 5 and 6.
4. Hewlett-Packard Application Note No. 63, "Spectrum Analysis," August 1968, and No. 150-4, "Spectrum Analysis-Noise Measurement," April 1974.
5. M. Schwartz, *Information Transmission, Modulation, and Noise,* McGraw-Hill, New York, 1970, Chap. 4.
6. C. Schiebold, "Theory and Design of the Delay Line Discriminator for Phase Noise Measurements," *Microwave Journal,* Vol. 26, No. 12, December 1983, p. 103.
7. J. B. Knorr, "Scattering Analysis of a Millimeter-Wave Scalar Network Analyzer," *IEEE Micro. Th. Tech.,* Vol. MTT-32, No. 2, February 1984, p. 183.
8. D. Rytting, "Appendix to an Analysis of Vectro Measurement Accuracy Enhancement Techniques," Hewlett-Packard Co., Santa Rosa, CA.

Part Four
Microwave Subsystems

21 | Microwave Receiver Front-End Design

21.1 CONFIGURATIONS OF MICROWAVE RECEIVERS

The configuration of microwave receivers is highly dependent on their application. For example, communication receivers may only require narrow bandwidth but must be tunable, whereas radar receivers have fixed frequency and moderate bandwidth (approximately the inverse of the pulse width). Electronic warfare (EW) receivers use very wide bandwidths to accommodate an uncertain emitter frequency. The characteristics of these three types of receivers are given in Table 21.1. Although these entries are true in general, numerous variations occur and must be considered on an individual basis. This is the job of system designers and engineers.

The most common receiver configuration for all these applications is the superheterodyne receiver (the tuned radio frequency receiver is sometimes used for fixed frequency applications). In the superheterodyne (1) the received signal may or may not be amplified and is shifted to an intermediate frequency by mixing in a nonlinear element (mixer). This IF is applied to the detection circuitry where the information is extracted. In some applications (particularly at millimeter waves) a single downconversion described above results in poor receiver performance due to simultaneous reception of the image frequency. To alleviate this effect, a double conversion superheterodyne receiver design may be necessary, as shown in Fig. 21.1(b).

Only the microwave (front-end) portion of the receiver is analyzed in this chapter. The low IF and detection circuitry is adequately described elsewhere for the three common applications—communications (2 and 3), radar, and EW (4–6).

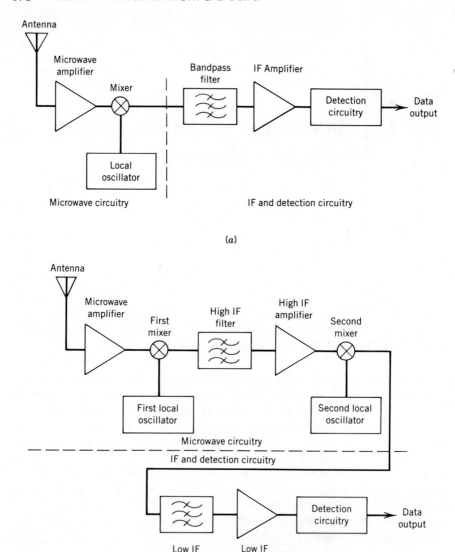

Figure 21.1. *Single and dual conversion superheterodyne receiver configurations. (a) Single conversion receiver; (b) dual conversion receiver.*

Two other receiver configurations are sometimes used. The tuned radio frequency (TRF) shown in Fig. 4.4 uses RF preselection to limit the bandwidth and tune the frequency. This receiver type is used for radar-warning applications because of its wide instantaneous bandwidth. The homodyne receiver uses some of the incoming signal to generate the LO signal along

Table 21.1   General characteristics of microwave receivers

Characteristic	Communications	Radar	Electronic Warfare
Typical applications	Point-to-point, Earth station (see Chapter 2)	Air surveillance, weather (see Chapter 3)	Radar warning, jammer receiver (see Chapter 4)
Frequency range	0.0005–60 GHz	1–100 GHz	1–100 GHz
Bandwidth of microwave circuitry	Several 100 MHz	Several megahertz	Several gegahertz
Sensitivity	Moderate	High	High
Dynamic range			
Total	Moderate	High	High
Spur-free	IMD products cause interference to other channels	Filtered for low IMD	Limits performance in dense signal environments

with the oscillator tuned to the IF, as shown in Fig. 21.2. The second mixer in the homodyne upconverts the fixed frequency oscillator operating at the IF to form the "pseudo-LO" for the downconverting mixer. The filter between the mixers removes the image frequency (RF–IF) generated by the upconverter. This receiver configuration could be used as a wideband EW receiver because the LO is generated by the incoming RF, but it has not

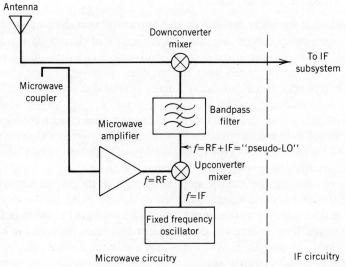

Figure 21.2. *Simplified diagram of a homodyne receiver configuration.*

been popular because simultaneous signals will not be received and the sensitivity and dynamic range are generally less than a superheterodyne receiver.

21.2 FIGURES OF MERIT FOR MICROWAVE RECEIVERS

The design of microwave receivers includes the analysis of the following parameters as a function of variables such as configuration, gain, filter location, and filter rejection/insertion loss:

Noise figure (noise factor or temperature)
Linear dynamic range
Spur-free dynamic range.

Noise factor (defined in Chapter 2) is the quotient of the signal-to-noise ratio (S/N) at the output of the receiver to the S/N at the input. The logarithm of the noise factor is the noise figure. In this chapter consider the noise figure at the output to be at the output end of the microwave portion in Fig. 21.1. Therefore, to compute parameters such as the bit error rate for a receiver, the effects of the IF and detection processing must be considered. In most receivers, the detection effects are small compared to the noise effects contributed by the microwave front end (in a typical EW receiver $\approx$ 2 dB is contributed by the detection circuitry).

21.2.1 Linear Dynamic Range

Linear dynamic range is defined as the allowable variation in input signal level between the receiver threshold (S/N = 1) and the signal level where the receiver is no longer linear (signal level where the gain of the receiver front end is reduced by 1 dB from the small or moderate signal level gain). This linear dynamic range is usually limited by the dynamic range of a component. By properly adjusting the gain distribution through the stages, the largest dynamic range can be achieved. In communication receivers, automatic gain control circuitry is used to expand the dynamic range by varying the gain of various stages. This intentional gain variation is not considered in the gain reduction for the 1-dB compression.

Linear dynamic range can be defined using the tangential sensitivity or the minimum discernable signal as the lower limit and the 1-dB compression point or overload point as the upper limit. The overload point is defined as the large signal level where the demodulator becomes poor or the bit error rate exceeds a given value. When quoting a dynamic range, the definition being used should be clearly stated.

21.2.2 Spur-Free Dynamic Range

Spur-free dynamic range is the input signal level range from the threshold level to the level where third- and second-order nonlinear effects equal the threshold level. Typically this dynamic range is less than the linear dynamic range. The common technique for parameterizing these effects is the intercept level to be described below. The spur-free dynamic range is readily increased by selectively filtering the spurs along the receiver chain; however, many times this is not possible because the spurs occur within the bandwidth of the signal being received.

21.2.3 Definition of Intercept Points

Distortion in amplifiers and undesirable mixing products in mixers are the chief source of spurs that limit the spur-free dynamic range of a receiver. A measure of the strength of these spurious signals is the intercept level. Intercept levels are fictitious extensions of the input-output power relationship for amplifiers and mixers. Devices are not operated at their intercept point level.

Using a development similar to the mixer development in Chapter 13, the output voltage of an amplifier having distortion is

$$V_{\text{out}} = a_0 + a_1 v_{\text{in}} + a_2 v_{\text{in}}^2 + a_3 v_{\text{in}}^3 + \cdots = \sum_{i=0}^{\infty} a_i v_{\text{in}}^i \qquad (21.1)$$

which is a Taylor series expansion. For practical applications, only the first few terms are important and are given special names (to be discussed below). The output voltage is dependent on the number of input signals, i.e., a single signal (sometimes referred to as a tone) with $v_{\text{in}} = V_1 \sin \omega_1 t$ or two signals (tones) with $v_{\text{in}} = V_1 \sin \omega_1 t + V_2 \sin \omega_2 t$.

The terms in Eq. 21.1 are now explained for the single and double signal cases.

21.2.3.1 a_0 and a_1 Terms. The a_0 term is a bias term. The term $a_1 v_{\text{in}}$ is the dominant linear relation in amplifiers (the output is a replica of the input).

21.2.3.2 a_2 Term. The $a_2 v_{\text{in}}^2$ term is the first distortion factor. For a single input signal it yields

$$V_{\text{out},2} = a_2(V_1 \sin \omega_1 t)^2 = \frac{a_2 V_1^2}{2} (1 - \cos 2\omega_1 t) \qquad (21.2)$$

The constant term in Eq. 21.2 offsets the bias in proportion to the input signal squared (this is the term used in a square law detector). The second term is the second harmonic distortion arising from curvature of the input–output characteristic (used to advantage in frequency doublers). For a two-signal input,

$$V_{out,2} = a_2[(V_1 \sin \omega_1 t)^2 + (V_2 \sin \omega_2 t)^2 + 2V_1 V_2 \sin \omega_1 t \cdot \sin \omega_2 t] \quad (21.3)$$

where the first two terms are second harmonic distortions, called inter-modulation distortion (IMD) when they arise from two input signals. For boardband amplifiers, these harmonics are not removed by filtering. Using the identify in Chapter 13, the last term becomes

$$V_{out,2}^1 = a_2 V_1 V_2 [\cos(\omega_1 - \omega_2)t - \cos(\omega_1 + \omega_2)t] \quad (21.4)$$

the familiar sum and difference frequency terms used in multipliers and mixers. In the superheterodyne receiver, the difference ($\omega_1 - \omega_2$) term is amplified in the IF circuitry. When used in this application, this term is not considered a distortion (it is the desired signal).

21.2.3.3 a_3 Term. The single signal input yields

$$V_{out,3} = a_3(V_1 \sin \omega_1 t)^3 = \frac{a_3 V_1^3}{4}(3 \sin \omega_1 t - \sin 3\omega_1 t) \quad (21.5)$$

which includes a term proportional to V_{in}^3 at the input frequency (a distortion of the $a_1 v_{in}$ term), which contributes to the 1-dB compression point.

The two signal input yields

$$V_{out,3} = a_3 \left((V_1 \sin \omega_1 t)^3 + (V_2 \sin \omega_2 t)^3 \right.$$
$$+ \frac{3a_3 V_1^2 V_2}{2}\{\sin \omega_2 t - \tfrac{1}{2}[\sin(2\omega_1 + \omega_2)t - \sin(2\omega_1 - \omega_2)t]\}$$
$$\left. + \frac{3a_3 V_1 V_2^2}{2}\{\sin \omega_1 t - \tfrac{1}{2}[\sin(2\omega_2 + \omega_1)t - \sin(2\omega_2 - \omega_1)t]\} \right) \quad (21.6)$$

The terms proportional to $V_1^2 V_2$ and $V_1 V_2^2$ cause "cross modulation" be-cause the amplitude of the ω_1 and ω_2 signals determine $V_{out,3}$. The famous third-order IMD signals ($2\omega_1 - \omega_2$ and $2\omega_2 - \omega_1$) are nearly the same frequency as ω_1 or ω_2 if $\omega_1 \approx \omega_2$, so they are difficult to filter. The $2\omega_1 + \omega_2$ and $2\omega_2 + \omega_1$ terms are close to the third harmonic term (if $\omega_1 \approx \omega_2$) and can be filtered out. These considerations lead to the requirements for filtering at various stages of the receiver to remove spurious signals. Note that the voltage of the third-order IMD is proportional to the third power of the total input voltage, so if both signals each increase by 1 dB, the IMD increases by 3 dB. Higher-order terms may also be considered analytically,

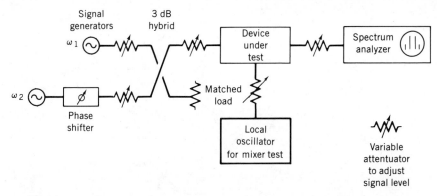

Figure 21.3. Test setup for measurement of intercepts of amplifiers or mixers.

but in practice the a_i ($i \geqslant 4$) are sufficiently small so they are masked by lower-order terms and can be neglected.

The values of the a_i can be inferred from measurements, but this procedure is somewhat complicated by the fact that several orders may contribute to a measurement (e.g., the third-order IMD has a fundamental component). For this reason a test setup as shown in Fig. 21.3 is used for the development of the intercept diagram rather than direct evaluation of the Taylor series coefficients. The spectrum analyzer is used as a frequency selective, calibrated power meter. The results for an amplifier are presented in Fig. 21.4, but mixer IFs provide results qualitatively similar as a function of RF level(s) for fixed LO level.

Using a single signal generator at a fixed frequency, the $V_{out,1} = a_1 v_{in}$ term can be plotted on log–log paper as shown in Fig. 21.4. The gain compression arises because of flat topping of the voltage waveforms, which is equivalent to diverting power into the harmonics of the input frequency. The $a_2 v_{in}^2$ terms lead to the second harmonic distortion (ω_1^2 terms) and second-order IMD ($\omega_1 + \omega_2$ and $\omega_1 - \omega_2$ in the two-signal tests). The plot in Fig. 21.4 shows the second-order IMD, which varies by ± 2 dB for ± 1 dB simultaneous change in the level of the two signals. The $a_1 v_{in}$ and $a_2 v_{in}^2$ plots intersect at the second-order intercept point. The slope of the $a_2 v_{in}^2$ plot is twice the $a_1 v_{in}$ plot. The second-order terms can be important for high dynamic range, broadband systems because for low signal levels, the second-order IMD exceeds the third-order IMD and limits the spur-free dynamic range.

The third-order IMD is also plotted in Fig. 21.4. Here the $2\omega_1 - \omega_2$ and $2\omega_2 - \omega_1$ terms are individually plotted (note the figure shows the power per signal). The third-order IMD power increases 3 dB if the two input signals increase 1 dB each. The intersection of the linear and third-order power plots is the third-order intercept point. The slope is three times that of the small signal gain line. Thus, given the small signal gain of the amplifier and the intercept points, the plots in Fig. 21.4 can be generated. These plots

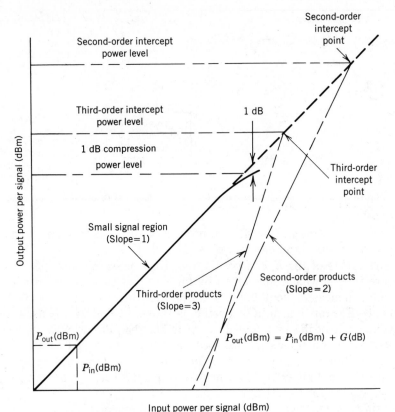

Figure 21.4. Intercept diagram for an amplifier.

apply to a single stage or an amplifier subsystem. When two amplifier stages are connected in cascade, the overall intercept point changes. This case is now analyzed.

21.2.3.4 Intercept Point of Cascaded Components. While considering the case of two or more components (amplifiers or mixers) in cascade, refer to

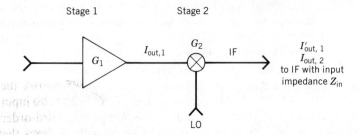

Figure 21.5. Calculation of the intercept for two cascade components. G_1 and G_2 ($= 1$/conversion loss) are power ratios, not expressed decibels.

Fig. 21.5. The third-order intercept will be computed for the cascade front end (the second-order intercept is done in Problem 5). The intercept power at the output of stage 1 is $I_{out,1}$, whereas $I'_{out,1}$ is the intercept referenced to the output of stage 2 ($I_{out,1}$ multiplied by the gain G_2). Since these calculations are needed for the spur-free dynamic range analysis, the reference plane is taken at the output of stage 2. For this analysis, all the intercepts must be referenced to the same plane, but any plane could have been selected.

Referring to Fig. 21.4, the two lines for the linear and third-order signal outputs can be represented by

$$\text{Linear:} \qquad \log P_{out} = \log G_1 + \log P_{in} \qquad (21.7)$$

$$\text{Third order:} \qquad \log P_{out} = \log C + 3(\log P_{in}) \qquad (21.8)$$

where C is a constant to be evaluated below. Evaluating these at the intercept point $I_{out} = P_{out}$, multiplying Eq. 21.7 by 3, and subtracting Eq. 21.8 yields an equation independent of P_{in}, namely,

$$\log C = 3(\log G_1) - 2(\log I_{out}) \qquad (21.9)$$

Writing this relation as a ratio,

$$C = \frac{G_1^3}{I_{out}^2} \qquad (21.10)$$

The amount of third-order distortion power at any level of input is then

$$P_{dist} = \left(\frac{G_1^3}{I_{out}^2}\right) P_{in}^3 = \left(\frac{P_{out}^3}{I_{out}^2}\right) \qquad (21.11)$$

by Eqs. 21.10 and 21.8 in ratio form. This distortion power appears across the input load Z_{in} of the IF subsystem. The distortion voltage is

$$V_{dist} = (P_{dist} Z_{in})^{1/2} = \frac{(P_{out}^3 Z_{in})^{1/2}}{I_{out}}$$

Since the phase is preserved through both the RF amplifier and mixer in most front ends, the distortion voltages must be added directly rather than incoherently like random noise. This is also the conservative approach and closely matches the experimental results. Using the intercepts at the output of the mixer, the total distortion voltage is (see Fig. 21.5)

$$V_{dist,T} = V_{dist,1} + V_{dist,2} = \left(\frac{1}{I'_{out,1}} + \frac{1}{I_{out,2}}\right) (P_{out}^3 Z_{in})^{1/2} \qquad (21.12)$$

The total output power in one signal of the third-order distortion is

$$P_{\text{dist,T}} = \frac{V_{\text{dist,T}}^2}{Z_{\text{in}}} = P_{\text{out}}^3 \left(\frac{1}{I_{\text{out,1}}'} + \frac{1}{I_{\text{out,2}}} \right)^2 = P_{\text{out}}^3 \left(\frac{1}{I_{\text{out,T}}} \right)^2 \quad (21.13)$$

This yields the desired result, namely,

$$I_{\text{out,T}} = \left(\frac{1}{I_{\text{out,1}}'} + \frac{1}{I_{\text{out,2}}} \right)^{-1} \quad \begin{array}{c} \text{(all intercepts evaluated} \\ \text{at the same plane)} \end{array} \quad (21.14)$$

which has the familiar form of the total resistance of two resistors in parallel. For example, if $I_{\text{out,1}}' = 10$ dBm, $G_2 = -10$ dB (10-dB conversion loss), $I_{\text{out,2}} = 13$ dBm, then $I_{\text{out,1}}' = 1$ mW and $I_{\text{out,2}} = 20$ mW and

$$I_{\text{out,T}} = \left(\frac{1}{1} + \frac{1}{20} \right)^{-1} = 0.95 \text{ mW} \rightarrow -0.22 \text{ dBm} \quad (21.15)$$

The intercept of cascade stages is always less than the lowest single intercept when referenced to the same plane. The cases for n stages and second-order distortion are developed in the problems.

Generally, the last gain stage in a cascade will limit the third-order IMD performance. This will occur if the output intercept of the previous stage exceeds the input intercept of the last stage. Considerations such as these lead to the distribution of gain through a receiver to provide the maximum overall spur-free dynamic range.

21.3   COMPONENTS IN MICROWAVE RECEIVER FRONT ENDS

21.3.1   Filters and Preamplifiers

The superheterodyne receiver is susceptible to reception of unwanted signals, called image signals, located at frequencies centered at the IF band of frequencies on either side of the LO, as shown in Fig. 21.6. The RF filter removes the image signals from the IF and reduces noise contributed by circuitry ahead of the mixer. The bandwidth of the RF filter is selected to allow reception of all desired signals, but it is usually less than an octave so second harmonics of signals are not received.

To compute the rejection performance required of the filter, simply determine the amount of image signal suppression required based on the receiver's detection criteria. For most receiver designs, symmetrical bandpass filters are used. Chebyshev filters (Chapter 9) are readily realizable and are specified by the relation (7)

$$\frac{V}{V_{\text{p}}} = \left\{ 1 + \left[\left(\frac{V_{\text{p}}}{V_{\text{v}}} \right)^2 - 1 \right] \cosh^2 \left[N \cosh^{-1} \left(\frac{X}{X_{\text{v}}} \right) \right] \right\}^{-1/2} \quad (21.16)$$

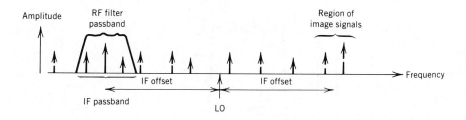

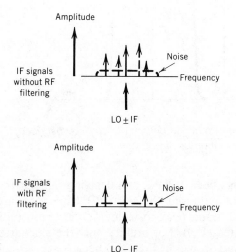

Figure 21.6. *Effect of RF filtering on image signals.*

where

V = output voltage at the normalized frequency x.

V_p = peak output voltage in the passband.

V_v = valley output voltage in the passband.

N = total number of resonators in the bandpass filter (number of re-active elements in a lowpass or highpass filter).

$X = (f_2 - f_1)/(f_1 f_2)^{1/2}$ = normalized frequency measured from center frequency $(f_1 f_2)^{1/2}$, where f_1 and f_2 are frequencies having the same attenuation. At the 3 dB attenuation point X_{3dB} = (3 dB bandwidth)/$(f_1 f_2)^{1/2}$.

X_v = value of X on the filter skirt where the output voltage equal V_v.

These parameters are also defined in Fig. 21.7.

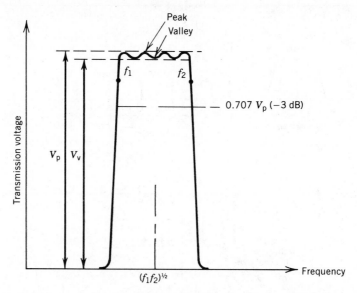

Figure 21.7. *Definition of terms to define a Chebyshev bandpass filter.* f_1 *and* f_2 *are frequencies where attenuation on skirt is equal.*

Equation 21.16 has been solved, and design graphs are readily available both in the passband and along the skirts (7). The filters are specified as a function of bandwidth and peak-to-valley voltage ratio (either linear or in decibels as $20 \log_{10}(V_p/V_v)$). For brevity only the skirt attenuation curves for $V_p/V_v = 1$ dB are shown in Fig. 21.8 because they yield N pole filters with reasonably steep skirts. Use these curves to determine the number of poles needed for a given skirt attenuation.

To attain the performance predicted in Fig. 21.8, the unloaded Q for each resonator (bandpass) or reactive element (lowpass) must exceed

$$Q_{min} = \frac{q_{min}(f_1 f_2)^{1/2}}{(\text{3-dB bandwidth})} \tag{21.17}$$

for the bandpass filter, and

$$Q_{min} = q_{min} \tag{21.18}$$

for the lowpass filter (8). Other aspects of these filters such as loss in the passband and phase dispersion are given in Ref. 7. The loss of the filter in its passband adds to the noise figure of the receiver. Therefore, the addition of the filter to eliminate image noise does not reduce the noise in the IF by 3 dB.

The need for an amplifier in the front end should be evaluated closely. For example, a high-gain amplifier will establish the noise figure but will also reduce the dynamic range. As a result the sensitivity and dynamic range are traded off and the amplifier has the effect of shifting the input signal level linear operating range but reducing the magnitude of the range. In some applications, such as Earth stations, this is highly desirable, but in others, such as EW receivers, it is undesirable. The complexity and cost of the amplifier may also be a key consideration. Only by trading off the system performance and limitations can the decision of having or not having a front-end amplifier be made.

21.3.2 Mixer

All superheterodyne receivers have at least one mixer. In a mixer the two groups of signals are the high-level local oscillator (LO) and accompanying noise and the low-to-moderate level RF (used as shorthand for microwave) signals. Because of the high-level LO, the IF contains many strong harmonics of the LO frequency. For a single ω_{RF} and ω_{LO}, the spurious signals will occur at $\omega_{out} = n\omega_{RF} \pm m\omega_{LO}$, where $n = 0, 1, 2, \ldots$ and $m = 0, 1, 2, \ldots$ etc. Depending on the type of mixer the magnitude of the response

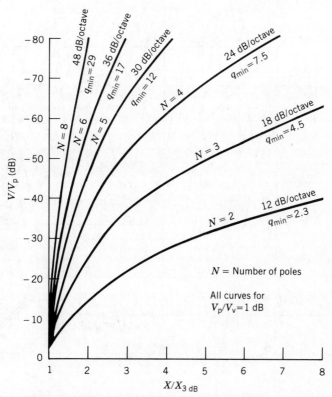

Figure 21.8. *Bandpass and lowpass filter skirt attenuation.*

for various n and m will be different but still present and therefore a source of interference.

To aid in the design of mixer circuits so that either no spurs or only higher-order (large m and n) spurious signals are present in the IF, the downconverter spurious chart in Fig. 21.9 has been developed (4). To simplify use of the chart for all frequencies, the high-frequency f_H is used to normalize the lower input frequency f_L (either the LO or RF frequency) and the IF output for the downconverter $f_H - f_L = f_{IF}$. The widest spurious-free bandwidth is centered at $L/H = 0.63$. Any larger RF bandwidth beyond 0.61 and 0.65 will yield other IF responses in the IF passband from 0.34 to 0.4. For wideband receivers $H - L$ can vary from 0.33 to 0.5 (zone B), which includes several higher-order spurs and is limited on one end by $H - L = L$ and on the other by $H - L = 2L - H$. The higher-order spurs that do occur in the band can be discerned by shifting the LO frequency by Δf_{LO} and noting whether the IF shifts by $n\Delta f_{LO}$. If $n > 1$, the receiver is

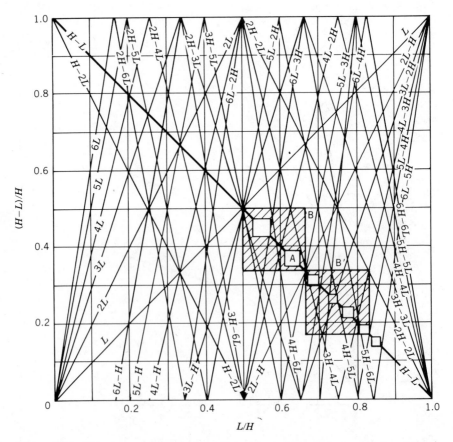

Figure 21.9. Downconverting mixer spurious chart. Region A, no spurious signals; regions B and B', weaker spurious signals.

hearing a spurious signal and should discount the information. Note that this technique assumes the receiver has a frequency measurement capability, and the shift will not discriminate against harmonics of the RF mixing with the fundamental LO signal.

Using a bounding region of the spur chart, some basic analysis can be used to derive the optimum frequency architecture for a receiver given certain assumptions. For example, if the IF is to be $2L - H < H - L < L$, where L is the RF signal and H is the LO (LO frequency above the RF), then in the limits of the inequalities, $L/H = 2/3$ and $L/H = 1/2$. These values agree with zone B. If a guard band (Δf) is desired between the bottom of the RF band and the top of the IF band, the equality

$$L - \Delta f = H - L \qquad (21.19)$$

applies. Also if the IF is not to be within $\Delta f'$ of the 2RF $-$ LO ($2L - H$) spur, then

$$H - L + \Delta f' = 2L - H \qquad (21.20)$$

These equations give relations for the IF and LO for a specified RF band. Specifically, Eq. 21.19 sets the LO frequency (H) since

$$L_{min} - \Delta f = H - L_{min} \quad \text{or} \quad H = 2L_{min} - \Delta f \qquad (21.21)$$

Given the value of the LO (assumed fixed), the maximum RF that can be received without spurious effects derives from Eq. 21.20, namely,

$$H - L_{max} + \Delta f' = 2L_{max} - H \quad \text{or} \quad L_{max} = \frac{(2H - \Delta f')}{3} \qquad (21.22)$$

Applying these relations to the lowest segment of a 18–40-GHz downconverter yields (for $L_{min} = 18$ GHz, $\Delta f = \Delta f' = 1$ GHz):

$$\text{LO frequency} = H = 2L_{min} - \Delta f \quad = 35 \text{ GHz} \qquad (21.23a)$$

and

$$\text{Maximum RF} = L_{max} = \frac{(2H - \Delta f')}{3} = 23 \text{ GHz} \qquad (21.23b)$$

The IF ranges from $H - L_{max}$ to $H - L_{min}$, so $\text{IF}_{min} = 12$ GHz and $\text{IF}_{max} = 17$ GHz. For frequencies between 23 and 40 GHz, the relation is applied again, except now the IF_{max} can exceed 17 GHz. For this case, the IF may limit the RF bandwidth depending on the specific design of the downconverter. Tradeoffs and analyses such as these are continued until the full frequency range is covered. Because LO power is difficult to generate at these frequencies, broadband receivers use a wideband IF to limit the number of fixed LO frequencies. The procedure yields a downconverter design as shown in Fig. 21.10. The channelizing filters are required to limit the bandwidth of the input signals to the higher-order spur region (zone B in Fig. 21.9).

The spurious effects chart does not predict the IMD produced by two or more signals whose difference frequency lies in the RF passband. Also the chart applies to the single-ended mixer design where suppression of some spurious signals due to cancellation in the circuitry surrounding the mixer is not present. In practice single-ended mixers are seldom used so a digression to other mixer designs is needed.

21.3.2.1 Types of Mixer Circuits. The following discussion presents a summary of the analysis of mixer circuits (6). Four mixer circuit types are readily commercially available. They are the single-ended, single-balanced or balanced, double-balanced, and image-rejection mixer types. The double-balanced design is the most popular and will be described in detail.

Double-Balanced Mixer. The double-balanced mixer (Fig. 21.11) is the most commonly used mixer having two baluns and four diodes. A balun is a transformer (implemented in microwaves in a transmission line configuration) designed to match a balanced transmission line to an unbalanced transmission line. The IF output is the difference in the current at the center tap of the LO balun.

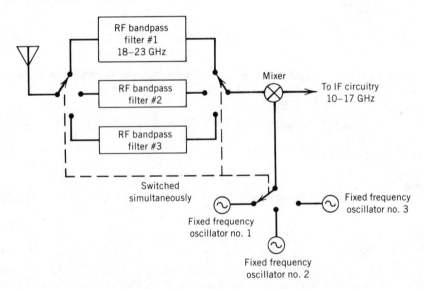

Figure 21.10. *Top-level design of a 18–40-GHz front end. FFO, fixed frequency oscillator.*

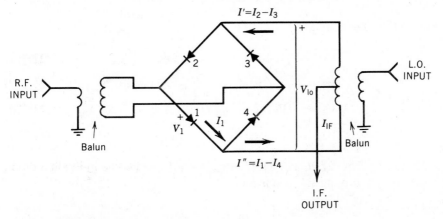

Figure 21.11. *Schematic of a double-balanced mixer circuit.*

The suppression of the harmonics can be seen by writing the relations for the currents in Fig. 21.11 and deriving the IF current. In diode j, the current is

$$I_j = I_s \left[\exp\left(\frac{eV_j}{nkT}\right) - 1 \right] \tag{21.24}$$

where I_s is the saturated back-biased current assumed equal for all the diodes, e is the electron charge, V_j is the voltage across the jth diode, n is a parameter near unity to account for the junction's carrier distribution, k is Boltzmann's constant (1.38×10^{-23} J/°K), and T is the absolute temperature. $V' = nkT/e = 25$ mV for most diodes at room temperature. The corresponding diode conductivity is

$$g_j = \frac{dI_j}{dV_j} = \left(\frac{I_s}{V'}\right) \exp\left(\frac{V}{V'}\right) \tag{21.25}$$

Since the LO voltage dominates the conductivity, first consider the effect of the LO and then consider the RF as a perturbation. Doing this for $V = V_{LO} \cos \omega_{LO} t$ yields

$$g_1 = \left(\frac{I_s}{V'}\right) \exp\left[\left(\frac{V_{LO}}{V'}\right) \cos \omega_{LO} t\right] = g_2 \tag{21.26a}$$

$$g_3 = \left(\frac{I_s}{V'}\right) \exp\left[\left(-\frac{V_{LO}}{V'}\right) \cos \omega_{LO} t\right] = g_4 \tag{21.26b}$$

and therefore the current in the diode is $I_j = g_j V_j$. The effect of the RF is now reintroduced as a perturbation so

$$V_1 = V_{LO} \cos \omega_{LO}t + V_{RF} \cos \omega_{RF}t \qquad (21.27a)$$

$$V_2 = V_{LO} \cos \omega_{LO}t - V_{RF} \cos \omega_{RF}t \qquad (21.27b)$$

$$V_3 = -V_{LO} \cos \omega_{LO}t - V_{RF} \cos \omega_{RF}t \qquad (21.27c)$$

$$V_4 = -V_{LO} \cos \omega_{LO}t + V_{RF} \cos \omega_{RF}t \qquad (21.27d)$$

Substituting Eqs. 21.26 and 21.27 into $I_j = g_j V_j$ yields the individual diode currents I_j. Forming the IF current

$$
\begin{aligned}
I_{IF} &= I'' - I' = I_1 - I_4 + I_3 - I_2 \\
&= 4V_{RF}\left(\frac{I_s}{V'}\right)\left\{\sinh\left[\left(\frac{V_{LO}}{V'}\right)\cos \omega_{LO}t\right]\right\}\cos \omega_{RF}t
\end{aligned}
\qquad (21.28)
$$

The lack of even LO harmonic terms becomes evident when the sinh term is expanded using the relation (6)

$$\sinh(A \cos \phi) = 2\sum_{k=0}^{\infty} I_{2k+1}(A)\cos(2k+1)\phi \qquad (21.29)$$

where $I_{2k+1}(A)$ are the modified Bessel functions of the first kind of order k. Since only small RF signals were considered, the suppression of the even harmonics of RF is not shown. A complete analysis including the RF in the g_j would yield this result. Similar analyses can be performed for other mixer types.

Summary of Mixer Circuits. The double-balanced mixer is the most commonly used type, but in some applications other circuit types may be optimum. The key parameters used for this decision are

Conversion loss
Harmonic suppression
Image rejection
LO/RF isolation
LO/IF isolation
Complexity
Amount of LO drive required for low conversion loss (proportional to the number of diodes)
Cost

A summary of these parameters is shown in Table 21.2. This table also

Table 21.2 Performance of Mixer Types

Type	No. of Diodes	RF Circuit	LO Circuit	IF Circuit	Harmonics/Images Suppressed	Isolation Port–Port
Single ended	1	Unbal	Unbal	Unbal with low-pass filter	None	Poor
Single balanced or balanced	2	90 or 180 hyb	90 or 180 hyb	0 or 180 hyb	RF − LO = 90/ IF = 0—complicated RF − LO = 180/ IF = 180 —even harmonics of LO	Good RF to LO
Double balanced	4	Balun	Balun	Unbal	Even LO Even RF	Good RF–LO–IF
Image rejection (two single-balanced mixers)	4	90 hyb	0 power divider	90 hyb (2 outputs)	One IF output for $\omega_{RF} > \omega_{LO}$ One IF output for $\omega_{RF} < \omega_{LO}$	Good RF–LO

Note: hyb = hybrid coupler (0, 90, or 180° phase shift between output ports); unbal = unbalanced.

includes information about the phase circuitry in the RF, LO, and IF circuits so that analyses similar to the above can be performed.

21.4 DESIGN OF A MICROWAVE RECEIVER FRONT END

Superheterodyne design is best facilitated with computer programs. Most microwave receiver manufacturers have internally developed, proprietary programs for this system analysis. A hand calculator will save time while computing the IMD levels for the system (9).

To begin the analysis, a block diagram of the cascaded elements should be prepared with a listing of the performance parameters of components and the system. Since this chapter deals only with the receiver front end, the IF and detection analysis is not presented. However, their influence is critical to the microwave design. For example, logarithmic amplifiers set the range of IF levels between -60 and 0 dBm. Thus the receiver gain distribution must be adjusted to present the correct signal levels at the IF input (10).

21.4.1 Communication Receiver Components

The example selected illustrates the procedures a receiver system designer can use to "optimize" the design. Consider the parameters shown in Fig. 21.12. A discussion of each of the components follows.

21.4.1.1 Filter. The input filter bandwidth has been selected to remove images, provide preselection ahead of the mixer to reduce spurious re-

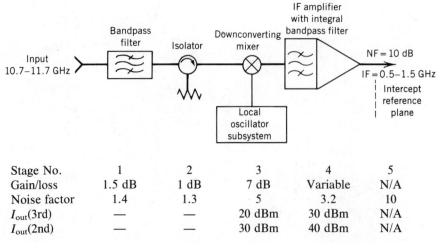

Stage No.	1	2	3	4	5
Gain/loss	1.5 dB	1 dB	7 dB	Variable	N/A
Noise factor	1.4	1.3	5	3.2	10
I_{out}(3rd)	—	—	20 dBm	30 dBm	N/A
I_{out}(2nd)	—	—	30 dBm	40 dBm	N/A

Figure 21.12. *Simplified terrestrial microwave link receiver design parameters.*

sponses, and yet allow reception of the full band without switched filters as shown in Fig. 21.10. The insertion loss of this filter is determined using the design techniques in Chapter 12 or the manufacturer's data. The number of poles of the filter is set by the amount of image suppression in the two possible image bands depending on the LO frequency of 12.2 or 10.2 GHz. The two image bands are then 12.7–13.7 GHz and 8.7–9.7 GHz. The upper image band is selected because the density of strong signals in the higher band is lower than in the lower band. Assuming that no local transmitters are operating in this band (12.7–13.7 GHz), the filter must suppress by 40 dB a signal located at 12.7 GHz. The 40 dB is selected because the minimum signals will be at ≈ -50 dBm (see below), the loss through the filter, isolator, and mixer is 10 dB, and the largest image signal level is not expected to exceed 0 dBm. The passband of the filter is assumed to extend to 12 GHz so that temperature drifts, etc., do not cause degradation of signals at 11.7 GHz. As a result, the filter must attenuate 40 dB in 700 MHz. This is unreasonably high for a filter with a moderate number of poles, so either the image suppression must decrease or the IF center frequency must be raised. For this design we select to reduce the image suppression to 20 dB, thus assuming that interfering signals will be below -20 dBm at the input.

21.4.1.2   Isolator. The isolator is used to present a good match to the RF input of the mixer. This assures that noise generated at the frequencies of LO $\pm$ IF that leaks through the LO–RF isolation of the mixer does not reflect directly back into the mixer. Use of an isolator here raises the noise figure by 0.5 dB but reduces the requirement to lower the LO noise. Also LO reradiation out the antenna is significantly reduced.

The bandwidth of the isolator must be 10.7–11.7 GHz. More bandwidth usually means more loss, but the unit must still have reasonable isolation at the LO frequency 12.2 GHz to be effective.

21.4.1.3   Mixer. A double-balanced mixer is selected for this application. This mixer type suppresses the even LO and RF harmonics, provides good LO–RF isolation, has a high third-order intercept (20 dBm), and only requires 10 dBm of LO drive. The input VSWR is also good (typically less than 2:1), so reflection losses at the input to the mixer can be neglected for this top-level analysis.

21.4.1.4   LO. Two options for the LO source should be considered. The equipment in Chapter 2 (see Fig. 2.6) was probably designed before the mid-1970s, so a temperature-stabilized transferred-electron (Gunn) source was used. This type of oscillator requires both a high Q cavity and a temperature oven for full frequency control ($\pm$ several MHz) from a fundamental oscillator. Considering both the heater power and the Gunn prime power (efficiencies range from 1 to 5%), the Gunn LO will use considerably more prime power than a dielectric-stabilized FET oscillator (abbreviated DSO).

Modern DSO designs maintain an accuracy of ± 1 MHz over temperature at 12.2 GHz with no external oven. Although at a terrestrial relay station prime power may not be a consideration, complexity, cost, and replacability should be considered.

21.4.1.5 IF Amplifier. The IF amplifier is included as part of the down-converter because it is the first gain stage and therefore plays a major role in setting the noise figure and signal levels in the receiver. Shown implicitly in this amplifier is input filtering to limit the IF bandwidth, thus reducing amplification of spurious mixer outputs. Because of the low IF frequency selected, second harmonic IF signals reappear in the IF output and can cause adjacent channel interference. The 1-GHz design shown here will be analyzed, but the effects of the second harmonic in the detection circuitry must be carefully considered. The second-order intercept of the IF amplifier must be measured if not given in the manufacturer's specifications. If the performance is not adequate, another option is to channelize the IF (separate into two channels, say 0.5–0.9 and 0.9–1.5 GHz).

It is assumed that the noise figure of the remainder of the receiver is 10 dB. A major design issue is selecting the gain of the IF amplifier.

21.4.2 Noise Figure

The noise factor of the receiver is computed using the cascade relation, namely,

$$F_T = F_1 + \sum_{i=2}^{N} \frac{F_i - 1}{\prod\limits_{j=1}^{i-1} G_j} \tag{21.30}$$

$$\text{NF} = 10 \log(F_T) \tag{21.31}$$

where F denotes the algebraic noise factor and G_j is the gain of the jth stage. Using the values in Fig. 21.12 where the stages are noted yields

$$F = F_1 + \frac{F_2 - 1}{G_1} + \frac{F_3 - 1}{G_1 G_2} + \frac{F_4 - 1}{G_1 G_2 G_3} + \frac{F_5 - 1}{G_1 G_2 G_3 G_4} \tag{21.32}$$

$$= 1.4 + \frac{0.3}{0.7} + \frac{4}{0.56} + \frac{2.2}{0.078} + \frac{9}{0.078 G_4} = 37.2 + \frac{9}{0.078 G_4}$$

A plot of NF versus gain G_4 of the IF amplifier is shown in Fig. 21.13a. For gains exceeding 10 dB, the overall noise figure is affected less than 1 dB by the IF amplifier.

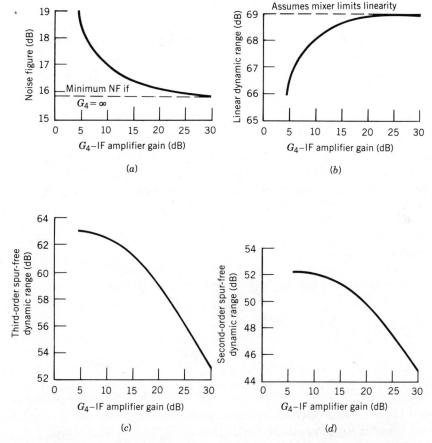

Figure 21.13. *Noise figure and dynamic range versus IF amplifier gain (G_4) for microwave downconverter in Fig 21.12.*

The linear dynamic range is determined by the input signal range from $S_i/N_i = 1$ ($N_i = F_T kTB$) to the 1-dB compression point. For this receiver front end with $G_4 = 20$ dB, $F_T = 38.4$, so

$$N_i = (38.4)(1.38 \times 10^{-23})(290)(10^9)$$
$$= 1.5 \times 10^{-10} \text{ W} \rightarrow -68.1 \text{ dBm} \tag{21.33}$$

Given the 1-dB compression point of approximately 0 dBm (10 dB below the LO drive level), the linear dynamic range is 68 dB. The linear dynamic range versus gain G_4 is shown in Fig. 21.13b. The spur-free dynamic ranges will be less than this number.

The overall gain of the front end is 10.5 dB from the RF input to the input of the IF subsystem with $G_4 = 20$ dB.

21.4.3   Third-Order Dynamic Range

The third-order dynamic range of the front end is computed at the reference plane shown in Fig. 21.10 assuming $G_4 = 20$ dB. Using Eq. 21.14,

$$I_{\text{out,T}} = \left(\frac{1}{I'_{\text{out,3}}} + \frac{1}{I_{\text{out,4}}} \right)^{-1} = \left(\frac{1}{10,000} + \frac{1}{1000} \right)^{-1}$$

$$= \frac{10,000}{11} = 909 \text{ mW} \rightarrow 29.6 \text{ dBm} \tag{21.34}$$

The third-order spur-free dynamic range is determined by the difference in input power level between the noise floor (S/N $= 1$) and the input level where the input signals produce spurs equal to this noise floor level. This level is readily computed using the relation in Eq. 21.11:

$$P_{\text{out}} = (P_{\text{dist}} I_{\text{out,T}}^2)^{1/3} \tag{21.35}$$

Where $P_{\text{dist}} = N_i G_1 G_2 G_3 G_4 = 1.8 \times 10^{-6}$ mW $=$ noise floor at the output reference plane. Computing the P_{out} level in decibels ($G_4 = 20$ dB) yields

$$P_{\text{out}}(\text{dBm}) = \frac{[P_{\text{dist}}(\text{dBm}) + 2I_{\text{out,T}}(\text{dBm})]}{3}$$

$$= \frac{[-57.4 + 2(39.6)]}{3} = -8.9 \text{ dBm} \tag{21.36}$$

Therefore two signals at -8.9 dBm at the output (-20 dBm at the input) result in third-order IMD signals at the noise floor at the output. The third-order spur-free dynamic range DR(3) for $G_4 = 20$ dB is, therefore,

$$\text{DR}(3) = P_{\text{out}} - P_{\text{dist}} = -8.9 - (-57.4) = 49 \text{ dB} \tag{21.37}$$

This is 19 dB less than the linear dynamic range.

The third-order spur-free dynamic range can also be plotted versus the gain of G_4. The results are shown in Fig. 21.13c. Remember that P_{dist} also varies with G_4. Note that increasing the IF gain decreases the dynamic range unless the intercept point of the components changes with gain. Clearly from Eq. 21.34 the intercept level of the IF amplifier needs to be raised to increase the dynamic range. This is accomplished by operating a more powerful unit having a higher I_{out} with the same gain and noise figure.

21.4.4   Second-Order Dynamic Range

Because the IF amplifier covers more than an octave, the second harmonic of strong IF signals below 0.75 GHz will also appear in the IF amplifier output. Referring to Eq. 21.3, the last term's amplitude (exceeds the two direct harmonic terms) is the first second-order spur to exceed the noise floor. As a result, the second-order intercept of the system must be known to estimate the magnitude of these signals. This value is typically 10 dB above the third-order values shown in Fig. 21.12 (also see Fig. 21.3). The second-order dynamic range versus gain G_4 is derived and plotted in Fig. 21.13*d* using the relations for the second-order intercept (see Problem 5).

Comparing Figs. 21.13*c* and 21.13*d* shows that the second-order dynamic range limits the receiver performance before the third-order. This can be avoided by increasing the intercept of the IF amplifier or adding filters to avoid the signals near the second harmonic from continuing into the IF system.

21.5 BROADBAND MICROWAVE RECEIVER FRONT-END DESIGN

For EW applications (see Chapter 4) the monitoring receivers have an instantaneous frequency coverage exceeding an octave. As a result spurious signals (at least of higher order) are inevitable. To retain a reasonable spur-free dynamic range, the designer has only two major components to work with—filters and the gain of amplifiers. All the remaining parameters are not readily modifiable to fit the situation. Thus for these receivers the third-order IMD usually limits performance. The use of high-level mixers (LO drive = 20 dBm or more) assists in raising the third-order intercept. Use of mixers with desirable harmonic suppression is also advised. For amplifiers, the use of high-power output stages assists with raising the intercepts. However, the first stage must retain a good noise figure or again the dynamic range will suffer.

When the microwave designer has achieved the "optimum" performance within complexity, weight, and cost constraints, the remainder must be done later in the receiver. Today the availability of fast-slewing LOs and tunable RF filters has allowed computer-controlled receivers to outperform the static or switched LO-filter designs of a few years back. Also the monolithic circuit technology has significantly reduced the number of subsystem interconnections, weight, and cost (only in large quantities) and now allows consideration of multiple, parallel channelized "optimum" designs having high probabilities of intercept.

The next major steps forward will be direct digitization of the incoming signal and then digital processing of the signals. Here again analog devices will undoubtedly set the level and introduce distortion before digitization. This approach is limited because the large dynamic range and frequency ranges require a significant number of bits in a period less than the highest information rate. Because of this situation, analog design techniques will be required for the foreseeable future.

PROBLEMS

1. Explain how the double downconversion shown in Fig. 21.1*b* eliminates the reception of images. What is the limitation on the frequency selection for the second (low) IF?

2. Estimate the bandpass filter rejection required between the two mixers in the homodyne receiver to avoid the "image." The microwave amplifier in Fig. 21.2 must have what characteristic to allow the receiver to have wide dynamic range?

3. Compute the amount of second-order distortion power at any input power level (the result is similar to Eq. 21.11).

4. Find the relation between the total third-order intercept $I_{out,T}$ and $I'_{out,1}$ and $I_{out,2}$ if the voltages add randomly.

5. Compute the relation between the total second-order intercept point and the individual second-order intercepts for two cascade components assuming coherent addition of voltages (the result is similar to Eq. 21.14).

6. Compute the total third-order intercept for N stages.

7. Compute the total second-order intercept for N stages.

8. Prepare a graph of the total third-order intercept (ordinate) for two cascaded stages where the abscissa is the input intercept of the last stage ± 10 dB. Assume the second stage gain is 10 dB.

9. For a superheterodyne receiver with LO = 11 GHz receiving RF from 9–10 GHz, determine the number of poles of a 1-dB ripple bandpass filter required to reject the image by 20, 30, 40, 50, and 60 dB.

10. Compute the numeric range $H-L$ for zone B′ of the downconverter spurious chart (Fig. 21.9).

11. List the spurious signals found in zone B of Fig. 21.9, and indicate which are not discernable by a LO frequency shift. Consider both LO = f_H and f_L.

12. Derive relations similar to Eqs. 21.16 to 21.19 for zone B′.

13. Complete the design of the RF filters in Fig. 21.10. Assume that the maximum IF frequency is 17 GHz.

14. Compute the frequency range of two IF channel filters to provide equal percentage bandwidth for each channel in Fig. 21.10. Estimate the second harmonic suppression assuming n = 2–6 pole bandpass filters with 1-dB ripple.

15. For the receiver in Fig. 21.12, compute the amount of additional IF

gain ($G_4 = 20$ dB) to provide -55 dBm to the input of a logarithmic amplifier when S/N $= 3$ dB.

16. Prepare a table of numbers to develop Fig. 21.13c. The column headings should be G_4 (dB), $I'_{out,3}$ (dBm), $I_{out,T}$ (dBm), P_{dist} (dBm), P_{out} (dBm) to cause the spurious signals to equal the noise floor, DR(3) dynamic range for third-order spurious signals.

17. Assuming the two signal levels differ by 10 dB, compute the levels of the two third-order spurious signals assuming $\omega_1 \approx \omega_2$.

18. Design a dual-conversion radar receiver front end suitable for air surveillance at 3.5 GHz with a 1 µs pulse length. Estimate its dynamic range, etc. Assume a $S_0/N_0 = 14$ dB is required at the output of the second (70 MHz) IF if the IF bandwidth is matched to the pulse length. Select reasonable component performance specifications. Compute minimum S_0.

19. Design a 2–16-GHz radar warning receiver front end suitable to receive 100-ns pulse lengths with $S_0/N_0 = 14$ dB in the appropriate bandwidth. Three octave bands, defined by filters, should be used. Estimate the sensitivity assuming antenna gain is 0 dB and linear dynamic range (see Table 13.2). Select reasonable component performance specifications and IF amplifier gain for a -40-dBm minimum signal at a video detector (TSS + 10 dB, see Table 13.2).

REFERENCES

1. G. Kennedy, *Electronic Communication Systems*, McGraw-Hill, New York, 1977.

2. J. C. Hancock, *An Introduction to the Principles of Communication Theory*, McGraw-Hill, New York, 1961.

3. J. M. Wozencraft and I. M. Jacobs, *Principles of Communication Engineering*, Wiley, New York, 1965.

4. M. I. Skolnik, ed., *Radar Handbook*, McGraw-Hill, New York, 1970, Chaps. 5, 16–23, and 39.

5. L. VanBrunt, *Applied ECM*, EW Engineering, Dunn Loring, VA 22027, 1978.

6. J. B. Tsui, *Microwave Receivers and Related Components*, NTIS, Springfield, VA, No. PB84-108711, Chap. 14.

7. *Reference Data for Radio Engineers*, 5th ed., H. W. Sams, 1968, Chap. 8.

8. M. Dishal, "Concerning the Minimum Number of Resonators and the Minimum Unloaded Q Needed in a Filter," *IRE Trans. Prof. Group Veh. Comm.*, Vol. PGVC-3, June 1953, p. 85.

9. F. W. Hauer, "Two Calculator Programs Expedite Front-End Design," *Microwaves*, November 1981, p. 81.

10. W. H. Hayward, *Introduction to Radio Frequency Design*, Prentice-Hall, Englewood Cliffs, NJ, 1982.

22 | Microwave Transmitter Design

22.1 INTRODUCTION

A transmitter is an RF amplifier. It receives its input from a waveform generator that produces the modulated RF to be amplified. The transmitter output is sent to the antenna. In some earlier systems the transmitter was a high-power oscillator and did not receive an RF input. However, these transmitters have limited application today. In some of today's systems, the transmitter provides a path from the antenna to the receiver; this path may include the receiver protector.

This chapter describes transmitters that operate in the 1–18 GHz range and that use a thermionic device as the amplifying element. In this device, the input RF is made to interact with an electron beam in such a manner that energy is coupled from the beam to the RF wave and the RF is amplified. Relatively low-power solid-state amplifiers that can be used as transmitters are discussed in Chapter 15.

22.2 BLOCK DIAGRAM

A simplified block diagram of a transmitter is shown in Fig. 22.1. The input level to the transmitter is typically in the -10 to $+10$ dBm range. The input may be CW or it may be gated on and off by the waveform generator; if it is gated, the RF may be either wider or narrower than the transmitter gate. These variables are a function of the system application and requirements.

The input RF assembly can have up to three functions. It can provide load matching for the waveform generator and/or input matching for the

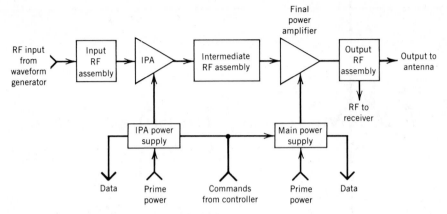

Figure 22.1. *Simplified block diagram of a transmitter.*

intermediate power amplifier (IPA); these functions are usually accomplished using isolators or circulators. The input RF assembly can also use directional couplers and RF level detecting circuits to provide built-in-test (BIT) or fault isolation test (FIT) data to the system or its interface.

The transmitter may have more than one stage of amplification depending on the transmitter input and output power level requirements. A single stage can have gain of 30–60 dB. Transmitters with output powers greater than 1000 W peak may have two or more intermediate amplifiers. If an IPA is included in the transmitter, an intermediate RF assembly may be used to provide load matching for the IPA, input matching for the next amplification stage, and BIT/FIT data. The final power amplifier (FPA) imparts the bulk of the energy to the RF wave.

The possible functions of the output RF assembly include load matching for the FPA, power monitoring for BIT/FIT, and RF fault protection via detection of either excessive reflected power coming back from the antenna or detection of waveguide arcs. This assembly can also include a high-power waveguide switch and an RF load so that the transmitter can be energized without radiating energy out from the antenna. And, in the case of a radar transmitter, this assembly can contain a circulator to route received energy coming from the antenna to the receiver via a receiver protector.

The IPA and FPA power supplies process the prime power into voltages and currents that can be used by the IPA and FPA. These supplies receive commands (for example, turn on or off) from the system and, in some cases, provide BIT/FIT data back to the system.

22.3 TRANSMITTER CHARACTERISTICS

Transmitters operating in the 1–18 GHz range can be categorized into three types—radar, communications, and ECM. Radar transmitters can be ground

Table 22.1 Characteristics of Transmitters

Type of Transmitter	Peak Output Power Range (kW)	Duty Cycle Range	Stability/Fidelity[a]	Bandwidth[b]	Reliability[a]	Maintainability[a]	Cooling[c]	Cost	Efficiency[a]	Packaging Density
Radar										
Ground based and shipborne	1–5,000	0.01–.5	M–H	M	M	M	L,A	M–H	L	L–M
Airborne	0.5–1,000	0.01–.5	M–H	M	L	L	L,A	M–H	M	H
Spaceborne	0.01–5	0.01–.5	M–H	M	H	L	L,R	H	H	H
Communications										
Line of sight	0.01–1	1	M–H	N–B	H	H	A	L	M	L–H
Tropospheric scatter	0.2–50	1	M–H	N–B	H	H	L,A	M	M	L–M
Satellite	0.01–1	1	M–H	N–B	H	H	L,R	H	H	H
ECM										
Ground based and shipborne	0.2–10	0.01–1	L	B	M	M	L,A	M	M	L–M
Airborne	0.2–2	0.01–1	L	B	L	L	L,A	M	M	H

[a] L, low; M, medium; H, high.
[b] N, narrow; M, medium; B, broad.
[c] L, liquid; A, air; R, radiation.

based (fixed or mobile), airborne, or spaceborne. Communications terminals can operate within sight of each other, over the horizon from each other (where the signal is "scattered" by the troposphere), or between ground and satellite or satellite and satellite. ECM transmitters are either ground based, shipborne, or airborne. Some general characteristics of these types of transmitters are given in Table 22.1. This is not a comprehensive listing of all transmitter types, just those that are the most common. All of the above transmitters can operate anywhere within the 1–18 GHz range, with the exception of troposheric scatter communications transmitters, which generally operate in the 1–5 GHz range. Other characteristics of transmitters too diverse to classify include type of prime power, environmental operating conditions (temperature, vibration, humidity, nuclear radiation, fungus, salt spray), EMI/EMC (electromagnetic interference and compatibility), gain, size, weight, and cost.

A transmitter design begins with the generation of a transmitter specification. The generation of this specification should be interactive between the system designer and the transmitter designer.

The interface between the electron tube and the rest of the transmitter (i.e., the modulator) can be a problem for the transmitter designer. The transmitter designer must suballocate the transmitter requirements to the tube designer. Again, this should be an interactive process if a cost-effective design is to result.

Understanding the use of transmitters in radar and communications systems provides a system perspective and an understanding of how the transmitter affects system performance and the transmitter stability required by the system.

22.4   TRANSMITTER REQUIREMENTS

22.4.1   Radar Transmitters

Different radar system applications require different radar transmitters. The transmitter requirements described below can occur singly or in combination.

22.4.1.1   MTI Systems. Moving target indicator (MTI) systems allow a moving target to be seen against a background of clutter. This is invariably done by subtracting one received pulse from the next (one pulse period later). If the target is stationary, as in the case of clutter, cancellation will occur. In the case of a moving target, owing to the change in range and phase and Doppler shift between successive pulses, cancellation will not occur and the resultant output will be "visible" against the cancelled clutter. The performance of such a system has been variously described in terms of "subclutter visibility," "cancellation ratio," and "improvement factor."

Cancellation ratio, which is more descriptive for the transmitter and will be used here, is the ratio in decibels of the energy residue after cancellation (subtraction) to the original pulse energy. Improvement factor is simply the reciprocal of this number. Both are obviously dependent on exact replication of one transmitted pulse to the next. This is termed pulse-to-pulse stability, and the transmitter contribution to it has four components, each contributing to the cancellation ratio as follows (1):

Amplitude $\quad 20 \log \dfrac{\Delta A}{A}$

Phase $\quad 20 \log \Delta \phi$

Timing $\quad 20 \log \dfrac{\sqrt{2}\Delta t}{t}$

Pulse width $\quad 20 \log \dfrac{\Delta t_0}{t_0}$

Transmitter frequency shift is not included here since it is assumed that most modern transmitters are of the MOPA (master oscillator, power amplifier) variety in which only phase variations occur.

If the pulse-to-pulse instabilities are random in time, they can be added in an RMS fashion to obtain the total effect.

It is generally found that if the phase requirement is met, the other instability requirements are also met. Also, intrapulse variations in amplitude and phase are of no consequence in MTI systems as long as they are repeated exactly from pulse to pulse.

22.4.1.2 Pulse Doppler Systems.

Pulse Doppler systems are the same in principle as MTI systems in that target velocity is used to discriminate a moving target against clutter. Instead of canceling the clutter, the Doppler frequency is measured directly by a bank of range-gated filters covering the frequency interval between half the PRF below the carrier and half the PRF above the carrier. Since unambiguous velocity is required by most systems, this means that the repetition rate of the radar must be at least twice that of the highest Doppler frequency to be detected. This is usually the first distinguishing feature of the pulse Doppler transmitter; repetition rates are typically in the range of 100–300 kHz, and the duty cycle can be as high as 50%.

The stability requirements of pulse Doppler transmitters are usually much more severe than for MTI transmitters owing to the higher system performance in terms of visibility against clutter that is required. The system designer therefore does not usually specify pulse-to-pulse stability per se but lumps the random terms together in an overall transmitter spectral density curve. Discrete instabilities, which appear as false targets in the receiver, are defined separately for amplitude and phase and in terms of sidebands relative to the carrier level. Their values are usually dependent on their frequency away from the carrier, and as noise is integrated by the filter and discretes are not, they are generally allowed to be higher than the pulse floor by the filter bandwidth (e.g., if the receiver filter is 100 Hz wide, the noise floor must be 20 dB below the discrete level). An example is shown in Fig. 22.2.

22.4.1.3 *Pulse Compression Systems.*

The main feature required of the transmitter in pulse compression systems is pulse fidelity (2). The important characteristics are ripple on top of the pulse and droop during the pulse. Ripple on top of the pulse produces unwanted amplitude and phase modulation (depending on the respective pushing figures), which give rise to time (range) sidelobes after compression (see Fig. 3.7). In the case of phase, the ratio of sidelobe level to mainlobe is given by $\phi/2$, where ϕ is the zero-to-peak deviation of the phase. Its distance from the main lobe is given by the reciprocal of its frequency or the number of ripples across the pulse divided by the pulse length. Amplitude modulation also gives rise to sidelobes whose ratio to the mainlobe is given by half the value of the modulation index. Distance from the mainlobe is the same as in the case of phase ripple. Phase is usually the main contributor to unwanted sidelobes, and the amplitude contribution is minimal.

Droop across the pulse primarily causes a power loss, which can usually

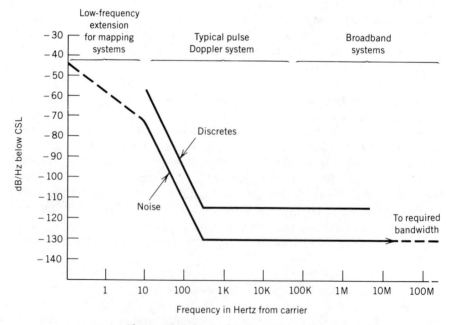

Figure 22.2. *Typical noise requirements.*

be compensated for by setting the nominal required level in the middle of the pulse. However, it also produces a phase shift across the pulse. The major component of the droop is generally linear, so no ill effects are produced except a frequency shift that gives rise to a time displacement after compression. If a quadratic component is present to any degree, this causes smearing (broadening) of the main lobe.

The system design engineer usually specifies all of the above parameters individually; i.e., amplitude and phase ripple and linear and quadratic droop.

22.4.1.4 *Broadband (High-Resolution) Systems.*

High-resolution systems (2) usually use pulse compression as above but with much greater time bandwidth products. The same requirements as above apply together with some additions. These are in the frequency domain; and because the bandwidths used in these systems are very large, the phase versus frequency and amplitude versus frequency characteristics of the power amplifiers must be taken into account. Any nonlinearity in these characteristics produces time sidelobes in exactly the same manner as the time domain instabilities above. As before these can have linear, quadratic, and sinusoidal (ripple) components, each producing its own effect on the compressed signal. In addition, the combination of these various distortions is usually much more complex than in the time domain (there can be more than one ripple frequency, for example), and it is usually impossible to determine their characteristics by simple measurements. Frequency weighting is also often used to reduce sidelobe levels, thus further complicating the problem so that the only true evaluation that can be made is by Fourier analysis.

These broadband parameters are not under control of the transmitter designer, and any specifications for them must be passed on to the tube designer directly. Since the effects of both the time and frequency domain distortions are the same, trades can be made between the two to achieve the overall sidelobe specification for the transmitter.

22.4.1.5 *Synthetic Aperture Systems.*

Synthetic aperture systems (3) integrate data in a coherent fashion over time intervals as long as a thousand pulse periods. Thus, high stability at very low frequencies (close to the carrier) is required. This is specified in terms of spectral noise density (as in the case of pulse Doppler systems), extending downward in frequency to 0.01–0.05 Hz (see Fig. 22.2). Difficulties in meeting this specification are usually associated with regulator noise and power tube flicker noise.

Another requirement for synthetic aperture radar transmitters is integrated sidelobe level, which is determined by integrating the noise spectrum of the transmitter from the lowest frequency of interest out to the bandwidth of the transmitted signal. Since these systems usually have high-resolution pulse compression, the bandwidth can be in the several hundred megahertz range.

22.4.2 Communications Transmitters

Communications transmitters are usually CW transmitters whose main requirement is linearity of the input/output characteristic. This is required to keep intermodulation products to a minimum when multiple signals are transmitted. Phase and amplitude linearity are not specified in terms of Fourier components as in broadband systems, but as a maximum rate of change per Hertz or the maximum slopes that are allowed in the phase and amplitude versus frequency characteristics. These parameters depend on the tube design and not the transmitter design.

Again, a transmitter is an amplifier and has two functions. It produces an output that is larger than its input, and it introduces noise (distortion). Just as distortion is of paramount importance to the designer of quality stereo amplifiers, noise is of concern to the designer of transmitters. Transmitter noise includes the random perturbations normally called noise and the periodic and/or predictable perturbations normally called distortion. "Stability" and "fidelity" are used synonymously; the more stable the transmitter, the less noise is introduced. The management of noise is included in the very core of transmitter technology. Noise in the transmitter can distort both the output amplitude and phase.

Noise considerations apply to both CW and pulsed transmitters. An understanding of the output spectrum of a pulsed system is necessary to the understanding of noise in the pulsed system.

22.5 TRANSMITTED SPECTRUM

Fourier analysis of the output of a simple pulsed transmitter gives the spectrum shown in Fig. 22.3. It consists of line spectra, with each line separated from the next by the pulse repetition frequency. Superimposed on the lines and forming an envelope for them is a waveform that can be described by a $(\sin X)/X$ function.

The actual expression for this envelope is

$$t_0 A \, \frac{\sin \pi f t_0}{\pi f t_0} \tag{22.1}$$

where

$$t_0 = \text{pulse width}$$

$$A = \text{pulse amplitude}$$

$$f = \text{frequency away from center frequency (fc)}$$

The power spectrum is the square of the amplitude spectrum.

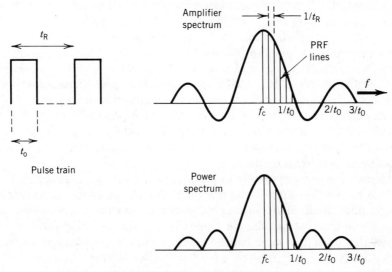

Figure 22.3. *Transmitted spectrum.*

For a video signal, the line appearing at f_c is the DC component. For a pulse modulated RF signal, the maximum or central spectral line corresponds to the carrier frequency. The separation between the lines is equal to the pulse repetition frequency. The first minimum on each side of the carrier occurs at a frequency equal to

$$f_c \pm \frac{1}{t_0} \qquad (22.2)$$

Succeeding minima occur at multiples of $1/t_0$ from the carrier.

One important parameter that is derived from the basic Fourier analysis is that the power in the central spectral line is given by

$$\left(\frac{A t_0}{t_R} \right)^2 \qquad (22.3)$$

where t_R is the interpulse period.

In a 1-ohm resistor, A^2 is equal to the peak power of the pulse. The power in the central spectral line therefore is equal to

$$\text{Peak power} \times (\text{duty cycle})^2 \qquad (22.4)$$

22.5.1 Sideband Reduction

Although the (sin X)/X characteristics of the spectrum depicted in Fig. 22.3 shows the sideband level gradually decreasing with frequency away from the carrier, this is generally not fast enough to avoid interference with other nearby transmitters operating in adjacent frequency channels. MIL-STD-469 Radar Engineering Design Requirements–Electromagnetic Compatibility places strict limits on allowable sideband levels at various distances from the carrier for all types of radiation. To meet these limits, it is necessary to tailor the naturally produced pulse modulated spectrum by other means.

Filtering is not feasible for modern systems that incorporate fast frequency agility because it is difficult to tune a high-power filter rapidly enough to follow the transmitter frequency excursions. Also, changes in pulse length and duty cycle require rapid changes in the filter bandwidth and skirt characteristics, another difficult task.

In practice it is impossible to produce an exactly rectangular pulse, and all real pulses have some finite rise time. This is due to the unavoidable, no matter how small, inductance and capacitance always present in the gating circuit and modulating tube electrode. This is shown in Fig. 22.4. Analysis shows that the actual modulating waveform always has a leading and trailing edge that can be described by a $\cos^2$ or "raised cosine" function.

When a pulse with these edge characteristics is subject to Fourier analysis, it is found to have greatly reduced sidelobe levels compared with those of the true rectangular pulse. The actual level and the rise and fall times of the pulse can be controlled by varying the value of the series inductance in Fig. 22.4. (*Note:* It is assumed that the capacitance is fixed.) The broadening of the mainlobe of the spectrum due to increased rise time, unless compensated, results in reduced radar range resolution.

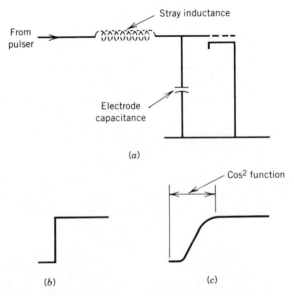

Figure 22.4. *Pulse rise waveforms of a typical modulator. (a) Tube stray reactances; (b) time response of input; (c) time response of output.*

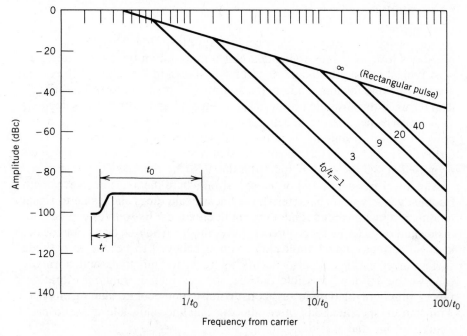

Figure 22.5. *Spectrum envelope of pulses with cos² edges.*

The actual power spectrum is given by the expression

$$G(f) = \frac{t_0}{\pi} \left(\frac{\sin \pi \alpha_0 \cos \pi \alpha_r}{\alpha_0 (1 - 4\alpha_r^2)} \right) \tag{22.5}$$

where

$$t_r = \text{rise time}$$

$$t_0 = \text{pulse width}$$

$$\alpha_r = f t_r$$

$$\alpha_0 = f t_0$$

Figure 22.5 gives the spectrum envelope for various ratios of pulse width to rise time for a pulse with cos² edges.

In practice, the sidelobe levels of the output pulse are actually lower than predicted by Fig. 22.5 because the pulse rise and fall times are modified by the transfer characteristic of the amplifying tube. This may be a three or five halves or cubic function, which results in the output pulse having higher-order (than squared) cosine function edges and even lower sidelobes.

22.6 NOISE

Noise is usually a concern for receiving systems. However, for many applications low noise requirements are also imposed on the transmitter.

A typical transmitter noise spectrum* is shown in Fig. 22.6. Two curves are shown, one for amplitude noise and one for phase noise. The ordinate is in terms of the absolute power in a convenient bandwidth such as decibels above 1 mW per Hertz. A better noise measure is the ratio of the noise power to the transmitted signal power, and the ordinate in this case is in terms of decibels per Hertz below the signal carrier or in the case of a gated or pulsed signal the central spectrum line (CSL).

The lower frequency end of the spectrum where the noise decreases with frequency is a typical characteristic of linear beam tubes and is due to signal modulation by unwanted random beam fluctuations. Its sources can be both inside and outside the tube. Flicker noise, the main source occurring inside, is an imperfectly understood phenomenon believed to be due to cathode surface irregularities. It also occurs in transistor junctions and normally follows the $1/f$ law. Possible outside sources include random vibration, power supply regulation/noise, acoustic noise, and cooling fluid turbulence. Transmitter specifications often allocate half the allowable noise to each source (inside and outside).

Because this noise is a true modulation mechanism, it is sometimes referred to as multiplicative noise as opposed to additive noise. Additive noise occurs in the high-frequency part of the spectrum and is generally referred to as the noise floor. It is additive because it is present all the time whether a signal is there or not and just adds to the output. In the case of AM noise, it can usually be calculated from the noise figure of the tube as follows:

$$\text{Noise from source} = kT = -174 \text{ dBm/Hz} + G \text{ (dB)} \qquad (22.6a)$$

$$\text{Noise from tube} = NF \qquad (22.6b)$$

$$\text{Total noise} = -174 + G \text{ (dB)} + NF \text{(dB) dBm/Hz} \qquad (22.6c)$$

where

$$k \text{ (Boltzmann's constant)} = 1.38 \times 10^{-23} \text{ J/°K}$$

$$T \text{ (temperature)} = 290°K$$

$$G = \text{Tube gain in decibels}$$

$$NF = \text{Noise figure in decibels}$$

PM noise cannot be treated in this fashion. In general it should be equal

---

* Figure 22.6 shows the noise spectrum associated with the central spectral line. In a pulsed system it is also repeated around all the other lines.

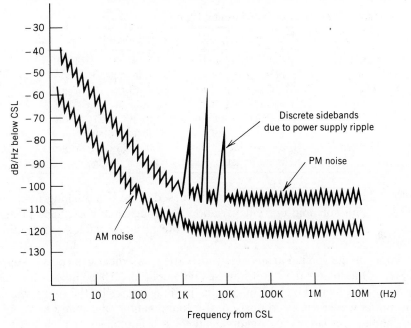

Figure 22.6. *Transmitter noise spectrum.*

to AM noise, but for linear beam tubes it is 15–20 dB higher because these tubes operate by velocity modulation of the beam, making the phase of the output more sensitive to disturbances than the amplitude.

22.7   UNWANTED OR SPURIOUS SIDEBANDS

22.7.1   Pushing Figures

In addition to noise, the spectrum of the transmitter also contains a number of unwanted or spurious discrete lines, also shown in Fig. 22.6. If these lines occur at frequencies away from the CSL corresponding to the harmonics of the power supply frequency, they are due to sidebands produced by unwanted modulation of the signal by the power supply ripple contained in the DC voltages and currents applied to the various electrodes of the transmitter tube. Variations in any of the tube voltages or current will cause the output RF amplitude and phase to vary. The amount by which this happens is a characteristic of both the electrode and the type of tube and is defined by the pushing figure. This is the change in phase or amplitude for a given change in the electrode voltage. The given change is usually 1 V or 1 A or 1% of the total. Two pushing figures thus exist for each electrode, one for phase variation and one for amplitude variation. Typical values of these parameters for a medium power (50 kW) traveling wave tube are shown in Table 22.2.

Table 22.2   Typical TWT Pushing Figures

Electrode	Phase Pushing Figure	Amplitude Pushing Figure
Cathode	40° per 1% of cathode voltage	0.5 dB per 1% of cathode voltage
Grid	6°–8° per 1% of grid voltage	0.1 dB per 1% of grid voltage
Collector	0.01° per 1% of collector voltage	0.01 dB per 1% of collector voltage
Filament	0.01° per 1% of filament voltage	0.001 dB per 1% of filament voltage

In addition, if the tube is solenoid focussed there is a pushing figure associated with solenoid current variation. Although many of these pushing figures can be calculated theoretically, more accurate values are obtained by measurement.

If the voltage applied to any electrode varies periodically (as power supply ripple), it modulates the output phase and amplitude. These modulation levels can be calculated by multiplying the pushing figures by the ripple values. Any ripple waveform is comprised of a fundamental, and many harmonics and modulation sidebands occur at all these frequencies, which must be evaluated individually.

22.7.2   Sideband Levels

Once the spurious modulation levels are known, the value of the sidebands they give rise to can be calculated.

22.7.2.1   Phase. Phase modulation at a single frequency gives rise to multiple sidebands. The sideband amplitudes are given by Bessel functions with θ, the peak phase deviation, as the argument. Thus,

$$J_0(\theta) = \text{carrier level}$$

$$J_1(\theta) = \text{first sideband}$$

$$J_2(\theta) = \text{second sideband}$$

$$J_3(\theta) = \text{third sideband}$$

Since θ is usually very small, only the first sideband is important. Using a reduction formula and approximation for small values of θ gives

$$\frac{J_1(\theta)}{J_0(\theta)} = \frac{\theta}{2} \tag{22.7}$$

where θ is in radians (4).

The ratio of a sideband to the carrier or CSL is thus given by $\theta/2$, where θ is the peak phase deviation previously calculated from the ripple value and the pushing figure. A composite expression therefore is

$$\frac{SB}{CSL} = 20 \log \frac{(E_T \times PF \times \pi)}{360} \text{ dB} \qquad (22.8)$$

where

$$
\begin{aligned}
SB \;\; &= \text{ sideband level} \\
CSL &= \text{ central spectral line level} \\
E_T \;\; &= \text{ peak ripple voltage} \\
PF \;\; &= \text{ pushing figure in degrees/Volt}
\end{aligned}
$$

Note that peak values, not RMS values, are used throughout.

22.7.2.2 *Amplitude.*
The expression for an amplitude modulated signal is given by

$$V = E_c \cos \omega_c t + \frac{m}{2}[E_c \cos(\omega_c + \omega_m)t] + \frac{m}{2}[E_c \cos(\omega_c - \omega_m)t] \qquad (22.9)$$

where

$$
\begin{aligned}
E_c \;\; &= \text{ amplitude of the unmodulated carrier} \\
\omega_c \;\; &= \text{ carrier frequency} \\
\omega_m &= \text{ modulating frequency} \\
m \;\; &= \text{ modulation index}
\end{aligned}
$$

This indicates a carrier signal at ω_c and two sidebands one above the carrier and the other below by an amount equal to the modulating frequency. The ratio of the single sideband to the carrier or CSL is given by $m/2$. m is the modulation index and is defined as

$$m = \frac{E_{max} - E_c}{E_c} = \frac{E_c - E_{min}}{E_c} = \frac{E_{max}}{E_c} - 1 \qquad (22.10)$$

where E_{max} and E_{min} are the maximum and minimum modulated carrier levels indicated in Fig. 22.7.

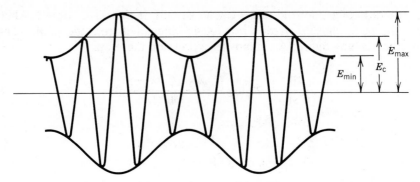

Figure 22.7. *Amplitude modulation. E_c, unmodulated carrier; m, modulation index; m = $(E_{max} - E_c)/E_c$; pushing figure, E_{max}/E_c.*

The ratio

$$\frac{E_{max}}{E_c}$$

can be related to the amplitude pushing figures as follows:

$$\frac{E_{max}}{E_c} = 10^{(\text{PFv}/20)} \qquad (22.11)$$

where PF is the pushing figure in decibels per Volt, and v is the modulating voltage.

$$M = \frac{E_{max}}{E_c} - 1 = 10^{(\text{PFv}/20)} - 1 \qquad (22.12)$$

so

$$\frac{\text{SSB}}{\text{CSL}} = 20 \log\left(\frac{10^{(\text{PFv}/20)} - 1}{2}\right) dB \qquad (22.13)$$

22.7.2.3   Ripple. To calculate the allowable ripple it is necessary to allocate the allowable phase deviation (θ) to the two electrodes. Two possible methods of doing this are as follows:

1. Divide θ between the two electrodes in the ratio of the pushing figures. This results in an equal percentage in obtaining given ripple attenuation in the two power supplies.
2. Divide θ in the ratio of a "difficulty factor" (again arbitrary) in obtaining given ripple attenuation in the two power supplies. For example, if the cathode supply is 25 kV and the grid only 500 V, it is

much more difficult to obtain a given ripple attenuation in the case of the cathode than the grid. Although this may seem unwieldy once the ratio gets beyond a 5 or 6 to 1, the effect of a greater ratio on the higher voltage supply is negligible but is considerable on the lower.

Once the allocation of θ is made, it is trivial to calculate the allowed ripple from the pushing figure.

22.7.3 Other Sources of Spurious Sidebands

In the same way that external mechanical disturbances produce random noise, they can also produce discrete sidebands if they are periodic. These include vibration, acoustic noise (from motors or generators), and coolant flow. Also, a random external excitation can excite a resonance in the tube and produce a modulation sideband at the resonant frequency if the mechanical structure involved is part of any beamforming or RF circuit. A tube model should be submitted to a vibration analysis early in the design phase to avoid this problem. If a resonance occurs in the frequency range of interest, modifications should be made either to eliminate it altogether or move its frequency out of range.

22.8 AM/PM AND AM/AM CONVERSION FACTORS

AM/PM and AM/AM conversion factors are, in effect, pushing figures for the RF input drive and describe the effect on the output of varying the drive. Typical values for a medium power TWT are

AM/PM	2°–4° per dB of drive change
AM/AM	0.2 dB change in output for 1 dB change in input drive

These numbers are for saturation drive levels only. Below saturation they both increase drastically. It is desirable to operate the tube in saturation to minimize their effect.

Since drive variations are not usually periodic, it is rare to observe sidebands from this source. However, in wideband systems in which phase and amplitude linearity is of prime importance, any change in drive level across the frequency band causes undesirable effects on the output.

22.9 TRANSMITTER BLOCK DIAGRAM

The block diagram of a typical transmitter is shown in Fig. 22.8. A picture of the AN/APG-68 airborne radar transmitter used in the F-16 aircraft is shown in Fig. 22.9.

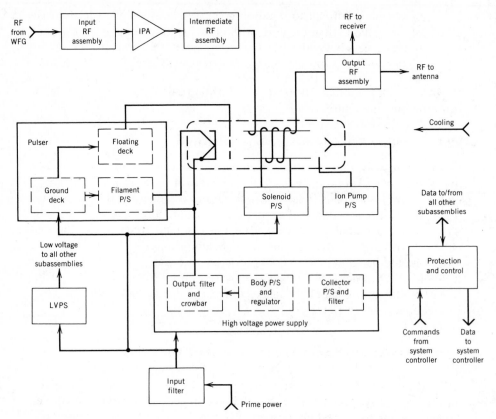

Figure 22.8. *Typical transmitter block diagram.* P/s = power supply, WFG = wideband function generator

The input filter has two purposes. First, it conditions the input power for use by the other subassemblies. Second, it isolates the power source from ripple currents drawn by the other subassemblies to accommodate the electromagnetic interference (EMI) and electromagnetic compatibility (EMC) requirements of the transmitter.

The high voltage power supply (HVPS) provides regulated high voltage power to the tube cathode, the collector(s), and (if included in the tube design) the anode. The HVPS can be one or more DC/DC converters, one or more "bulk" supplies (a voltage step-up transformer with a rectifying bridge), or a combination of these. There are several approaches to reduce the ripple voltages on the tube electrodes to enhance the stability of the transmitter. Passive LC filters are commonly used. A series regulator can be included on the input or the output of the supplies. If the transmitter is pulsed, novel shunt regulators can be used to control the cathode voltage droop during the pulse.

The pulser is used to modulate a tube electrode to turn the tube on and off. The modulated tube electrode can be a shadow grid, an intercepting grid, a focus electrode, or a modulation anode. Typically, these tube electrodes are driven relative to the tube cathode; this means that the controlling circuitry must be referenced to the tube cathode. As a consequence, the pulser usually includes the tube filament supply because it also is referenced to the tube cathode. The pulser usually includes some "ground deck" (referenced to chassis ground) circuitry to provide power to the "floating deck" (referenced to the tube cathode), to provide commands to the floating deck, and to receive health and status telemetry from the floating deck. The pulser usually needs to swing the tube electrode both positive and negative relative to the cathode (to turn the tube on and off, respectively). To reduce phase errors, the positive supply is usually well regulated, and the voltage pulse is sometimes clamped within the pulser and/or physically on the tube.

The low voltage power supply (LVPS), if included, provides low voltage power to the other transmitter subassemblies. If size and weight are of concern, the transmitter may consume low voltage power from the system level supplies. Incorporation of an LVPS within the transmitter adds an extra level of isolation between the high voltages within the transmitter and sensitive low voltage circuits within the system.

If the tube incorporates a solenoid to focus the electron beam, a current regulated solenoid supply is needed.

The protection and control unit (PCU) serves two functions. First, it

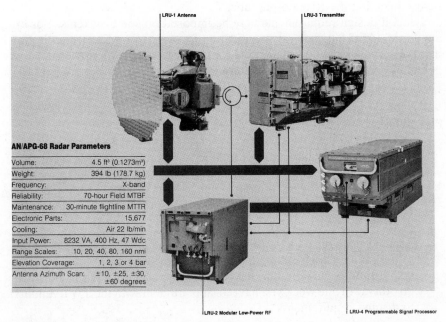

AN/APG-68 Radar Parameters

Volume:	4.5 ft³ (0.1273m³)
Weight:	394 lb (178.7 kg)
Frequency:	X-band
Reliability:	70-hour Field MTBF
Maintenance:	30-minute flightline MTTR
Electronic Parts:	15,677
Cooling:	Air 22 lb/min
Input Power:	8232 VA, 400 Hz, 47 Wdc
Range Scales:	10, 20, 40, 80, 160 nmi
Elevation Coverage:	1, 2, 3 or 4 bar
Antenna Azimuth Scan:	±10, ±25, ±30, ±60 degrees

Figure 22.9. F-16 radar (APG-68) with callouts of key subassemblies. LRU = line replaceable unit, MTBF = mean time between failures, MTTR = mean time to repair. (Photo/data courtesy of Westinghouse Electric Corp.)

senses anomalous performance within the transmitter and (typically) turns off the faulty supply. This action helps to prevent the fault from propagating, both within the supply and to the tube. Second, the PCU controls the operation of the transmitter; the PCU receives commands from the system and sequences the various power supplies within the transmitter on/off in the sequence needed by the tube and the system. If the transmitter has a BIT/FIT requirement, the PCU typically buffers the data received from the other transmitter subassemblies for the system.

If the tube incorporates an ion pump to maintain a vacuum, an ion pump power supply is needed. Typically, this is a 3-kV, 1-mA supply.

Packaging is an integral part of the transmitter design. The package must protect the user from the high voltages within the transmitter and must protect the high-voltage circuits from the environment (low atmospheric pressure, salt spray). If the environment is relatively benign, the package may be a simple cabinet; if the environment is not conducive to high-voltage management, the package may include a fluid (liquid or gas) chamber for the high voltage circuits.

If the transmitter stability requirements are so demanding that voltage ripple reduction techniques are insufficient, RF phase and amplitude corrections can be made within the transmitter. If ripple voltages less than 0.001% (tens of millivolts of ripple on tens of kilovolts) are dictated, the designer can use RF correction techniques whereby the output RF is compared to the input RF and phase and amplitude are adjusted dynamically before injecting the RF into the tube.

General design requirements may be imposed on some systems. For example, MIL-E-5400 (general specification for Aerospace Electronic Equipment) may be applied to transmitters for the air force. A comprehensive text containing more details for the electrical design of transmitters is Ref. 1.

PROBLEMS

1. Generate a block diagram of an approach to implement phase correction around a single stage TWT transmitter. The scheme should correct for distortions that are both short compared to a pulse width and long compared to the period. *Hint:* A delay line and a sample and hold may be needed.

2. Assume that you are a transmitter designer. You have received a listing of preliminary transmitter requirements indicating that a TWT should be used; unfortunately, this list contains only electrical parameters and no others. In one week you have an appointment with the TWT vendors to discuss requirements.

   (a) List those questions to which you would try to find answers before you meet with the TWT vendor.

(b)   List all the topics you want to discuss with the TWT vendors when they arrive.

3.   Assume that the transmitter requirements are such that two TWTs must be used in series to achieve the required gain. Discuss how your single stability requirement for a pulse doppler system can be suballocated to each stage, and provide some mathematical basis.

4.   Why should airborne ECM transmitters have relatively low reliability and maintainability?

5.   An MTI transmitter has the following pulse-to-pulse characteristics:
      The amplitude varies by 1%.
      The phase varies by 30°.
      The period varies by 0.0002%.
      The pulse width varies by 0.0001%.
      Assume these errors are random in nature.
      (a)   What is the largest contributor to the instability of the transmitter?
      (b)   What are the total errors of the transmitter?

6.   Consider a pulse doppler radar system operating with a PRF of 100 kHz, a pulse width of 5 µs, a center frequency of 10 GHz, and a peak output power of 5 kW.
      (a)   Sketch the spectrum of the transmitted pulse.
      (b)   Calculate the amplitude of the central spectral line.
      (c)   Calculate the frequency of the first minimum.
      (d)   What is the maximum unambiguous velocity?
      (e)   Suppose the transmitter has a noisy power supply in it and distorts the transmitted signal in such a manner that an extra spectral line is generated every 30 kHz. What does this do to the system performance?

7.   A TWT has the following pushing figures:

Electrode	Phase	Amplitude
Grid	6°/%	0.1 dB/%
Cathode	35°/%	0.4 dB/%

The peak grid voltage ripple is 0.03% and the peak cathode voltage ripple is 0.02%. Both ripples are at 2400 Hz and are synchronized with each other. Assume their effects on the output phase and amplitude are additive. The transmitter operates at a duty cycle of 0.5 and has an average output power of 2 kW. Calculate
      (a)   the size of the phase sideband relative to the central spectral line.
      (b)   the size of the amplitude sideband relative to the central spectral line.

8. Assume the requirements for a transmitter are to have phase sidebands of -60 dBc and amplitude sidebands of -70 dBc if these distortions occur at 1200 Hz. Also assume that it is three times as difficult to reduce ripple in the cathode supply as it is to reduce ripple in the grid supply. For a TWT with the pushing figures in Problem 7 and for a transmitter operating at a duty cycle of 0.4 with a peak output power of 50 kW, calculate the maximum allowable voltage ripple on the grid and cathode supplies if the ripples are in synchrony at 1200 Hz and their effects on the output phase and amplitude are additive.

9. A TWT has the following pushing figures:

Electrode	Phase	Amplitude
Grid	8°/%	0.3 dB/%
Cathode	40°/%	0.5 dB/%
Solenoid	2°/%	0.2 dB/%

The power supplies have the following ripples, all in synchrony at 2400 Hz:

Electrode	Ripple
Grid	0.001%
Cathode	0.007%
Solenoid	0.003%

The transmitter operates at a duty cycle of 0.45 and has a peak output power of 1 kW. Calculate the size of the phase and amplitude sidebands relative to the central spectral line. Assume the effects of the ripple on output phase and amplitude are additive.

10. Assume the requirements for a transmitter are to have phase and amplitude sidebands of -65 and -75 dBc at 10 kHz, respectively. All of the power supplies within the transmitter are DC/DC converters operating at 10 kHz. The TWT has a solenoid. Assume the collector and filament pushing figures are such that the ripples on these supplies can be ignored. Assume that the difficulty factors for removing ripples from the grid, cathode, and solenoid supplies are 1, 2, and 4, respectively. Calculate the allowed ripples on these supplies if the transmitter operates at a duty cycle of 50%, has a peak output power of 100 W, and the TWT has the pushing figures of Problem 9. Assume the effects of the ripples on the output phase and amplitude are additive.

REFERENCES

1. M. I. Skolnik (ed.), *Radar Handbook*, McGraw-Hill Book Co., New York, 1970, pp. 17–48.

2. Ibid., pp. 20–27.

3. Ibid., Chap. 23.

4. G. Watson, *Theory of Bessel Functions*, Cambridge University Press, 1958, p. 17.

General Reference

G. W. Ewell, *Radar Transmitters*, McGraw-Hill Book Co., New York, 1981.

Index